7.1 Making and Selling Markers and T-Shirts

Using a Graph to Solve a Linear System

7

Learning By Doing Lesson Map

Get Ready

Objectives

In this lesson, you will:

- Analyze cost and income equations.
- Graph cost and income equations on the same graph.
- Find the break-even point graphically.

Key Terms

- income
- profit
- point of intersection
- break-even point

Indiana's Academic Standards

Standard 3 Pairs of Linear Equations and Inequalities

A1.3.1

Understand the relationship between a solution of a pair of linear equations in two variables and the graphs of the corresponding lines and solve pairs of linear equations in two variables by graphing, substitution or elimination.

A1.3.3

Solve problems that can be modeled using pairs of linear equations in two variables, interpret the solutions, and determine whether the solutions are reasonable.

Lesson Overview

Within the context of this lesson, students will be asked to:

- Compare and analyze cost and income equations graphically and algebraically.
- Graph cost and income equations on the same graph.
- Find a break-even point graphically.

Essential Questions

The following key questions are addressed in this lesson:

1. What is income?
2. What is profit?
3. What is a point of intersection on a graph?
4. What is a break-even point?

7.2 Time Study

Graphs and Solutions of Linear Systems

Learning By Doing Lesson Map

Get Ready

Objectives

In this lesson, you will:

- Determine the number of solutions of a linear system.
- Identify parallel and perpendicular lines.

Key Terms

- systems of linear equations
- linear system
- solution
- point of intersection
- parallel lines
- perpendicular lines
- reciprocals

Indiana's Academic Standards

Standard 3 Pairs of Linear Equations and Inequalities

A1.3.1

Understand the relationship between a solution of a pair of linear equations in two variables and the graphs of the corresponding lines and solve pairs of linear equations in two variables by graphing, substitution or elimination.

A1.3.3

Solve problems that can be modeled using pairs of linear equations in two variables, interpret the solutions, and determine whether the solutions are reasonable.

Lesson Overview

Within the context of this lesson, students will be asked to:

- Graph systems of linear equations.
- Determine the number of solutions of linear systems of equations.
- Identify parallel and perpendicular lines.

Essential Questions

The following key questions are addressed in this lesson:

1. What is a system of linear equations?
2. What is a solution of an equation?
3. What is a point of intersection?
4. What are parallel lines?
5. What are perpendicular lines?
6. What are reciprocals?

7.3 Hiking Trip

Using Substitution to Solve a Linear System

7

Learning By Doing Lesson Map

Get Ready

Objective

In this lesson, you will:

- Solve linear systems by using substitution.

Key Terms

- standard form of a linear equation
- substitution method

Indiana's Academic Standards

Standard 3 Pairs of Linear Equations and Inequalities

A1.3.1

Understand the relationship between a solution of a pair of linear equations in two variables and the graphs of the corresponding lines and solve pairs of linear equations in two variables by graphing, substitution or elimination.

A1.3.3

Solve problems that can be modeled using pairs of linear equations in two variables, interpret the solutions, and determine whether the solutions are reasonable.

Lesson Overview

Within the context of this lesson, students will be asked to:

- Write equations in standard form.
- Write equations in slope-intercept form.
- Use substitution to solve systems of linear equations.

Essential Questions

The following key questions are addressed in this lesson:

1. What is the standard form of an equation?
2. What do the letters A, B, and C represent in the standard form of an equation?
3. What is substitution?
4. What does it mean to solve a system of linear equations?

7.4 Basketball Tournament

Using Linear Combinations to Solve a Linear System

Learning By Doing Lesson Map

Get Ready

Objective

In this lesson, you will:

- Solve a linear system by using linear combinations.

Key Terms

- standard form of a linear equation
- linear combination
- linear combinations method

Indiana's Academic Standards

Standard 3 Pairs of Linear Equations and Inequalities

A1.3.1

Understand the relationship between a solution of a pair of linear equations in two variables and the graphs of the corresponding lines and solve pairs of linear equations in two variables by graphing, substitution or elimination.

A1.3.3

Solve problems that can be modeled using pairs of linear equations in two variables, interpret the solutions, and determine whether the solutions are reasonable.

Lesson Overview

Within the context of this lesson, students will be asked to:

- Write equations in standard form.
- Multiply equations in systems of equations by constants to get pairs of like terms that are opposites of each other.
- Use the linear combinations method to solve systems of linear equations.

Essential Questions

The following key questions are addressed in this lesson:

1. What is the standard form of a linear equation?
2. What do the letters *A*, *B*, and *C* represent in the standard form of an equation?
3. What is a linear combination?
4. What is the linear combinations method?
5. What does it mean to solve a system of linear equations?

7.5 Finding the Better Paying Job

Using the Best Method to Solve a Linear System, Part I

7

Learning By Doing Lesson Map

Get Ready

Objective

In this lesson, you will:

- Solve a linear system by using an algebraic method.

Key Terms

- linear system
- inequality

Indiana's Academic Standards

Standard 3 Pairs of Linear Equations and Inequalities

A1.3.1

Understand the relationship between a solution of a pair of linear equations in two variables and the graphs of the corresponding lines and solve pairs of linear equations in two variables by graphing, substitution or elimination.

A1.3.3

Solve problems that can be modeled using pairs of linear equations in two variables, interpret the solutions, and determine whether the solutions are reasonable.

Lesson Overview

Within the context of this lesson, students will be asked to:

- Write a system of two linear equations for the weekly salary from different companies.
- Choose a method to solve their system of linear equations.
- Add a third linear equation for a constant function to their system of linear equations.
- Analyze their system of linear equations.

Essential Questions

The following key questions are addressed in this lesson:

1. What is a linear system?
2. How can we analyze the system of linear equations using a restriction and an inequality?
3. What is the best method to choose how to solve a linear system of equations?
4. How can you choose the best method to use to solve a linear system?

7.6 World Oil: Supply and Demand

Using the Best Method to Solve a Linear System, Part 2

Learning By Doing Lesson Map

Get Ready

Objective

In this lesson, you will:

- Solve a linear system by using an algebraic method.

Key Term

- linear system

Indiana's Academic Standards

Standard 3 Pairs of Linear Equations and Inequalities

A1.3.1

Understand the relationship between a solution of a pair of linear equations in two variables and the graphs of the corresponding lines and solve pairs of linear equations in two variables by graphing, substitution or elimination.

A1.3.3

Solve problems that can be modeled using pairs of linear equations in two variables, interpret the solutions, and determine whether the solutions are reasonable.

Lesson Overview

Within the context of this lesson, students will be asked to:

- Write a linear system of equations that uses large numbers to model the supply and demand of world oil.
- Choose an algebraic method to solve the linear system of equations.
- Solve the linear system of equations.

Essential Questions

The following key questions are addressed in this lesson:

1. What is a linear system?
2. How can you solve a linear system?
3. How can you choose which algebraic method to use to solve a linear system?

7

7.7 Picking the Better Option

Solving Linear Systems

Learning By Doing Lesson Map

Get Ready

Objective

In this lesson, you will:

- Use a system of linear equations to solve a problem.

Key Term

- break-even point

Indiana's Academic Standards

Standard 3 Pairs of Linear Equations and Inequalities

A1.3.1

Understand the relationship between a solution of a pair of linear equations in two variables and the graphs of the corresponding lines and solve pairs of linear equations in two variables by graphing, substitution or elimination.

A1.3.3

Solve problems that can be modeled using pairs of linear equations in two variables, interpret the solutions, and determine whether the solutions are reasonable.

Lesson Overview

Within the context of this lesson, students will be asked to:

- Write a system of linear equations to model a situation for the cost of producing bicycles under different production plans.
- Solve and graph the system of linear equations.
- Write and solve related systems of linear equations.

Essential Questions

The following key questions are addressed in this lesson:

1. What is a break-even point?
2. What are production costs?
3. What is a system of linear equations?

7.8 Video Arcade

Writing and Graphing an Inequality in Two Variables

Learning By Doing Lesson Map

Get Ready

Objectives

In this lesson, you will:

- Write an inequality in two variables.
- Graph an inequality in two variables.

Key Terms

- linear inequality in two variables
- inequality symbol
- linear equation
- coordinate plane
- half-plane

Indiana's Academic Standards

Standard 2 Linear Functions, Equations and Inequalities

A1.2.6

Graph a linear inequality in two variables.

Lesson Overview

Within the context of this lesson, students will be asked to:

- Write an inequality in two variables to model a situation that involves the points available after playing arcade games.
- Graph an inequality in two variables.
- Shade the appropriate half-plane for an inequality.
- Determine the type of line, solid or dashed, that should be used to graph various linear inequalities.
- Graph various linear inequalities.

Essential Questions

The following key questions are addressed in this lesson:

1. What is a linear equation?
2. What is a linear inequality?
3. What is a linear combination?
4. What are the symbols of inequality and what is the meaning of each?
5. What is a half-plane?
6. How can you graph a linear inequality?

7.9 Making a Mosaic

Solving Systems of Linear Inequalities

7

Learning By Doing Lesson Map

Get Ready

Objectives

In this lesson, you will:

- Write a system of linear inequalities.
- Graph a system of linear inequalities.
- Identify solutions of a system of linear inequalities.

Key Terms

- system of linear equations
- linear inequality
- system of linear inequalities

Indiana's Academic Standards

Standard 3 Pairs of Linear Equations and Inequalities

A1.3.2

Graph the solution set for a pair of linear inequalities in two variables with and without technology and use the graph to find the solution set.

Lesson Overview

Within the context of this lesson, students will be asked to:

- Write a system of linear equations to represent the number of bags of glass and metallic tiles to buy.
- Write a system of linear inequalities that better represents the number of bags of glass and metallic tiles that can be bought.
- Solve the systems of linear equations.
- Solve and graph the system of linear inequalities.

Essential Questions

The following key questions are addressed in this lesson:

1. What is a mosaic?
2. Why would an art group donate time and work to create an art project for a school?
3. Why would a foundation or company donate the money to pay the expenses?
4. What is a system of equations?
5. What is a system of inequalities?
6. What does it mean to solve a system of linear inequalities?

8.1 Web Site Design

Introduction to Quadratic Functions

Learning By Doing Lesson Map

Get Ready

Objectives

In this lesson, you will:

- Graph quadratic functions.
- Identify coefficients in quadratic functions.
- Evaluate quadratic functions.

Key Terms

- rate of change
- curve
- quadratic function
- evaluate

Indiana's Academic Standards

Standard 5 Quadratic Equations and Functions

A1.5.1

Graph quadratic functions.

A1.5.3

Solve problems that can be modeled using quadratic equations, interpret the solutions, and determine whether the solutions are reasonable.

Lesson Overview

Within the context of this lesson, students will be asked to:

- Graph quadratic functions.
- Identify the coefficients of the terms in a quadratic function and label them as *a*, *b*, and *c*.
- Evaluate quadratic functions for given values.

Essential Questions

The following key questions are addressed in this lesson:

1. What is a web site design?
2. What is a logo?
3. How can you calculate a rate of change?
4. What is a quadratic function?
5. How do you evaluate a quadratic function?

8.2 Satellite Dish

Parabolas

8

Learning By Doing Lesson Map

Get Ready

Objectives

In this lesson, you will:

- Graph quadratic functions.
- Find the line of symmetry of a parabola.
- Find the vertex of a parabola.
- Identify the maximum or minimum value of a function.

Key Terms

- parabola
- line of symmetry
- vertical line
- vertex
- minimum
- maximum

Indiana's Academic Standards

Standard 1 Relations and Functions

A1.1.2

Identify the domain and range of relations represented by tables, graphs, words, and equations.

Standard 5 Quadratic Equations and Functions

A1.5.1

Graph quadratic functions.

A1.5.3

Solve problems that can be modeled using quadratic equations, interpret the solutions, and determine whether the solutions are reasonable.

Lesson Overview

Within the context of this lesson, students will be asked to:

- Graph quadratic functions.
- Find the line of symmetry of a parabola graphically and algebraically.
- Find the vertex of a parabola graphically and algebraically.
- Identify the maximum or minimum value of a quadratic function.

Essential Questions

The following key questions are addressed in this lesson:

1. What is a satellite dish?
2. What is a parabola? What does a parabola represent?
3. What is a vertex, a line of symmetry, a maximum, and a minimum value?

8.3 Dog Run

Comparing Linear and Quadratic Functions

Learning By Doing Lesson Map

Get Ready

Objectives

In this lesson, you will:

- Use linear and quadratic functions to model a situation.
- Determine the effect on the area of a rectangle when its length or width doubles.

Key Terms

- linear function
- quadratic function

Indiana's Academic Standards

Standard 5 Quadratic Equations and Functions

A1.5.1

Graph quadratic functions.

A1.5.3

Solve problems that can be modeled using quadratic equations, interpret the solutions, and determine whether the solutions are reasonable.

A1.5.5

Sketch and interpret linear and non-linear graphs representing given situations and identify independent and dependent variables.

Lesson Overview

Within the context of this lesson, students will be asked to:

- Create tables of values and graph linear and quadratic functions to model situations.
- Identify the effects on the area of a rectangle when the length or width is doubled.

Essential Questions

The following key questions are addressed in this lesson:

1. What is a dog run?
2. What shape is the dog run for this situation?
3. What is a linear function? What does the graph of a linear function look like?
4. What is a quadratic function? What does the graph of a quadratic function look like?
5. What is meant by the "maximum possible area"?

8.4 Guitar Strings and Other Things

Square Roots and Radicals

Learning By Doing Lesson Map

Get Ready

Objectives

In this lesson, you will:

- Evaluate the square root of a perfect square.
- Approximate a square root.

Key Terms

- square root
- positive square root
- negative square root
- principal square root
- radical symbol
- radicand
- perfect square

Indiana's Academic Standards

Standard 5 Quadratic Equations and Functions

A1.5.3

Solve problems that can be modeled using quadratic equations, interpret the solutions, and determine whether the solutions are reasonable.

Lesson Overview

Within the context of this lesson, students will be asked to:

- Evaluate the square root of a perfect square.
- Approximate the square root of values that are not perfect squares.

Essential Questions

The following key questions are addressed in this lesson:

1. What is a square root?
2. What is a perfect square?
3. How can you approximate a square root?
4. How can you identify a radicand?
5. How can you evaluate a square root?

8

8.5 Tent Designing Competition

Solving by Factoring and Extracting Square Roots

Learning By Doing Lesson Map

Get Ready

Objectives

In this lesson, you will:

- Solve a quadratic equation by factoring.
- Solve a quadratic equation by extracting square roots.

Key Terms

- parabola
- intercepts
- pi

Indiana's Academic Standards

Standard 5 Quadratic Equations and Functions

A1.5.1

Graph quadratic functions.

A1.5.2

Solve quadratic equations in the real number system with real number solutions by factoring, by completing the square, and by using the quadratic formula.

A1.5.3

Solve problems that can be modeled using quadratic equations, interpret the solutions, and determine whether the solutions are reasonable.

A1.5.4

Analyze and describe the relationships among the solutions of a quadratic equation, the zeros of a quadratic function, the *x*-intercepts of the graph of a quadratic function, and the factors of a quadratic expression.

Lesson Overview

Within the context of this lesson, students will be asked to:

- Solve a quadratic function by factoring and setting the factors of the function equal to zero.
- Solve a quadratic function by extracting square roots.

Essential Questions

The following key questions are addressed in this lesson:

1. What is a tent design?
2. What is a parabolic shape for a tent?
3. How can you calculate the intercepts for a quadratic equation?
4. What is pi?
5. How can you solve a quadratic function?

8.6 Kicking a Soccer Ball

Using the Quadratic Formula to Solve Quadratic Equations

Learning By Doing Lesson Map

Get Ready

8

Objectives

In this lesson, you will:

- Solve a quadratic equation by using the quadratic formula.
- Find the value of the discriminant.

Key Terms

- quadratic formula
- discriminant

Indiana's Academic Standards

Standard 5 Quadratic Equations and Functions

A1.5.2

Solve quadratic equations in the real number system with real number solutions by factoring, by completing the square, and by using the quadratic formula.

A1.5.3

Solve problems that can be modeled using quadratic equations, interpret the solutions, and determine whether the solutions are reasonable.

Lesson Overview

Within the context of this lesson, students will be asked to:

- Solve a quadratic function by using the quadratic formula.
- Find the value of the discriminant.
- Identify the number of roots for a quadratic equation based on the sign of the discriminant.

Essential Questions

The following key questions are addressed in this lesson:

1. What is the quadratic formula?
2. What is a solution to a quadratic equation?
3. When can you use the quadratic formula?
4. How can you use the quadratic formula to find solutions for quadratic equations?
5. What is a discriminant and why is it important?

8.7 Pumpkin Catapult

Using a Vertical Motion Model

Learning By Doing Lesson Map

Get Ready

Objective

In this lesson, you will:

- Write and use a vertical motion model.

Key Term

- vertical motion model

Indiana's Academic Standards

Standard 5 Quadratic Equations and Functions

A1.5.1

Graph quadratic functions.

A1.5.2

Solve quadratic equations in the real number system with real number solutions by factoring, by completing the square, and by using the quadratic formula.

A1.5.3

Solve problems that can be modeled using quadratic equations, interpret the solutions, and determine whether the solutions are reasonable.

Lesson Overview

Within the context of this lesson, students will be asked to:

- Write and use a vertical motion model to represent the height of a pumpkin as a function of time.
- Write a quadratic equation to model the vertical motion to represent the height of a pumpkin as a function of horizontal distance.
- Graph a vertical motion model that represents a real-life situation.

Essential Questions

The following key questions are addressed in this lesson:

1. What is a pumpkin catapult?
2. What is a vertical motion model?
3. How can you calculate vertical height if given time or horizontal distance?
4. How can you calculate time or horizontal distance if given vertical height?

8.8 Viewing the Night Sky

Using Quadratic Functions

Learning By Doing Lesson Map

Get Ready

8

Objective

In this lesson, you will:

- Analyze a quadratic function that models the shape of an object.

Key Terms

- axis of symmetry
- vertex
- domain
- range

Indiana's Academic Standards

Standard 1 Relations and Functions

A1.1.2

Identify the domain and range of relations represented by tables, graphs, words, and equations.

Standard 5 Quadratic Equations and Functions

A1.5.1

Graph quadratic functions.

A1.5.3

Solve problems that can be modeled using quadratic equations, interpret the solutions, and determine whether the solutions are reasonable.

Lesson Overview

Within the context of this lesson, students will be asked to:

- Graph quadratic functions.
- Analyze two quadratic functions that model the shape of a telescope lens.
- Evaluate quadratic functions for given values.

Essential Questions

The following key questions are addressed in this lesson:

1. What is a telescope?
2. What is the vertex for a telescope lens?
3. How can you find the domain and the range for a telescope lens?
4. What is the quadratic function used to find the lens height?
5. How can you evaluate a quadratic function for given values?

9.1 The Museum of Natural History

Powers and Prime Factorization

Learning By Doing Lesson Map

Get Ready

Objectives

In this lesson, you will:

- List the factors of numbers.
- Identify prime and composite numbers.
- Write the prime factorizations of numbers.

Key Terms

- factors
- prime number
- composite number
- prime factorization

Indiana's Academic Standards

Standard 1 Number Sense and Computation

8.1.3

Use the laws of exponents for integer exponents and evaluate expressions with negative integer exponents.

Lesson Overview

Within the context of this lesson, students will be asked to:

- Find and list the whole number factors of given numbers.
- Classify whole numbers as either prime, composite, or neither.
- Find and write the prime factorization of given numbers.
- Find and list any common whole number factors for given pairs of numbers.

Essential Questions

The following key questions are addressed in this lesson:

1. What is a museum of natural history?
2. What is a prime number?
3. What is a composite number?
4. What is a factor?
5. How can you write the prime factorization for a given number?

9.2 Bits and Bytes

Multiplying and Dividing Powers

Learning By Doing Lesson Map

Get Ready

Objectives

In this lesson, you will:

- Write numbers as powers.
- Multiply powers.
- Divide powers.

Key Terms

- power
- exponent
- product
- quotient

Materials

- calculator

Indiana's Academic Standards

Standard 4 Polynomials

A1.4.1

Use the laws of exponents for variables with exponents and multiply, divide, and find powers of variables with exponents.

Lesson Overview

Within the context of this lesson, students will be asked to:

- Work with whole numbers of large magnitude.
- Write large numbers as powers of the same base.
- Multiply and divide large numbers.
- Develop a rule to multiply large numbers by finding the product of powers of the same base.
- Develop a rule to divide large numbers by finding the quotient of powers of the same base.

Essential Questions

The following key questions are addressed in this lesson:

1. What is computer memory space?
2. What is a quotient?
3. What is a product?
4. What is a power?
5. What is an exponent?

9

9.3 As Time Goes By

Zero and Negative Exponents

Learning By Doing Lesson Map

Get Ready

Objectives

In this lesson, you will:

- Write a number as a power.
- Evaluate powers with positive, negative, and zero exponents.

Key Terms

- positive exponent
- negative exponent
- zero exponent

Indiana's Academic Standards

Standard 4 Polynomials

A1.4.1

Use the laws of exponents for variables with exponents and multiply, divide, and find powers of variables with exponents.

Lesson Overview

Within the context of this lesson, students will be asked to:

- Use unit analysis to find equivalent times.
- Write numbers as powers.
- Evaluate powers that have positive exponents, powers that have negative exponents, and powers that have an exponent of zero.
- Evaluate the products and the quotients of powers with various exponents.

Essential Questions

The following key questions are addressed in this lesson:

1. What is a computer operation or communication?
2. What is a positive exponent?
3. What is a negative exponent?
4. What is a power with an exponent of zero?
5. How can you evaluate powers with various exponents?

9.4 Large and Small Measurements

Scientific Notation

Learning By Doing Lesson Map

Get Ready

Objectives

In this lesson, you will:

- Write numbers in scientific notation.
- Write numbers in standard form.

Key Terms

- standard form
- scientific notation

Indiana's Academic Standards

Standard 4 Polynomials

A1.4.1

Use the laws of exponents for variables with exponents and multiply, divide, and find powers of variables with exponents.

Lesson Overview

Within the context of this lesson, students will be asked to:

- Write very large and very small numbers that are represented in standard form by using scientific notation.
- Write numbers that are represented in scientific notation by using standard form.

Essential Questions

The following key questions are addressed in this lesson:

1. What is scientific notation?
2. When does it help to write numbers in scientific notation?
3. How can you write a number in scientific notation?
4. What is standard form?
5. How can you write a number in standard form?

9

9.5 The Beat Goes On

Properties of Powers

Learning By Doing Lesson Map

Get Ready

Objectives

In this lesson, you will:

- Use the power of a power property.
- Use the power of a product property.
- Use the power of a quotient property.

Key Terms

- power
- product
- quotient

Materials

- Poster paper for the closure activity

Indiana's Academic Standards

Standard 4 Polynomials

A1.4.1

Use the laws of exponents for variables with exponents and multiply, divide, and find powers of variables with exponents.

Lesson Overview

Within the context of this lesson, students will be asked to:

- Calculate the lengths of radii when given diameters.
- Calculate the area of circular drumheads.
- Develop and apply the power of a power property.
- Develop and apply the power of a product property.
- Develop and apply the power of a quotient property.

Essential Questions

The following key questions are addressed in this lesson:

1. What is a drumhead?
2. How can you calculate the area of a circle?
3. How can you calculate the length of a radius if given the length of the diameter of a circle?
4. What is the power of a power property?
5. What is the power of a product property?
6. What is the power of a quotient property?

9.6 Sailing Away

Radicals and Rational Exponents

Learning By Doing Lesson Map

Get Ready

Objectives

In this lesson, you will:

- Find the *n*th root of a number.
- Write an expression in radical form.
- Write an expression in rational exponent form.

Key Terms

- cube root
- index
- *n*th root
- radicand
- rational exponent
- radical

Indiana's Academic Standards

Standard 4 Polynomials

A1.4.1

Use the laws of exponents for variables with exponents and multiply, divide, and find powers of variables with exponents.

Standard 6 Rational and Radical Expressions and Equations

A1.6.3

Simplify radical expressions involving square roots.

Lesson Overview

Within the context of this lesson, students will be asked to:

- Evaluate expressions to solve equations.
- Solve for the *n*th root of numbers.
- Develop and apply a rule for writing expressions in radical form.
- Develop and apply a rule for writing expressions in rational exponent form.

Essential Questions

The following key questions are addressed in this lesson:

1. What is a root?
2. What is a radical?
3. What is a rational exponent?
4. How can you write a radical as a power?
5. How can you write a power as a radical?

9

10.1 Water Balloons

Polynomials and Polynomial Functions

Learning By Doing Lesson Map

Get Ready

Objectives

In this lesson, you will:

- Identify terms and coefficients of polynomials.
- Classify polynomials by the number of terms.
- Classify polynomials by degree.
- Write polynomials in standard form.
- Use the Vertical Line Test to determine whether equations are functions.

Key Terms

- polynomial
- term
- coefficient
- degree
- monomial
- binomial
- trinomial
- standard from
- Vertical Line Test

Indiana's Academic Standards

Standard 1 Relations and Functions

A1.1.1

Determine whether a relation represented by a table, graph, words or equation is a function or not a function and translate among tables, graphs, words and equations.

Lesson Overview

Within the context of this lesson, students will be asked to:

- Identify if an expression is a polynomial.
- Identify terms and coefficients of polynomials.
- Classify polynomials by the number of terms as well as by the highest degree term.
- Write polynomials in standard form.
- Apply the Vertical Line Test to graphs of equations to determine whether the graphed equations are functions.

Essential Questions

The following key questions are addressed in this section:

1. What is a polynomial?
2. What is vertical motion?
3. How can you determine the coefficient of a term?
4. How can you determine the degree of a term? How can you determine the degree of a polynomial?
5. How can you determine whether a graphed equation is a function?
6. How can you write a function in standard form?
7. What is a monomial? What is a binomial? What is a trinomial?

10.2 Play Ball!

Adding and Subtracting Polynomials

Learning By Doing Lesson Map

Get Ready

Objectives

In this lesson, you will:

- Add polynomials.
- Subtract polynomials.

Key Terms

- combine like terms
- distributive property
- add
- subtract

Indiana's Academic Standards

Standard 4 Polynomials

A1.4.2

Add, subtract and multiply polynomials and divide polynomials by monomials.

Lesson Overview

Within the context of this lesson, students will be asked to:

- Evaluate two polynomials for the same x-value, then find the sum of the results.
- Add two polynomials, then evaluate the sum for a given value of x.
- Compare the results for the two sums.
- Add polynomials.
- Subtract polynomials.
- Simplify expressions requiring addition and subtraction of polynomials.

Essential Questions

The following key questions are addressed in this section:

1. What is the distributive property?
2. What are like terms?
3. How can you combine like terms?
4. How can you add polynomials?
5. How can you subtract polynomials?

10.3 Se Habla Español

Multiplying and Dividing Polynomials

Learning By Doing Lesson Map

Get Ready

Objectives

In this lesson, you will:

- Use an area model to multiply polynomials.
- Use distributive properties to multiply polynomials.
- Use long division to divide polynomials.

Key Terms

- area model
- distributive property
- divisor
- dividend
- remainder

Indiana's Academic Standards

Standard 4 Polynomials

A1.4.2

Add, subtract and multiply polynomials and divide polynomials by monomials.

Lesson Overview

Within the context of this lesson, students will be asked to:

- Evaluate polynomials for a given value of x and then multiply the results.
- Multiply the polynomials, then evaluate the product for the same given value of x.
- Compare the answer from each process and identify them as resulting in the same value.
- Follow a similar procedure for division of polynomials.
- Use long division to divide polynomials.
- Students will identify the divisor, dividend, and remainder when dividing polynomials.

Essential Questions

The following key questions are addressed in this section:

1. What is an area model?
2. What is the distributive property and how can it help you to multiply polynomials?
3. How can you multiply polynomials?
4. What is long division?
5. How can you use long division to divide polynomials?

10.4 Making Stained Glass

Multiplying Binomials

Learning By Doing Lesson Map

Get Ready

Objectives

In this lesson, you will:

- Use the FOIL pattern to multiply binomials.
- Use formulas to find special products.

Key Terms

- FOIL pattern
- square of a binomial sum
- square of a binomial difference

Indiana's Academic Standards

Standard 4 Polynomials

A1.4.2

Add, subtract and multiply polynomials and divide polynomials by monomials.

Lesson Overview

Within the context of this lesson, students will be asked to:

- Multiply binomials.
- Develop patterns and formulas for multiplying binomials.
- Use the FOIL Pattern to multiply binomials efficiently.
- Use formulas to find the products of the most common types of special binomials.

Essential Questions

The following key questions are addressed in this section:

1. What is a binomial?
2. What is the product of two binomials?
3. How can you use the FOIL Pattern to multiply two binomials?
4. What special binomial products are common?
5. What patterns for multiplying binomials exist?

10.5

Suspension Bridges

Factoring Polynomials

Learning By Doing Lesson Map

Get Ready

Objectives

In this lesson, you will:

- Factor a polynomial by factoring out a common factor.
- Factor a polynomial of the form $x^2 + bx + c$.
- Factor a polynomial of the form $ax^2 + bx + c$.

Key Terms

- factor
- linear factor
- trinomial
- FOIL pattern

Indiana's Academic Standards

Standard 4 Polynomials

A1.4.3

Factor common terms from polynomials and factor quadratic expressions.

Lesson Overview

Within the context of this lesson, students will be asked to:

- Solve a quadratic equation using the quadratic formula.
- Factor out a common monomial factor to solve a quadratic function.
- Factor trinomials of the form $x^2 + bx + c$.
- Factor trinomials of the form $ax^2 + bx + c$.

Essential Questions

The following key questions are addressed in this section:

1. What is a suspension bridge?
2. What is a factor?
3. What is a linear factor?
4. How can you factor a trinomial of the form $x^2 + bx + c$?
5. How can you factor a trinomial of the form $ax^2 + bx + c$?

10.6 Swimming Pools

Rational Expressions

Learning By Doing Lesson Map

Get Ready

Objectives

In this lesson, you will:

- Find the domains of rational expressions.
- Simplify rational expressions.
- Add, subtract, multiply, and divide rational expressions.

Key Terms

- rational expression
- domain
- excluded value
- restricting the domain

Indiana's Academic Standards

Standard 6 Rational and Radical Expressions and Equations

A1.6.1

Add, subtract, multiply, divide, reduce, and evaluate rational expressions with polynomial denominators. Simplify rational expressions with linear and quadratic denominators, including denominators with negative exponents.

Lesson Overview

Within the context of this lesson, students will be asked to:

- Represent a ratio as a rational function.
- Determine the domains for rational expressions.
- Simplify rational expressions.
- Add, subtract, multiply, and divide rational expressions.

Essential Questions

The following key questions are addressed in this section:

1. What is a rational expression?
2. How are rational expressions similar to fractions? How are they different?
3. How can you add rational expressions? How can you subtract rational expressions? How can you multiply rational expressions? How can you divide rational expressions?
4. How can you restrict the domain for rational expressions?
5. Why must you restrict the domain for rational expressions?

10

11.1 Your Best Guess

Introduction to Probability

Learning By Doing Lesson Map

Get Ready

Objective

In this lesson, you will:

- Find the probability of an event.
- Find the odds in favor of an event.
- Find the odds against an event.

Key Terms

- outcomes
- sample space
- event
- probability
- probability of the event
- favorable outcome
- odds in favor
- odds against
- complementary events

Materials

- Coins for each student to toss in Problem 1
- Number cubes for each student to roll in Problem 1

Lesson Overview

Within the context of this lesson, students will be asked to:

- Simulate guessing for a true or false question with the toss of a coin and guessing for a 6 choice multiple-choice question with the rolling of a number cube.
- Calculate probabilities for events.
- Calculate the probability of a complementary event.
- Calculate the odds in favor of an event occurring.
- Calculate the odds against an event occurring

Essential Questions

The following key questions are addressed in this section:

1. What is an event?
2. What is a sample space?
3. What is a probability?
4. What are odds?

11.2 What's in the Bag

Theoretical and Experimental Probabilities

Learning By Doing Lesson Map

Get Ready

Objective

In this lesson, you will:

- Find theoretical probabilities.
- Find experimental probabilities.
- Compare theoretical and experimental probabilities.

Key Terms

- theoretical probability
- experimental probability
- experiment
- trial
- success

Materials

- 30 small slips of paper for each student
- 1 bag or container for each student
- A marker, pen, or pencil for each student

Lesson Overview

Within the context of this lesson, students will be asked to:

- Calculate the theoretical probability for an experiment.
- Prepare and conduct an experiment.
- Calculate experimental probabilities.
- Compare theoretical and experimental probabilities.
- Consider changes to the experiment that would likely result in experimental probabilities that would more closely approximate the theoretical probabilities.

Essential Questions

The following key questions are addressed in this section:

1. What is an experiment?
2. What is a trial for an experiment?
3. What is a success?
4. What is a theoretical probability?
5. What is an experimental probability?

11.3 A Brand New Bag

Using Probabilities to Make Predictions

Learning By Doing Lesson Map

Get Ready

Objective

In this lesson, you will:

- Use experimental probabilities to make predictions.

Key Terms

- trial
- experimental probability

Materials

- 30 small slips of paper for each student
- 1 bag or container for each student
- A marker, pen, or pencil for each student

Lesson Overview

Within the context of this lesson, students will be asked to:

- Prepare and conduct an experiment.
- Calculate experimental probabilities.
- Use experimental probabilities to make predictions about the contents of the bag or container.
- Consider changes to the experiment that would likely result in experimental probabilities that would more closely approximate the actual probabilities.

Essential Questions

The following key questions are addressed in this section:

1. What is an experiment?
2. What is a trial for an experiment?
3. What is an experimental probability?

11.4

Fun with Number Cubes

Graphing Frequencies of Outcomes

Learning By Doing Lesson Map

Get Ready

Objective

In this lesson, you will:

- Use a line plot to graph frequencies of outcomes.
- Find and compare probabilities.

Key Terms

- line plot

Materials

- A pair of 6-sided number cubes for each student

Lesson Overview

Within the context of this lesson, students will be asked to:

- Create a line plot for data based on theoretical probabilities.
- Create a line plot for the data from an actual experiment.
- Calculate and compare theoretical and experimental probabilities.

Essential Questions

The following key questions are addressed in this section:

1. What is a line plot?
2. What type of information can be determined from a line plot?
3. How can the shape of a line plot for theoretical data differ from the shape of a line plot for experimental data?

11.5 Going to the Movies

Counting and Permutations

Learning By Doing Lesson Map

Get Ready

Objective

In this lesson, you will:

- Find the number of permutations of n objects.
- Simplify expressions that involve factorials.

Key Terms

- tree diagram
- Fundamental Counting Principle
- permutation
- factorial

Materials

- The students may need a calculator for the warm up exercises.

Lesson Overview

Within the context of this lesson, students will be asked to:

- Calculate the number of ways that 5 friends can sit beside each other in a row.
- Calculate the number of permutations of a given number n of objects.
- Simplify expressions that involve factorials.

Essential Questions

The following key questions are addressed in this section:

1. What is a tree diagram?
2. How can a tree diagram help you to find the number of possible outcomes for a situation?
3. What is a permutation?
4. How can you calculate the number of permutations of n objects?
5. What is a factorial?

11.6 Going Out for Pizza

Permutations and Combinations

Learning By Doing Lesson Map

Get Ready

Objective

In this lesson, you will:

- Find the number of permutations of n objects taken r at a time.
- Find the number of combinations of n objects taken r at a time.

Key Terms

- permutations
- permutation of n distinct objects taken r at a time
- combinations
- combination of n distinct objects taken r at a time

Materials

- The students may need a calculator for the warm up exercises.

Lesson Overview

Within the context of this lesson, students will be asked to:

- Calculate the number of permutations of n objects taken r objects at a time.
- Calculate the number of permutations of n objects taken without concern for order r at a time.
- Determine the relationship between permutations and combinations.

Essential Questions

The following key questions are addressed in this section:

1. What is a permutation?
2. What is a combination?
3. Is a permutation or a combination an ordered arrangement of objects?
4. Is a permutation or a combination a group of objects without concern for order?
5. How can you calculate the number of permutations? How can you calculate the number of combinations?

11.7 Picking Out Socks

Independent and Dependent Events

Learning By Doing Lesson Map

Get Ready

Objective

In this lesson, you will:

- Find the probability of independent events.
- Find the probability of dependent events.
- Use a tree diagram to find probabilities.

Key Terms

- compound events
- independent event
- dependent event
- tree diagram

Materials

- You may want to bring in a spinner from a game to help students understand Investigate Problem 1 Question 1 more easily.

Lesson Overview

Within the context of this lesson, students will be asked to:

- Create organized lists and tree diagrams to represent the random selection of socks from a drawer with replacement and without replacement.
- Calculate probabilities of compound events for independent events.
- Calculate probabilities of compound events for dependent events.
- Use a tree diagram and an organized list to more easily calculate probabilities.

Essential Questions

The following key questions are addressed in this section:

1. What is a tree diagram?
2. What is a compound event?
3. What are independent events?
4. What are dependent events?

11.8 Probability on the Shuffleboard Court

Geometric Probabilities

Learning By Doing Lesson Map

Get Ready

Objective

In this lesson, you will:

- Find and use geometric probabilities.

Key Terms

- triangle
- parallelogram
- trapezoid
- congruent
- area
- geometric probability

Lesson Overview

Within the context of this lesson, students will be asked to:

- Compare areas and find ratios of areas.
- Calculate the geometric probabilities for a shuffleboard court.
- Use the concept of congruent shapes and areas to find geometric probabilities.
- Identify the geometric shapes of a triangle, parallelogram, and a trapezoid.

Essential Questions

The following key questions are addressed in this section:

1. What is a geometric probability?
2. What is a triangle? What is a parallelogram? What is a trapezoid?
3. What are congruent figures?
4. What is area?

11.9 Game Design

Geometric Probabilities and Fair Games

Learning By Doing Lesson Map

Get Ready

Objective

In this lesson, you will:

- Use geometric probabilities to find values in a game.
- Determine whether a game is fair.
- Alter the rules of a game so that it is fair.

Key Terms

- geometric probability
- fair game

Materials

- You may want to create a spinner to simulate the spinner in Problem 1.
- You will need a pair of number cubes for each group of 2 students in the class.

Lesson Overview

Within the context of this lesson, students will be asked to:

- Calculate geometric probabilities.
- Calculate the expected values for the outcomes of a game of chance.
- Compare probabilities to determine whether the possible outcomes are equally likely and whether the game is fair.
- Alter the rules of a game that are unfair to develop a related game that is fair.

Essential Questions

The following key questions are addressed in this section:

1. What is a geometric probability?
2. What is a fair game?
3. When is a game determined to be fair?

12.1

Taking the PSAT

Measures of Central Tendency

Learning By Doing Lesson Map

Get Ready

Objectives

In this lesson, you will:

- Create a stem-and-leaf plot.
- Determine the distribution of a data set.
- Find the mean, median, and mode of a data set.
- Compare the mean and median for different distributions.

Key Terms

- stem-and-leaf plot
- distribution
- measure of central tendency
- median
- mode
- mean

Indiana's Academic Standards

Standard 7 Data Analysis

A1.7.1

Organize and display data using appropriate methods to detect patterns and departures from patterns. Summarize the data using measures of center (mean, median) and spread (range, percentiles, variance, standard deviation). Compare data sets using graphs and summary statistics.

A1.7.2

Distinguish between random and non-random sampling methods, identify possible sources of bias in sampling, describe how such bias can be controlled and reduced, evaluate the characteristics of a good survey and well-designed experiment, design simple experiments or investigations to collect data to answer questions of interest, and make inferences from sample results.

Lesson Overview

Within the context of this lesson, students will be asked to:

- Create a stem-and-leaf plot.
- Determine the shape of the distribution of data.
- Calculate the mean, median, and mode for a data set and recognize these as measures of central tendency.
- Analyze and compare measures of central tendency for various distributions.

Essential Questions

The following key questions are addressed in this lesson:

1. What is a measure of central tendency?
2. What is a distribution?
3. How can you calculate the mean, median, and mode for a set of data?
4. How can you identify the shape of the distribution for a set of data?
5. How can you create a stem-and-leaf plot for a set of data?

12.2 Compact Discs

Collecting and Analyzing Data

Learning By Doing Lesson Map

Get Ready

Objectives

In this lesson, you will:

- Collect and organize data.
- Find the mean and median of a data set.
- Determine how data values affect the mean and median of a data set.

Key Terms

- survey
- mean
- median
- sample size

Materials

- Small pieces of paper for the motivating activity
- Additional copies of the pages of this lesson on paper or transparency for each group if having the students conduct the survey and then each contributing data from some of their reviewers to create a combined set of data for the group

Indiana's Academic Standards

Standard 7 Data Analysis

A1.7.2

Distinguish between random and non-random sampling methods, identify possible sources of bias in sampling, describe how such bias can be controlled and reduced, evaluate the characteristics of a good survey and well-designed experiment, design simple experiments or investigations to collect data to answer questions of interest, and make inferences from sample results.

Lesson Overview

Within the context of this lesson, students will be asked to:

- Conduct a survey of ratings for music CDs by students.
- Organize the data into a chart.
- Calculate the mean and median of the data.
- Determine how data values affect the mean and median of the data set.
- Discuss how sample size can affect the results of a survey.

You may want to do the motivating activity and assign part (B) and Questions 1 through 3 of Problem 1 as a homework assignment the day or two before you do this activity in class.

Essential Questions

The following key questions are addressed in this lesson:

1. What is a survey?
2. Why should we conduct surveys?
3. How can you conduct a survey?
4. What is a sample size and why is it important?
5. How can you calculate the mean and median for survey data?

12.3 Breakfast Cereals

Quartiles and Box-and-Whisker Plots

Learning By Doing Lesson Map

Get Ready

Objectives

In this lesson, you will:

- Find the range and extremes of a data set.
- Find the first, second, and third quartiles of a data set.
- Represent a data set graphically by using a box-and-whisker plot.
- Identify outliers in a data set.
- Find percentiles of a data set.
- Find the IQR of a data set.

Key Terms

- range
- extreme
- median
- quartile
- *n*th percentile
- box-and-whisker plot
- outlier
- percentile
- interquartile range

Indiana's Academic Standards

Standard 7 Data Analysis

A1.7.1

Organize and display data using appropriate methods to detect patterns and departures from patterns. Summarize the data using measures of center (mean, median) and spread (range, percentiles, variance, standard deviation). Compare data sets using graphs and summary statistics.

Lesson Overview

Within the context of this lesson, students will be asked to:

- Calculate the range and interquartile range for a given set of data.
- Calculate quartiles and percentiles for a given set of data.
- Construct a box-and-whisker plot for a given set of data.
- Identify the effect of an outlier in a data set on a box-and-whisker plot.

Essential Questions

The following key questions are addressed in this lesson:

1. What is a range for data?
2. What is an interquartile range?
3. How do you construct a box-and-whisker plot?
4. How do you find the quartiles for data?
5. How do you find a given percentile or find the percentile for a given number?
6. What are outliers?

12.4 Home Team Advantage?

Sample Variance and Standard Deviation

Learning By Doing Lesson Map

Get Ready

Objectives

In this lesson, you will:

- Use a line plot to represent a data set.
- Find the deviation of a data set.
- Find the sample variance of a data set.
- Find the sample standard deviation of a data set.

Key Terms

- line plot
- deviation
- absolute deviation
- mean absolute deviation
- sample variance
- sample standard deviation

Indiana's Academic Standards

Standard 7 Data Analysis

A1.7.1

Organize and display data using appropriate methods to detect patterns and departures from patterns. Summarize the data using measures of center (mean, median) and spread (range, percentiles, variance, standard deviation). Compare data sets using graphs and summary statistics.

Lesson Overview

Within the context of this lesson, students will be asked to:

- Create a line plot to represent a set of data.
- Calculate the deviations for each value in a data set.
- Calculate the square of each deviation, the sum of the squares and divide the sum by one less than the number of sample values to calculate the sample variance of a set of data.
- Calculate the sample standard deviation of a data set by finding the positive square root of the sample variance for the data set.

Essential Questions

The following key questions are addressed in this lesson:

1. What is a line plot?
2. What is a deviation?
3. How can you calculate the sample variance for a set of data?
4. How can you calculate the standard deviation for a set of data?
5. What do the deviations, the sample variance, and the sample standard deviation represent?

13.1 Solid Carpentry

The Pythagorean Theorem and Its Converse

Learning By Doing Lesson Map

Get Ready

Objectives

In this lesson, you will:

- Use the Pythagorean Theorem to find the side length of a right triangle.
- Use the converse of the Pythagorean Theorem to determine whether a triangle is a right triangle.

Key Terms

- legs
- Pythagorean Theorem
- hypotenuse
- converse
- converse of the Pythagorean Theorem

Materials

- graph paper
- rulers
- scissors

Indiana's Academic Standards

Standard 3 Data Analysis and Probability

8.3.3

Explain why the Pythagorean Theorem is valid using a variety of methods and use the Pythagorean Theorem and its converse to calculate lengths of line segments.

Lesson Overview

Within the context of this lesson, students will be asked to:

- Verify the Pythagorean Theorem visually with constructions.
- Identify the legs and hypotenuse of a right triangle.
- Apply the Pythagorean Theorem.
- Apply the converse of the Pythagorean Theorem.

Essential Questions

The following key questions are addressed in this lesson:

1. What is a carpenter and what does a carpenter do?
2. What is the Pythagorean Theorem and how can you use it?
3. What is the converse of the Pythagorean Theorem and how can you use it?
4. What are the parts of a right triangle?

13.2 Location, Location, Location

The Distance and Midpoint Formulas

Learning By Doing Lesson Map

Get Ready

Objectives

In this lesson, you will:

- Find the distance between two points in the coordinate plane.
- Find the midpoint between two points in the coordinate plane.

Key Terms

- Distance Formula
- midpoint
- Midpoint Formula

Indiana's Academic Standards

Standard 3 Data Analysis and Probability

8.3.3

Explain why the Pythagorean Theorem is valid using a variety of methods and use the Pythagorean Theorem and its converse to calculate lengths of line segments.

Lesson Overview

Within the context of this lesson, students will be asked to:

- Calculate the distance between 2 points by creating right triangles and using the Pythagorean Theorem to solve for the distance.
- Develop and apply the Distance Formula.
- Develop and apply the Midpoint Formula.

Essential Questions

The following key questions are addressed in this lesson:

1. What is a map?
2. What is a distance?
3. What is a midpoint?
4. What is a formula?

13.3 "Old Mathematics"

Completing the Square and Deriving the Quadratic Formula

Learning By Doing Lesson Map

Get Ready

Objectives

In this lesson, you will:

- Solve quadratic equations by completing the square.
- Derive the Quadratic Formula.

Key Terms

- perfect square trinomial
- factor
- complete the square
- Quadratic Formula

Indiana's Academic Standards

Standard 5 Quadratic Equations and Functions

A1.5.2

Solve quadratic equations in the real number system with real number solutions by factoring, by completing the square, and by using the quadratic formula.

Lesson Overview

Within the context of this lesson, students will:

- Represent polynomials with the area of geometric figures.
- Complete squares geometrically and algebraically.
- Use the method of completing the square to solve quadratic equations.
- Develop the Quadratic Formula from completing the square.

Essential Questions

These key questions are addressed in the lesson:

1. What is a perfect square trinomial?
2. What is a factor?
3. How can you complete the square to solve a quadratic equation?
4. When should you use the Quadratic Formula to solve a quadratic equation?

13.4 Learning to Be a Teacher

Vertex Form of a Quadratic Equation

Learning By Doing Lesson Map

Get Ready

Objectives

In this lesson, you will:

- Write quadratic equations in standard form.
- Write quadratic equations in factored form.
- Write quadratic equations in vertex form.

Key Terms

- standard form
- factored form
- vertex form

Indiana's Academic Standards

Standard 5 Quadratic Equations and Functions

A1.5.4

Analyze and describe the relationships among the solutions of a quadratic equation, the zeros of a quadratic function, the x-intercepts of the graph of a quadratic function, and the factors of a quadratic expression.

Lesson Overview

Within the context of this lesson, students will be asked to:

- Write equations in standard form, in factored form, and in vertex form when given one of these forms of the equation.
- Graph quadratic functions.
- Identify the direction a parabola will open based on the coefficient of the x^2 term for a quadratic equation written in standard form.
- Identify the x-intercepts of a parabola for a quadratic equation written in factored form.
- Identify the vertex of a parabola for a quadratic equation written in vertex form.

Essential Questions

The following key questions are addressed in this lesson:

1. What is a quadratic equation?
2. What will the graph of a quadratic equation look like?
3. How can you write a quadratic equation in standard form?
4. How can you write a quadratic equation in factored form?
5. How can you write a quadratic equation in vertex form?

13.5 Screen Saver

Graphing by Using Parent Functions

Learning By Doing Lesson Map

Get Ready

Objectives

In this lesson, you will:

- Use the graph of a parent function to describe the graph of a function.
- Identify the parent function given a function.
- Write equations of functions based on the graphs of functions.

Key Terms

- parent function
- reflection
- translation

Materials

- Colored pencils
- Transparency graph grid overlays

Indiana's Academic Standards

Standard 5 Quadratic Equations and Functions

A1.5.1

Graph quadratic functions.

Lesson Overview

Within the context of this lesson, students will be asked to:

- Graph a parent function and related functions.
- Describe the graph of a function in relation to the graph of the parent function.
- Identify the parent function of a given function.
- Write the equations of functions based on the graphs of functions.

Essential Questions

The following key questions are addressed in this lesson:

1. What is a parent function?
2. What is a translation?
3. What is a reflection?

13.6 Science Fair

Introduction to Exponential Functions

Learning By Doing Lesson Map

Get Ready

Objectives

In this lesson, you will:

- Write and graph exponential functions.
- Identify translations of exponential functions.

Key Terms

- exponential function

Indiana's Academic Standards

Standard 6 Exponential and Logarithmic Functions

A2.6.1

Analyze, describe, and sketch graphs of exponential functions by examining intercepts, zeros, domain and range, and asymptotic and end behavior.

Lesson Overview

Within the context of this lesson, students will be asked to:

- Write and graph an exponential function to model exponential growth.
- Evaluate and graph various exponential functions.
- Identify transformations of exponential functions.
- Identify the appropriate domain and range for exponential functions.

Essential Questions

The following key questions are addressed in this lesson:

1. What is an exponential function?
2. What are the requirements for an exponential function?
3. How can you evaluate an exponential function?
4. How can you graph an exponential function?

13.7

Money Comes and Money Goes

Exponential Growth and Decay

Learning By Doing Lesson Map

Get Ready

Objectives

In this lesson, you will:

- Write and use an exponential growth model.
- Write and use an exponential decay model.

Key Terms

- simple interest
- compound interest
- exponential growth model
- exponential growth
- growth rate
- growth factor
- exponential decay model
- exponential decay
- decay rate
- decay factor

Indiana's Academic Standards

Standard 6 Exponential and Logarithmic Functions

A2.6.4

Solve problems that can be modeled using exponential and logarithmic equations, interpret the solutions, and determine whether the solutions are reasonable using technology as appropriate.

Lesson Overview

Within the context of this lesson, students will be asked to:

- Write, evaluate, and graph an exponential growth model to represent the balance in a savings account after interest is accrued.
- Write, evaluate, and graph an exponential decay model to represent the value of a car after it is purchased.
- Use the models to predict values.

Essential Questions

The following key questions are addressed in this lesson:

1. What is exponential growth?
2. What is exponential decay?
3. What is simple interest?
4. What is compound interest?

13.8 Camping

Special Topic: Logic

Learning By Doing Lesson Map

Get Ready

Objectives

In this lesson, you will:

- Prove a statement using a direct proof.
- Prove a statement using an indirect proof.
- Find a counterexample.

Key Terms

- logical reasoning
- proof
- direct proof
- counterexample
- indirect proof

Lesson Overview

Within the context of this lesson, students will be asked to:

- Construct direct proofs to prove algebraic relationships.
- Construct an indirect proof to prove an algebraic statement.
- Develop counterexamples to disprove algebraic relationships.

Essential Questions

The following key questions are addressed in this lesson:

1. What is logical reasoning?
2. What is a proof?
3. What is a direct proof?
4. What is an indirect proof?
5. What is a counterexample?

ALGEBRA I

Teacher's Implementation Guide

VOLUME 2

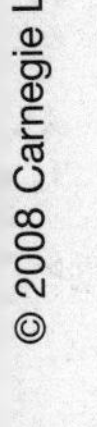

Pittsburgh, PA
Phone 888.851.7094
Fax 412.690.2444

www.carnegielearning.com

Acknowledgements

We would like to thank those listed below who helped to prepare the Cognitive Tutor® ***Algebra I*** Teacher's Implementation Guide.

William S. Hadley
Jessica Pflueger
Kathy Dickensheets
Michele Covatto
The Carnegie Learning Development Team

Algebra was used in a variety of ways to design the building and surrounding grounds on the front cover. Architects designed the sides of the pyramid with slopes that were structurally sound and pleasing to the eye. Stone masons calculated the number of bricks needed for the sidewalk and designed the pattern in which they were laid. Landscapers mixed the fertilizer for the grass and shrubbery with the correct ratio of water and chemicals. As you work through the Cognitive Tutor Algebra I text and software, you will see additional opportunities for using Algebra in your everyday activities.

ISBN-13 978-1-932409-63-5
ISBN-10 1-932409-63-7
Teacher's Implementation Guide, Volume 2

Printed in the United States of America
1-2006-VH
2-2006-VH
3-2007-VH
4-4/2008-HPS

Dear Student,

You are about to begin an exciting adventure using mathematics, the language of science and technology. As you sit in front of a computer screen or video game, ride in an automobile, fly in a plane, talk on a cellular phone, or use any of the tools of modern society, realize that mathematics was critical in its invention, design, and production.

The workplace today demands that employees be technologically literate, work well in teams, and be self-starters. At Carnegie Learning, we have designed a mathematics course that uses state-of-the-art computer software with collaborative classroom activities.

As you use the Cognitive Tutor® ***Algebra I*** software, it actually learns about you as you learn about mathematics. As you work, you will receive "just-in-time" instruction so that you are always ready for the next problem. In the classroom, you will work with your peers to solve real-world problem situations. Working in groups, you will learn to use multiple representations to analyze questions and write or present your answers.

Throughout the entire process, your teacher will be a facilitator and guide in support of your learning. As a result, you will become a self-sufficient learner, moving through the software and Student Text at your own rate and discovering solutions to problems that you never thought were possible to solve.

Throughout this year, have fun while Learning by Doing!

The Cognitive Tutor® *Algebra I* Development Team

Contents

Contents

Contents

Collaborative Classroom

As you begin the process of planning for the school year, you will want to give serious consideration to how your classroom is structured. Early research on teaching and learning has revealed that what happens in the classroom in the first three days determines the environment for the entire year. This insight is important as you begin to think about your classroom and the Cognitive Tutor® ***Algebra I*** curriculum. An effective implementation of the curriculum is most likely to occur in the collaborative classroom, a classroom in which knowledge is shared.

Carnegie Learning's philosophy—Learning by Doing®—captures the belief that students develop understanding and skill by taking an active role in their environment. Furthermore, it is Carnegie Learning's belief that effective communication and collaboration are essential skills for the successful learner. It is through dialogue and discussion of different strategies and perspectives that students become knowledgeable independent learners. These beliefs can be realized in the collaborative classroom.

Defining a Collaborative Classroom

A collaborative classroom is an environment in which knowledge and authority are shared between the teacher and the students. In a collaborative classroom, teachers are facilitators and students are active participants. All students, not segregated by ability level, interest, or achievement, benefit from the environment created in the collaborative classroom.

Teachers in the collaborative classroom combine their extensive knowledge about teaching and learning, content, and skills with the informal and formal knowledge, strategies, and individual experiences of their students. The collaborative classroom differs from the traditional classroom in which the teacher is seen as an information giver (Tinzmann, M.B.; Jones, B.F.; Fennimore, T.F.; Bakker, J; Fine, C.; and Pierce, J., 1990).

Characteristics of the Collaborative Classroom

The collaborative classroom is identified by discussion, with in-depth accountable talk and two-way interactions, whether among members of the whole class or small groups. It is a well-structured environment in which questioning and dialogue are valued and appropriate parameters are set so that active learning can occur. Careful planning by the teacher ensures that students can work together to attain individual and collective goals and to develop learning strategies.

In the collaborative classroom, students are encouraged to take responsibility for their learning through monitoring and reflective self-evaluation. The collaborative classroom is one in which teachers spend more time in true academic interactions as they guide students to search for information and help students to share what they know. As facilitators, teachers have the opportunity to provide the correct amount of help to individual students by providing appropriate hints, probing questions, feedback, and help in clarifying thinking or the use of a particular strategy.

Collaborative Learning versus Cooperative Learning

Two types of learning occur in the collaborative classroom; collaborative learning, which focuses on interaction, and cooperative learning, which is a structure of interaction that helps students to accomplish a goal or end product. While these two forms of learning are often described and used interchangeably, differences do exist. The significant difference between collaborative and cooperative learning environments is the amount of control that the teacher exercises in setting goals and providing choice. For instance, in a collaborative classroom, students are positioned to set their own goals and choose activities, whereas in the cooperative learning environment, the teacher directs these activities. (Ten Panitz, 1996, http://www.city.londonmet.ac.uk/deliberations/collab.learning/panitz2.html)

Learning in the Collaborative Classroom

Critical to teaching and learning in the collaborative-cooperative environment is the ability to define the responsibilities of the teacher and students. For effective collaboration and cooperative teamwork, teachers and students must agree to certain responsibilities that support the learning process. The table below reflects the parallel responsibilities of teachers and students.

Effective Collaboration and Cooperative Teamwork

Teacher Responsibilities	Student Responsibilities
Monitor student behavior.	Develop the skills to work cooperatively.
Provide assistance when needed.	Learn to talk and discuss problems with each other in order to accomplish the group goal.
Answer questions only when they are group questions.	Ask for help only after each person in the group has considered the problem and the group has a question for the teacher.
Interrupt the process to reinforce cooperative skill or to provide direct instructions to all students.	Believe that all members of the group work together toward a common goal. Understand that the success or failure of the group is to be shared by all members.
Provide closure for the lesson.	Reflect on the work of the group.
Evaluate the group process by discussing the actions of the group members.	Appreciate that working together is a process and encourage each group member to interact and relate to the rest of the group members.
Help students to become individually accountable for learning and reinforce this understanding regularly.	Realize that each member must contribute as much as he or she can to the group goal. Understand that the success of the group is dependent on the individual work of each member of the group. Understand that group members are individually accountable for their own learning.

What the Collaborative Classroom is Not

It must be agreed upon by the teacher and the students that the collaborative classroom is not one in which students:

- Work in small groups on a problem or group of problems without direction or individual responsibility.
- Work individually while sitting in a group working on problems.
- Work without conversation or interaction regarding the method or process being used to solve the problem.
- Allow one member of the group to do all of the work while others sit passively.

Shaping the Collaborative Classroom

To ensure that the spirit and purpose of the collaborative classroom is clear from the onset of school, you will want to engage your students in a collaborative activity on the first day. In doing so, you can accomplish two important goals. First, students immediately understand the importance and value of working together, and secondly, students quickly move into their role as active participants.

The activity "Facts in Five," as you may have experienced in training, is an activity designed to meet these goals. Another popular activity with students is known as "Broken Squares" (Spencer Kagan: Cooperative Learning©). In this activity, members of the team are each given several pieces of a broken square. The pieces belong to different squares. Students must create the whole square by taking turns giving each other one piece. No one may speak during the activity, that is, no one can ask for what he or she needs. This activity is perfect for teaching sensitivity and the importance of communication.

During the first few days of class, it is extremely important that expectations and the "rules of the game" be defined. The best approach is to have the students work together in small groups to generate the guidelines for teamwork (See Lesson: Creating Collaborative Classroom Guidelines on page xvi).

As a guidepost for identifying the elements for successful group interactions, we suggest reviewing the "Ten guidelines for students doing group work in mathematics" written by Anne E. Brown for the CLUME Project (http://www.uwplatt.edu/clume/tenguide.htm). Brown developed these guidelines after viewing the video and audio tapes of more than a dozen group sessions of her students. This list reflects the apparent actions critical to the success or failure of the group. In summary, the guidelines state the following:

1. Groups should be formed quickly and members of the group should sit together, facing each other, and get to work quickly. Members should call each other by first name. Members should not engage in "off-task" discussion. Everyone should be encouraged to participate.

2. All instructions should be read aloud so that everyone is aware of the expectations of the assignment.

3. Members of the group should listen to each other and not interrupt. Comments or questions should be acknowledged and responded to by other group members.
4. Members of the group should not accept being confused. If a member of the group does not understand the information that is presented, this person should ask someone to paraphrase or re-phrase what was said.
5. Members of the groups should ask for clarification if a word is used in a way that is confusing.
6. The members of the group should work together on the same problem and check for agreement frequently.
7. Members of the group should explain their reasoning by "thinking out loud" and ask others to do the same. This helps everyone to relate the information being presented to what they already know.
8. Members of the group should monitor the group's progress and be aware of time constraints so that all members of the group meet the goals of the assignment.
9. If the group gets stuck, the members of the group should review and summarize what they have done so far. The group can then ask for questions to find errors or missing connections to help the group's work to proceed.
10. Members of the group should engage in questioning, the engine that drives mathematical investigation.

Group Work in the Collaborative Classroom

If we expect students to work well in groups, they will need to understand what it means to learn collaboratively and how it will benefit them. A good description of collaborative learning used by many of our teachers is:

Collaborative learning is a process in which each individual contributes personal knowledge and skill with the intent of improving his or her learning accomplishments along with those of others.

Students should be aware that one of the most important goals of collaborative learning is to create a "community of learners." They should understand that the community will grow and thrive only if all members of the group are active participants. Students must also understand that their role in the classroom will be different than what they may have experienced in other classes and so will the teacher's role!

You will want to introduce the features of a collaborative classroom to your students. Important characteristics of the collaborative classroom include:

- Shared responsibility
- Choice
- Discussion about how we learn from what is right as well as what is wrong
- Working in groups, whether as an entire class or as several small groups

Finally, you will want students to understand the goals and expectations of a collaborative classroom:

- Students learn collaboratively to gain greater individual proficiency.
- Groups "sink or swim" together.
- EVERYONE suggests, questions, and encourages.
- Group members are responsible for each other's learning.
- All group members bring valued talents and information to the task at hand.

Getting Started in the Collaborative Classroom

When problems and investigations in the text require that students work in groups, you will want to structure the groups. When problems and investigations in the text require that students work individually, it is possible to maintain a collaborative classroom where students are free to communicate with each other and to share information.

To form groups initially, you may want to set arbitrary groups and make changes as you observe students. One suggestion for structuring groups is to think about having two types of groups, long-term groups and short-term groups. The long-term groups, or home groups, stay together for the entire school year and sit together in class. Long-term groups enable students to build trust and confidence and to learn how to negotiate with each other to derive success. On the other hand, the short-term groups are randomly assigned for specific tasks. Short-term groups allow students to develop the ability to work with many different people. Clearly, how you arrange the groups will depend on how to best meet your students' needs.

Most importantly, you want to make sure that students are respectful of one another at all times. The success of the group depends on cooperation, which can be achieved only if students accept one another and value the contributions of others.

If you have students who do not want to work in groups, do not force the issue. Allow those students to work alone. It is important that the student who is working alone understands that the teacher is not a member of his or her group. After these students find that they cannot talk with others and that those who are sharing information are progressing more easily, they will naturally gravitate to a group.

You want to structure the success of the group experience, so it is important to use guidelines and timelines. Although you will want the students to come up with the operating guidelines, timelines are probably better left to you to determine.

After the groups are formed, you may want to have one person from each group be designated as a facilitator. Some responsibilities of the facilitator include:

- Obtaining and returning all materials
- Communicating information from the teacher to the group
- Handing in the completed assignments for the group

Success while working collaboratively depends upon every group member working on every part of the problem, so you may find that you do not want to assign roles such as recorder or reader to group members.

Students working together should generate noise and movement in the room. Some have defined this attribute as "controlled chaos." To ensure that the group work remains in control, you will want to monitor group interactions and check for understanding of the task at hand. You may also want to ask students to complete parts of the problem or investigation, stop and discuss the work done, summarize the main points of the task, and then continue. This works well when the problem or investigation is lengthy.

Because groups will work at different paces, you might want to prepare some additional tasks or extensions of the problem or investigation for those groups who finish quickly.

Facilitating Groups in the Collaborative Classroom

Facilitating the group process is critical. As noted earlier, you should only answer a question posed by the group rather than by individual students. You may also restrict the number of questions that a group can ask, being generous the first few times that students work in groups. When a group asks questions, answer by redirecting with guiding questions such as:

- What does your group think?
- How did you arrive at that answer?
- How does this relate to past activities?
- What work have you done so far?
- What do you know about the problem?
- What do you need to figure out?
- What materials might help you to figure this out?
- Are there other parts of the problem that you can do first?

Other tips to consider as you manage your collaborative classroom include:

- Provide additional instruction to those struggling with a task.
- Listen carefully and value diversity of thought that often provides instructional opportunities.
- Balance learning with working effectively. Remember that no one is on task 100% of the time.
- Deal with conflict constructively.
- Ask students to sign-off on other group members' papers to acknowledge that everyone understands the group's results.

Holding the groups accountable for an end product, such as a presentation, will add further value to the learning activity. As you have surely discovered, when you truly understand a concept or idea yourself, then you are able to explain that concept or idea to someone else.

Presentations and Discussions in the Collaborative Classroom

To successfully close or wrap-up a problem or investigation with a presentation and discussion, students must know exactly what you expect from them. You should also make sure that students know that you will hold the entire group accountable for the presentation. (This helps to ensure that students will hold each other accountable.)

Some suggestions for facilitating the presentation process include:

- Choose presenters in a group to ensure that all students have the opportunity to present.
- Require that students defend and talk about their solutions.
- Hold all students accountable by asking questions of group members who are not presenting.
- Ask presenters to make connections and generalizations and extend concepts.
- Allow groups time to process feedback and to celebrate their achievements.

To bring closure to the group work and presentation process, engage students in discussion or have them keep learning journals. Some suggestions for summary wrap-up questions include:

- What was something that you learned from this problem?
- What were the mathematical concepts that you applied in solving this problem?
- About what concepts do you still have questions?
- What are three things that your group did well?
- What is at least one thing that your group could do even better the next time?

Checklist of Teacher-Directed and Learner-Centered Classrooms

To understand where you are in the transition process from creating a teacher-directed classroom to creating a learner-centered classroom, you may use the criteria below to evaluate your classroom (Courtesy of Jacquelyn Snyder, Jan Sinopoli, and Vince Vernachhio, Pittsburgh Public Schools). Use your initial evaluation as a baseline measure and check yourself at regular intervals throughout the school year.

Teacher-Directed Classroom	Learner-Centered Classroom
The teacher directs all classroom activity.	The teacher facilitates classroom activity.
Each activity is dependent on the teacher.	Most activities require only guidance by the teacher.
The teacher is in the front of the room instructing the entire class using the blackboard or overhead most of the time.	The teacher walks around the classroom during all activities, watching and listening to student-to-student discourse.
The teacher models examples of the lesson objective and directs students to practice similar problems found in the text or on handouts designed by the teacher.	The teacher monitors the students to keep them on task, while the students actively work together on an activity.
Students are seated in rows, working as a class with the teacher at the front of the class or working independently.	The students are typically paired or grouped to work together while the teacher facilitates the process.
The teacher presents the material while students watch and take notes.	The teacher systematically brings the class together on several occasions, assuring that the mathematics of the lesson is understood.
The students work independently as the teacher tries to help each student individually.	Students are required to make presentations, explaining their progress within the activity.
The teacher completely answers the problem for the student when he or she is having difficulty.	If a student is having difficulty understanding something, even after consulting with his or her group members, the teacher asks the group leading questions to guide them to the desired outcome.
The teacher does the thinking and the work.	The students do the thinking and the work.
The teacher asks low-level or fill-in-the-blank types of questions that can be answered with a single number or in a word or two.	The teacher asks thought-provoking questions that required students to explain their thinking and processes.
The majority of classroom discourse is teacher-to-student discourse.	The majority of discourse is student-to-student discourse.
The teacher encourages students to memorize rules, procedures, and formulas.	The teacher encourages students to construct knowledge. Prior knowledge is assessed as new concepts emerge.

Lesson: Creating Collaborative Classroom Guidelines

Preparing for the Lesson

- Arrange the class seats in groups of three or four such that students face each other. Position the desks in such a way that students need only do a half turn of their heads if you call their attention to the front of the room.
- Give poster boards to each group.
- Give colorful markers to each group.

Expected Student Growth

- Students will gain experience in working cooperatively, listening and respecting the ideas of others, and coming to a consensus regarding the final product.
- Students will learn how to share power with the teacher.

Initiating the Activity

- Ask students if they have ever worked in groups.
- Ask students to think about good and not-so-good group experiences.
- Have students make lists of things that happen in groups or things that they think should happen in groups to have a group work more productively to complete a task.
- Direct students to develop social guidelines for group work in class. The guidelines should be phrased positively and refer to observable behavior. Lists of guidelines should not be too long.

Facilitating the Activity

- Monitor student behavior.
- Offer assistance only if necessary. For instance, students may be making their lists too long.
- Interrupt the process to reinforce cooperative skills or to provide directions.

Student Presentations

- Have all groups present their guidelines.
- Have students determine which guidelines are similar and record those.
- Have students look at the remainder of the guidelines and determine which should be included in the list of guidelines. Students should be able to justify their choices. As all students will be using these guidelines, there should be consensus on the final list.

Lesson Closure

- Indicate that the final list will be generated and every member in the class must agree by signing off on the list. By doing so, students have agreed to honor the list of guidelines and will be held accountable.
- Indicate that groups not adhering to the guidelines may have their group grades reduced.

Presentation Rubric

The rubric below can be used to help you score group presentations to the entire class. The presentation scores, which range from 1 to 5, are detailed in the rubric. It is a good idea to copy the Presentation Rubric and distribute it to the class so that students understand how they are scored.

Score	Description
5	You earn a 5 for your presentation if your presentation is nearly perfect. Your mathematics must be correct with only a very minor flaw (not having to do with the main idea of the problem). Your public speaking skills must also be perfect or quite close to perfect. You must look at your audience. Your must present yourself well and not make distracting gestures or hand motions during the presentation. Your rate of speech must be neither too fast nor too slow.
4	You earn a 4 for your presentation if you miss one thing within the mathematical content of your presentation. OR You earn a 4 for your presentation if there is one thing that you do not do very well within the public speaking part of the presentation.
3	You earn a 3 for your presentation if you can complete the problem, but your public speaking skills are poor. This score means that you do not make eye contact, you speak inaudibly, your mumble your words, etc. OR You earn a 3 for your presentation if you have some content knowledge and make one major error, as well as omit one of the important aspects of good public speaking.
2	You earn a 2 for your presentation if you stand up for your presentation but really have very little content knowledge. This score means that you are unable to complete the problem and your speaking skills are poor.
1	You earn a 1 for your presentation for being willing to stand up and try to present.
0	You earn a 0 for your presentation if you refuse to stand up and try to present.

Looking Ahead to Chapter 7

7

Focus In Chapter 7, you will learn to write, graph, and solve systems of equations both graphically and algebraically. You will also learn how to write, graph, and solve systems of linear inequalities, as well as identify solutions to linear inequalities.

Chapter Warm-up

Answer these questions to help you review skills that you will need in Chapter 7.

Use the distributive property to simplify each expression.

1. $3(5x + 7)$

$15x + 21$

2. $-2(10 + 4y)$

$-20 - 8y$

3. $-\frac{1}{6}(3x - 12)$

$-\frac{1}{2}x + 2$

Write each linear equation in standard form.

4. $y = -4x + 7$

$4x + y = 7$

5. $2y - 4 = 3x + 23$

$-3x + 2y = 27$

6. $-8y = \frac{3}{5}x + \frac{1}{5}$

$3x + 40y = -1$

Read the problem scenario below.

A bicycle company is trying to determine the number of bikes that they have sold. The company began in 1995. In the year 1997, the company sold a total of 285 bikes, and in the year 2000, the company sold a total of 684 bikes. Assume that the number of bikes sold is a linear function of the time in years since 1995.

7. Find the linear function that describes the total number of bikes sold as a function of the time in years since 1995.

$f(x) = 133x + 19$

8. Use the linear function to find the total number of bikes sold in 2010.

$f(15) = 133(15) + 19 = 2014$ bikes

Key Terms

income ■ p. 299
profit ■ p. 299
point of intersection ■ p. 306, 309
break-even point ■ p. 306
system of linear equations ■ p. 309
linear system ■ p. 309
solution ■ p. 309
parallel lines ■ p. 311
perpendicular lines ■ p. 314
reciprocals ■ p. 314
standard form of a linear equation ■ p. 315
substitution method ■ p. 317
linear combinations method ■ p. 326
linear combination ■ p. 326
inequality ■ p. 332, 345
linear inequality in two variables ■ p. 345
inequality symbol ■ p. 345
linear equation ■ p. 347
coordinate plane ■ p. 347
half-plane ■ p. 347
system of linear inequalities ■ p. 353

CHAPTER

7

Systems of Equations and Inequalities

The earliest known bricks were made of mud. Today, most bricks are made of clay or ground shale. In Lesson 7.2, you will compare the number of bricks that can be laid by a novice bricklayer and an experienced bricklayer.

7

7.1

Making and Selling Markers and T-Shirts

Using a Graph to Solve a Linear System

7

Learning By Doing Lesson Map

Get Ready

Objectives

In this lesson, you will:

- Analyze cost and income equations.
- Graph cost and income equations on the same graph.
- Find the break-even point graphically.

Key Terms

- income
- profit
- point of intersection
- break-even point

NCTM Content Standards

Grades 9–12 Expectations

Algebra Standards

- Analyze functions of one variable by investigating rates of change, intercepts, zeros, asymptotes, and local and global behavior.
- Write equivalent forms of equations, inequalities, and systems of equations and solve them with fluency—mentally or with paper and pencil in simple cases and using technology in all cases.
- Use symbolic algebra to represent and explain mathematical relationships.
- Draw reasonable conclusions about a situation being modeled.
- Approximate and interpret rates of change from graphical and numerical data.

Measurement Standards

- Make decisions about units and scales that are appropriate for problem situations involving measurement.
- Apply informal concepts of successive approximation, upper and lower bounds, and limit in measurement situations.

Lesson Overview

Within the context of this lesson, students will be asked to:

- Compare and analyze cost and income equations graphically and algebraically.
- Graph cost and income equations on the same graph.
- Find a break-even point graphically.

Essential Questions

The following key questions are addressed in this lesson:

1. What is income?
2. What is profit?
3. What is a point of intersection on a graph?
4. What is a break-even point?

Show The Way

Warm Up

7

Place the following questions or an applicable subset of these questions on the board before students enter class. Students should begin working as soon as they are seated.

Evaluate each expression. Simplify your result, if possible.

1. $\frac{11}{12} - \frac{2}{3}$ $\frac{1}{4}$

2. $\frac{11}{12} \div \frac{2}{3}$ $\frac{11}{8}$

3. $\frac{5}{24} + \frac{1}{3}$ $\frac{13}{24}$

4. $\frac{9}{2} - \frac{5}{2}$ 2

5. $\frac{15}{28} \cdot \frac{4}{5}$ $\frac{3}{7}$

6. $\frac{8}{13} \div \frac{4}{39}$ 6

7. $\frac{4}{15} + \frac{7}{5}$ $\frac{5}{3}$

8. $\frac{9}{4} \cdot \frac{16}{7}$ $\frac{36}{7}$

9. $\frac{49}{50} - \frac{8}{25}$ $\frac{33}{50}$

10. $\frac{5}{10} \div \frac{1}{5}$ $\frac{5}{2}$

11. $\frac{3}{4} + \frac{9}{5}$ $\frac{51}{20}$

12. $\frac{-18}{7} \cdot \frac{21}{72}$ $\frac{-3}{4}$

Motivator

Begin the lesson with the motivator to get students thinking about the topic of the upcoming problem. This lesson is about sales of markers and T-shirts. The motivating questions are about selling products. Ask the students the following questions to get them interested in the lesson.

- Why would a person or company produce items to sell?
- What items have you helped to sell to raise money for your teams or clubs?
- What kind of T-shirts might someone buy in a large quantity and try to sell to earn a profit?
- What names have we learned for the amount of money that is spent to produce something to sell?
- What names have we learned for the amount of money that is received for something we sell?

Explore Together

Problem 1

Students will create equations for income and for cost and use their equations to calculate profit.

Problems 1 and 2 are similar mathematically.

Grouping

Ask for a student volunteer to read the Scenario and Problem 1 aloud. Have a student restate the problem. Ask for another student volunteer to read Problem 2 aloud. Have a student restate the problem. Pose the Guiding Questions below to verify student understanding.

Because the mathematics in Problems 1 and 2 is similar, assign Problem 1 parts (A) through (D) and Questions 1 through 6 of Investigate Problem 1 to half of the students. Assign Problem 2 parts (A) through (D) and Questions 1 through 6 of Investigate Problem 2 to the other half of the students. Have the students work in small groups to complete the questions assigned to them. Have students discuss and present these questions before assigning Question 7.

Take Note

Remember that the **profit** is the amount of money that is left from sales (income) after the costs are subtracted.

Guiding Questions

- What is income?
- What is a production cost?
- What is profit?
- What does a positive profit represent?
- What does a negative profit represent?
- What is constant in the marker situation? What is changing?
- What is constant in the T-shirt situation? What is changing?

SCENARIO You have a part-time job at a company that makes and sells color art markers. As part of your job, you are studying the company's production costs. The markers are made one color at a time. It costs \$2 to manufacture each marker and there is a \$100 set-up cost for each color. You are also studying the **income,** or the amount of money that the company earns, from the sales of the markers. The company sells the markers to office and art supply stores for \$3 per marker.

Problem 1 Making and Selling Markers

A. Write an equation that gives the production cost in dollars to make one color of marker in terms of the number of markers produced. Be sure to describe what your variables represent. Use a complete sentence in your answer.

The equation is $y = 2x + 100$ where x is the number of markers produced and y is the production cost in dollars.

B. Write an equation that gives the income in dollars in terms of the number of markers sold. Be sure to describe what your variables represent. Use a complete sentence in your answer.

The equation is $y = 3x$ where x is the number of markers sold and y is the income in dollars.

C. Find the production cost to make 80 markers of the same color. Show all your work and use a complete sentence in your answer.

$y = 2(80) + 100 = 260$; The production cost for 80 markers is \$260.

Find the income from selling the 80 markers that you made. Show all your work and use a complete sentence in your answer.

$y = 3(80) = 240$; The income for 80 markers is \$240.

Find the profit from the sale of the 80 markers that you made. Show all your work and use a complete sentence in your answer.

Profit: $240 - 260 = -20$; The profit is –\$20.

Explore Together

7

Problem 1

Students will evaluate their income and cost equations for 100 markers.

Common Student Errors

Students will often subtract the smaller number from the larger number to get a positive profit even if the cost is larger than the income. The students may need to be refocused on the meaning of profit as income minus cost.

Investigate Problem 1

Grouping

Students will be working in small groups to complete these questions.

Students will complete a table of values for the cost and for the income for various marker production amounts.

Common Student Errors

Students may have difficulty determining appropriate bounds for their graph. You may need to pose the Guiding Questions below to help them refine their understanding.

Guiding Questions

- What is the smallest possible number of markers that could be produced? What is the smallest possible number of markers that could be sold?
- Will the number of markers made and sold be represented by an x-value or a y-value on the graph?
- What is the largest reasonable number of markers that could be made and sold?
- Will that number be the lower or upper bound? Is it a bound for x or for y?
- What is the smallest amount of money possible for the cost of making the markers?
- What is the smallest amount of money possible for the sale of the markers?
- Will the amount of money be represented by the x-value or the y-value on the graph?

Problem 1 Making and Selling Markers

D. Find the production cost to make 100 markers of the same color. Show all your work and use a complete sentence in your answer.

$y = 2(100) + 100 = 300$; The production cost for 100 markers is $300.

Find the income from selling the 100 markers that you made. Show all your work and use a complete sentence in your answer.

$y = 3(100) = 300$; The income for 100 markers is $300.

Find the profit if 100 markers are made and sold. Show all your work and use a complete sentence in your answer.

Profit: $300 - 300 = 0$; The profit is $0.

Investigate Problem 1

1. Complete the table of values that shows the production cost and income for different numbers of markers of the same color.

Quantity Name	Number of markers	Product cost	Income
Unit	markers	dollars	dollars
Expression	x	$2x + 100$	$3x$
	0	100	0
	20	140	60
	30	160	90
	35	170	105
	55	210	165
	125	350	375
	200	500	600
	400	900	1200

- What is the largest reasonable amount of money for the cost of producing the markers?
- What is the largest reasonable income from selling the markers?
- What upper bound will you use for the y-values?

Explore Together

Investigate Problem 1

Students will graph both the cost and income equations in the same coordinate plane. Students will then use their graph to answer questions comparing cost and income.

Grouping

Students will be working in small groups to complete these questions.

Key Formative Assessments

- How can you interpret the part of the graph where the cost line is above the income line?
- How can you interpret the part of the graph where the cost line is below the income line?
- Summarize the meaning of one line being above another line for an interval of markers.
- Summarize the meaning of one line being below another line for an interval of markers.
- Summarize the meaning of the point where both of the lines have the same value for both the number of markers and the amount of money.

Investigate Problem 1

7

2. Create a graph of both the production cost and income equations on the grid below. Use the bounds and intervals below. Be sure to label your graph clearly.

Variable quantity	Lower bound	Upper bound	Interval
Markers	0	150	10
Money	0	450	30

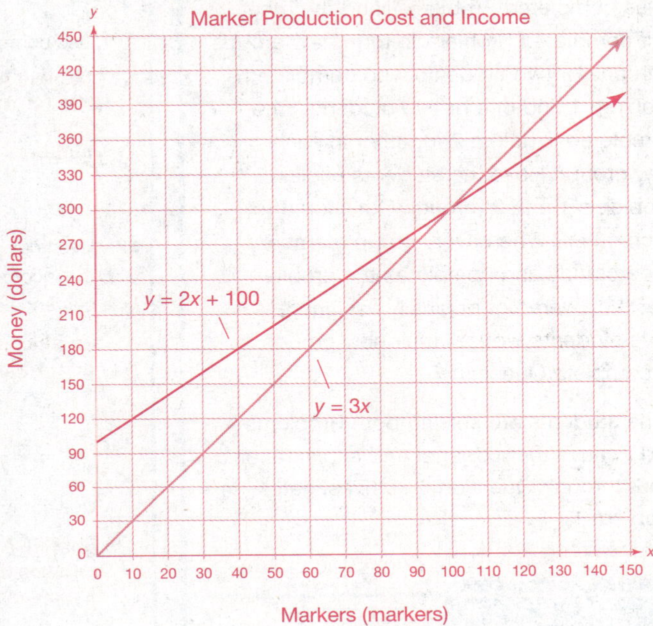

3. Use your graph to determine the numbers of markers for which the production cost is greater than the income. Use a complete sentence in your answer.

The production cost is greater than the income when less than 100 markers are produced and sold.

Use complete sentences to explain how you found your answer.

Sample Answer: First determine where on the graph the line for the production cost is above the line for the income. Then look at the *x*-axis and determine the corresponding numbers of markers.

Explore Together

7

Investigate Problem 1

Students will answer questions to determine the break-even point informally for the marker sale situation.

Grouping

Call the class back together. If half of the class completed Problem 1 and the other half completed Problem 2, form new groups with one or two students who completed Problem 1 and one or two students who completed Problem 2 together in each new group. Have the students explain Problems 1 and 2 through Question 6 to each other. When every group member understands and completes the problem that they were not originally assigned, have students work in their new groups to complete Question 7.

If the students are solving both Problems 1 and 2, have them debrief and explain their work through Question 6 of Investigate Problem 1.

Take Note

Whenever you see the share with the class icon, your group should prepare a short presentation to share with the class that describes how you solved the problem. Be prepared to ask questions during other groups' presentations and to answer questions during your presentation.

Investigate Problem 1

4. Use your graph to determine the numbers of markers for which the income is greater than the production cost. Use a complete sentence in your answer.

The income is greater than the production cost when more than 100 markers are produced and sold.

Use complete sentences to explain how you found your answer.

Sample Answer: First determine where on the graph the line for the income is above the line for the production cost. Then look at the *x*-axis and determine the corresponding numbers of markers.

5. Use your graph to determine the number of markers for which the income is equal to the production cost. Use a complete sentence in your answer.

The income and production cost are equal when 100 markers are produced and sold.

Use complete sentences to explain how you found your answer.

Sample Answer: Determine where the line for the income crosses the line for the production cost. Then look at the *x*-axis and determine the corresponding number of markers.

6. Describe the numbers of markers that must be sold in order for your profit to be at least $0. Use complete sentences to explain how you found your answer.

Sample Answer: One hundred markers or more must be sold to make a profit. When 100 markers are sold, the profit is $0. According to the graph, when more than 100 markers are sold, the income is greater than the production cost, so the profit is greater than $0.

Explore Together

Problem 2

Students will solve a problem similar mathematically to Problem 1 but with the new context of selling T-shirts.

Students will calculate income and cost values for various numbers of T-shirts. Students will calculate the break-even point informally in part (D) of Problem 2.

Grouping

If students are all completing Problem 1 and then will all complete Problem 2, call the class back together to have the students discuss and present their work for parts (A) through (D) of Problem 1 and Questions 1 through 6 of Investigate Problem 1.

Ask for a student volunteer to read Problem 2 aloud. Have a student restate the problem. Pose the Guiding Questions below to verify student understanding. Have students work in small groups to complete parts (A) through (D) of Problem 2 and Questions 1 through 6 of Investigate Problem 2 if you have not already assigned these questions.

If the students are solving both Problems 1 and 2, have them discuss and present their work through Question 6 of Investigate Problem 1.

Guiding Questions

- How is this problem different from Problem 1?
- How is this problem similar to Problem 1?
- What is constant in this situation?
- What is varying in this situation?
- What units will be needed in this situation?

Take Note

Remember that the profit is the amount of money that is left from sales (income) after the costs are taken out.

Problem 2 Making and Selling T-Shirts

Your work at the marker company has inspired you to start your own business. You decide to design and sell customized T-shirts. The company that supplies your T-shirts charges you $7.50 for each T-shirt and a set-up cost of $22.50 for a new design. You decide to sell the T-shirts for $8.25 each.

A. Write an equation that gives the production cost in dollars to make one design of T-shirt in terms of the number of T-shirts made. Be sure to describe what your variables represent. Use a complete sentence in your answer.

The equation is $y = 7.50x + 22.50$ where x is the number of T-shirts made and y is the production cost in dollars.

B. Write an equation that gives the income (the amount of money that you earn) in dollars in terms of the number of T-shirts sold. Be sure to describe what your variables represent. Use a complete sentence in your answer.

The equation is $y = 8.25x$ where x is the number of T-shirts sold and y is the income in dollars.

C. Find the production cost to make 15 T-shirts in the same design. Show all your work and use a complete sentence in your answer.

$y = 7.50(15) + 22.50 = 135$; The production cost for 15 T-shirts is $135.

Find the income from selling the 15 T-shirts that you made. Show all your work and use a complete sentence in your answer.

$y = 8.25(15) = 123.75$; The income for 15 T-shirts is $123.75.

Find the profit from the sale of the 15 T-shirts that you made. Show all your work and use a complete sentence in your answer.

Profit: $123.75 - 135 = -11.25$; The profit for 15 T-shirts is –$11.25.

D. Find the production cost to make 30 T-shirts in the same design. Show all your work and use a complete sentence in your answer.

$y = 7.50(30) + 22.50 = 247.50$; The production cost for 30 T-shirts is $247.50.

Find the income from selling the 30 T-shirts that you made. Show all your work and use a complete sentence in your answer.

$y = 8.25(30) = 247.50$; The income for 30 T-shirts is $247.50.

Find the profit if 30 T-shirts are made and sold. Show all your work and use a complete sentence in your answer.

Profit: $247.50 - 247.50 = 0$; The profit for 30 T-shirts is $0.

Investigate Problem 2

1. Complete the table of values on the next page that shows the production cost and income for different numbers of T-shirts in the same design.

Common Student Errors

Some students have difficulty understanding the difference between cost, income, and profit.

Explore Together

Investigate Problem 2

7

Students will create a table of values for cost and income in Question 1.

Students will graph both the cost and income equations in the same coordinate plane. Students will use their graphs to answer questions comparing cost and income.

Grouping

Students will be working in small groups to complete these questions.

Key Formative Assessments

- Is it possible to sell zero shirts?
- Would the cost to produce the shirts be different if one person ordered 15 shirts or if 15 different people ordered 1 shirt each?
- Would the income from the shirts be different if one person ordered 15 shirts or if 15 different people ordered 1 shirt each?
- In the table of values in Question 1, does the 400 shirts represent one person buying 400 shirts, or does it represent 400 total shirts being produced and sold?
- What is the meaning of the 25 in the *Number of T-shirts* column?
- What is the smallest amount of money possible for the sale of the shirts?

Investigate Problem 2

Quantity Name	Number of T-shirts	Product cost	Income
Unit	T-shirts	dollars	dollars
Expression	x	$7.50x + 22.50$	$8.25x$
	0	22.50	0.00
	20	172.50	165.00
	25	210.00	206.25
	30	247.50	247.50
	100	772.50	825.00
	200	1522.50	1650.00
	400	3022.50	3300.00

2. Create a graph of both the production cost and income equations on the grid below. Use the bounds and intervals below. Be sure to label your graph clearly.

Variable quantity	Lower bound	Upper bound	Interval
T-shirts	0	45	3
Money	0	375	25

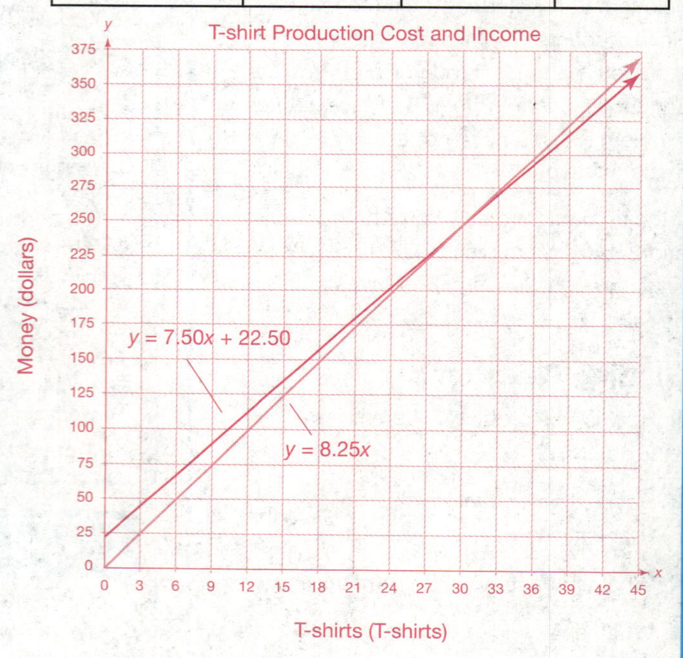

Explore Together

Investigate Problem 2

Students will informally calculate the break-even point in Question 5 and then will formally learn the meaning of a break-even point and point of intersection from a graph in Question 7.

Guiding Questions

- What is the smallest possible number of T-shirts that could be made? What is the smallest possible number of T-shirts that could be sold?
- Will the number of T-shirts made and sold be represented by an *x*-value or by a *y*-value on the graph?
- What is the largest reasonable number of T-shirts that could be made and sold?
- Will that number be the lower or upper bound? Is it a bound for *x* or for *y*?
- What is the smallest amount of money possible for the cost of making the T-shirts?
- What is the smallest amount of money possible for the sale of the T-shirts?
- Will the amount of money be represented by an *x*-value or *y*-value on the graph?
- What is the largest reasonable amount of money for the cost of producing the T-shirts?
- What is the largest reasonable income from selling the T-shirts?
- What upper bound will you use for the *y*-values?
- How can you determine from the graph where the cost to produce the T-shirts is more than the income from selling the T-shirts? In what interval (of the number of T-shirts) is this?
- How can you determine from the graph when the cost for selling the T-shirts is more than the cost to produce the T-shirts? In what interval (of the number of T-shirts) is this?

Investigate Problem 2

3. Use your graph to determine the numbers of T-shirts for which the production cost is greater than the income. Use a complete sentence in your answer.

The production cost is greater than the income when less than 30 T-shirts are produced and sold.

Use complete sentences to explain how you found your answer.

Sample Answer: First determine where on the graph the line for the production cost is above the line for the income. Then look at the *x*-axis and determine the corresponding numbers of T-shirts.

4. Use your graph to determine the numbers of T-shirts for which the income is greater than the production cost. Use a complete sentence in your answer.

The income is greater than the production cost when more than 30 T-shirts are produced and sold.

Use complete sentences to explain how you found your answer.

Sample Answer: First determine where on the graph the line for the income is above the line for the production cost. Then look at the *x*-axis and determine the corresponding numbers of T-shirts.

5. Use your graph to determine the number of T-shirts for which the income is equal to the production cost. Use a complete sentence in your answer.

The income and production cost are equal when 30 T-shirts are produced and sold.

Use complete sentences to explain how you found your answer.

Sample Answer: Determine where the line for the income crosses the line for the production cost. Then look at the *x*-axis and determine the corresponding number of T-shirts.

Explore Together

Investigate Problem 2

7

Grouping

Ask for a student volunteer to read Question 7 aloud. Have a student restate the problem.

Have the students work in small groups to complete Question 7. Then call the class back together to have the students discuss and present their work for Question 7.

Just the Math

Ask the students to discuss the meaning of the break-even point and the point of intersection. Pose the Guiding Questions below to verify student understanding.

Guiding Questions

- What is a point of intersection?
- What is a break-even point?
- Why is the word *point* in the term *break-even point*?

Key Formative Assessments

- What is income? What is profit?
- What is a production cost?
- What is a point of intersection?
- What was the meaning of the point of intersection of the graph for Problem 1 about markers?
- What was the meaning of the point of intersection of the graph for Problem 2 about T-shirts?
- If two lines are graphed for other situations that do not represent cost and income, could the point of intersection still be meaningful? What could it represent instead of a break-even point?

Investigate Problem 2

6. Describe the numbers of T-shirts that must be sold in order for your profit to be at least $0. Use complete sentences to explain how you found your answer.

Sample Answer: Thirty T-shirts or more must be sold. When 30 T-shirts are sold, the profit is $0. According to the graph, when more than 30 T-shirts are sold, the income is greater than the production cost, so the profit is greater than $0.

7. **Just the Math: Break-Even Point** When two graphs cross (or intersect) each other, the point where they cross is called a **point of intersection.** When one line represents the production cost of an item and the other line represents the income from selling the item, the x-coordinate of this point is called the break-even point. What is the **break-even point** for making and selling markers? Use a complete sentence in your answer.

The break-even point for marker sales is 100 markers.

What is the company's profit at the break-even point? Show all your work and use a complete sentence in your answer.

Profit: 300 – 300 = 0; The profit is $0.

What is the break-even point for making and selling T-shirts? Use a complete sentence in your answer.

The break-even point for T-shirt sales is 30 T-shirts.

What is your profit from the T-shirts at the break-even point? Show your work and use a complete sentence in your answer.

Profit: 247.50 – 247.50 = 0; The profit is $0.

7

Wrap Up

Close

- Review all key terms and their definitions. Include the terms *income, profit, point of intersection,* and *break-even point*.
- You may also want to review any other vocabulary terms that were discussed during the lesson, which may include *revenue* and *production cost*.
- Remind the students to write the key terms and their definitions in the notes section of their notebooks. You may also want the students to include examples.
- Ask the students what other situations they can think of where the point of intersection of two lines graphed would be meaningful.
- Ask the students how they can find a point of intersection for two lines graphically and how they could find the point of intersection of two lines algebraically.

Ties to the Cognitive Tutor Software

In the Cognitive Tutor software, students represent systems of equations in a table, with two functions of x. This representation encourages students to equate the expressions in the dependent variable columns and set up a single equation with one unknown. This is a simpler method than setting up two equations with two unknowns, at least when the equations would naturally be in slope-intercept form.

Follow Up

Assignment

Use the Assignment for Lesson 7.1 in the Student Assignments book. See the Teacher's Resources and Assessments book for answers.

Assessment

See the Assessments provided in the Teacher's Resources and Assessments book for Chapter 7.

Open-Ended Writing Task

Ask the students to graph a third equation on their graph from this lesson in Problem 1 for the amount of profit for making and selling the markers and then in Problem 2 for the amount of profit for making and selling T-shirts. Have them predict what the graph of each equation will look like before they graph them. Then have them compare what they expected to what they found.

Reflections

7

Insert your reflections on the lesson as it played out in class today.

What went well?

____

____

What did not go as well as you would have liked?

____

____

How would you like to change the lesson in order to improve the things that did not go well and capitalize on the things that did go well?

____

____

____

____

____

Notes

7.2

Time Study

Graphs and Solutions of Linear Systems

7

Learning By Doing Lesson Map

Get Ready

Objectives

In this lesson, you will:

- Determine the number of solutions of a linear system.
- Identify parallel and perpendicular lines.

Key Terms

- systems of linear equations
- linear system
- solution
- point of intersection
- parallel lines
- perpendicular lines
- reciprocals

NCTM Content Standards

Grades 9–12 Expectations

Algebra Standards

- Generalize patterns using explicitly defined and recursively defined functions.
- Analyze functions of one variable by investigating rates of change, intercepts, zeros, asymptotes, and local and global behavior.
- Interpret representations of functions of two variables.
- Use symbolic algebra to represent and explain mathematical relationships.
- Draw reasonable conclusions about a situation being modeled.
- Approximate and interpret rates of change from graphical and numerical data.

Measurement Standards

- Make decisions about units and scales that are appropriate for problem situations involving measurement.

Lesson Overview

Within the context of this lesson, students will be asked to:

- Graph systems of linear equations.
- Determine the number of solutions of linear systems of equations.
- Identify parallel and perpendicular lines.

Essential Questions

The following key questions are addressed in this lesson:

1. What is a system of linear equations?
2. What is a solution of an equation?
3. What is a point of intersection?
4. What are parallel lines?
5. What are perpendicular lines?
6. What are reciprocals?

Show The Way

7

Warm Up

Place the following questions or an applicable subset of these questions on the board before students enter class. Students should begin working as soon as they are seated.

Decide whether each point is a solution of the equation $y = 15x + 280$. Answer yes or no.

1. (5, 280) **No**

2. (5, 375) **No**

3. (12, 470) **No**

4. (30, 730) **Yes**

5. (100, 1780) **Yes**

6. (2, 310) **Yes**

The ordered pairs represent the function $y = 10{,}400 - 52x$. Complete each ordered pair.

7. (0, ____) **10,400**

8. (1, ___) **10,348**

9. (5, ____) **10,140**

10. (50, ____) **7800**

11. (150, ____) **2600**

12. (___ , 0) **200**

Motivator

Begin the lesson with the motivator to get students thinking about the topic of the upcoming problem. This lesson is about the number of bricks being set. The motivating questions are about brick setting.

Ask the students the following questions to get them interested in the lesson.

- What does it mean to set bricks?
- How many bricks do you think you could set in an hour if you were to try to make a small wall for a decoration?
- Why is mortar layered between bricks?
- How many years do you think it takes most brick setters to become very skilled at brick setting?
- How might a person learn to set bricks?

Explore Together

Problem 1

Students will write and analyze systems of linear equations with different slopes and y-intercepts.

Grouping

Working as a class, ask for a student volunteer to read the Scenario and Problem 1 aloud. Have a student restate the problem. Pose the Guiding Questions below to verify student understanding.

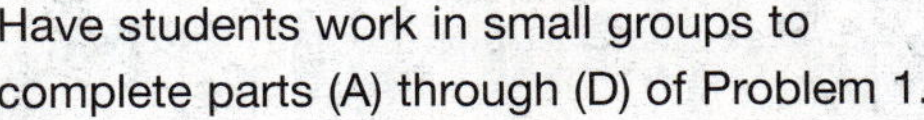

Have students work in small groups to complete parts (A) through (D) of Problem 1.

Guiding Questions

- What is a novice?
- What does it mean if a worker is an experienced worker?
- What is the rate of brick setting for the novice worker?
- What is the rate of brick setting for the experienced worker?
- Are these rates constant?
- What is changing in this problem?
- Where would we include the number of bricks already set in the equation for the brick setters?
- What is meant by the phrase, "in terms of"?
- Why are we asked to use x and y for this situation rather than variables such as b for the number of bricks and t for the amount of time?
- Who currently has more bricks set?
- How can a novice have more bricks set than the experienced brick setter?

SCENARIO A process engineer is performing a time study on a construction site. As part of the study, the work rates of a novice (beginner) bricklayer and a more experienced bricklayer are being recorded. At the beginning of the study, the novice had put 1510 bricks into place and was setting the bricks in place at a rate of thirty eight bricks per hour. The experienced worker started the job after the novice and had put 960 bricks into place so far and was setting the bricks in place at a rate of sixty bricks per hour.

7

Problem 1 The Novice and the Pro

A. For each worker, write an equation that gives the total number of bricks y set in place in terms of the time x in hours after the beginning of the time study.

Novice worker: $y = 38x + 1510$;
Experienced worker: $y = 60x + 960$

B. After eight hours of the time study, how many bricks in all will each worker have set into place? Show all your work and use complete sentences in your answer.

$y = 38(8) + 1510 = 1814$; $y = 60(8) + 960 = 1440$;
The novice worker will have set 1814 bricks and the experienced worker will have set 1440 bricks.

Which worker has set more bricks into place after eight hours of the time study? Use a complete sentence in your answer.

The novice worker has set more bricks.

C. After forty hours of the time study, how many bricks in all will each worker have set into place? Show all your work and use complete sentences in your answer.

$y = 38(40) + 1510 = 3030$; $y = 60(40) + 960 = 3360$;
The novice worker will have set 3030 bricks and the experienced worker will have set 3360 bricks.

Which worker has set more bricks into place after forty hours of the time study? Use a complete sentence in your answer.

The experienced worker has set more bricks.

Explore Together

Problem 1

Call the class back together to have the students discuss and present their work for parts (A) through (D) of Problem 1.

Common Student Errors

Students may not understand the significance of their answer in part (D) of Problem 1. You may need to ask them to identify the meaning of their answer to help them understand its importance.

Investigate Problem 1

Students will informally calculate the solution to a system of linear equations. Students will also graph a system of linear equations.

Grouping

Ask for a student volunteer to read Question 1 aloud. Pose the Guiding Questions below to verify student understanding.

Have students work in small groups to complete Questions 1 through 4.

Guiding Questions

- What two quantities are being studied in each of the equations that you wrote in part (A) of Problem 1?
- What is the best unit to measure the time for this situation?
- What is the most appropriate unit to measure the number of bricks set?
- What is the smallest possible amount of time that can be spent?
- What is the smallest possible number of bricks that can be set?
- What is the largest reasonable amount of time that one worker is likely to spend setting bricks?
- What is the largest reasonable number of bricks that will likely be set by one worker?

Problem 1 The Novice and the Pro

D. Find the number of hours that the time study would need to run in order for each worker to set a total of 2460 bricks. Show all your work and use complete sentences in your answer.

$2460 = 38x + 1510$ $\quad$ $2460 = 60x + 960$

$950 = 38x$ $\quad$ $1500 = 60x$

$x = 25$ $\quad$ $x = 25$

It will take both the novice worker and the experienced worker 25 hours.

Investigate Problem 1

1. Create a graph of both equations on the grid below. First, choose your bounds and intervals. Be sure to label your graph clearly.

Variable quantity	Lower bound	Upper bound	Interval
Time	0	120	8
Bricks	0	7500	500

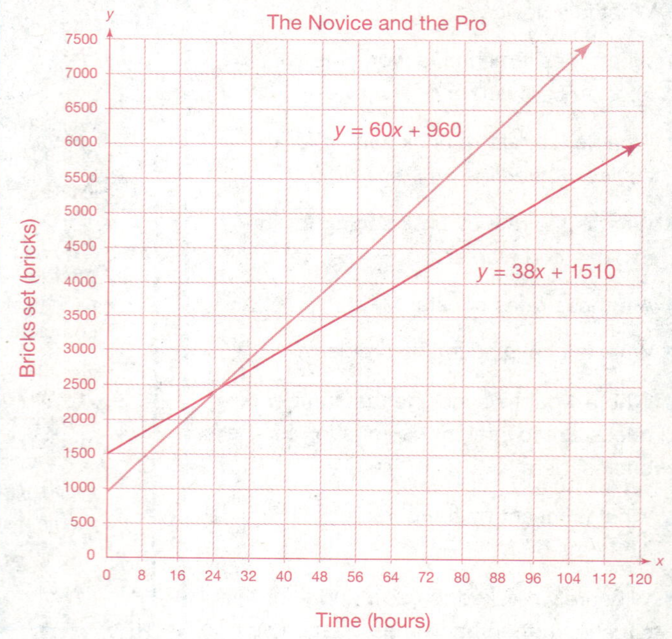

2. Find the amount of time that it will take in the time study for the number of bricks set by each worker to be the same. Use a complete sentence to explain how you found your answer.

Sample Answer: It will take about 25 hours. In part (D), it was determined that it took both workers 25 hours to set 2460 bricks.

- How can you use the answers to the last four Guiding Questions to develop your graph?

Note To complete Question 2, students may remember their work in part (D) of Problem 1 or may use their graphs.

Explore Together

Investigate Problem 1

Students will analyze their system of linear equations and the meaning of various parts of each equation in the system.

Grouping

Call the class back together to have the students discuss and present their work for Questions 1 through 4.

Key Formative Assessments

- What is the meaning of the x-values and the y-values in each equation?
- What is the meaning of the point of intersection on your graph in Question 1?
- Summarize what you notice about your graph.

Just the Math

The concepts of systems of linear equations and the solution of a system of linear equations are introduced formally in Questions 5 and 6.

Take Note

Recall that to algebraically verify that an ordered pair is a solution of an equation, substitute the values given by the ordered pair for x and y in the equation. These values should give you a true statement.

Grouping

Ask for a student volunteer to read Questions 5 and 6 aloud. Pose the Guiding Questions below to verify student understanding. Have students work in small groups to complete Questions 5 and 6. Then discuss these questions together as a class.

Guiding Questions

- What is a system of linear equations?
- What is a solution of a linear system?
- Is it possible to have a solution of a system of more than two linear equations? What might such a solution look like on a graph?

Investigate Problem 1

3. What does the slope of each line represent in this problem situation? Use a complete sentence in your answer.

Sample Answer: The slope indicates the number of bricks set by each worker in one hour.

Which worker sets bricks faster? How do you know? Use a complete sentence in your answer.

Sample Answer: The experienced worker sets bricks faster because this worker sets 60 bricks per hour and the novice worker only sets 38 bricks per hour.

4. What does the y-intercept of each line represent in this problem situation? Use a complete sentence in your answer.

Sample Answer: The y-intercept indicates the number of bricks that have already been set.

How do the y-intercepts of the lines compare? What does this mean in the problem situation? Use complete sentences in your answer.

Sample Answer: The y-intercept of the novice worker is greater than the y-intercept of the experienced worker. The novice worker has set more bricks so far.

5. Just the Math: Systems of Linear Equations

In this lesson and in Lesson 7.1, you considered the graphs of two linear equations together. When you do this, you form a **system of linear equations** or a **linear system.** Write the linear system represented by the graph in Problem 1.

$y = 38x + 1510$ **and** $y = 60x + 960$

6. Just the Math: Solution of a Linear System

The **solution** of a linear system is an ordered pair (x, y) that is a solution to *both* equations in the system. Graphically, the solution is the **point of intersection** of the system. What is the solution of the linear system in this problem situation? Use your graph to help you. Write your answer using a complete sentence.

The solution is the ordered pair (25, 2460).

Algebraically, verify that the ordered pair is a solution of your system. Remember that the ordered pair needs to be a solution of both equations.

$2460 \stackrel{?}{=} 38(25) + 1510$ $\quad$ $2460 \stackrel{?}{=} 60(25) + 960$

$2460 \stackrel{?}{=} 950 + 1510$ $\quad$ $2460 \stackrel{?}{=} 1500 + 960$

$2460 = 2460$ $\quad$ $2460 = 2460$

Explore Together

Problem 2

Students will create a system of linear equations in which the slopes are equal.

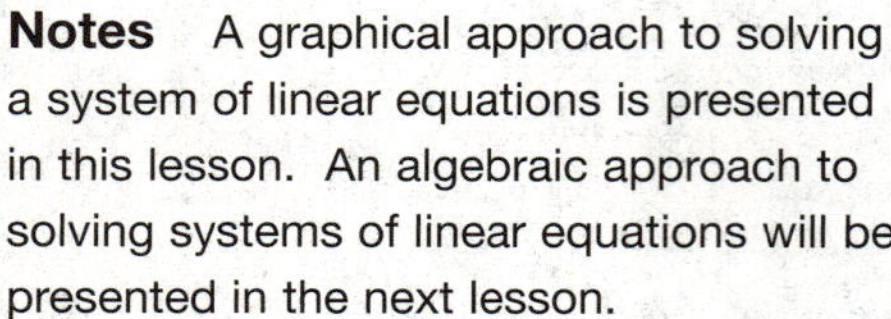

Notes A graphical approach to solving a system of linear equations is presented in this lesson. An algebraic approach to solving systems of linear equations will be presented in the next lesson.

Grouping

Ask for a student volunteer to read Problem 2 aloud. Have a student restate the problem. Pose the Guiding Questions below to verify student understanding.

Have students work in small groups to complete parts (A) through (C) of Problem 2. Then call the class back together to have the students discuss and present their work.

Guiding Questions

- How is this situation different from the situation in Problem 1?
- How is this situation similar to the situation in Problem 1?
- Which equation from Problem 1 will be used again in this problem, the equation for the number of bricks set by the novice or the experienced brick setter?
- What is the rate to set bricks for the experienced brick setter discussed in Problem 1? What is the rate for the second experienced brick setter in Problem 2?
- The rate to set bricks is the same for both experienced workers. How will this fact affect the equations for the number of bricks set?
- What are appropriate bounds for the amount of time and the number of bricks set for this problem?
- What do you predict the graph of this system of linear equations will look like when you graph it in part (C)?

Problem 2 The Pros

A. Another experienced bricklayer is having her time recorded as a part of the time study. At the beginning of the study, this worker had set 600 bricks so far and can set 60 bricks in one hour. Write an equation that gives the total number of bricks y set in place in terms of the time x in hours after the beginning of the time study.

$y = 60x + 600$

B. Write a linear system that shows the total number of bricks set in terms of time for both experienced workers.

$y = 60x + 960$ and $y = 60x + 600$

C. Create a graph of the linear system on the grid below. First, choose your bounds and intervals. Be sure to label your graph clearly.

Variable quantity	Lower bound	Upper bound	Interval
Time	0	120	8
Bricks	0	7500	500

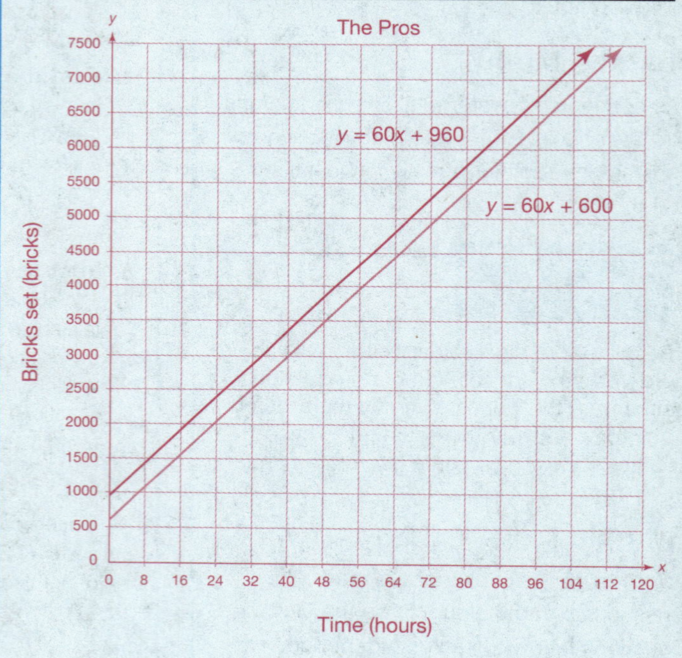

Explore Together

Investigate Problem 2

Students will answer questions about a system of linear equations with equal slopes and use this information to learn about parallel lines.

Grouping

Ask for a student volunteer to read Question 1 aloud. Have students work in small groups to complete Questions 1 through 4 of Investigate Problem 2. Then call the class back together to have the students discuss and explain their work for Questions 1 through 4.

Just the Math

Students will be formally introduced to parallel lines.

Grouping

Ask for a student volunteer to read Question 5 aloud. Have a student restate the problem. Pose the Guiding Questions below to verify student understanding.

Have the students work together in small groups to complete Questions 5 through 7. Then call the class back together to have the students discuss and explain their work for Questions 5 through 7.

Guiding Questions

- What does the phrase "in the same plane" mean?
- What are parallel lines?
- What do parallel lines look like when they are graphed?
- If the two experienced workers who set bricks at the same rate had started their work at the same time, how would the number of bricks set compare at all times for the two workers?
- Because the two experienced workers did not start at the same time, but they work at the same rate, will the second experienced worker be able to catch up to the number of bricks set by the first experienced worker?

Investigate Problem 2

1. What does the slope of each line represent in this problem situation? Use a complete sentence in your answer.

Sample Answer: The slope indicates the number of bricks set by each worker in one hour.

Which worker sets bricks faster? How do you know? Use a complete sentence in your answer.

Sample Answer: The workers set the bricks at the same rate, 60 bricks per hour, because the slope of each line is 60.

2. What does the y-intercept of each line represent in this problem situation? Use a complete sentence in your answer.

Sample Answer: The y-intercept indicates the number of bricks that have already been set.

How do the y-intercepts of the lines compare? What does this mean in the problem situation? Use complete sentences in your answer.

Sample Answer: The y-intercept of the first experienced worker is greater than the y-intercept of the second experienced worker. The first experienced worker has set more bricks so far.

3. Does there appear to be any point of intersection of the lines?

No.

4. Use complete sentences to describe how the lines are related to each other.

Sample Answer: The lines are the same distance from one another.

5. Just the Math: Parallel Lines The lines that you graphed in part (C) are *parallel lines*. Two lines in the same plane are **parallel** to each other if they do not intersect. What can you conclude about the slopes of parallel lines? Use a complete sentence in your answer.

The slopes of parallel lines are equal.

6. Does the linear system for the two experienced workers have a solution? Use complete sentences to explain your reasoning.

Sample Answer: Because the lines for the equations are parallel, they never intersect. So, there is no solution.

7. Will the two experienced workers ever set the same number of bricks during the time study? Use complete sentences to explain your reasoning.

Sample Answer: They will never set the same number of bricks because the lines for the equations never intersect.

Explore Together

Investigate Problem 2

Just the Math

Students will create a table of values for two linear functions that are equivalent.

Grouping

Ask for a student volunteer to read Question 8 aloud. Have a student restate the problem. Pose the Guiding Questions below to verify student understanding.

Have students work in small groups to complete Question 8. Then call the class back together to have the students discuss and explain their work for Question 8.

Guiding Questions

- Which situation had one solution of the system of linear equations in this lesson? What was the solution of that system?
- Which situation had no solution of the system of linear equations in this lesson? Why did that system have no solution?
- What does it mean for a system to have one solution? What does it mean for a system to have no solution?
- Would it be possible for a system of two linear equations to have exactly two different solutions?

Key Formative Assessments

- What possible number of solutions exists for a system of two linear equations?
- What does it mean for a system of linear equations to have an infinite number of solutions?
- What is the relationship between the two lines if the system of linear equations has an infinite number of solutions?

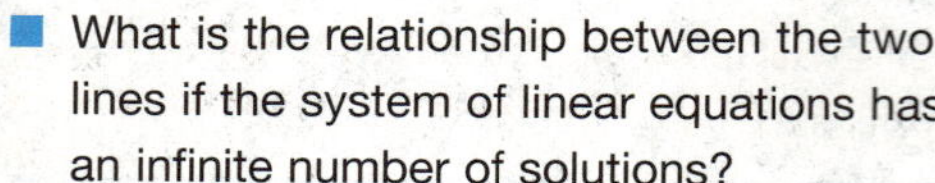

Problem 3

Grouping

Ask for a student volunteer to read part (A) of Problem 3 aloud. Have a student restate the problem. After students understand the situation, have them work in small groups to complete part (A).

Investigate Problem 2

8. **Just the Math: Number of Solutions of a Linear System** So far in this lesson, we have seen a linear system with one solution and a linear system with no solution. Use complete sentences to describe the graphs of these kinds of linear systems.

Sample Response: The graph of a linear system with one solution contains intersecting lines. The graph of a linear system with no solution contains parallel lines.

Consider the following linear system:

$y = 2x - 4$ and $y = -2(2 - x)$.

Complete the table of values for this linear system.

Expression	x	$2x - 4$	$-2(2 - x)$
	–5	**–14**	**–14**
	0	**–4**	**–4**
	5	**6**	**6**
	10	**16**	**16**
	12	**20**	**20**
	15	**26**	**26**

What can you conclude about the number of solutions of this linear system? Use a complete sentence in your answer.

Sample Answer: There are an infinite number of solutions.

Because every point on the graph of $y = 2x - 4$ is on the graph of $y = -2(2 - x)$, we can say that this system has an *infinite number* of solutions. Use a complete sentence to explain why you think this is true.

Sample Answer: A line is made up of an infinite number of points and the lines have the same graph, so the number of solutions is infinite.

Problem 3 When Is the Job Done?

A. The experienced bricklayer who sets bricks at a rate of 60 bricks per hour and has set 960 bricks so far must set approximately 20,000 additional bricks before the job is done. Write an equation that gives the total number of bricks y left to set in terms of the time x in hours after the beginning of the time study.

$y = 20,000 - 60x$

Explore Together

Problem 3

Students will compare two equations for the same situation with a different focus. One equation for the bricklayer is solved for the number of bricks already set. The second equation for the same bricklayer is solved for the number of bricks remaining to be set.

Grouping

Ask for a student volunteer to read part (B) of Problem 3 aloud. Have a student restate the problem. Pose the Guiding Questions below to verify student understanding. Then, have students work in small groups to complete part (B).

Guiding Questions

- What is the meaning of the first equation in your system of linear equations?
- What is the meaning of the second equation in your system of linear equations?
- What is the y-intercept of the first equation?
- What is the y-intercept of the second equation?
- Do you expect the graphs of the lines in this system to intersect?
- How many solutions to this system do you expect to find?

Investigate Problem 3

Grouping

Have the students work in small groups to complete Questions 1 through 5.

Common Student Errors

Students may be confused with the graph for this system of linear equations. It will be new to many students to have two different equations to model the same situation with a different focus.

Problem 3 When Is the Job Done?

B. Form a linear system with the equation in part (A) and the equation from Problem 1, part (A) that gives the total number of bricks set by this worker in terms of the time after the beginning of the time study.

$y = 20{,}000 - 60x$ and $y = 60x + 960$

Investigate Problem 3

1. Create a graph of the linear system on the grid below. First, choose your bounds and intervals. Be sure to label your graph clearly

Variable quantity	Lower bound	Upper bound	Interval
Time	0	375	25
Bricks	0	22,500	1500

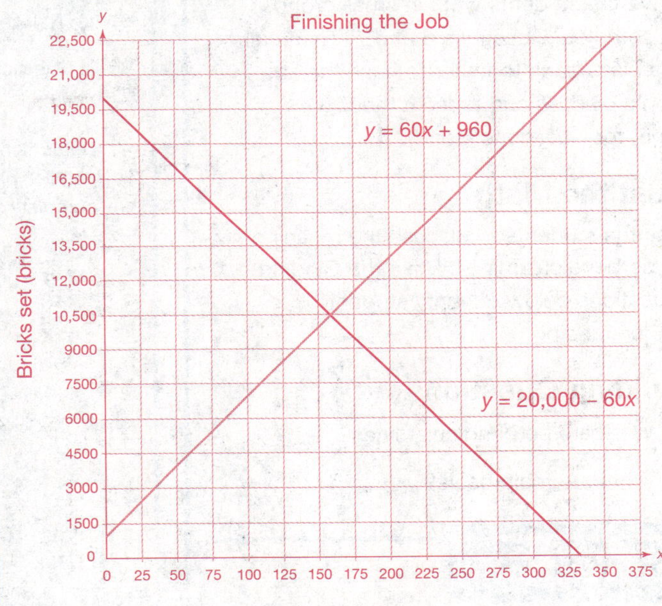

2. What does the point of intersection of the lines represent?

   Sample Answer: The number of hours it takes for the number of bricks set to equal the number of bricks left to set.

3. Compare the slopes of the lines. Use a complete sentence in your answer.

   Sample Answer: One slope is positive, one is negative.

Explore Together

7

Investigate Problem 3

Students will be formally introduced to the terminology for perpendicular lines.

Common Student Errors

Students may need to have help understanding the term perpendicular lines.

Grouping

Call the class back together to have the students discuss and present their work for Questions 1 through 5. Then ask for a student volunteer to read Question 6 aloud. Have a student restate the problem.

Have the students work in small groups to complete Questions 6 and 7. Then call the class back together to have the students discuss and present their work for Questions 6 and 7.

Just the Math

Ask the students to discuss the meaning of perpendicular lines. Pose the Guiding Questions below to verify student understanding.

Guiding Questions

- What are perpendicular lines?
- What are reciprocals?

Take Note

Two numbers are **reciprocals** if their product is –1.

Key Formative Assessments

- What are the possible numbers of solutions of a system of two linear equations?
- What is a system of linear equations?
- What are parallel lines and what do we know about the slopes of parallel lines?
- What are perpendicular lines and what do we know about the slopes of perpendicular lines?

Investigate Problem 3

4. Are the lines *perpendicular*? That is, do they intersect at a right angle?

No.

5. Consider the graph of your linear system and the equations of the lines. What do you notice about the slopes of perpendicular lines?

Sample Answer: One slope is positive and one slope is negative.

6. Just the Math: Perpendicular Lines A property of **perpendicular lines** is that the product of their slopes must be –1. So, this means that the slopes must have opposite signs and must be **reciprocals** of each other. For instance, the lines $y = -3x + 4$ and $y = \frac{1}{3}x + 1$ are perpendicular because $-3\left(\frac{1}{3}\right) = -1$. Algebraically show that the lines in your graph in Question 1 are *not* perpendicular. Show your work.

$60(-60) = -3600 \neq -1$

7. Determine whether the graphs of each pair of equations are parallel, perpendicular, or neither. Show your work and use a complete sentence to explain your reasoning.

$y = \frac{2}{3}x + 4$ and $y = -\frac{3}{2}x + 1$

$\frac{2}{3}\left(-\frac{3}{2}\right) = -1$; The lines are perpendicular because the product of the slopes is –1.

$y = 5x - 4$ and $y = -5x + 4$

$5(-5) = -25$; The lines are neither parallel nor perpendicular because the slopes are not equal and the product of the slopes is not –1.

$y = 4x$ and $y = \frac{1}{4}x - 2$

$4\left(\frac{1}{4}\right) = 1$; The lines are neither parallel nor perpendicular because the slopes are not equal and the product of the slopes is not –1.

$y = -1.8x + 15$ and $y = 6 - 1.8x$

The lines are parallel because the slopes are the same.

Wrap Up

Close

- Review all key terms and their definitions. Include the terms *system of linear equations, linear system, solution, point of intersection, parallel lines, reciprocals, perpendicular lines,* and *negative reciprocals*.
- You may also want to review any other vocabulary terms that were discussed during the lesson, which may include *brick setting, in terms of, plane, catch up, novice worker*, and *experienced worker*.
- Remind the students to write the key terms and their definitions in the notes section of their notebooks. You may also want the students to include examples.
- Ask the students to summarize all the mathematics in this lesson.
- Divide the students into six small groups. Assign one of these key terms to each of the six groups: *linear equations* and *linear system; solution; point of intersection; parallel lines; reciprocals; and perpendicular lines*. Have each group discuss and then present a review of the key term that they were assigned.
- Have the students work together as a class. One student will give a slope, another student will give the slope of a line parallel to the line with the original slope, and a third student will give the slope of a line perpendicular to a line with the original slope.

Ties to the Cognitive Tutor Software

In the Cognitive Tutor software, most students will use algebraic methods (setting up and solving an equation) to find the point of intersection in a problem with systems of linear equations. However, some students will plot the graph first, and read the intersection point from the graph to complete the table.

Follow Up

Assignment

Use the Assignment for Lesson 7.2 in the Student Assignments book. See the Teacher's Resources and Assessments book for answers.

Assessment

See the Assessments provided in the Teacher's Resources and Assessments book for Chapter 7.

Open-Ended Writing Task

Ask the students to create a vocabulary practice worksheet for the key terms in this lesson. They should include an answer key as well as the original worksheet. This activity can be done individually or in a small group.

7

Reflections

Insert your reflections on the lesson as it played out in class today.

What went well?

What did not go as well as you would have liked?

How would you like to change the lesson in order to improve the things that did not go well and capitalize on the things that did go well?

Notes

7.3

Hiking Trip

Using Substitution to Solve a Linear System

7

Learning By Doing Lesson Map

Get Ready

Objective

In this lesson, you will:

- Solve linear systems by using substitution.

Key Terms

- standard form of a linear equation
- substitution method

NCTM Content Standards

Grades 9–12 Expectations

Algebra Standards

- Interpret representations of functions of two variables.
- Understand the meaning of equivalent forms of expressions, equations, inequalities, and relations.
- Write equivalent forms of equations, inequalities, and systems of equations and solve them with fluency—mentally or with paper and pencil in simple cases and using technology in all cases.
- Use symbolic algebra to represent and explain mathematical relationships.
- Draw reasonable conclusions about a situation being modeled.

Measurement Standards

- Make decisions about units and scales that are appropriate for problem situations involving measurement.

Lesson Overview

Within the context of this lesson, students will be asked to:

- Write equations in standard form.
- Write equations in slope-intercept form.
- Use substitution to solve systems of linear equations.

Essential Questions

The following key questions are addressed in this lesson:

1. What is the standard form of an equation?
2. What do the letters A, B, and C represent in the standard form of an equation?
3. What is substitution?
4. What does it mean to solve a system of linear equations?

Show The Way

Warm Up

7

Place the following questions or an applicable subset of these questions on the board before students enter class. Students should begin working as soon as they are seated.

Write the reciprocal of each number.

	Original Number	Reciprocal		Original Number	Reciprocal
1.	$\frac{3}{2}$	$\frac{2}{3}$	**2.**	$-\frac{1}{3}$	-3
3.	$-\frac{9}{4}$	$-\frac{4}{9}$	**4.**	$\frac{1}{6}$	6

Complete the table below.

Slope of original line	Slope of a line parallel to original line	Slope of a line perpendicular to original line
5. $\frac{3}{2}$	$\frac{3}{2}$	$\frac{-2}{3}$
6. $-\frac{1}{3}$	$-\frac{1}{3}$	3
7. $-\frac{9}{4}$	$-\frac{9}{4}$	$\frac{4}{9}$
8. -6	-6	$\frac{1}{6}$

Motivator

Begin the lesson with the motivator to get students thinking about the topic of the upcoming problem. This lesson is about a hiking trip. The motivating questions are about students' experiences with hiking.

Ask the students the following questions to get them interested in the lesson.

- What is a hiking trip?
- What is trail mix?
- Have you ever eaten trail mix?
- What was in the trail mix you have eaten?
- Have you ever gone hiking? If so, where have you hiked?
- Have you hiked with a group, club, or troop?

Explore Together

Problem 1

Students will write and analyze a system of linear equations that has one equation written in standard form and the other in slope-intercept form.

Take Note

A linear equation is in **standard form** if it is written as $Ax + By = C$, where A, B, and C are constants and A and B are not both zero.

Grouping

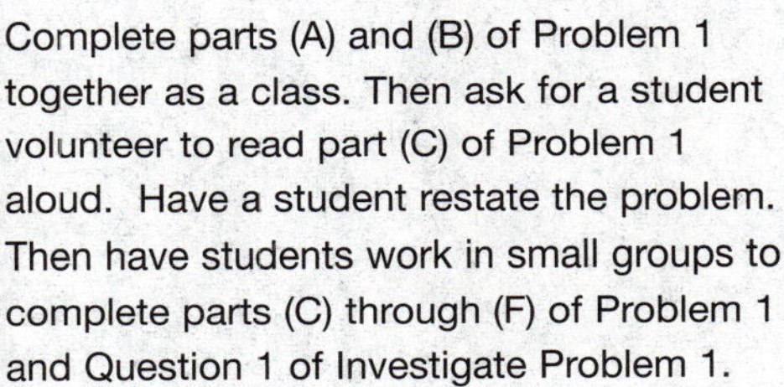

Ask for a student volunteer to read the Scenario and Problem 1 aloud. Have a student restate the problem. Pose the Guiding Questions below to verify student understanding.

Complete parts (A) and (B) of Problem 1 together as a class. Then ask for a student volunteer to read part (C) of Problem 1 aloud. Have a student restate the problem. Then have students work in small groups to complete parts (C) through (F) of Problem 1 and Question 1 of Investigate Problem 1.

Guiding Questions

- What is trail mix?
- What is constant in this situation?
- What is variable in this situation?
- What does x represent in this situation?
- What does y represent in this situation?
- What is a solution of an equation?
- What does it mean for values to satisfy an equation?

Common Student Errors

Students may have difficulty with the arithmetic in these questions, especially the multiplication.

SCENARIO The Outdoor Club at school is going on a hiking trip and is making trail mix as part of the food that they will take. The trail mix will be made up of nuts and dried fruits, such as raisins, dried cherries, and banana chips. The nuts cost \$4.50 per pound and the dried fruits cost \$3.25 per pound. The group can spend \$15 on the trail mix.

7

Problem 1 Making Trail Mix

A. Write an equation in standard form that relates the numbers of pounds of nuts and dried fruits that can be bought for \$15. Use x to represent the number of pounds of nuts and y to represent the number of pounds of dried fruits that can be bought.

$4.50x + 3.25y = 15$

B. The group agreed to have one and a half times as much dried fruits as nuts in the mix. Write an equation in x and y as defined in part (A) that represents this situation.

$y = 1.5x$

C. Will two pounds of nuts and three pounds of dried fruits satisfy both of your equations? Show all your work.

$4.50(2) + 3.25(3) \stackrel{?}{=} 15$

$9 + 9.75 \stackrel{?}{=} 15$

$18.75 \neq 15$

$3 \stackrel{?}{=} 1.5(2)$

$3 = 3$

No.

D. Will two and one quarter pounds of nuts and one and a half pounds of dried fruits satisfy both of your equations? Show all your work.

$4.50(2.25) + 3.25(1.5) \stackrel{?}{=} 15$

$10.125 + 4.875 \stackrel{?}{=} 15$

$15 = 15$

$1.5 \stackrel{?}{=} 1.5(2.25)$

$1.5 \neq 3.375$

No.

Explore Together

7

Problem 1

Students will graph both equations in the same coordinate plane and then try to estimate the point of intersection from their graphs.

Grouping

Students will be working together in small groups to complete these questions.

Common Student Errors

Students may have difficulty choosing appropriate bounds for their graphs in part (E) of Problem 1. Pose the Guiding Questions below to verify student understanding.

Guiding Questions

- What two quantities are being studied in the equation that you wrote in part (A) of this problem?
- What two quantities are being studied in the equation that you wrote in part (B) of this problem?
- What is the quantity that you called the x-value?
- What is the quantity that you called the y-value?
- If you increase the amount of nuts in your trail mix, what will happen to the amount of dried fruits that you can put in the trail mix?
- If you decrease the amount of nuts in your trail mix, what will happen to the amount of dried fruits that you can put in the trail mix?
- What is the smallest possible amount of nuts that you can buy? What is the smallest possible amount of dried fruits that you can buy?
- What is the largest reasonable quantity of nuts that you can buy? What is the largest reasonable quantity of dried fruits that you can buy?

Problem 1 Making Trail Mix

E. Create a graph of both equations on the grid below. First, choose your bounds and intervals. Be sure to label your graph clearly.

Variable quantity	Lower bound	Upper bound	Interval
Nuts	0	7.5	0.5
Dried fruits	0	7.5	0.5

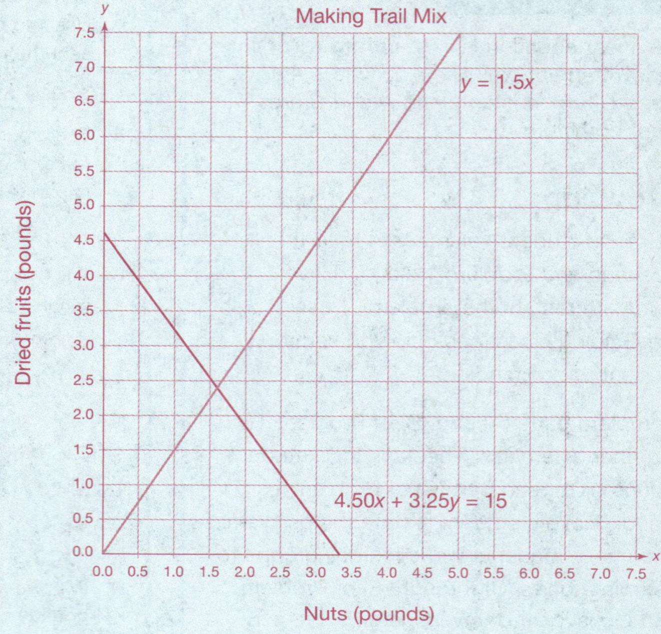

F. Can you determine the solution of this linear system exactly from your graph? Use a complete sentence to explain your answer.

Sample Answer: No, because the point does not fall exactly on the grid lines.

Common Student Errors

Some students may incorrectly believe that they can estimate the point of intersection accurately from their graph.

Explore Together

Investigate Problem 1

Grouping

Call the class back together to have the students discuss and present their work for parts (C) through (F) of Problem 1 and Question 1 of Investigate Problem 1.

Just the Math

Students will be formally introduced to the substitution method for solving a system of linear equations in Question 2. The term *substitution* in the mathematical context will be new to some students. Use the Guiding Questions below to help them.

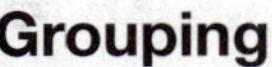

Grouping

Ask for a student volunteer to read Question 2 aloud. Have a student restate the problem. Pose the Guiding Questions below to verify student understanding.

Complete Question 2 together as a class. Then have students work in small groups to complete Questions 3 and 4.

Guiding Questions

- If we substitute margarine for butter in a recipe that we are cooking, what does that mean?
- If we substitute a make-up version of a test for a student who is absent, what does that mean?
- What does it mean to substitute for a variable in an equation?
- How can you substitute for a variable in an equation?

Take Note

It does not matter which equation of the linear system that you use to find the value of y. You could have used the equation $4.50x + 3.25y = 15$ to find the value of y.

Investigate Problem 1

1. Estimate the point of intersection from your graph.

Sample Answer: (1.6, 2.4)

Check your point in each of the equations. Is your point the solution of the linear system?

Sample Answer:

$$4.50(1.6) + 3.25(2.4) \stackrel{?}{=} 15$$
$$7.2 + 7.8 \stackrel{?}{=} 15$$
$$15 = 15$$
$$2.4 \stackrel{?}{=} 1.5(1.6)$$
$$2.4 = 2.4$$

Yes.

2. Just the Math: Substitution Method In many systems it is difficult to determine the solution from the graph, so there is an algebraic method for finding the solution. Consider the linear system for this problem situation:

$$4.50x + 3.25y = 15$$
$$y = 1.5x.$$

Because y is equal to $1.5x$, we can substitute $1.5x$ for y in the first equation.

$$4.50x + 3.25y = 15$$
$$4.50x + 3.25(1.5x) = 15$$

You now have an equation in x only. Solve this equation for x. Show all your work.

$$4.50x + 3.25(1.5x) = 15$$
$$4.50x + 4.875x = 15$$
$$9.375x = 15$$
$$x = 1.6$$

Now that you have the x-value of the solution, find the y-value by substituting your result for x into the equation $y = 1.5x$. Show all your work.

$y = 1.5(1.6) = 2.4$

So, the solution to the linear system is (1.6, 2.4). Is this solution confirmed by your graph?

Yes.

3. Interpret the solution of the linear system in the problem situation. Use a complete sentence in your answer.

Sample Answer: The group should buy 1.6 pounds of nuts and 2.4 pounds of dried fruits for the trail mix.

Explore Together

Investigate Problem 1

Students will have to determine the total amount of trail mix by finding the sum of the x- and y-values.

Call the class back together to have the students discuss and explain their work for Questions 3 and 4 from Investigate Problem 1.

Problem 2

Students will write and analyze a system of linear equations with equivalent slopes. They will attempt to solve a system of linear equations that has no solution.

Grouping

Ask for a student volunteer to read Problem 2 aloud. Have a student restate the problem. Pose the Guiding Questions below to verify student understanding. The units for time are especially important in this problem.

Have students work in small groups to complete parts (A) through (E) of Problem 2. Then call the class back together to have the students discuss and present their work for parts (A) through (E).

Guiding Questions

- How is this situation different from the situation in Problem 1? How is this situation similar to the situation in Problem 1?
- How will the equations that you write for this new situation be the same as each other?
- What will be different between the two equations that you write for the situation in Problem 2?
- What are the units for the rate at which each group is hiking?
- What are the units given for the amount of time that the first group hikes before the second group begins to hike? How can that amount of time be converted to hours rather than minutes to match the rest of the information?

Investigate Problem 1

4. How many pounds of trail mix will the club have? Use a complete sentence to explain how you found your answer.

4 pounds; Add the number of pounds of nuts and the number of pounds of dried fruits to get the total number of pounds of trail mix.

Problem 2 Hiking the Trail

The Outdoor Club splits up into two smaller groups to hike the trail. The first group leaves the beginning of the trail and hikes at a rate of 2.5 miles per hour. The second group leaves 30 minutes later and hikes at a rate of 2.5 miles per hour.

A. Write an equation for the first group that gives the distance hiked y in miles in terms of the amount of time x in hours that the group has been hiking.

$y = 2.5x$

B. How far will the first group have traveled after 30 minutes of hiking? Show your work and use a complete sentence in your answer.

$y = 2.5(0.5) = 1.25$; **The first group will have traveled 1.25 miles.**

C. Write an equation for the second group that gives the distance hiked y in miles in terms of the amount of time since the first group started hiking x.

$y = 2.5x - 1.25$

D. How far will each group have traveled 45 minutes after the first group started hiking? Show all your work and use a complete sentence in your answer.

$y = 2.5(0.75) = 1.875$; $y = 2.5(0.75) - 1.25 = 0.625$; **The first group will have traveled 1.875 miles, and the second group will have traveled 0.625 mile.**

How far will each group have traveled after 2 hours? Show all your work and use a complete sentence in your answer.

$y = 2.5(2) = 5$; $y = 2.5(2) - 1.25 = 3.75$; **The first group will have traveled 5 miles, and the second group will have traveled 3.75 miles.**

E. Will the second group catch up to the first group? Use complete sentences to explain your reasoning.

Sample Answer: No, because they left after the first group, and they are hiking at the same rate as the first group.

Common Student Errors

Students may struggle to create the equations for this problem and may not remember to convert the 30-minute advanced start for the first group to hours.

Explore Together

Investigate Problem 2

Students will graph and will attempt to solve their system of linear equations.

Grouping

Ask for a student volunteer to read Question 1 from Investigate Problem 2 aloud. Have a student restate the problem. Then have students work in small groups to complete Questions 1 through 5.

Key Formative Assessments

- What would a solution of this system of linear equations represent?
- What is the relationship between the graphs of these equations?
- If your brother is 4 years younger than you, will he ever be the same age as you are at the same time as you are that age?
- Is there any way that a number can be equal to the value of itself decreased by another non-zero number? For instance, is there any number that we could say is equal to that same number subtracted by 4?
- Why can't the second group ever catch up to the first group of hikers if they continue hiking at the same rate?
- How many solutions did we find for systems of linear equations with parallel lines in the previous lesson about the brick setters?
- What is the meaning of the solution you found in Question 1?

Investigate Problem 2

1. Solve the linear system by using the substitution method. First, write your system below.

   $y = 2.5x$

   $y = 2.5x - 1.25$

   Next, because you have an expression for *y* in terms of *x*, substitute your expression for *y* from the first equation into the second equation.

   $2.5x = 2.5x - 1.25$

   Now solve the equation for *x*. What is the result? Use a complete sentence in your answer.

   Sample Answer: The result doesn't make sense because 0 is not equal to –1.25.

2. Create a graph of your linear system on the grid below. First, choose your bounds and intervals. Be sure to label your graph clearly.

Variable quantity	Lower bound	Upper bound	Interval
Time	0	7.5	0.5
Distance	0	7.5	0.5

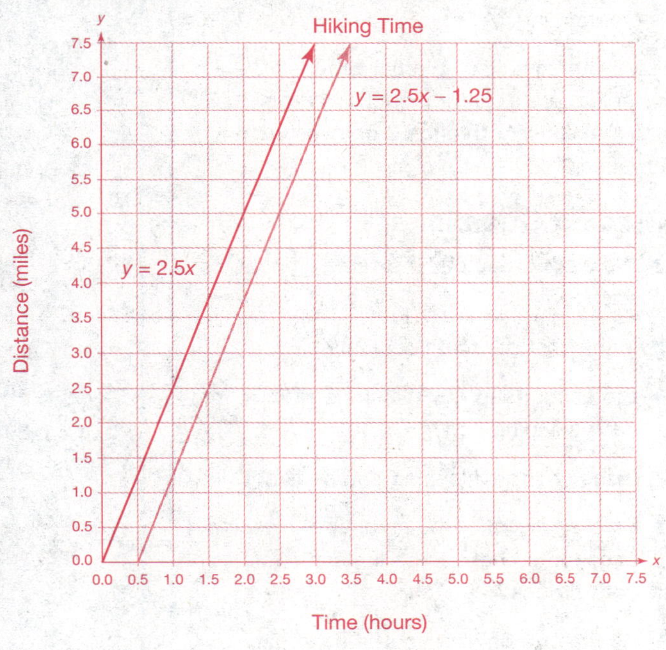

7

Explore Together

7

Investigate Problem 2

Grouping

Call the class back together to have the students discuss and present their work for Questions 1 through 5 of Investigate Problem 2.

Problem 3

Students will create and solve another system of linear equations. This linear system will most likely begin with both equations written in standard form.

Grouping

Ask for a student volunteer to read Problem 3 aloud. Have a student restate the problem. Pose the Guiding Questions below to verify student understanding.

Have students work in small groups to complete parts (A) through (C) of Problem 3 and Questions 1 through 3 of Investigate Problem 3.

Guiding Questions

- Did each group pay the same amount for each tent?
- Is the cost per tent constant?
- Did each group pay the same amount for each sleeping bag?
- Is the cost per sleeping bag constant?
- Do the values for the number of sleeping bags rented have to be the same for both groups?
- Do the groups have to rent an equal number of sleeping bags and tents?
- Would each group most likely rent more tents or more sleeping bags?
- What other information is given in this problem?

Investigate Problem 2

3. What is the relationship between the lines in the graph? Use a complete sentence in your answer.

The lines are parallel.

4. What is the solution of the linear system? Use a complete sentence in your answer.

Sample Answer: The linear system has no solution.

5. What is the result when you try to algebraically solve a linear system that has no solution? Use a complete sentence in your answer.

Sample Answer: The result is something that does not make sense.

Problem 3 Camping

Another community group joins the Outdoor Club at the campsite. The new group has rented six tents and twenty four sleeping bags for $186. The Outdoor Club rented from the same place and rented eight tents and thirty sleeping bags for $236. Each tent costs the same, and each sleeping bag costs the same.

A. For each group, write an equation in standard form for this problem situation. Use x to represent the cost of one tent in dollars and use y to represent the cost of one sleeping bag in dollars.

Community group: $6x + 24y = 186$;
Outdoor Club: $8x + 30y = 236$

B. Without solving the linear system, interpret the solution of the linear system in part (A). Use a complete sentence in your answer.

The solution will give the cost of one sleeping bag and one tent.

C. Can you tell from looking at the equations whether the linear system has a solution? Use a complete sentence to explain your reasoning.

Sample Answer: No, because you cannot directly determine the slopes or y-intercepts from the equations in this form.

Explore Together

Investigate Problem 3

Students will solve this system of linear equations using the substitution method.

Common Student Errors

Many students will not quickly recognize that the difference between the equations in Problems 1 and 2 and the equations in Problem 3 is the form of the equations. You may have to remind the students of slope-intercept form and standard form using the Guiding Questions below.

Take Note

Whenever a product involves a sum, such as $4(x + 3)$, you must use the distributive property to simplify: $4(x + 3) = 4(x) + 4(3)$.

Guiding Questions

- What three forms have we learned for writing an equation?
- What is slope-intercept form?
- What is standard form?
- What is point-slope form?
- Make a list of each individual equation you have written so far in this lesson. What form of equation is each one?
- How many solutions to this linear system do you expect to find?
- What is the substitution method of solving a system of linear equations?
- Why is this linear system more difficult to solve using the substitution method than the linear systems in Problems 1 and 2?

Grouping

Call the class back together to have the students discuss and present their work for parts (A) through (C) of Problem 3 and Questions 1 through 3 of Investigate Problem 3. Then ask for a student volunteer to read Question 4 aloud. Have a student restate the problem. Complete Questions 4 through 6 together as a class.

Investigate Problem 3

1. How does this linear system differ from the linear systems that you wrote in Problems 1 and 2? Use complete sentences in your answer.

   Sample Answer: In Problems 1 and 2, at least one equation was in slope-intercept form. In this linear system, neither equation is in slope-intercept form.

2. To solve this linear system by using the substitution method, what do you think you would have to do first? Use a complete sentence in your answer.

   Sample Answer: I need to write one of the equations in slope-intercept form.

3. Write the equation for the community group in slope-intercept form. Show all your work.

   Sample Answer: $6x + 24y = 186$

   $24y = -6x + 186$

   $y = -0.25x + 7.75$

4. Now, use the substitution method to solve the linear system. Begin by substituting your expression from Question 3 for y in terms of x into the equation for the Outdoor Club.

   $8x + 30(\boxed{-0.25x + 7.75}) = 236$

   Now solve this equation for x. Show all your work.

   $8x + 30(-0.25x + 7.75) = 236$

   $8x - 7.5x + 232.5 = 236$

   $0.5x = 3.5$

   $x = 7$

   Finally, find the value for y. Show all your work.

   Sample Answer: $y = -0.25(7) + 7.75 = 6$

5. Check your answer algebraically. Show all your work.

   $6(7) + 24(6) \stackrel{?}{=} 186$

   $42 + 144 \stackrel{?}{=} 186$

   $186 = 186$

   $8(7) + 30(6) \stackrel{?}{=} 236$

   $56 + 180 \stackrel{?}{=} 236$

   $236 = 236$

6. Interpret the solution of the linear system in the problem situation. Use complete sentences in your answer.

   The tents are $7 each, and the sleeping bags are $6 each.

7

Explore Together

Investigate Problem 3

7

Students will practice solving systems of linear equations using the substitution method.

Common Student Errors

Students may forget to write one of the equations in slope-intercept form to use the substitution method. Remind the students to refocus them as necessary.

Grouping

Ask for a student volunteer to read Question 7 aloud. Have a student restate the problem. Pose the Guiding Questions below to verify student understanding.

Have the students work in small groups to complete Question 7. Then call the class back together to have the students discuss and present their work for Question 7.

Guiding Questions

- How can you solve a system of linear equations using the substitution method?
- What are the important ideas to remember when solving a linear system using the substitution method?
- What errors do you think people could make if they are not careful in solving linear systems this way?

Key Formative Assessments

- What is the substitution method?
- How can you solve a system of linear equations using the substitution method?
- What are the most important steps in solving a linear system with the substitution method?
- Why do we make the restriction that A and B cannot both equal zero for the standard form of an equation?

Investigate Problem 3

7. If possible, solve each linear system by using the substitution method. Show all your work and use a complete sentence in your answer. Then check your answer algebraically.

$4x + 3y = 10$
$y = 2x$

$4x + 3y = 10$
$4x + 3(2x) = 10$
$4x + 6x = 10$
$10x = 10$
$x = 1$

$y = 2(1) = 2$
The solution is (1, 2).

$4(1) + 3(2) \stackrel{?}{=} 10$
$4 + 6 \stackrel{?}{=} 10$
$10 = 10$

$2 \stackrel{?}{=} 2(1)$
$2 = 2$

$y = 2x - 1$
$y = 3x + 1$

$y = 2x - 1$
$3x + 1 = 2x - 1$
$x = -2$

$y = 2(-2) - 1 = -4 - 1 = -5$
The solution is $(-2, -5)$.

$-5 \stackrel{?}{=} 2(-2) - 1$
$-5 \stackrel{?}{=} -4 - 1$
$-5 = -5$

$-5 \stackrel{?}{=} 3(-2) + 1$
$-5 \stackrel{?}{=} -6 + 1$
$-5 = -5$

$8x - 2y = 7$
$2x + y = 4$

$y = -2x + 4$

$8x - 2(-2x + 4) = 7$
$8x + 4x - 8 = 7$
$12x = 15$
$x = 1.25$

$y = -2(1.25) + 4 = 1.5$
The solution is (1.25, 1.5).

$8(1.25) - 2(1.5) \stackrel{?}{=} 7$
$10 - 3 \stackrel{?}{=} 7$
$7 = 7$

$2(1.25) + 1.5 \stackrel{?}{=} 4$
$2.5 + 1.5 \stackrel{?}{=} 4$
$4 = 4$

$6x + 3y = 5$
$y = -2x + 1$

$6x + 3y = 5$
$6x + 3(-2x + 1) = 5$
$6x - 6x + 3 = 5$
$3 \neq 5$

There is no solution.

Wrap Up

Close

- Review all key terms and their definitions. Include the terms *standard form of a linear equation* and *substitution method*.
- You may also want to review any other vocabulary terms that were discussed during the lesson, which may include *slope-intercept form, point-slope form, in terms of,* and *satisfy an equation*.
- Remind the students to write the key terms and their definitions in the notes section of their notebooks. You may also want the students to include examples.
- Ask the students to summarize all the mathematics in this lesson.
- Divide the students into several small groups. Have each group develop a new system of linear equations that can be solved using the substitution method. Then have them solve their own linear system. Each group should write their linear system on a piece of paper with the solution written on the back. They will leave that paper on one desk in a set of desks. Finally, have the students rotate clockwise around the room to solve each of the linear systems and check their work with the solution on the back of the paper. If any group disagrees with the solution written by the group that created the linear system, they should notify you.

Ties to the Cognitive Tutor Software

Students should understand that there are different methods for solving systems of equations, and they need to make choices about which methods are appropriate in a particular situation. In the Cognitive Tutor software, students can choose to solve through linear combination or graphically.

Follow Up

Assignment

Use the Assignment for Lesson 7.3 in the Student Assignments book. See the Teacher's Resources and Assessments book for answers.

Assessment

See the Assessments provided in the Teacher's Resources and Assessments book for Chapter 7.

Open-Ended Writing Task

Ask the students to write a summary of the steps needed to solve a system of linear equations using the substitution method.

7

Reflections

Insert your reflections on the lesson as it played out in class today.

What went well?

What did not go as well as you would have liked?

How would you like to change the lesson in order to improve the things that did not go well and capitalize on the things that did go well?

Notes

7.4

Basketball Tournament

Using Linear Combinations to Solve a Linear System

7

Learning By Doing Lesson Map

Get Ready

Objective

In this lesson, you will:

- Solve a linear system by using linear combinations.

Key Terms

- standard form of a linear equation
- linear combination
- linear combinations method

NCTM Content Standards

Grades 9–12 Expectations

Algebra Standards

- Interpret representations of functions of two variables.
- Understand the meaning of equivalent forms of expressions, equations, inequalities, and relations.
- Write equivalent forms of equations, inequalities, and systems of equations and solve them with fluency—mentally or with paper and pencil in simple cases and using technology in all cases.
- Use symbolic algebra to represent and explain mathematical relationships.
- Draw reasonable conclusions about a situation being modeled.

Lesson Overview

Within the context of this lesson, students will be asked to:

- Write equations in standard form.
- Multiply equations in systems of equations by constants to get pairs of like terms that are opposites of each other.
- Use the linear combinations method to solve systems of linear equations.

Essential Questions

The following key questions are addressed in this lesson:

1. What is the standard form of a linear equation?
2. What do the letters *A*, *B*, and *C* represent in the standard form of an equation?
3. What is a linear combination?
4. What is the linear combinations method?
5. What does it mean to solve a system of linear equations?

Show The Way

7

Warm Up

Place the following questions or an applicable subset of these questions on the board before students enter class. Students should begin working as soon as they are seated.

Solve each system of equations by using the substitution method.

1. $x - y = 3$ **(5, 2)**

$2x + y = 12$

2. $x + 3y = 14$ **(−1, 5)**

$4y = x + 21$

3. $3x = y - 20$ **(0, 20)**

$x + y = 20$

4. $x = 2y - 2$ **(4, 3)**

$2x + 3y = 17$

5. $7x + y = 20$ **(2, 6)**

$3x + 2y = 18$

6. $x - y = 6$ **(7, 1)**

$5x + y = 36$

Motivator

Begin the lesson with the motivator to get students thinking about the topic of the upcoming problem. This lesson is about a basketball tournament. The motivating questions are about the details of a basketball tournament.

Ask the students the following questions to get them interested in the lesson.

- What is a basketball tournament?
- Why would tickets be sold in advance and at the door?
- Have you ever played in a basketball tournament?
- How did your team do?
- Why might a team need to sell snacks to participate in a tournament?
- Did your team sell anything to raise money to travel or pay the entrance fee for a tournament?

Explore Together

Problem 1

Students will write and solve a system of equations using the linear combinations method.

Grouping

Ask for a student volunteer to read the Scenario and Problem 1 aloud. Have a student restate the problem. Pose the Guiding Questions below to verify student understanding.

Have students work together in small groups to complete parts (A) through (C) of Problem 1. Then call the class back together to have the students discuss and present their work for parts (A) through (C).

Guiding Questions

- What information is given about the total number of tickets sold?
- What does x represent in this situation?
- What are the units for x?
- What does y represent in this situation?
- What are the units for y?
- Which value will be greater, x or y?
- Will the difference $x - y$ be a positive number or negative number?
- What is the format to write an equation in standard form?
- How could we interpret a solution for the system of equations in this situation?

Investigate Problem 1

Grouping

Ask for a student volunteer to read Question 1 aloud. Have a student restate the problem. Have students work together in small groups to complete Questions 1 and 2 of Investigate Problem 1.

Common Student Errors

The initial equations may be difficult for some students to formulate. The Guiding Questions above should help to guide them through any difficulties.

SCENARIO Your school hosted a basketball tournament. Tickets were sold before the tournament and at the door. More tickets were bought before the tournament than were bought at the door. In fact, there was a difference of 84 tickets between the two kinds of tickets sold. A total of 628 tickets were sold.

7

Problem 1 Ticket Sales

A. Write an equation in standard form that represents the total number of tickets sold. Use x to represent the number of tickets sold before the tournament and use y to represent the number of tickets sold at the door.

$x + y = 628$

B. Write an equation in standard form that represents the difference in the numbers of tickets sold.

$x - y = 84$

C. How are these equations different? How are they the same? Use complete sentences in your answer.

Sample Answer: The equations are both in standard form. The coefficients of x are the same, and the coefficients of y are opposites.

Investigate Problem 1

1. Write the linear system for this problem situation below.

$x + y = 628$

$x - y = 84$

Now, add the equations together.

$$\begin{array}{r} x + y = 628 \\ + \; x - y = 84 \\ \hline 2x \quad\;\; = 712 \end{array}$$

Solve the resulting equation. Use a complete sentence in your answer.

The solution is $x = 356$.

7

Explore Together

Investigate Problem 1

Students will verify their solutions to the system of equations in Problem 1 both algebraically and graphically.

Common Student Errors

Question 1 will challenge many students because it is a new way to approach solving a linear system of equations. Expect the need to refocus any students who are frustrated by this question.

Grouping

Call the class back together to have the students discuss and present their work for Questions 1 and 2. Then ask for a student volunteer to read Questions 3 and 4 aloud. Have a student restate the questions. Have students work in small groups to complete Questions 3 through 6.

Key Formative Assessments

- How did we solve this system of linear equations?
- How was this method of solving a linear system different from solving a linear system using substitution?
- What is the general strategy to solving a system of equations using the linear combinations method?
- Explain the meaning of the phrase "linear combination."
- What relationship must exist for two terms in the system to use the linear combinations method to solve the system?
- If opposite terms do not already exist, do you think there is a way we might alter the system to create an equivalent system with a pair of opposite terms?

Investigate Problem 1

Now find the value for y by substituting your value for x into one of the original equations.

$356 + y = 628$

$y = 272$

What is the solution of your linear system? Use a complete sentence in your answer.

The solution is (356, 272).

2. Check your solution algebraically.

$356 + 272 \stackrel{?}{=} 628$ $\quad$ $356 - 272 \stackrel{?}{=} 84$

$628 = 628$ $\quad$ $84 = 84$

3. Check your solution by creating a graph of your linear system on the grid below. First, choose your bounds and intervals. Be sure to label your graph clearly.

Variable quantity	Lower bound	Upper bound	Interval
Presale tickets	0	450	30
Door tickets	0	450	30

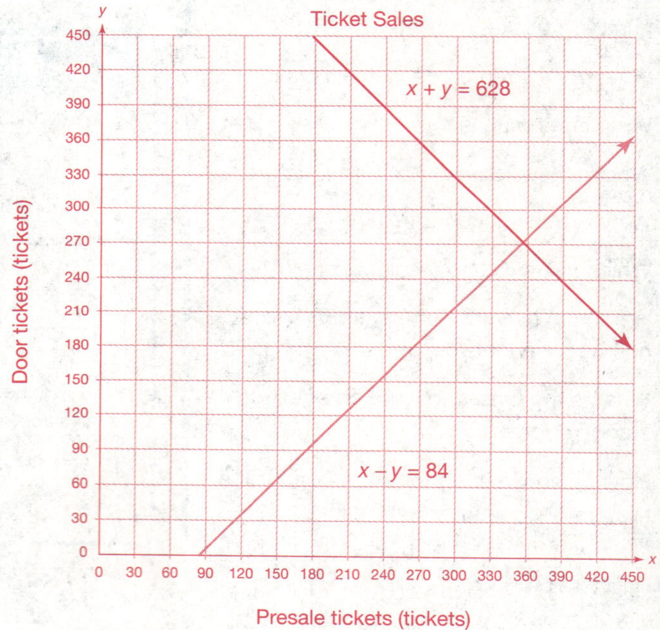

Explore Together

Investigate Problem 1

Grouping

Call the class back together to have the students discuss and present their work for Questions 3 through 6.

Problem 2

Students will write and solve another system of equations using the linear combinations method. (This can also be referred to as the elimination method.)

Grouping

Ask for a student volunteer to read Problem 2 aloud. Pose the Guiding Questions below to verify student understanding.

Have students work in small groups to complete parts (A) through (C) of Problem 2. Then call the class back together to have the students debrief their work for parts (A) through (C).

Guiding Questions

- How is this situation different from the situation in Problem 1? How is it similar to Problem 1?
- What information is given about the total number of tins of snacks sold?
- How much money was earned with the snack fund-raiser?
- What does x represent in this situation? What does y represent in this situation?
- What is the unit for x? What is the unit for y?
- Without solving, are we given enough information to know if the x- or y-value will be greater?
- What is the format to write an equation in standard form?
- How can we interpret a solution for the system of equations for this situation?

Investigate Problem 1

4. Interpret the solution of the linear system in the problem situation. Use a complete sentence in your answer.

   Three hundred fifty six tickets were sold before the tournament and 272 tickets were sold at the door.

5. What effect did adding the equations together have? Use complete sentences in your answer.

   Sample Answer: Adding the equations together eliminated one of the variables.

6. Describe how the coefficients of y in the original system are related. Use a complete sentence in your answer.

   Sample Answer: They are opposites.

Problem 2 Traveling to the Tournament

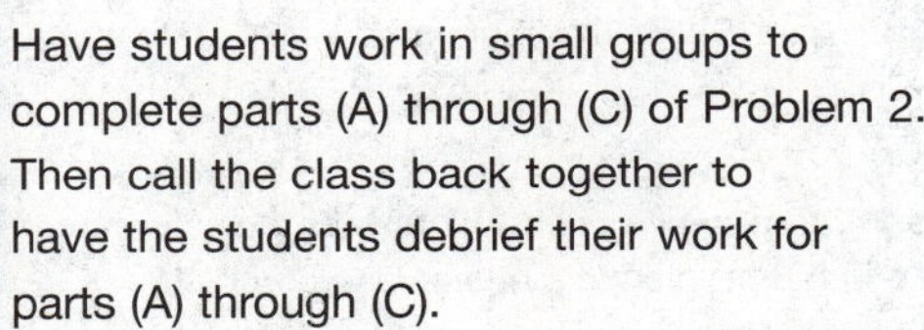

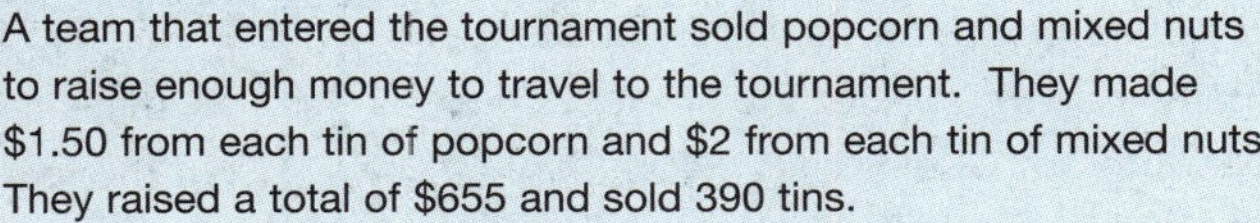

A team that entered the tournament sold popcorn and mixed nuts to raise enough money to travel to the tournament. They made \$1.50 from each tin of popcorn and \$2 from each tin of mixed nuts. They raised a total of \$655 and sold 390 tins.

A. Write an equation in standard form that represents the total amount of money raised. Use x to represent the number of tins of popcorn sold and use y to represent the number of tins of nuts sold.

   $1.50x + 2y = 655$

B. Write an equation in standard form that represents the total number of tins sold.

   $x + y = 390$

C. How are these equations different? How are they the same? Use complete sentences in your answer.

   Sample Answer: The equations are both in standard form. The coefficients of x and y are different.

Investigate Problem 2

1. Multiply each side of the equation that represents the total number of tins sold by –2. Show your work.

   $-2(x + y) = -2(390)$

   $-2x - 2y = -780$

Investigate Problem 2

Grouping

Have students work in small groups to complete Questions 1 through 5 of Investigate Problem 2.

Explore Together

Investigate Problem 2

7

Take Note

When a variable is multiplied by a number, the number is called the **coefficient.**

Students will continue to solve a system of linear equations. Students will need to use multiplication to rewrite the system so that it can be solved using the linear combinations method.

Common Student Errors

The process of multiplying to create equations with opposite terms will be challenging for many students. After the students have time to attempt Questions 1 through 5 in small groups, you will need to call the class back together to have the students discuss and present their work for Questions 1 through 5. This is likely to take a considerable amount of discussion to verify student understanding.

Grouping

After students understand Questions 1 through 5, have the students work in small groups to complete Questions 6 through 8 of Investigate Problem 2.

Call the class back together to have the students discuss and present their work for Questions 6 through 8. Then ask for a student volunteer to read Question 9 aloud. Have a student restate the problem. Complete Question 9 together as a class.

Just the Math

The formal introduction to solving a system of equations using the linear combinations method is given in Question 9. Pose the Guiding Questions on the next page about this concept to verify student understanding.

Investigate Problem 2

2. Write a linear system from the equation in part (A) and the equation in Question 1.

$$1.50x + 2y = 655$$
$$-2x - 2y = -780$$

3. How do the coefficients of the equations in your linear system compare? Use complete sentences in your answer.

Sample Answer: The coefficients of y are opposites. The coefficients of x are of opposite sign.

4. Add the equations in your linear system together. Then simplify the result. Show your work.

$$\begin{aligned} 1.50x + 2y &= 655 \\ -2x - 2y &= -780 \\ \hline -0.5x &= -125 \\ x &= 250 \end{aligned}$$

5. What does the result in Question 4 represent? Use a complete sentence in your answer.

The result represents the number of tins of popcorn sold.

6. Find the value for y by substituting your value for x into the original equation from part (B). Show your work.

$$250 + y = 390$$
$$y = 140$$

7. What is the solution of the linear system? Interpret the solution of the linear system in the problem situation. Use complete sentences in your answer.

The solution is (250, 140). Two hundred fifty tins of popcorn were sold and 140 tins of nuts were sold.

8. Check your solution algebraically. Show all your work.

$$1.50(250) + 2(140) \stackrel{?}{=} 655$$
$$375 + 280 \stackrel{?}{=} 655$$
$$655 = 655$$

$$250 + 140 \stackrel{?}{=} 390$$
$$390 = 390$$

9. Just the Math: Linear Combinations Method
The method you used to solve the linear systems in Problems 1 and 2 is called the **linear combinations method.** A **linear combination** is an equation that is the result of adding two equations to each other. The goal of adding the equations together is to get an equation in one variable. Then you can find the value of one variable and use it to find the value of the other variable.

Explore Together

Investigate Problem 2

Students will solve systems of equations using the linear combinations method. To solve the linear system, students will need to multiply to get opposite common terms first.

Guiding Questions

- What is the linear combinations method of solving a linear system?
- How can multiplying one or both of the equations in a system by a constant enable you to use the linear combinations method?
- What relationship needs to exist for a pair of common terms in order to use the linear combinations method?

Grouping

Have students work in small groups to complete Questions 10 through 12.

Common Student Errors

Students will frequently make arithmetic errors when solving systems of equations using the linear combinations method.

Students will also often struggle to find an appropriate constant by which to multiply an equation, or appropriate constants by which to multiply both equations in order to find common terms that are opposites.

Notes Question 10 asks only for the first step that would be required to solve the system of equations using the linear combinations method. You may want to have the students find the solutions to each of these four systems for additional practice as part of a homework assignment or in small groups.

Investigate Problem 2

In many cases, one (or both) of the equations in the system must be multiplied by a constant so that when the equations are added together, the result is an equation in one variable. For instance, consider the system

$4x + 2y = 3$
$5x - 3y = 1.$

What is the least common multiple of 2 and 3?

6

What do you have to multiply 2 by to get 6? What do you have to multiply 3 by to get 6?

Multiply 2 by 3 to get 6 and multiply 3 by 2 to get 6.

So, multiply the first equation by 3 and multiply the second equation by 2. Complete the steps below.

$3(4x + 2y) = 3(3)$ ⇨ $\boxed{12}\,x + \boxed{6}\,y = \boxed{9}$

$2(5x - 3y) = 2(1)$ ⇨ $\boxed{10}\,x - \boxed{6}\,y = \boxed{2}$

Now, solve the new linear system. Show all your work and use a complete sentence in your answer.

$$12x + 6y = 9$$
$$+\ 10x - 6y = 2$$
$$22x = 11$$
$$x = 0.5$$

$$4(0.5) + 2y = 3$$
$$2 + 2y = 3$$
$$2y = 1$$
$$y = 0.5$$

The solution is (0.5, 0.5).

10. For each linear system below, describe the first step you would take to solve the system by using the linear combinations method. Identify the variable that will be solved for when you add equations. Use complete sentences in your answer.

$4x - 3y = 8$ and $2x - 3y = 1$
Sample Answer: Multiply the first equation by –1 to find the value of *x*.

$3x + 4y = 2$ and $2x - y = 4$
Sample Answer: Multiply the second equation by 4 to find the value of *x*.

$6x + 5y = 1$ and $3x + 4y = 2$
Sample Answer: Multiply the second equation by –2 to find the value of *y*.

$8x + 3y = 2$ and $-7x + 4y = 5$
Sample Answer: Multiply the first equation by 7 and the second equation by 8 to find the value of *y*.

Explore Together

7

Investigate Problem 2

Students will practice solving systems of equations using the linear combinations method.

Grouping

Students will be working in small groups together to complete these questions.

Key Formative Assessments

- What ways have we used to solve systems of linear equations in this chapter?
- Is it worthwhile to be able to solve linear systems using the substitution method?
- Is it worthwhile to be able to solve linear systems using the linear combinations method?
- Is there frequently one method that would be more efficient to use than the other for solving a linear system?
- Does the method that is most efficient change for various linear systems?
- What would make the substitution method easier to use?
- What would make the linear combinations method easier to use?

Grouping

Call the class back together to have the students discuss and present their work for Questions 10 through 12.

Common Student Errors

Students may need more practice with choosing between methods of solving systems of linear equations. This skill is difficult for many students.

Investigate Problem 2

11. Solve each linear system using linear combinations. Show all your work.

$-5x + 2y = -10$
$3x - 6y = -18$

$-15x + 6y = -30$
$+\ 3x - 6y = -18$
$-12x = -48$
$x = 4$
$-5(4) + 2y = -10$
$-20 + 2y = -10$
$2y = 10$
$y = 5$
The solution is (4, 5).

$7x - 4y = -3$
$2x + 5y = -7$

$35x - 20y = -15$
$+\ 8x + 20y = -28$
$43x = -43$
$x = -1$
$7(-1) - 4y = -3$
$-7 - 4y = -3$
$-4y = 4$
$y = -1$
The solution is (–1, –1).

12. Describe the kinds of linear systems for which you would use the substitution method you learned in the last lesson to solve the system. Describe the kinds of linear systems for which you would use the linear combinations method to solve the system. Use complete sentences in your answer.

Answers will vary but should include the following observations:

Linear systems that have variables whose coefficients are 1 or –1 are easier to solve using substitution.

Linear systems written in standard form with coefficients other than 1 or –1 are easier to solve using linear combinations.

Wrap Up

Close

- Review all key terms and their definitions. Include the terms *standard form of a linear equation, linear combination,* and *linear combinations method*.
- You may also want to review any other vocabulary terms that were discussed during the lesson, which may include *opposites, opposite terms,* and *common terms.*
- Remind the students to write the key terms and their definitions in the notes section of their notebooks. You may also want the students to include examples.
- Ask the students to summarize all the mathematics in this lesson.
- Divide the students into several small groups. Have each group develop a new system of linear equations that can be solved using the linear combinations method. Then have them solve their own system. Each group should write their system on a piece of paper with the solution written on the back. They will leave that paper on a set of desks. Finally, have the students rotate clockwise around the room to solve each of the systems and check their work with the solution on the back of the paper. If any group disagrees with the solution written by the group that created the system, they should notify you.

7

Ties to the Cognitive Tutor Software

Students should understand that there are different methods for solving systems of equations, and they need to make choices about which methods are appropriate in a particular situation. In the Cognitive Tutor software, students can choose to solve through linear combination or graphically.

Follow Up

Assignment

Use the Assignment for Lesson 7.4 in the Student Assignments book. See the Teacher's Resources and Assessments book for answers.

Assessment

See the Assessments provided in the Teacher's Resources and Assessments book for Chapter 7.

Open-Ended Writing Task

Ask the students to write a summary of the steps to solve a system of linear equations using the linear combinations method. Students should also compare and contrast solving linear systems using the substitution method with solving linear systems using the linear combinations method.

Reflections

7

Insert your reflections on the lesson as it played out in class today.

What went well?

____

____

What did not go as well as you would have liked?

____

____

How would you like to change the lesson in order to improve the things that did not go well and capitalize on the things that did go well?

____

____

____

____

____

Notes

7.5

Finding the Better Paying Job

Using the Best Method to Solve a Linear System, Part I

7

Learning By Doing Lesson Map

Get Ready

Objective

In this lesson, you will:

- Solve a linear system by using an algebraic method.

Key Terms

- linear system
- inequality

NCTM Content Standards

Grades 9–12 Expectations

Algebra Standards

- Interpret representations of functions of two variables.
- Understand the meaning of equivalent forms of expressions, equations, inequalities, and relations.
- Write equivalent forms of equations, inequalities, and systems of equations and solve them with fluency—mentally or with paper and pencil in simple cases and using technology in all cases.
- Use symbolic algebra to represent and explain mathematical relationships.
- Draw reasonable conclusions about a situation being modeled.

Measurement Standards

- Make decisions about units and scales that are appropriate for problem situations involving measurement.

Lesson Overview

Within the context of this lesson, students will be asked to:

- Write a system of two linear equations for the weekly salary from different companies.
- Choose a method to solve their system of linear equations.
- Add a third linear equation for a constant function to their system of linear equations.
- Analyze their system of linear equations.

Essential Questions

The following key questions are addressed in this lesson:

1. What is a linear system?
2. How can we analyze the system of linear equations using a restriction and an inequality?
3. What is the best method to choose how to solve a linear system of equations?
4. How can you choose the best method to use to solve a linear system?

Show The Way

Warm Up

7

Place the following questions or an applicable subset of these questions on the board before students enter class. Students should begin working as soon as they are seated.

Solve each system of equations by using the linear combinations method.

1. $x - y = 3$ **(5, 2)**

$2x + y = 12$

2. $x + 3y = 14$ **(–1, 5)**

$4y = x + 21$

3. $3x = y - 20$ **(0, 20)**

$x + y = 20$

4. $x = 2y - 2$ **(4, 3)**

$2x + 3y = 17$

5. $7x + y = 20$ **(2, 6)**

$3x + 2y = 18$

6. $x - y = 6$ **(7, 1)**

$5x + y = 36$

7. Compare your solutions of the six linear systems above to your solutions of the Warm Up in Lesson 7.4, where you used the substitution method to solve the same linear systems. Which systems were easier to solve using substitution? Which systems were easier to solve using linear combinations? Write your observations using complete sentences.

Student answers will vary but should include the systems that they thought were easier to solve by using substitution and linear combinations.

Motivator

Begin the lesson with the motivator to get students thinking about the topic of the upcoming problem. This lesson is about helping a friend to choose a company for which to work. The motivating questions are job offers made by companies.

Ask the students the following questions to get them interested in the lesson.

- Have you or someone you know ever had more than one company offer you or them a job at the same time?
- Did the companies pay the same way and the same amount?
- What is a commission?
- Have you ever worked for a company that paid a salary based on commission?
- What was the commission rate?

Explore Together

Problem 1

Students will write a system of linear equations and then choose a method to solve their system of equations.

Grouping

Working as a class, ask for a student volunteer to read the Scenario and Problem 1 aloud. Have a student restate the problem. Pose the Guiding Questions below to verify student understanding.

Have students work together in small groups to solve parts (A) through (E) of Problem 1.

Guiding Questions

- What is commission?
- What are total sales?
- What does the phrase "in terms of" mean?
- What does x represent in this situation?
- What are the units for x?
- What does y represent in this situation?
- What are the units for y?
- What will a solution for this system of equations represent?
- How can you write 10% as a decimal value?
- How can you write 20% as a decimal value?
- If you are given a total sales amount for either weekly salary equation, how can you calculate the weekly salary amount?
- If you are given the weekly salary amount for either weekly salary equation, how can you calculate the total sales amount?

Notes Students are not instructed to write their system of equations using the variables x and y. However, most will probably do so without being told. If you prefer, you can instruct them to use these variables.

SCENARIO A friend of yours interviewed for two different sales positions at competing companies. One of the companies, Stellar, pays $500 per week plus a 10% commission on the total sales per week in dollars. The other company, Lunar, pays $200 per week plus a 20% commission on the total sales per week in dollars.

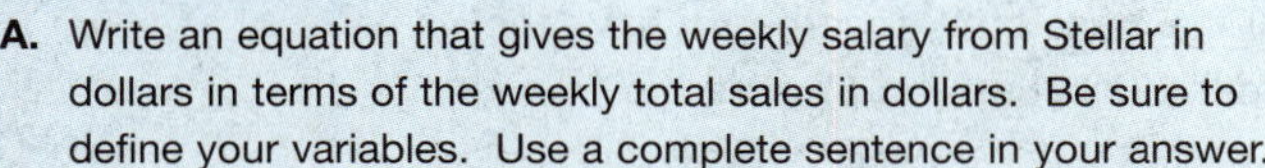

A. Write an equation that gives the weekly salary from Stellar in dollars in terms of the weekly total sales in dollars. Be sure to define your variables. Use a complete sentence in your answer.

Sample Answer: The equation is $y = 0.1x + 500$, where x is the weekly total sales in dollars and y is the weekly salary in dollars.

B. Write an equation that gives the weekly salary from Lunar in dollars in terms of the weekly total sales in dollars. Be sure to define your variables. Use a complete sentence in your answer.

Sample Answer: The equation is $y = 0.2x + 200$, where x is the weekly total sales in dollars and y is the weekly salary in dollars.

C. Find the salary from Stellar if the total sales are $1200 in one week. Show your work and use a complete sentence in your answer.

$y = 0.1(1200) + 500 = 620$; The salary from Stellar is $620.

Find the salary from Lunar if the total sales are $1200 in one week. Show your work and use a complete sentence in your answer.

$y = 0.2(1200) + 200 = 440$; The salary from Lunar is $440.

D. Find the total sales from Lunar if the weekly salary was $1200. Show your work and use a complete sentence in your answer.

$1200 = 0.2x + 200$

$1000 = 0.2x$

$5000 = x$ **The total sales from Lunar were $5000.**

Explore Together

7

Problem 1

Students will continue to analyze their system of equations for the weekly salary.

Grouping

Call the class back together to have the students discuss and present their work for parts (A) through (E) of Problem 1.

Investigate Problem 1

Grouping

Ask for a student volunteer to read Question 1 aloud. Have a student restate the problem. Pose the Guiding Questions below to verify student understanding.

Have students work in small groups to solve Questions 1 through 5.

Guiding Questions

- What algebraic methods have you learned to solve a system of linear equations?
- How can you decide which method you want to use to solve the system of linear equations in this situation?
- Is it possible to use the substitution method to solve this system?
- Is it possible to use the linear combinations method to solve this system?

Notes Students should all get the same values as answers to these questions, but their solution methods will vary. Sample solution methods are given.

Problem 1 Comparing Salaries

E. The salary from Stellar for one week is $540. Find the salary at Lunar if the total sales at Lunar are the same as the total sales at Stellar for this week. Show all your work and use a complete sentence in your answer.

Sales at Stellar:

$540 = 0.1x + 500$

$40 = 0.1x$

$400 = x$

Salary at Lunar:

$y = 0.2(400) + 200$

$y = 80 + 200$

$y = 280$

The salary at Lunar is $280.

Investigate Problem 1

1. Use an algebraic method to determine whether the salary from Lunar will ever be the same as the salary at Stellar. Show all your work and use a complete sentence in your answer.

Sample Answer:

$0.1x + 500 = 0.2x + 200$

$0.1x + 300 = 0.2x$

$300 = 0.1x$

$3000 = x$

The salaries will be the same when the total sales are $3000.

If the salaries will be the same, what will the salaries be? Show all your work and use a complete sentence in your answer.

Sample Answer: $0.1(3000) + 500 = 300 + 500 = 800$; The salaries will be $800.

2. Which method did you use to find the answer to Question 1? Use a complete sentence to explain your choice.

Sample Answer: I used the substitution method because both equations were in slope-intercept form.

3. Check your solution by creating a graph of your linear system on the grid on the next page. First, choose your bounds and intervals. Be sure to label your graph clearly.

Variable quantity	Lower bound	Upper bound	Interval
Sales	0	4500	300
Salary	0	1200	80

Explore Together

Investigate Problem 1

Students will analyze their system of equations both graphically and using a table of values.

Grouping

Students will be working together in small groups in order to solve these questions.

Notes Students will need to have an upper bound for sales of at least \$3000 and an upper bound for salary of at least \$800 in order for their graphs to be meaningful.

The graph shown in Question 3 has two lines for Question 3 as well as the line $y = 750$ that will be added by the students in Question 6.

Sample answers are shown in the table in Question 4.

Key Formative Assessments

- How did you choose to solve your linear system of equations?
- How can you use your graph to verify your solution?
- For what interval of sales is the salary from Stellar greater?
- For what interval of sales is the salary from Lunar greater?
- What does the point of intersection represent in this situation?
- What is the point of intersection for this graph?

Guiding Questions

- What is the smallest possible total sales amount that your friend could have for either company?
- What is the largest reasonable total sales amount that your friend could have for either company?
- What bounds would be appropriate for the total sales amount?
- What is the smallest possible salary that your friend could earn from either company?

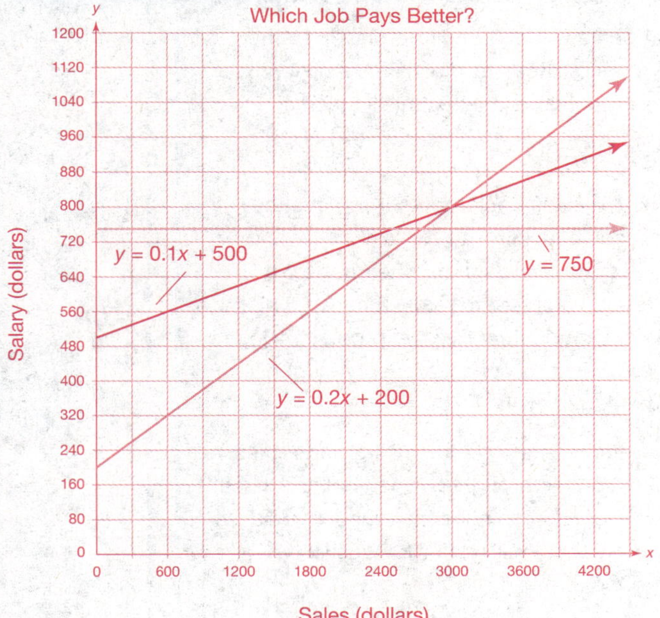

Is your solution confirmed by your graph?

Yes.

4. Complete the table of values that shows the salaries from both companies for different sales amounts.

Quantity Name	Total sales	Stellar salary	Lunar salary
Unit	dollars	dollars	dollars
Expression	x	$0.1x + 500$	$0.2x + 200$
	0	500	200
	100	510	220
	200	520	240
	500	550	300
	1000	600	400
	2500	750	700
	5000	1000	1200
	8000	1300	1800
	10,000	1500	2200

- What is the largest reasonable salary that your friend could earn from either company?
- What bounds would be appropriate for the salary?

7

Explore Together

7

Investigate Problem 1

Students will add a third equation to their system of linear equations in Question 6. This function is a constant function.

Grouping

Call the class back together to have the students discuss and explain their work for Questions 1 through 5. Then ask for a student volunteer to read Question 6 aloud. Have a student restate the problem. Pose the Guiding Questions to verify student understanding.

Have students work in small groups to solve Questions 6 through 8. Then call the class back together to have the students discuss and present their work for Questions 6 through 8.

Guiding Questions

- How is the pay structure for Solar different than the pay structure from Stellar and Lunar?
- How are the pay structures similar, if at all?
- How can you add the graph of this equation to your graph in Question 3?
- What will the graph of this equation look like?

Key Formative Assessments

- If your friend chooses to work for Stellar, will their total salary depend on the total amount of sales? What if they work for Lunar? What if they work for Solar?
- Which of the three companies would you recommend to your friend to choose?
- Why would you recommend that company?
- What is the meaning of the inequality that you wrote in Question 8?
- Is there one solution to the system of all three linear equations for this situation?
- Is it possible to have a solution for a system of three linear equations?

Investigate Problem 1

5. Which company would you recommend to your friend? Why? Use complete sentences in your answer.

Answers will vary, but students should identify the sales for which company's salary is greater than the other company's salary.

6. Your friend interviews at a third company, Solar. Solar pays a salary of $750 per week with no commissions. Write an equation that gives the salary in dollars in terms of the total sales in dollars. Then add the graph of this equation to your graph in Question 3.

$y = 750$

7. Describe the conditions for which the salary from Solar is better than the salaries at Stellar and Lunar. Show all your work and use complete sentences in your answer.

Solar and Stellar:

$750 = 0.1x + 500$

$250 = 0.1x$

$2500 = x$

Solar is better than Stellar when the sales are less than $2500.

Solar and Lunar:

$750 = 0.2x + 200$

$550 = 0.2x$

$2750 = x$

Solar is better than Lunar when the sales are less than $2750.

8. Your friend takes the job with Stellar and wants to earn at least $975 each week. Write an **inequality** that represents this situation.

$0.1x + 500 \geq 975$

Solve the inequality. Then use a complete sentence to explain what the solution means in the problem situation.

$0.1x + 500 \geq 975$

$0.1x \geq 475$

$x \geq 4750$

Your friend's total sales must be at least $4750 each week.

Wrap Up

7

Close

- Review all key terms and their definitions; include the *standard form of a linear system* and *inequality*.
- You may also want to review any other vocabulary terms that were discussed during the lesson, which may include *commission* and *in terms of*.
- Remind the students to write the key terms and their definitions in the notes section of their notebooks. You may also want the students to include examples.
- Ask the students to summarize the two ways to solve systems of linear equations. Have them compare and contrast the solution methods.
- Have the students summarize how they can choose the method to use to solve a system of linear equations.

Ties to the Cognitive Tutor Software

Students should understand that there are different methods for solving systems of equations, and they need to make choices about which methods are appropriate in a particular situation. In the Cognitive Tutor software, students can choose to solve by linear combination or graphically.

Follow Up

Assignment

Use the Assignment for Lesson 7.5 in the Student Assignments book. See the Teacher's Resources and Assessments book for answers.

Assessment

See the Assessments provided in the Teacher's Resources and Assessments book for Chapter 7.

Open-Ended Writing Task

Ask the students to write a paragraph explaining which method of solving systems of linear equations they most prefer to use. Be sure to have the students explain why they most prefer the method that they prefer.

Reflections

7

Insert your reflections on the lesson as it played out in class today.

What went well?

What did not go as well as you would have liked?

How would you like to change the lesson in order to improve the things that did not go well and capitalize on the things that did go well?

Notes

7.6

World Oil: Supply and Demand

Using the Best Method to Solve a Linear System, Part 2

Learning By Doing Lesson Map

Get Ready

Objective

In this lesson, you will:

- Solve a linear system by using an algebraic method.

Key Term

- linear system

NCTM Content Standards

Grades 9–12 Expectations

Algebra Standards

- Write equivalent forms of equations, inequalities, and systems of equations and solve them with fluency—mentally or with paper and pencil in simple cases and using technology in all cases.
- Use symbolic algebra to represent and explain mathematical relationships.
- Judge the meaning, utility, and reasonableness of the results of symbol manipulations, including those carried out by technology.
- Draw reasonable conclusions about a situation being modeled.

Lesson Overview

Within the context of this lesson, students will be asked to:

- Write a linear system of equations that uses large numbers to model the supply and demand of world oil.
- Choose an algebraic method to solve the linear system of equations.
- Solve the linear system of equations.

Essential Questions

The following key questions are addressed in this lesson:

1. What is a linear system?
2. How can you solve a linear system?
3. How can you choose which algebraic method to use to solve a linear system?

Show The Way

7

Warm Up

Place the following questions or an applicable subset of these questions on the board before students enter class. Students should begin working as soon as they are seated.

Choose either substitution or the linear combinations method to solve each system of linear equations. Then solve the linear system. State the method that you chose.

1. $3x - 2y = 9$ **(5, 3)**

$-3x + y = -12$

2. $2x + 2y = 4$ **(7, –5)**

$2y = x - 17$

3. $y = x - 9$ **(12, 3)**

$-x + 10y = 18$

4. $7 = 6y - 7x$ **(–1, 0)**

$-2x + 3y = 2$

5. $6x + 2y = 56$ **(10, –2)**

$-x + y = -12$

6. $4x - 3y = -9$ **(–6, –5)**

$y = 4x + 19$

Motivator

Begin the lesson with the motivator to get students thinking about the topic of the upcoming problem. This lesson is about the supply and demand of world oil. The motivating questions are about the amount of oil that is available in the world and the amount of oil that is used.

Ask the students the following questions to get them interested in the lesson.

- What uses do we have for oil?
- Do you think that the amount of oil used increases or decreases each year?
- Do you think the amount of oil located and available increases or decreases each year?
- What do you think can be done to reduce the amount of oil used in the world?
- Do you think it is important to reduce the amount of oil that we use?

Explore Together

Problem 1

Students will calculate rates of change and use them to write and solve a system of equations to analyze the oil supply and demand.

Grouping

Ask for a student volunteer to read the Scenario, part (A), and part (B) of Problem 1 aloud. Have a student restate the situation. Pose the Guiding Questions below to verify student understanding.

Complete parts (A) and (B) of Problem 1 together as a class. Then ask for a student volunteer to read part (C) of Problem 1 aloud. Have a student restate the problem. Complete part (C) together as a class.

Guiding Questions

- Is the amount of oil being produced increasing or decreasing?
- How can the amount of oil being produced increase when oil is being used?
- What is meant in part (B) of Problem 1 by the phrase, "number of years since 1965"?
- How many years since 1965 is this year?
- What does *x* represent in this situation?
- What does *y* represent in this situation?
- Is it at all reasonable to assume that demand for oil will be constant over time?

Note This problem considers the cumulative amount of oil produced since 1965, and the total amount of oil used during the same time period.

SCENARIO In 2003, there were approximately 28,179.4 million barrels of oil being produced in the world. In 1965, there were approximately 15,856 million barrels of oil being produced in the world.

Problem 1 Supply and Demand

A. What is the rate of change in the amount of oil being produced in millions of barrels per year from 1965 to 2003? Show all your work and use a complete sentence in your answer. Round your answer to the nearest tenth, if necessary.

$$\frac{28{,}179.4 - 15{,}856}{2003 - 1965} = \frac{12{,}323.4}{38} = 324.3$$

The rate of change is 324.3 million barrels per year.

B. The amount of oil being produced is called the *supply* of oil. Write an equation that gives the supply in millions of barrels in terms of the number of years since 1965. Assume that the rate of change in the supply is the same as the rate of change from 1965 to 2003. Be sure to define your variables. Show all your work and use complete sentences in your answer.

Sample Answer: Let x represent the number of years since 1965 and let y represent the supply in millions of barrels. Then, the graph of the equation passes through the point (0, 15,856) and has a slope of 324.3. The equation is $y = 324.3x + 15{,}856$.

C. The amount of oil that the world uses is called the *demand* for oil. In 1965, the demand was approximately 15,179 million barrels per year and was increasing at a rate of 360.1 million barrels per year. Write an equation that gives the demand in millions of barrels in terms of the number of years since 1965. Be sure to define your variables and use a complete sentence in your answer.

The equation is $y = 360.1x + 15{,}179$ where x is the number of years since 1965 and y is the demand in millions of barrels.

Explore Together

7

Investigate Problem 1

Students will solve the equations that they wrote for various values.

Grouping

Ask for a student volunteer to read Question 1 aloud. Have a student restate the problem. Pose the Guiding Questions below to verify student understanding.

Have students work in small groups to complete Questions 1 through 11 of Investigate Problem 1. Be sure that the students are writing in complete sentences.

Guiding Questions

- What is given in Question 1?
- What is unknown in Question 1?
- How can you solve to find the amount of oil produced in the year 1971?
- How many years after 1965 is the year 1971?
- How can you determine the demand for oil between the year 1965 and any given year?
- How can you determine the year if you are given the demand for oil?
- How can you determine the year if you are given the supply of oil?
- What questions do you still have about this problem situation?

Notes The answers to these questions assume that the time period is the number of years since the beginning of the year 1965. Therefore, if the amount of time is 2.5 years, it represents 2.5 years since the beginning of 1965, or halfway through 1967.

Investigate Problem 1

1. In which year was the supply 18,000 million barrels? Show all your work and use a complete sentence in your answer.

$18{,}000 = 324.3x + 15{,}856$

$2144 = 324.3x$

$6.6 \approx x$

$1965 + 6.6 = 1971.6$

In late 1971, the supply was 18,000 million barrels.

2. In which year will the supply be 30,000 million barrels? Show all your work and use a complete sentence in your answer.

$30{,}000 = 324.3x + 15{,}856$

$14{,}144 = 324.3x$

$43.6 \approx x$

$1965 + 43.6 = 2008.6$

In late 2008, the supply should be 30,000 million barrels.

3. Find the demand in 1975. Show all your work and use a complete sentence in your answer.

$1975 - 1965 = 10$

$y = 360.1(10) + 15{,}179 = 18{,}780$

In 1975, the demand was 18,780 million barrels of oil.

4. Find the demand in 2010. Show all your work and use a complete sentence in your answer.

$2010 - 1965 = 45$

$y = 360.1(45) + 15{,}179 = 31{,}383.5$

In 2010, the demand will be 31,383.5 million barrels of oil.

5. In which year was the demand 18,000 million barrels of oil? Show all your work and use a complete sentence in your answer.

$18{,}000 = 360.1x + 15{,}179$

$2821 = 360.1x$

$7.8 \approx x$

$1965 + 7.8 = 1972.8$

In late 1972, the demand was 18,000 million barrels of oil.

Explore Together

Investigate Problem 1

Students will continue to solve their equations to determine time and amount of oil for both supply and demand.

Common Student Errors

The numbers in this problem are difficult for students to envision and to work with. Remind students to be careful, to be sure that their answers are logical, and to check them with the other members of their groups.

Students may have difficulty interpreting some of their answers. It is especially important for students to realize that in the questions in which the students must solve for the year, they should write a sentence in their answer to explain any solutions that have decimal values. For instance, if the student calculates the year to be $x = 4.9$, they must recognize that the answer represents almost 5 years after 1965, or more specifically, it will be late in the year 1969.

Grouping

Call the class back together to have the students discuss and present their work for Questions 1 through 11 of Investigate Problem 1.

Key Formative Assessments

- How can you check to see whether your hypothesis in Question 11 is correct?
- Is there another way to check?
- Use your equation to predict the demand for oil this year.
- Use your other equation to predict the supply of oil this year.

7

Investigate Problem 1

6. In which year will the demand be 40,000 million barrels of oil? Show all your work and use a complete sentence in your answer.

$40{,}000 = 360.1x + 15{,}179$

$24{,}821 = 360.1x$

$68.9 \approx x$

$1965 + 68.9 = 2033.9$

In late 2033, the demand will be 40,000 million barrels of oil.

7. Find the supply 25 years after 1965. Show all your work and use a complete sentence in your answer.

$y = 324.3(25) + 15{,}856 = 23{,}963.5$;
Twenty five years after 1965, the supply was 23,963.5 million barrels of oil.

8. In which year was the demand 10,000 million barrels of oil? Show all your work and use a complete sentence in your answer.

$10{,}000 = 360.1x + 15{,}179$

$-5179 = 360.1x$

$-14.4 \approx x$

$1965 - 14.4 = 1950.6$

In late 1950, the demand was 10,000 million barrels of oil.

9. Find the year in which the supply was 10,000 million barrels. Show all your work and use a complete sentence in your answer.

$10{,}000 = 324.3x + 15{,}856$

$-5856 = 324.3x$

$-18.1 \approx x$

$1965 - 18.1 = 1946.9$

In late 1946, the supply was 10,000 million barrels of oil.

10. Write the linear system that represents the supply and the demand since 1965.

Sample Answer: $y = 324.3x + 15{,}856$ and $y = 360.1x + 15{,}179$

11. Do you think that the supply was ever the same as the demand? Use what you know about the equations of a linear system to explain your answer.

Sample Answer: Yes. Because the equations do not have the same slope, they cannot represent parallel lines. The graphs of the equations must either be perpendicular lines or intersecting lines and so will have a value for which the equations are equal.

Explore Together

7

Investigate Problem 1

Students will create a table of values to compare the supply of oil and the demand for oil in Question 12. Then, students will solve the system of equations for the world oil situation in Question 13. They will choose an algebraic method to solve the system.

Grouping

Ask for a student volunteer to read Question 12 aloud. Have a student restate the problem. Then have students work in small groups to complete Question 12.

Call the class back together to have the students discuss and check their answer to Question 12.

Have students work in small groups to complete Question 13.

Common Student Errors

Students may struggle to choose a method to solve the linear system of equations.

Notes Sample answers are shown in the table in Question 12. The answers to Question 13 should be the same for all students, but the solution and method used to get the solution will vary.

Investigate Problem 1

12. Complete the table of values below that shows the supply and the demand for different numbers of years.

	Years since 1965	Supply	Demand
Quantity Name			
Unit	years	million barrels	million barrels
Expression	x	$324.3x + 15{,}856$	$360.1x + 15{,}179$
	1	16,180.3	15,539.1
	5	17,477.5	16,979.5
	10	19,099.0	18,780.0
	25	23,963.5	24,181.5
	30	25,585.0	25,982.0
	40	28,828.0	29,583.0
	50	32,071.0	33,184.0

13. Use the table to decide whether the supply was ever the same as the demand. Use a complete sentence to explain your reasoning.

Sample Answer: Yes, the amounts will be the same because the table shows that sometime between 10 and 25 years after 1965, the demand goes from being less than the supply to greater than the supply.

If so, determine the number of years that will pass before the supply and demand will be equal. Show all your work and use a complete sentence in your answer.

Sample Answer:

$$324.3x + 15{,}856 = 360.1x + 15{,}179$$
$$324.3x + 677 = 360.1x$$
$$677 = 35.8x$$
$$18.9 \approx x$$

Approximately 18.9 years after 1965, the supply and demand are equal.

How did you find your answer? Use a complete sentence to explain.

Sample Answer: I used the substitution method.

What was the amount of oil when the supply and demand were equal? Show all your work and use a complete sentence in your answer.

Sample Answer: $y = 360.1(18.9) + 15{,}179 \approx 21{,}985$;
The amount of oil was about 21,985 million barrels.

Explore Together

Investigate Problem 1

Students will graph their system of equations to verify their solution.

Grouping

Ask for a student volunteer to read Question 14 aloud. Have a student restate the problem. Then have students work in small groups to complete Questions 14 and 15.

Call the class back together to have the students discuss and present their work for Questions 14 and 15.

Key Formative Assessments

- What does each equation in your system of equations represent?
- What is the relationship between the graphs of these equations?
- Where do the graphs of your equations intersect?
- What is the meaning of the point of intersection of your graph?
- What is the meaning of the solution you found in Question 13?
- Do you think your solution to Question 13 accurately shows when the world oil supply would not be able to meet the world oil demand?

Investigate Problem 1

14. Check your estimate by creating a graph of your linear system on the grid below. First, choose your bounds and intervals. Be sure to label your graph clearly.

Variable quantity	Lower bound	Upper bound	Interval
Time	0	45	3
Oil	15,000	29,000	1000

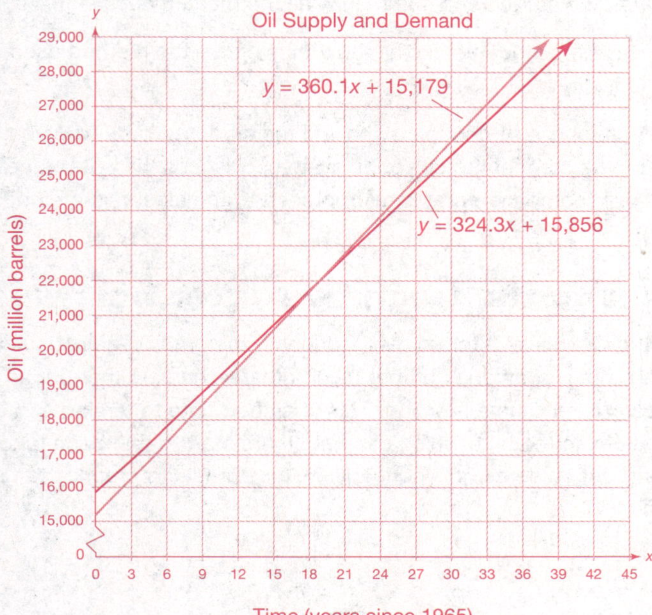

Is your estimate confirmed by your graph?

Yes.

15. Do you think that your equations are accurate models for your data? Use complete sentences to explain your reasoning.

Answers will vary.

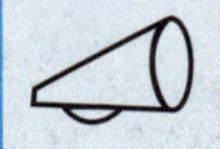

7

Wrap Up

Close

7

- Review all key terms and their definitions. Include the term *standard form of a linear system*.
- You may also want to review any other vocabulary terms that were discussed during the lesson, which may include *world oil supply, world oil demand,* and *consumption*.
- Remind the students to write the key terms and their definitions in the notes section of their notebooks. You may also want the students to include examples.
- Ask the students to summarize all the mathematics in this lesson.
- Have the students explain how they chose the method used to solve their system of equations in Question 13.
- Summarize the lesson by having the students work with you to compare all the ways suggested by the students to solve the system. Discuss the advantages and disadvantages of each approach. Also, compare a graphical approach to the algebraic methods used.

Ties to the Cognitive Tutor Software

In the Cognitive Tutor software, shading the graph requires students to think of a relation such as "greater than" not just as one number being greater than another but as one region of the plane representing the area greater than a function. This is a good time to remind students of the discussion of function and relations in the text, which can help them start to think of a function as something that divides the graph. For most students, this concept will take time and practice to fully develop.

Follow Up

Assignment

Use the Assignment for Lesson 7.6 in the Student Assignments book. See the Teacher's Resources and Assessments book for answers.

Assessment

See the Assessments provided in the Teacher's Resources and Assessments book for Chapter 7.

Open-Ended Writing Task

Ask the students to write a paragraph of their opinion of what factors will affect the world oil supply and demand over the next century, and whether the factors will cause the world oil supply and demand to increase or decrease over time in the future.

Reflections

Insert your reflections on the lesson as it played out in class today.

What went well?

What did not go as well as you would have liked?

How would you like to change the lesson in order to improve the things that did not go well and capitalize on the things that did go well?

Notes

7.7

Picking the Better Option

Solving Linear Systems

7

Learning By Doing Lesson Map

Get Ready

Objective

In this lesson, you will:

- Use a system of linear equations to solve a problem.

Key Term

- break-even point

NCTM Content Standards

Grades 9–12 Expectations

Algebra Standards

- Interpret representations of functions of two variables.
- Understand the meaning of equivalent forms of expressions, equations, inequalities, and relations.
- Write equivalent forms of equations, inequalities, and systems of equations and solve them with fluency—mentally or with paper and pencil in simple cases and using technology in all cases.
- Use symbolic algebra to represent and explain mathematical relationships.
- Draw reasonable conclusions about a situation being modeled.

Measurement Standards

- Make decisions about units and scales that are appropriate for problem situations involving measurement.

Lesson Overview

Within the context of this lesson, students will be asked to:

- Write a system of linear equations to model a situation for the cost of producing bicycles under different production plans.
- Solve and graph the system of linear equations.
- Write and solve related systems of linear equations.

Essential Questions

The following key questions are addressed in this lesson:

1. What is a break-even point?
2. What are production costs?
3. What is a system of linear equations?

Show The Way

7

Warm Up

Place the following questions or an applicable subset of these questions on the board before students enter class. Students should begin working as soon as they are seated.

Find the slope of the line passing through each pair of points.

1. (1, 2) and (2, 4) $m = 2$

2. (4, 6) and (2, 3) $m = \frac{3}{2}$

3. (–10, –1) and (8, 8) $m = \frac{1}{2}$

4. (–3, –1) and (0, –1) $m = 0$

5. (5, –4) and (–4, 8) $m = -\frac{4}{3}$

6. (3, 0) and (–7, 5) $m = -\frac{1}{2}$

Motivator

Begin the lesson with the motivator to get students thinking about the topic of the upcoming problem. This lesson is about designing and building a new product. The motivating questions are about their knowledge of product design and production.

Ask the students the following questions to get them interested in the lesson.

- When a new product is designed and created, what types of expenses would you expect to exist?
- Who might pay for those expenses?
- What is a prototype?
- What types of products are created by first making a prototype before mass-producing the actual products?
- Do you think that many changes are made after a prototype is developed and before the actual products are manufactured?
- How can a company change the production costs of a product?

Explore Together

Problem 1

Students will write and analyze a system of equations to compare manufacturing costs for new bicycles.

Grouping

Ask for a student volunteer to read the Scenario and Problem 1 aloud. Have a student restate the problem. Pose the Guiding Questions below to verify student understanding.

Have students work together in small groups to complete parts (A) through (E) of Problem 1.

Guiding Questions

- Why are the costs so great to design a new product and build a prototype?
- Is one of the plans more expensive for every possible number of bicycles to be made?
- How can you determine which plan would be less expensive to manufacture bicycles?
- What is constant for the first plan?
- What is changing for the first plan?
- What is constant for the second plan?
- What is changing for the second plan?

Common Student Errors

Students may try to create only one equation rather than a system of equations to model this situation.

Grouping

Call the class back together to have the students discuss and present their work for parts (A) through (E).

SCENARIO The Bici Bicycle Company is planning on making a low price ultra-light bicycle. There are two different plans being considered for building this bicycle. The first plan includes a cost of $125,000 to design and build a prototype bicycle. The materials and labor costs for each bike made under the first plan will be $225. The second plan includes a cost of $100,000 to design and build the prototype. The materials and labor costs for each bike made under the second plan will be $275.

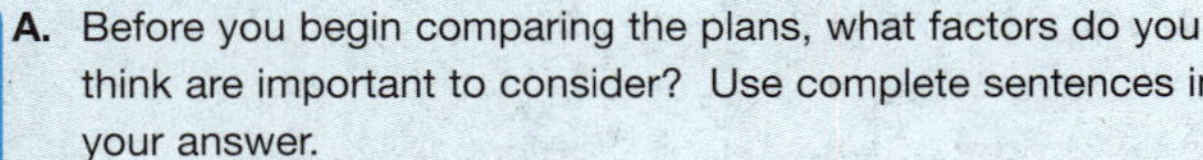

Problem 1 Which Plan Is the Better Plan?

A. Before you begin comparing the plans, what factors do you think are important to consider? Use complete sentences in your answer.

Answers will vary. Sample Answer: The number of bicycles made, the total cost to produce the bicycles, and the selling price of each bicycle should be considered.

B. For each plan, write an equation that gives the total cost in dollars in terms of the total number of bicycles made. Be sure to define your variables.

Use x for the number of bicycles made and use y for the total cost in dollars.

Plan 1: $y = 225x + 125{,}000$; Plan 2: $y = 275x + 100{,}000$

C. For each plan, what do the slope and y-intercept of the graph of the equation represent in the problem situation? Use complete sentences in your answer.

Sample Answer: In each plan, the slope represents the additional cost of making one more bicycle. In each plan, the y-intercept represents the cost before making a single bicycle.

D. Will there be a number of bicycles for which the total costs are the same? How do you know? Use a complete sentence to explain your reasoning.

Sample Answer: Yes, because the slopes of the lines are different, making the lines not parallel to each other.

E. Describe the different methods you can use to find the number of bicycles for which the total costs are the same. Use a complete sentence in your answer.

Sample Answer: You could graph the system and find the point of intersection or you could use the method of substitution to find the answer.

Explore Together

Investigate Problem 1

Students will create a table of values to compare the costs for the two plans for various numbers of bicycles made.

Grouping

Ask for a student volunteer to read Question 1 aloud. Have a student restate the problem. Then have students work together in small groups to complete Questions 1 and 2 of Investigate Problem 1.

Common Student Errors

Students may have difficulty evaluating the equations to complete the table of values. The numbers are very large and students do not easily comprehend such large values. This is because they do not have any way to connect those values to anything with which they are familiar.

Grouping

Call the class back together to have the students discuss and present their work for Questions 1 and 2.

Key Formative Assessments

- For which plan is it less expensive to design and build a prototype?
- For which plan is it less expensive to build each bike, not including the design costs?
- What relationship did you find between the plans in your table?
- What do you predict a graph of your system of linear equations would look like?

Grouping

Ask for a student volunteer to read Question 3 aloud. Pose the Guiding Questions on the next page to verify student understanding.

Have students work in small groups to complete Questions 3 through 5. Call the class back together to have the students discuss and present their work for Questions 3 through 5.

Investigate Problem 1

1. Complete the table of values that shows the total cost of both plans for different numbers of bicycles.

Quantity Name	Bicycles made	Plan 1 cost	Plan 2 cost
Unit	bicycles	dollars	dollars
Expression	x	$225x + 125,000$	$275x + 100,000$
	0	125,000	100,000
	50	136,250	113,750
	100	147,500	127,500
	200	170,000	155,000
	500	237,500	237,500
	1000	350,000	375,000
	1500	462,500	512,500

2. Can you determine from your table the number of bicycles for which the total costs are the same? If so, describe the numbers of bicycles for which the first plan is better and the numbers of bicycles for which the second plan is better. If not, use an algebraic method to answer the question. Then describe the numbers of bicycles for which each plan is the better plan.

Answer methods will vary. The table may be sufficient to answer the question. Sample Answer:

$275x + 100,000 = 225x + 125,000$

$50x + 100,000 = 125,000$

$50x = 25,000$

$x = 500$

$y = 275(500) + 100,000$

$= 237,500$

Plan 1 is better when more than 500 bicycles are made.
Plan 2 is better when less than 500 bicycles are made.

3. Create a graph of your linear system on the grid on the next page to verify your answer. First, choose your bounds and intervals. Be sure to label your graph clearly.

Variable quantity	Lower bound	Upper bound	Interval
Bicycles	0	1500	100
Total cost	0	600,000	40,000

Explore Together

Investigate Problem 1

Students will graph their system of linear equations.

Guiding Questions

- What is the smallest number of bicycles that can be produced? What is the largest number of bicycles that can be produced?
- What is the largest reasonable amount of money for the cost to produce the bicycles?

Common Student Errors

Students may have difficulty determining the labels for the graph axes. It may help to connect this problem to those that the students have already solved in the Cognitive Tutor software.

Take Note

Recall that the *break-even point* is the x-coordinate of the point where the graph of the cost intersects the graph of the income.

Key Formative Assessments

- What is the independent variable in both equations?
- What does the x-value represent in both equations? What does the y-value represent in each equation?
- What do the y-values for each equation have in common?
- What is the point of intersection of your graph? What is the meaning of the point of intersection?
- What label can you use for the y-axis that will represent the y-values in both equations?
- What units will you use for your x-axis? What units will you use for your y-axis?
- What is meant by the selling price of a bicycle?

Investigate Problem 1

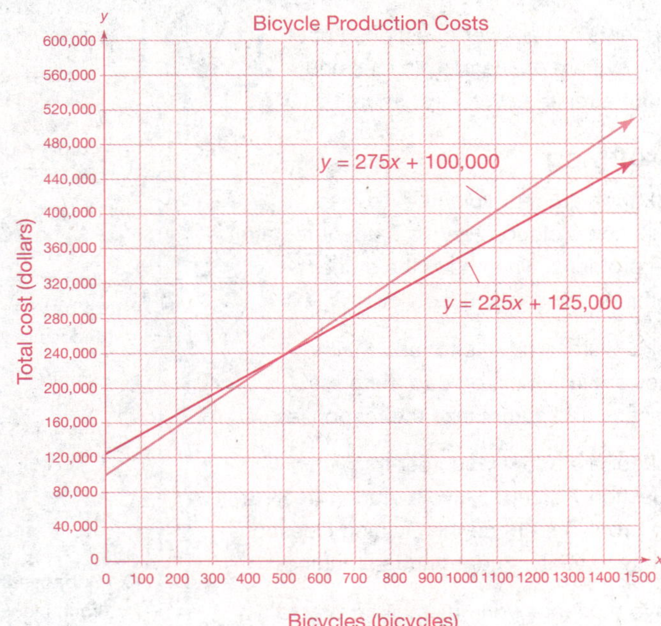

4. Now consider the selling price of the bikes. Suppose that the company wants to sell the bikes for $525 each. Write an equation that gives the total earnings in dollars in terms of the number of bicycles sold.

   $y = 525x$

5. For each plan, determine the break-even point. Show all your work. Use complete sentences in your answer.

   Plan 1:

   $y = 225x + 125{,}000$

   $525x = 225x + 125{,}000$

   $300x = 125{,}000$

   $x \approx 416.67$

   Plan 2:

   $y = 275x + 100{,}000$

   $525x = 275x + 100{,}000$

   $250x = 100{,}000$

   $x = 400$

   The break-even point for Plan 1 is 417 bicycles, and the break-even point for Plan 2 is 400 bicycles.

- What is a break-even point? How can you calculate the break-even point?
- Will the break-even point be the same for each plan? Is the break-even point related to the point of intersection in your graph in Question 3?

Explore Together

Investigate Problem 1

7

Students will answer questions about their system of linear equations and about a related system of equations.

Grouping

Ask for a student volunteer to read Question 6 aloud. Have a student restate the problem. Then have the students complete Question 6 individually.

Ask several students to read their answers for Question 6 to the class. Discuss and summarize the responses.

Ask for a student volunteer to read Question 7 aloud. Have a student restate the problems. Pose the Guiding Questions below to verify student understanding.

Have students work in small groups to complete Question 7. Call the class back together to have the students discuss and explain their work for Question 7.

Guiding Questions

- Would a company design and build a new product if they did not expect to earn a profit?
- How would you decide which plan is better to use?
- Will the plan you recommend be the same for all possible numbers of bicycles? Explain.

Grouping

Ask for a student volunteer to read Question 8 aloud. Have a student restate the problem. Pose the Guiding Questions below to verify student understanding. Have students work in small groups to complete Questions 8 and 9.

Guiding Questions

- How does the situation in Question 8 differ from the previous situation?
- How might the costs be reduced by $22.50 for each bicycle?

Investigate Problem 1

6. Use a complete sentence to explain what the break-even point means in this situation.

Sample Answer: The break-even point is the number of bicycles that need to be sold for the profit to be $0. Any number of bicycles greater than the break-even point will create a positive profit.

7. Use the results from Questions 2 and 5 to describe the numbers of bicycles for which each plan is better. Use complete sentences in your answer.

Sample Answer: The better plan is the plan with the lesser production cost and the greater profit. Plan 1 should be used if the company expects to sell more than 500 bicycles. Plan 2 should be used if the company expects to sell between 400 and 500 bicycles.

8. Suppose that the company wants to change the total costs of each plan. The company wants to reduce the material and labor costs under each plan by $22.50. For each plan, write the new equation for the total cost in terms of the number of bicycles made. Then use your equations to determine the numbers of bicycles for which each plan is better. Use complete sentences in your answers.

Plan 1: $y = 202.50x + 125{,}000$

Plan 2: $y = 252.50x + 100{,}000$

$$202.50x + 125{,}000 = 252.50x + 100{,}000$$
$$125{,}000 = 50x + 100{,}000$$
$$25{,}000 = 50x$$
$$500 = x$$

Plan 1 is better for more than 500 bicycles, and Plan 2 is better for less than 500 bicycles.

Explore Together

Investigate Problem 1

Students will consider related systems of linear equations and choose the more cost effective plan under the new conditions.

Grouping

Call the class back together to have the students discuss and present their work for Questions 8 and 9.

Ask for a student volunteer to read Question 10 aloud. Have a student restate the problem. Pose the Guiding Questions below to verify student understanding.

Have students work in small groups to complete Questions 10 through 13.

Guiding Questions

- What is different in the conditions given in Question 10?
- Are the design and prototype costs the same as the original situation, or have they changed?
- Is the cost to build each bicycle the same as in the original situation?
- Have the costs to build the bicycles increased or decreased from the original situation?
- Did the cost to build each bicycle decrease more than the selling price decreased, or less than the selling price decreased?

Investigate Problem 1

7

9. Does this answer surprise you? Why or why not? Use complete sentences to explain your answer.

Sample Answer: No, because the cost per bicycle for each plan changed by the same amount.

10. The company also decides to reduce the selling price under each plan to $450. For each new plan, determine the break-even point. Show all your work. Use complete sentences in your answer.

Plan 1:

$y = 202.50x + 125{,}000$

$450x = 202.50x + 125{,}000$

$247.50x = 125{,}000$

$x \approx 505$

Plan 2:

$y = 252.50x + 100{,}000$

$450x = 252.50x + 100{,}000$

$197.50x = 100{,}000$

$x \approx 506$

The break-even point for Plan 1 is 505 bicycles, and the break-even point for Plan 2 is 506 bicycles.

11. Use the results from Questions 8 and 10 to describe the numbers of bicycles for which each plan is better. Use complete sentences in your answer.

Sample Answer: Plan 1 is the only plan that should be used only if the company expects to sell more than 505 bicycles.

Explore Together

7

Investigate Problem 1

Students will verify their solution to a system of linear equations by using a graph.

Grouping

Students will be working in small groups to complete Questions 12 and 13.

Key Formative Assessments

- How does your answer to Question 11 compare to your answer to Question 7?
- How did you use a graph to verify your solution to Question 11?
- With the initial conditions, by using which plan would your company lose the least amount of money if it sold less than 500 bikes? What about with the new conditions?

Investigate Problem 1

12. Create a graph of the linear system you found in Question 8 on the grid below. First, choose your bounds and intervals. Be sure to label your graph clearly.

Variable quantity	Lower bound	Upper bound	Interval
Bicycles	0	1500	100
Total cost	0	600,000	40,000

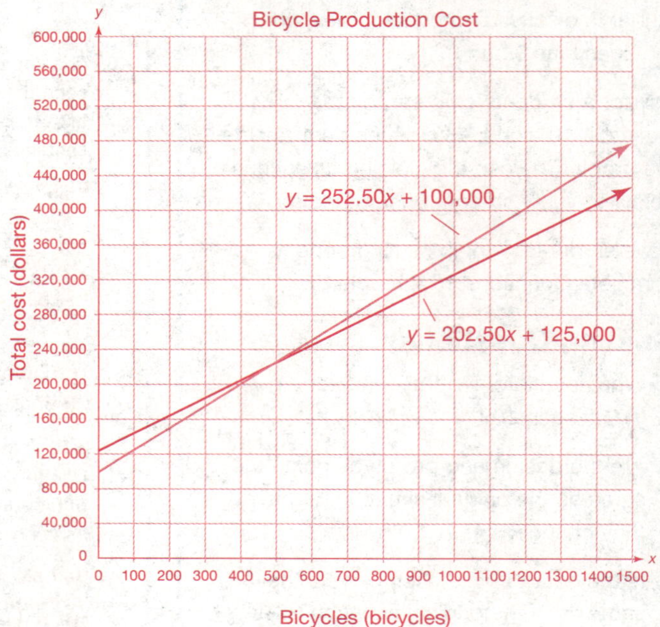

13. Estimate the point of intersection from your graph. Use complete sentences to compare the points of intersection that you found algebraically and graphically.

Sample Answer: The point of intersection is (500, 226,250). It does not matter which method I use, I find the same point of intersection. So, I can use a graph to verify my algebraic answer or I can use algebra to verify my graphical answer.

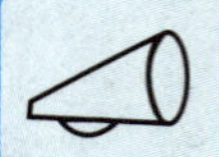

Wrap Up

Close

- Review all key terms and their definitions. Include the term *break-even point*.
- You may also want to review any other vocabulary terms that were discussed during the lesson, which may include *production costs, independent variable, dependent variable, profit,* and *point of intersection*.
- Remind the students to write the key terms and their definitions in the notes section of their notebooks. You may also want the students to include examples.
- Ask the students to compare and contrast the various situations in this lesson. Ask them to summarize how they chose a plan under each of the conditions given.

Ties to the Cognitive Tutor Software

In the Cognitive Tutor software, shading the graph requires students to think of a relation such as "greater than" not just as one number being greater than another but as one region of the plane representing the area greater than a function. This is a good time to remind students of the discussion of function and relations in the text, which can help them start to think about a function as something that divides the graph. For most students, this concept will take time and practice to fully develop.

Follow Up

Assignment

Use the Assignment for Lesson 7.7 in the Student Assignments book. See the Teacher's Resources and Assessments book for answers.

Assessment

See the Assessments provided in the Teacher's Resources and Assessments book for Chapter 7.

Open-Ended Writing Task

Ask the students to think of a product that could be produced for which there might be different possible production costs. Have them write a paragraph explaining two possible production plans and their associated costs. The students should write a system of linear equations to model each of the plans and then solve the system. The students should also explain the meaning of their solution. Ask the students to choose which production plan they would recommend and explain their choice.

7

Reflections

Insert your reflections on the lesson as it played out in class today.

What went well?

____

____

What did not go as well as you would have liked?

____

____

How would you like to change the lesson in order to improve the things that did not go well and capitalize on the things that did go well?

____

____

____

____

____

Notes

7.8 Video Arcade

Writing and Graphing an Inequality in Two Variables

7

Learning By Doing Lesson Map

Get Ready

Objectives

In this lesson, you will:

- Write an inequality in two variables.
- Graph an inequality in two variables.

Key Terms

- linear inequality in two variables
- inequality symbol
- linear equation
- coordinate plane
- half-plane

NCTM Content Standards

Grades 9–12 Expectations

Algebra Standards

- Generalize patterns using explicitly defined and recursively defined functions.
- Interpret representations of functions of two variables.
- Use symbolic algebra to represent and explain mathematical relationships.
- Identify essential quantitative relationships in a situation and determine the class or classes of functions that might model the relationships.
- Draw reasonable conclusions about a situation being modeled.

Lesson Overview

Within the context of this lesson, students will be asked to:

- Write an inequality in two variables to model a situation that involves the points available after playing arcade games.
- Graph an inequality in two variables.
- Shade the appropriate half-plane for an inequality.
- Determine the type of line, solid or dashed, that should be used to graph various linear inequalities.
- Graph various linear inequalities.

Essential Questions

The following key questions are addressed in this lesson:

1. What is a linear equation?
2. What is a linear inequality?
3. What is a linear combination?
4. What are the symbols of inequality and what is the meaning of each?
5. What is a half-plane?
6. How can you graph a linear inequality?

Show The Way

7

Warm Up

Place the following questions or an applicable subset of these questions on the board before students enter class. Students should begin working as soon as they are seated.

Complete each statement using >, <, or =.

1. $-4 \; \boxed{<} \; 3$
2. $0 \; \boxed{>} \; -7$
3. $50.013 \; \boxed{>} \; 50.009$
4. $\frac{12}{5} \; \boxed{>} \; 2.2$
5. $38\% \; \boxed{>} \; 0.039$
6. $75\% \; \boxed{=} \; \frac{3}{4}$
7. $\frac{27}{30} \; \boxed{=} \; 0.9$
8. $1\% \text{ of } 23 \; \boxed{<} \; 3$
9. $25\% \text{ of } 20 \; \boxed{=} \; 5$

Complete each statement using ≥ or ≤.

10. $0.6\% \; \boxed{\geq} \; 0.0006$
11. $7028\% \; \boxed{\geq} \; 70$
12. $-4.009 \; \boxed{\leq} \; -4.0039$
13. $\frac{2}{3} \text{ of } 36 \; \boxed{\geq} \; 22$
14. $3\% \; \boxed{\leq} \; 3$
15. $5\% \; \boxed{\leq} \; 0.5$

Motivator

Begin the lesson with the motivator to get students thinking about the topic of the upcoming problem. This lesson is about a video arcade. The motivating questions are about students' experiences in a video arcade.

Ask the students the following questions to get them interested in the lesson.

- What is a video arcade?
- Have you ever gone to a video arcade?
- How did you pay for each game? Did you use tokens, coins, or cards with point values?
- What is your favorite game to play at a video arcade?

Explore Together

Problem 1

Grouping

Ask for a student volunteer to read the Scenario and Problem 1 aloud. Have a student restate the problem. Pose the Guiding Questions below to verify student understanding.

Have students complete parts (A) through (C) of Problem 1 individually. Then discuss their solutions together as a class.

Guiding Questions

- Do you think the cost of a point is more or less than one dollar? Why?
- Why would different games require a different number of points from the card to play?
- What is constant in this situation?
- What is varying in this situation?

Take Note

Recall that an **inequality** is a statement that is formed by placing an **inequality symbol** (<, >, ≤, ≥) between two expressions.

Take Note

The forms of a linear inequality in two variables are:

$Ax + By < C$
$Ax + By > C$
$Ax + By \le C$
$Ax + By \ge C$

Grouping

Ask for a student volunteer to read part (D) of Problem 1 aloud. Have a student restate the problem. Pose the Guiding Questions at the right to verify student understanding.

Have students work together in small groups to complete parts (D) through (F) of Problem 1.

SCENARIO Your cousin's graduation party is at a restaurant that has a large video arcade. Each person at the party receives a card with fifty points on it to play the games in the arcade. One of your favorite games, a driving game, uses twelve card points per game. Another game that you like, a basketball game, uses eight points per game.

7

Problem 1 Playing Games

A. Can you play three driving games and two basketball games and not go over the number of points on the card? Show your work.

$12(3) + 8(2) = 52 > 50$; No.

B. Can you play two driving games and three basketball games and not go over the number of points on the card? Show your work.

$12(2) + 8(3) = 48 < 50$; Yes.

C. Can you play one driving game and four basketball games and not go over the number of points on the card? Show your work.

$12(1) + 8(4) = 44 < 50$; Yes.

D. Write an expression that represents the total number of points used by playing x driving games and y basketball games.

$12x + 8y$

E. What restrictions must be placed on this expression so that you do not go over the number of points on the card? Use a complete sentence in your answer.

The value of the expression must be less than or equal to 50.

F. One form of a **linear inequality in two variables** can be written as $Ax + By \le C$. Write an inequality in two variables that represents this problem situation.

$12x + 8y \le 50$

Investigate Problem 1

1. Complete the table on the next page that shows different numbers of driving and basketball games played and the numbers of points used.

Guiding Questions

- What does x represent in this situation? What does y represent in this situation?

Explore Together

Problem 1

Grouping

Call the class back together to have the students discuss and present their work for parts (D) through part (F) of Problem 1.

Investigate Problem 1

Students will analyze and graph the inequality that they wrote in part (F) of Problem 1.

Grouping

Ask for a student volunteer to read Questions 1 and 2 aloud. Have a student restate the problems. Then have the students work together in small groups to complete Questions 1 and 2.

Common Student Errors

Students may be confused with the third column in the table. It is uncomfortable for them to have a situation in which the number of points used is greater than the amount available.

The bounds and intervals for the graph are fixed to be sure that the students are able to focus on the graphical representation of their inequality.

Grouping

Call the class back together to have students discuss and present their work for Questions 1 and 2.

Key Formative Assessments

- What is the meaning of each "x" on your graph?
- What is the meaning of each point that does not have an "x" on your graph?
- Where are the points that do not have an "x" located compared to the points that have an "x"?
- Is that relationship true for every point compared to every "x"?
- What other relationships do you notice on your graph, if any?

Investigate Problem 1

Quantity Name	Driving games	Basketball games	Points used
Unit	games	games	points
	0	5	40
	1	3	36
	2	3	48
	2	4	56
	3	2	52
	3	3	60
	4	0	48
	4	1	56

2. Create a graph of the data in the table on the grid below. If the number of points used in a row does not exceed the card's points, draw a point for the numbers of games. If the number of points used does exceed the card's points, draw an "x" for the numbers of games. Use the bounds and intervals given below. Label your graph clearly.

Variable quantity	Lower bound	Upper bound	Interval
Driving game	0	7.5	0.5
Basketball game	0	7.5	0.5

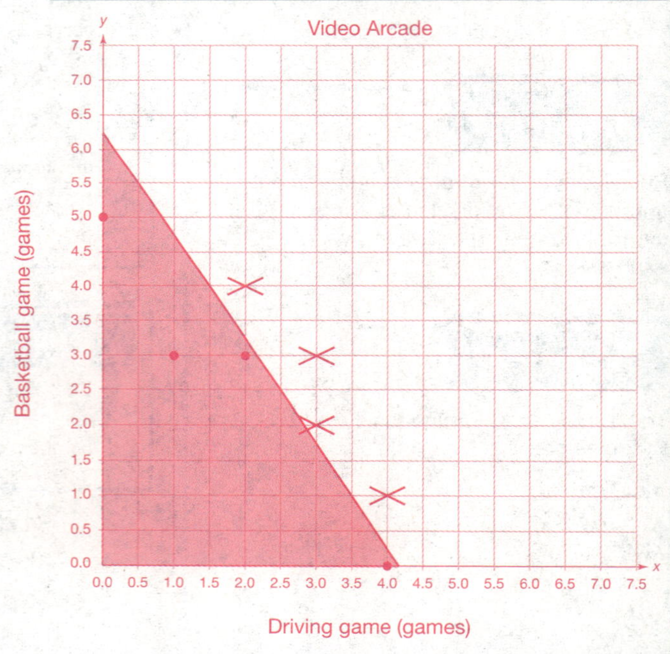

- How could you graph the set of points where the number of points used for x driving games and y basketball games is exactly 50 points?
- Is it possible to spend exactly 50 points on only driving and basketball games?

Explore Together

Investigate Problem 1

Grouping

Ask for a student volunteer to read Question 3 aloud. Have a student restate the problem. Then have students work together in small groups to complete Questions 3 and 4. Finally, call the class back together to have the students discuss and present their work for Questions 3 and 4.

Take Note

A **linear equation** in two variables is an equation in which each of the variables is raised to the first power (such as x, rather than x^2) and, when in simplest form, each variable only appears once.

Just the Math

Students will be formally introduced to the concept of a linear inequality in Question 5 and the concept of a half-plane in Question 6.

Ask for a student volunteer to read Question 5 aloud. Have a student restate the problem. Pose the Guiding Questions below to verify understanding.

Have the students complete Question 5 individually. Discuss their results as a class.

Guiding Questions

- What is a linear inequality?
- In the situation for this problem with the arcade points, is it possible to use more points than are available on the card? Is it possible to use fewer points than are available on the card?
- If it were possible, would it be acceptable to use the same number of points as on the card?

Grouping

Ask for a student volunteer to read the first part of Question 6 aloud. Have a student restate the problem. Pose the Guiding Questions at the right to verify student understanding. Have the students complete the first part of Question 6 individually. Discuss their results as a class.

Guiding Questions

- What is a plane?
- Why is the graph of a linear inequality called a half-plane?

Investigate Problem 1

3. Write an equation that represents the number of driving games x and the number of basketball games y that can be played for exactly 50 points. Then add the graph of this equation to your graph in Question 2.

$12x + 8y = 50$

4. What do you notice about your graph? Use a complete sentence in your answer.

Sample Answer: The x's fall on one side of the line and the points fall on the other side of the line.

5. Just the Math: Linear Inequality Shade the side of the graph that contains all of the points. This graph is the graph of the *linear inequality* $12x + 8y \le 50$. A linear inequality is the same as a linear equation except that an inequality symbol ($<$, $>$, $\le$, or $\ge$) is used instead of an equals sign. How do the solutions of the linear equation $12x + 8y = 50$ differ from the solutions of the linear inequality $12x + 8y \le 50$? Use complete sentences in your answer.

Sample Answer: The solutions of a linear equation form a line. The solutions of a linear inequality include half of the grid.

6. Just the Math: Graphs of Linear Inequalities The graph of a linear inequality is a **half-plane,** or half of a **coordinate plane**. A line, given by the inequality, divides the plane into two half-planes and the inequality symbol tells you which half-plane contains all the solutions. If the symbol is $\le$ or $\ge$, the graph includes the line. If the symbol is $<$ or $>$, the graph does not include the line and is represented by a dashed line. For which inequalities below would you include the line? Which inequalities below would you represent by using a dashed line? Write your answers using complete sentences.

$y > -6 - x$ $\quad$ $2x + 3y \ge 4$ $\quad$ $x + 5y \le 10$

$3x + 12y > 5$ $\quad$ $y \ge -x + 2$ $\quad$ $x - y < 3$

Sample Answer: Include the line for the inequalities $2x + 3y \ge 4$, $x + 5y \le 10$, and $y \ge -x + 2$. Use a dashed line for the inequalities $y > -6 - x$, $3x + 12y > 5$, and $x - y < 3$.

Explore Together

Investigate Problem 1

Students will consider and graph linear inequalities.

Grouping

Ask for a student volunteer to read the second part of Question 6 aloud. Have the students complete this part individually.

Ask for a student volunteer to read the third part of Question 6 aloud. Have a student restate the problem. Pose the Guiding Questions below to verify student understanding.

Have the students draw the graph in Question 6 individually. Then discuss the graph together as a class.

Guiding Questions

- How can more than one point be a solution to a linear inequality?
- How does shading represent the half-plane that is the solution to the linear inequality?
- How many points are included in a half-plane?
- How does the author suggest for you to decide which is the appropriate region to shade in the half-plane?
- Why might the author have suggested that we use the point (0, 0) to decide where to shade the graph?
- Would the same strategy work for any other points on the plane?
- Would this strategy work if the point (0, 0) were part of the line?
- If the point (0, 0) were on the line, how could you determine which half-plane to shade?
- What other strategy could you use to decide where to shade?

Investigate Problem 1

Consider the linear inequality $y < 4x + 3$. The line that divides the plane is given by $y = 4x + 3$. Should this line be a solid line or a dashed line? Use a complete sentence to explain. Then draw the correct type of line on the grid below.

Because the inequality symbol is <, the line is not included in the graph, and so is represented by a dashed line.

After you draw the correct type of line, you need to decide which half-plane contains all the solutions, because this is the half-plane that you will shade. To make your decision, consider the point (0, 0). If (0, 0) is a solution, then the half-plane that contains (0, 0) contains all the solutions and should be shaded. If (0, 0) is not a solution, then the half-plane that does not contain (0, 0) contains all the solutions and should be shaded.

Is (0, 0) a solution? Show your work.

$0 \overset{?}{<} 4(0) + 3$

$0 < 3$

Yes.

Now shade the correct half-plane on the grid below.

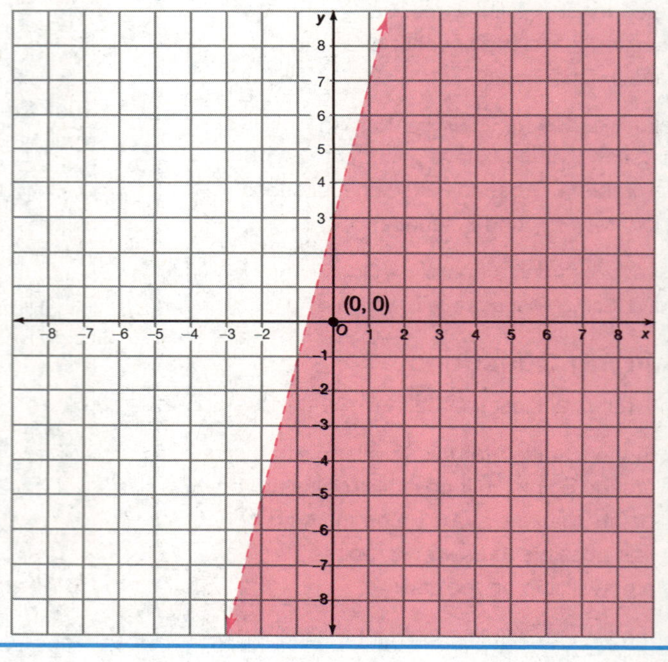

Explore Together

Investigate Problem 1

Students will practice graphing linear inequalities.

Grouping

Have students work together in small groups to complete Question 7.

Call the class back together to have the students discuss and present their work for Question 7. Give a transparency copy of this page or a transparency copy of blank grids to students and have them copy their graph for one of the inequalities onto the transparency overlay. Use these to help the students see appropriate graphs.

Common Student Errors

Many students will struggle to understand graphing linear inequalities. It will help to have the students summarize the process and the meaning of graphing linear inequalities by using the Key Formative Assessments below.

Key Formative Assessments

- How are linear equations and linear inequalities similar?
- How are linear equations and linear inequalities different?
- What does the graph of a linear equation look like?
- What does the graph of a linear inequality look line?
- How are the graphs of linear equations and linear inequalities similar? How are they different?

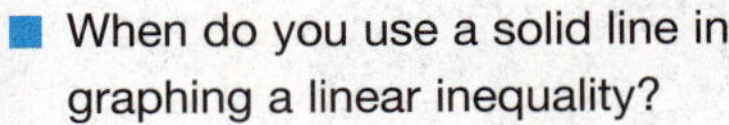

- When do you use a solid line in graphing a linear inequality?
- When do you use a dashed line in graphing a linear inequality?
- Why does a graph of a linear inequality with the < or > symbol have a dashed line while a linear inequality with the symbol ≤ or ≥ has a solid line?

Investigate Problem 1

7. Graph each linear inequality.

$y > x + 2$

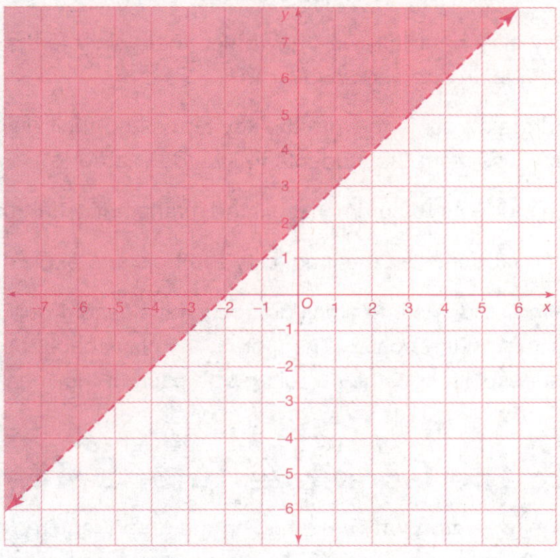

$y \leq -x + 3$

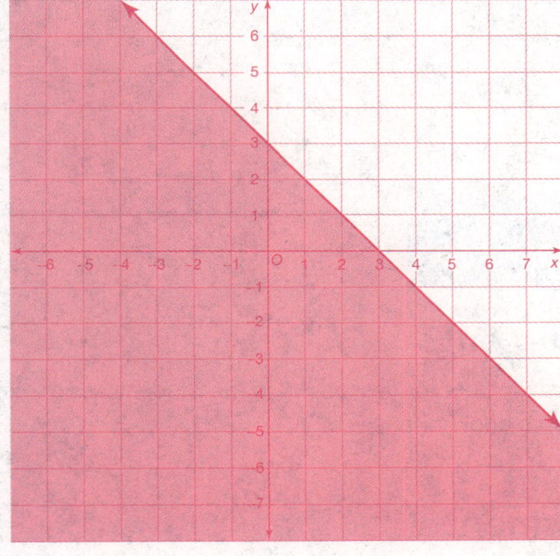

7

Wrap Up

Close

- Review all key terms and their definitions. Include the terms *standard form of a linear equation, linear inequality in two variables, inequality symbol, coordinate plane,* and *half-plane*.
- You may also want to review any other vocabulary terms that were discussed during the lesson, which may include *arcade* and *shading*.
- Remind the students to write the key terms and their definitions in the notes section of their notebooks. You may also want the students to include examples.
- Ask the students to summarize all the mathematics in this lesson.
- Have the students list the inequality symbols and explain the meaning of each.
- Divide the students into small groups. Have each group develop a linear inequality. List each of the inequalities on the board. Then, bring the class back together and have a discussion in which the students explain how they would graph each of the linear inequalities.

Ties to the Cognitive Tutor Software

In the Cognitive Tutor software, shading the graph requires students to think of a relation such as "greater than" not just as one number being greater than another but as one region of the plane representing the area greater than a function. This is a good time to remind students of the discussion of function and relations in the text, which can help them start to think of a function as something that divides the graph. For most students, this concept will take time and practice to fully develop.

Follow Up

Assignment

Use the Assignment for Lesson 7.8 in the Student Assignments book. See the Teacher's Resources and Assessments book for answers.

Assessment

See the Assessments provided in the Teacher's Resources and Assessments book for Chapter 7.

Open-Ended Writing Task

Ask the students to write out a summary of the steps to graph a linear inequality. They should include four different examples with their summary that show and explain the graph of a linear inequality that involves each of the following inequality symbols: $<$, $>$, $\leq$, and $\geq$.

Reflections

Insert your reflections on the lesson as it played out in class today.

What went well?

What did not go as well as you would have liked?

How would you like to change the lesson in order to improve the things that did not go well and capitalize on the things that did go well?

Notes

7

7

7.9

Making a Mosaic

Solving Systems of Linear Inequalities

7

Learning By Doing Lesson Map

Get Ready

Objectives

In this lesson, you will:

- Write a system of linear inequalities.
- Graph a system of linear inequalities.
- Identify solutions of a system of linear inequalities.

Key Terms

- system of linear equations
- linear inequality
- system of linear inequalities

NCTM Content Standards

Grades 9–12 Expectations

Algebra Standards

- Interpret representations of functions of two variables.
- Understand the meaning of equivalent forms of expressions, equations, inequalities, and relations.
- Write equivalent forms of equations, inequalities, and systems of equations and solve them with fluency—mentally or with paper and pencil in simple cases and using technology in all cases.
- Use symbolic algebra to represent and explain mathematical relationships.
- Draw reasonable conclusions about a situation being modeled.

Measurement Standards

- Make decisions about units and scales that are appropriate for problem situations involving measurement.

Lesson Overview

Within the context of this lesson, students will be asked to:

- Write a system of linear equations to represent the number of bags of glass and metallic tiles to buy.
- Write a system of linear inequalities that better represents the number of bags of glass and metallic tiles that can be bought.
- Solve the systems of linear equations.
- Solve and graph the system of linear inequalities.

Essential Questions

The following key questions are addressed in this lesson:

1. What is a mosaic?
2. Why would an art group donate time and work to create an art project for a school?
3. Why would a foundation or company donate the money to pay the expenses?
4. What is a system of equations?
5. What is a system of inequalities?
6. What does it mean to solve a system of linear inequalities?

Show The Way

Warm Up

7

Place the following questions or an applicable subset of these questions on the board before students enter class. Students should begin working as soon as they are seated.

Find the equation of the line passing through the given points and parallel to the given line.

1. (1, 2); parallel to $y = 2x + 4$ $y = 2x$

2. (4, 6); parallel to $y = -x - 2$ $y = -x + 10$

3. (–9 , –1); parallel to $y = \frac{2}{3}x$ $y = \frac{2}{3}x + 5$

4. (–3, –1); parallel to $y = 5$ $y = -1$

Find the equation of the line passing through the given points and perpendicular to the given line.

5. (–6, –4); perpendicular to $y = -\frac{1}{2}x + 2$ $y = 2x + 8$

6. (3, 0); perpendicular to $y = -\frac{3}{2}x - 8$ $y = \frac{2}{3}x - 2$

Motivator

Begin the lesson with the motivator to get students thinking about the topic of the upcoming problem. This lesson is about an art project donated to the school. The motivating questions are about art projects that students have encountered.

Ask the students the following questions to get them interested in the lesson.

- What is a mosaic?
- Is there a local art group or organization that might be able to donate an art project to our school?
- What expenses would a group have to pay to create an art project for our school?
- What company or foundation might we be able to ask to donate the money to pay the expenses?
- What kind of art project would you recommend that would be appropriate for our school?

Explore Together

Problem 1

Students will write and solve a system of equations to determine the amount of each type of tile to buy for a mosaic.

Grouping

Working as a class, ask for a student volunteer to read the Scenario and Problem 1 aloud. Have a student restate the problem. Pose the Guiding Questions below to verify student understanding.

Have students work together in small groups to complete parts (A) through (E) of Problem 1.

Guiding Questions

- What is a mosaic?
- What is a tile?
- Why might the artists want to buy different types of tiles?
- Why would metallic tiles cost more per bag than glass tiles?
- What is constant in this situation?
- What is variable in this situation?
- What does x represent in this situation?
- What does y represent in this situation?

Common Student Errors

Students may have difficulty solving the system of equations because they may not be as familiar with the numbers in this situation as the numbers in other situations.

Grouping

Call the class back together to have the students discuss and present their work for parts (A) through (E) of Problem 1.

Notes In Lesson 7.7, students worked with linear inequalities. In Lesson 7.8, students reviewed solving a system of linear equations. In this lesson, students will combine these concepts to solve a system of linear inequalities.

SCENARIO A local arts group is donating a mural that will be placed at the entrance to your school. The mural will be 6 feet tall and 12 feet wide. The group has calculated that they will need approximately 110 bags of tiles to complete the project. The mural will be made of glass and metallic tiles. Each bag of glass tiles costs \$10 and each bag of metallic tiles costs \$18. Another group has donated \$1500 for the purchase of the tiles.

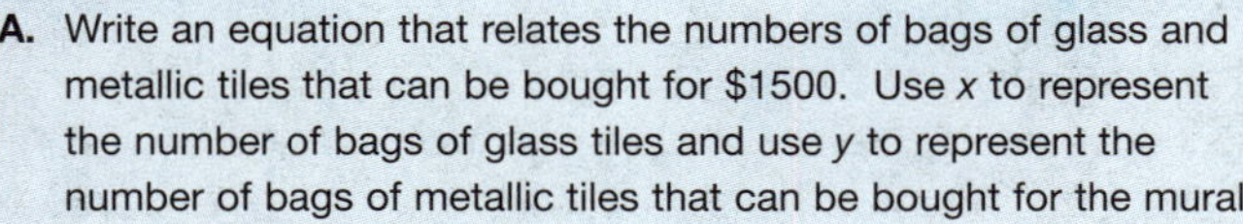

Problem 1 Getting the Tiles

A. Write an equation that relates the numbers of bags of glass and metallic tiles that can be bought for \$1500. Use x to represent the number of bags of glass tiles and use y to represent the number of bags of metallic tiles that can be bought for the mural.

$10x + 18y = 1500$

B. Write an equation that relates the numbers of bags of glass and metallic tiles to the total number of bags of tiles needed for the project.

$x + y = 110$

C. What does the solution of the linear system formed by the equations in part (A) and part (B) represent?

Sample Answer: The number of bags of each kind of tile that can be bought for \$1500 so that there are 110 bags of tiles.

D. Solve the linear system. Show all your work and use a complete sentence in your answer.

$y = 110 - x$

$10x + 18(110 - x) = 1500$

$10x + 1980 - 18x = 1500$

$-8x + 1980 = 1500$

$-8x = -480$

$x = 60$

$y = 110 - 60 = 50$

The solution is (60, 50).

E. What does the solution mean in the problem situation? Use a complete sentence in your answer.

The group should buy 60 bags of glass tiles and 50 bags of metallic tiles.

7

Explore Together

Investigate Problem 1

7

Students will analyze their system of equations to determine alternate possible solutions to the situation for the tiles. Students will then write an inequality that better represents the situation than the equation.

Grouping

Ask for a student volunteer to read Question 1 aloud. Have a student restate the problem. Then have students work together in small groups to complete Questions 1 through 5. Pose the Guiding Questions below to verify student understanding.

Guiding Questions

- Why might the group of artists want to have extra tile?
- Why might the group of artists want to have more money remaining after they buy the bags of tile?

Common Student Errors

These questions may seem very simple to some students. Careless errors may occur if the students rush through these questions. Have the students work together in small groups to prevent such errors.

Investigate Problem 1

1. Suppose that the group wants to buy 75 bags of glass tiles and 35 bags of metallic tiles. Is this enough tile?

   75 + 35 = 110; Yes.

   Can the group afford this assortment of tile? Show all your work and use a complete sentence to explain your reasoning.

   10(75) + 18(35) = 1380; Yes, because $1380 is less than $1500, the group can afford this assortment of tile.

2. Suppose that the group wants to buy 90 bags of glass tiles and 25 bags of metallic tile. Is this enough tile?

   90 + 25 = 115; Yes.

   Can the group afford this assortment of tile? Show all your work and use a complete sentence to explain your reasoning.

   10(90) + 18(25) = 1350; Yes, because $1350 is less than $1500, the group can afford this assortment of tile.

3. Suppose that the group wants to buy 80 bags of glass tiles and 38 bags of metallic tiles. Is this enough tile?

   80 + 38 = 118; Yes.

   Can the group afford this assortment of tile? Show all your work and use a complete sentence to explain your reasoning.

   10(80) + 18(38) = 1484; Yes, because $1484 is less than $1500, the group can afford this assortment of tile.

4. Does the group have to spend all of the money to get enough tile? Use a complete sentence to explain your reasoning.

   Sample Answer: No. For instance, if they buy 80 bags of glass tiles and 38 bags of metallic tiles, it only costs $1484 for 118 bags of tile.

   Write an inequality that represents the amounts of money the group can spend on x bags of glass tiles and y bags of metallic tiles.

   $10x + 18y \leq 1500$

Explore Together

Investigate Problem 1

Students will write and analyze a system of inequalities that better represents the situation for buying tiles than the system of equations.

Grouping

Call the class back together to have the students discuss and present their work for Questions 1 through 5.

Just the Math

Students will be formally introduced to systems of linear inequalities.

Grouping

Ask for a student volunteer to read Question 6 aloud. Have a student restate the problem. Pose the Guiding Questions below to verify student understanding.

Have students work in small groups to complete Questions 6 and 7. Then call the class back together to have the students discuss and present their work for Questions 6 and 7.

Guiding Questions

- What is a system of linear equations?
- How can you solve a system of linear equations?
- What is a system of linear inequalities?
- How might you solve a system of linear inequalities?
- What conditions must be met in order for a point to be a solution of a system of linear inequalities?

7

Investigate Problem 1

5. Can the group buy more bags of tiles than is needed and not spend all the money? Use a complete sentence to explain your reasoning.

Sample Answer: Yes. For instance, if they buy 80 bags of glass tiles and 38 bags of metallic tiles, it only costs $1484 for 118 bags of tiles.

Write an inequality that represents the total numbers of bags of tiles they can use to complete the project.

$x + y \geq 110$

6. Just the Math: System of Linear Inequalities
Together, the linear inequalities in Questions 4 and 5 form a **system of linear inequalities.** Write the system of linear inequalities below.

$10x + 18y \leq 1500$ **and** $x + y \geq 110$

What do you think it means to be a solution of a system of linear inequalities? Use a complete sentence in your answer.

Sample Answer: A solution is an ordered pair that yields a true statement when substituted into each inequality in the system.

Determine whether the numbers of bags of tiles given in Questions 1 through 3 are solutions of your system of inequalities. Show all your work.

(75, 35)
$10(75) + 18(35) \overset{?}{\leq} 1500$
$750 + 630 \overset{?}{\leq} 1500$
$1380 \leq 1500$
$75 + 35 \overset{?}{\geq} 110$
$110 \geq 110$

(90, 25)
$10(90) + 18(25) \overset{?}{\leq} 1500$
$900 + 450 \overset{?}{\leq} 1500$
$1350 \leq 1500$
$90 + 25 \overset{?}{\geq} 110$
$115 \geq 110$

(80, 38)
$10(80) + 18(38) \overset{?}{\leq} 1500$
$800 + 684 \overset{?}{\leq} 1500$
$1484 \leq 1500$
$80 + 38 \overset{?}{\geq} 110$
$118 \geq 110$

Yes, the values are solutions of the system of inequalities.

7. How many solutions do you think a system of linear inequalities can have? Use complete sentences to explain your reasoning.

Sample Answer: Many. The solutions of a linear inequality are represented as a half-plane. As long as the half-planes overlap, there will be many solutions.

Explore Together

7

Investigate Problem 1

Students will graph their system of linear inequalities and shade the appropriate half-planes.

Grouping

Ask for a student volunteer to read Question 8 aloud. Have a student restate the problem. Pose the Guiding Questions below to verify student understanding.

Have students work together in small groups to complete Questions 8 through 10. Then call the class back together to have the students discuss and present their work for Questions 8 through 10.

Guiding Questions

- What is the smallest possible number of bags of glass tiles that can be purchased?
- What is the smallest possible number of bags of metallic tiles that can be purchased?
- How can we determine the largest possible number of bags of glass tiles that can be purchased? How can we determine the largest possible number of bags of metallic tiles?
- When do we graph an inequality using a solid line?
- When do we graph an inequality using a dashed line?
- How can you determine which half-plane to shade for the graph of an inequality?
- How is the graph of a system of linear inequalities similar to the graph of a system of linear equations?

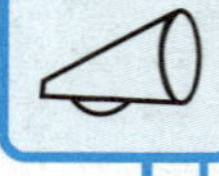

- How does the graph of a system of linear inequalities differ from the graph of a system of linear equations?
- How does the solution of a system of linear equations differ from the solution of a system of linear inequalities?

Investigate Problem 1

8. Create a graph of your system of inequalities on the grid below. Use a different color pen or pencil for each inequality. First, choose your bounds and intervals. Be sure to label your graph clearly.

Variable quantity	Lower bound	Upper bound	Interval
Glass tiles	0	150	10
Metallic tiles	0	150	10

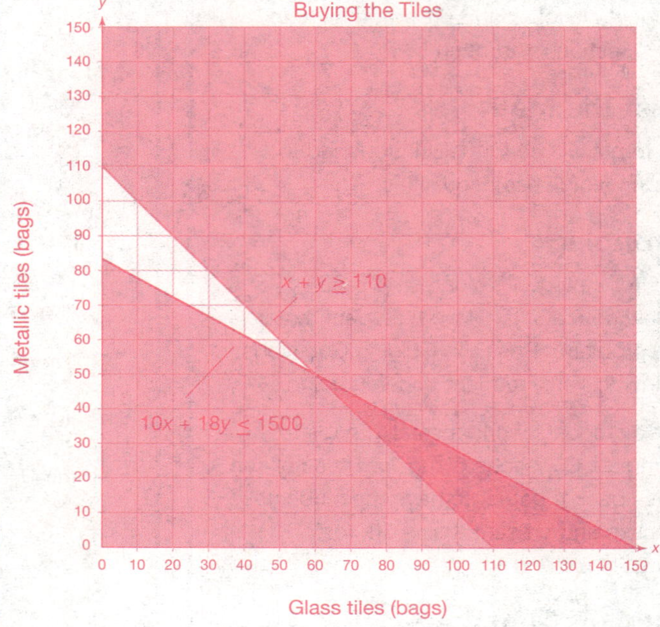

9. What part of the graph do you think represents the solution of the system of linear inequalities? Use a complete sentence in your answer.

The solution is the part where the shaded areas and lines overlap.

10. Identify three different solutions of the system of inequalities. What do these solutions represent in the problem situation? Use complete sentences in your answer.

Answers will vary. The solutions represent the different numbers of bags of each kind of tile that can be bought.

Wrap Up

7

Close

- Review all key terms and their definitions. Include the terms *standard form of a system of linear equations, linear inequality*, and *system of linear inequalities*.
- You may also want to review any other vocabulary terms that were discussed during the lesson, which may include *half-plane, linear, solution of a system of linear equations,* and *solution of a system of linear inequalities.*
- Remind the students to write the key terms and their definitions in the notes section of their notebooks. You may also want the students to include examples.
- Ask the students to discuss how the situation would change if the cost of the bags of glass tiles were raised to $12 per bag and the cost of the bags of metallic tiles were raised to $25 per bag. Have the students write, solve, and graph the new system of inequalities.

Ties to the Cognitive Tutor Software

In the Cognitive Tutor software, shading the graph requires students to think of a relation such as "greater than" not just as one number being greater than another but as one region of the plane representing the area greater than a function. This is a good time to remind students of the discussion of function and relations in the text, which can help them start to think of a function as something that divides the graph. For most students, this concept will take time and practice to fully develop.

Follow Up

Assignment

Use the Assignment for Lesson 7.9 in the Student Assignments book. See the Teacher's Resources and Assessments book for answers.

Assessment

See the Assessments provided in the Teacher's Resources and Assessments book for Chapter 7.

Open-Ended Writing Task

Ask the students to write a paragraph to answer the following question. The mosaic is 6 feet by 12 feet and needs to have 110 bags of tiles. How many bags of tiles per square foot are needed?

7

Reflections

Insert your reflections on the lesson as it played out in class today.

What went well?

What did not go as well as you would have liked?

How would you like to change the lesson in order to improve the things that did not go well and capitalize on the things that did go well?

Notes

7

Looking Ahead to Chapter 8

Focus In Chapter 8, you will graph, solve, and analyze quadratic functions using methods such as factoring, extracting square roots, and the quadratic formula. You will also learn to find the vertex, and minimum and maximum values of a parabola.

Chapter Warm-up

Answer these questions to help you review skills that you will need in Chapter 8.

8

Find the square of each number.

1. 3 **9**

2. 11 **121**

3. –4 **16**

4. –8.64 **74.65**

5. 6.4 **40.96**

6. 12.5 **156.25**

Evaluate each expression when $x = 3$.

7. $4x^2 - 6$

$4(3)^2 - 6$ 30

8. $5x^2 + 8x + 1$

$5(3)^2 + 8(3) + 1$ 70

9. $2x^2 + 9x$

$2(3)^2 + 9(3)$ 45

Read the problem scenario below.

You and your friends are playing soccer in a field by your house. You mark each corner of the field with a flag. Given the distances below, what is the area of your soccer field? Be sure to include the correct units in your answers. Write your answers using a complete sentence.

10. The length of the field is 150 feet, and the width is 50 feet.

The area of the field is 7500 square feet.

11. The length of the field is 75 yards, and the width is 20 yards.

The area of the field is 1500 square yards.

12. The length of the field is 80 meters, and the width is 25 meters.

The area of the field is 2000 square meters.

Key Terms

CHAPTER

8

Quadratic Functions

8

When a musician plucks a guitar string, the string vibrates and transmits its vibration through the guitar. The sound is amplified and you hear a musical note. In Lesson 8.4, you will use an equation to find the tension of the string and wave speed of the vibrations.

8.1 Web Site Design

Introduction to Quadratic Functions

Learning By Doing Lesson Map

Get Ready

Objectives

In this lesson, you will:

- Graph quadratic functions.
- Identify coefficients in quadratic functions.
- Evaluate quadratic functions.

Key Terms

- rate of change
- curve
- quadratic function
- evaluate

NCTM Content Standards

Grades 9–12 Expectations

Algebra Standards

- Generalize patterns using explicitly and recursively defined functions.
- Analyze functions of one variable by investigating rates of change, intercepts, zeros, asymptotes, and local and global behavior.
- Interpret representations of functions of two variables.
- Use symbolic algebra to represent and explain mathematical relationships.
- Identify essential quantitative relationships in a situation and determine the class or classes of functions that might model the relationships.
- Draw reasonable conclusions about a situation being modeled.

Measurement Standards

- Understand and use formulas for the area, surface area, and volume of geometric figures, including cones, spheres, and cylinders.

Lesson Overview

Within the context of this lesson, students will be asked to:

- Graph quadratic functions.
- Identify the coefficients of the terms in a quadratic function and label them as a, b, and c.
- Evaluate quadratic functions for given values.

Essential Questions

The following key questions are addressed in this lesson:

1. What is a web site design?
2. What is a logo?
3. How can you calculate a rate of change?
4. What is a quadratic function?
5. How do you evaluate a quadratic function?

Show The Way

8

Warm Up

Place the following questions or an applicable subset of these questions on the board before students enter class. Students should begin working as soon as they are seated.

Find the slope of the line that contains each pair of points.

1. (2, 7), (–3, –3)

$m = 2$

2. (–10, –4), (1, 7)

$m = 1$

3. (0, –7), (–2, 1)

$m = -4$

4. (3, –4), (–7, –10)

$m = \frac{3}{5}$

5. (–2, 2), (2, –9)

$m = -\frac{11}{4}$

6. (1, –2), (4, 6)

$m = \frac{8}{3}$

7. (–3, –8), (–9, –13)

$m = \frac{5}{6}$

8. (–1, –5), (2, –5)

$m = 0$

9. (6, 3), (10, 15)

$m = 3$

10. (17, 1), (17, –4)

m is undefined

11. (–2, –6), (–6, –5)

$m = -\frac{1}{4}$

12. (2, –3), (–7, –9)

$m = \frac{2}{3}$

Motivator

Begin the lesson with the motivator to get students thinking about the topic of the upcoming problem. This lesson is about creating an animation for a web site. The motivating questions are about a web site design.

Ask the students the following questions to get them interested in the lesson.

- What is a web site?
- What web sites have you used that were well designed?
- What do you like about the web site for our school?
- What makes a web site interesting to you?

Explore Together

Problem 1

Students will graph and evaluate quadratic functions in this lesson.

Grouping

Ask for a student volunteer to read the Scenario and Problem 1 aloud. Have a student restate the problem. Pose the Guiding Questions below to verify student understanding. Have students work together in small groups to complete part (A) of Problem 1.

Guiding Questions

- What information is given in this problem?
- What is a frame for the animation?
- What shape is each frame of the logo?
- How can you determine the length of the side of each frame?
- How can you calculate the area of each logo?
- Will the length increase or decrease with each frame? Will the area increase or decrease with each frame?
- Will the length and area change at the same rate?
- What is meant by the term square inches?

Common Student Errors

Students may need to be reminded how to calculate the area of a square.

Students sometimes don't understand the meaning of square units. Check for understanding of the concept of square units as the number of squares of a constant size that would be needed to cover the shape.

Call the class back together to have the students discuss and explain their work for part (A) of Problem 1.

SCENARIO Your brother is a graphic artist who works at a company that creates and maintains web sites. One of his jobs is to make art pieces that are put together to form movies (or animations) that are played on different pages of the web site. Each art piece is a frame of the animation. When the frames are displayed one after another, movement can be shown, and the animation is created. His current project is for a web site for a sporting goods company.

Problem 1 Creating an Animation

Your brother's first job on his current project is to create the frames for an animation of the company's logo that will play on the web site's main page. Some of the frames for the animation are shown. When the frames are displayed one after another, the logo will appear to grow.

The initial side length of the square logo is 1 inch. The side length grows by one inch in each frame.

A. Complete the table of values that shows the side length of the logo, the area of the logo, and the corresponding frame numbers. Copy the columns of the table into the correct columns in the margins of pages 361 and 362.

Quantity Name	Frame	Length	Area
Unit	numbers	inches	square inches
Expression	x	x	x^2
	1	1	1
	2	2	4
	3	3	9
	4	4	16
	5	5	25

8

Explore Together

Problem 1

Students will investigate the growth rates of the length of the logo and of the area of the logo.

Grouping

Ask for a student volunteer to read part (B) of Problem 1 aloud.
Have a student restate the problem.
Complete part (B) of Problem 1 together as a whole class.

Have students work together in small groups to complete parts (C) through (G) of Problem 1.

Key Formative Assessments

- How did you determine the rate of growth of the length from one frame to another?
- How did you determine the rate of growth of the area from one frame to another?
- Which was a constant rate of change?
- Why do you think the length has a constant rate of change?
- What was the rate of change of the length in each part? Include the units.
- Was the rate of change of the area also constant?
- Why do you think that the rate of change of the area varies?
- Is there a pattern in how the rate of growth of the area increases?
- What are the units for the rate of change of the area?

Common Student Errors

Students may incorrectly assume that the rates of change are constant for both the length and the area. Also, students may have difficulty finding the units for the rates of change. These units are an important part of the understanding. So, spend time to verify the students' understanding of the units.

8

Problem 1 Creating an Animation

B. How does the length grow as the frame number increases? How does the area grow as the frame number increases? Use complete sentences in your answer.

Sample Answer: The length grows by one inch as the frame number increases by one. The area grows by many square inches as the frame number increases by one.

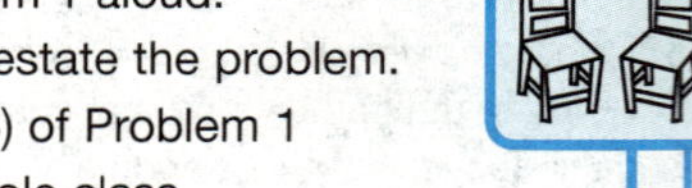

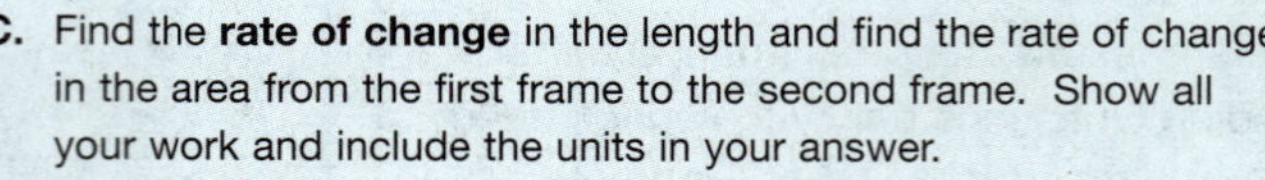

C. Find the **rate of change** in the length and find the rate of change in the area from the first frame to the second frame. Show all your work and include the units in your answer.

Length: $\frac{2-1}{2-1} = \frac{1}{1}$; **1 inch per frame**

Area: $\frac{4-1}{2-1} = \frac{3}{1}$; **3 square inches per frame**

D. Find the rate of change in the length and find the rate of change in the area from the second frame to the third frame. Show all your work and include the units in your answer.

Length: $\frac{3-2}{3-2} = \frac{1}{1}$; **1 inch per frame**

Area: $\frac{9-4}{3-2} = \frac{5}{1}$; **5 square inches per frame**

E. Find the rate of change in the length and find the rate of change in the area from the third frame to the fourth frame. Show all your work and include the units in your answer.

Length: $\frac{4-3}{4-3} = \frac{1}{1}$; **1 inch per frame**

Area: $\frac{16-9}{4-3} = \frac{7}{1}$; **7 square inches per frame**

F. Find the rate of change in the length and find the rate of change in the area from the fourth frame to the fifth frame. Show all your work and include the units in your answer.

Length: $\frac{5-4}{5-4} = \frac{1}{1}$; **1 inch per frame**

Area: $\frac{25-16}{5-4} = \frac{9}{1}$; **9 square inches per frame**

Explore Together

Investigate Problem 1

Students will create a scatter plot of the data that they found in part (A) in order to graph the length of the logo as a function of the frame number.

Grouping

Call the class back together to have the students discuss and present their work for parts (C) through (G).

Quantity Name	Frame	Length
Unit	numbers	inches
Expression	x	x
	1	1
	2	2
	3	3
	4	4
	5	5

Ask for a student volunteer to read Question 1 aloud. Have a student restate the problem. Pose the Guiding Questions below to verify student understanding. Have students work together in small groups to complete Questions 1 through 5.

Guiding Questions

- What is a scatter plot?
- How can you graph a scatter plot?
- To plot the length as a function of the frame number, which variable is the independent variable? Which variable is the dependent variable?
- Which variable should be graphed on the horizontal axis? Which variable should be graphed on the vertical axis?
- What is the unit for the frame number in this situation?
- What is the unit for the length in this situation?
- What bounds are appropriate for the frame number?

Problem 1 Creating an Animation

G. What do you notice about the rates of change in the length with respect to the frame number? What do you notice about the rates of change in the area with respect to the frame area? Use a complete sentence in your answer.

The rates of change that involve the length are the same, and the rates of change that involve the area are different.

Investigate Problem 1

1. Create a scatter plot of the length as a function of the frame number on the grid below. First, choose your bounds and intervals. Be sure to label your graph clearly.

Variable quantity	Lower bound	Upper bound	Interval
Frame	0	7.5	0.5
Length	0	7.5	0.5

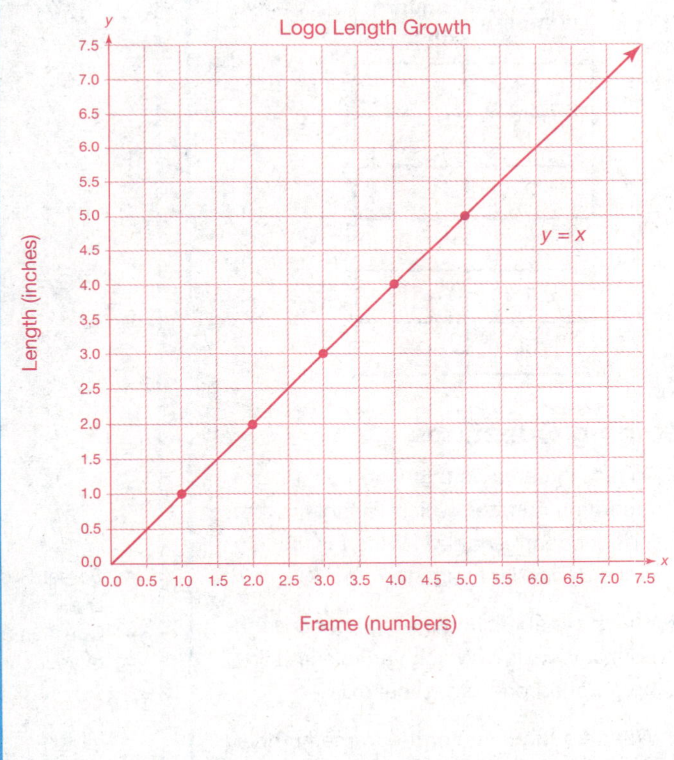

- What interval is appropriate for the frame number?
- What bounds are appropriate for the length?
- What interval is appropriate for the length?

Explore Together

Investigate Problem 1

Take Note

A **curve** can be a straight line or a curved line.

8

Students will create a scatter plot of the quadratic data of the area as a function of the frame number.

Grouping

Students will be working in small groups to complete these questions.

Quantity Name	Frame	Area
Unit	numbers	square inches
Expression	x	x^2
	1	1
	2	4
	3	9
	4	16
	5	25

Guiding Questions

- To plot the area as a function of the frame number, which variable is the independent variable? Which variable is the dependent variable?
- Which variable should be graphed on the horizontal axis? Which variable should be graphed on the vertical axis?
- What are the units for the frame number in this situation?
- What are the units for the area in this situation?
- What bounds are appropriate for the frame number?
- What interval is appropriate for the frame number?

Investigate Problem 1

2. Draw the curve that best fits the data on your graph in Question 1. Use a complete sentence to describe the shape of your curve.

The curve is a straight line.

3. Create a scatter plot of the area as a function of the frame number on the grid below. First, choose your bounds and intervals. Be sure to label your graph clearly.

Variable quantity	Lower bound	Upper bound	Interval
Frame	0	15	1
Area	0	120	8

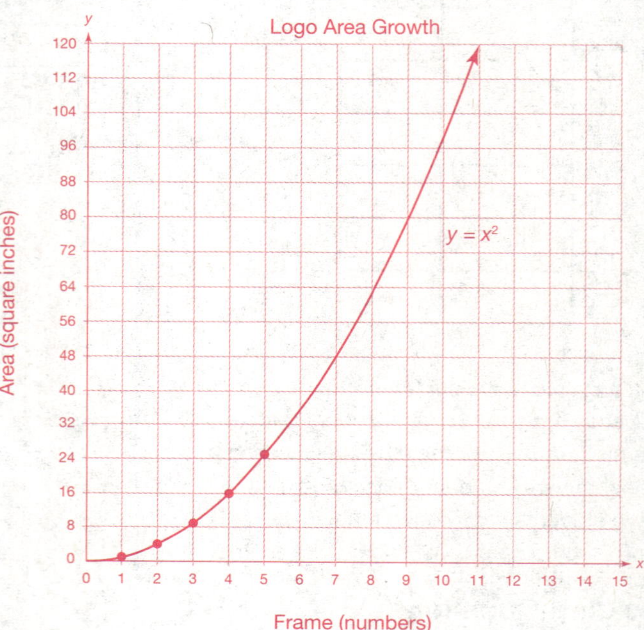

4. Sometimes, the curve that best fits the data is not a straight line. Draw the curve that best fits the data on your graph in Question 3. Use a complete sentence to describe the shape of your curve.

Sample Answer: The curve is shaped like half of a "U."

- What bounds are appropriate for the area?
- What interval is appropriate for the area?
- What is a curve?

Explore Together

Investigate Problem 1

Students have graphed a linear function comparing frame number and length and then graphed the quadratic function comparing frame number and area. The students will now write equations in Question 5 to model the situations and to identify characteristics of quadratic functions.

Grouping

Call the class back together to have the students discuss and present their work for Questions 1 through 5.

Just the Math

Quadratic functions will be formally introduced in Question 6. The standard form for a quadratic function is used.

Ask for a student volunteer to read Question 6 aloud. Have a student restate the problem. Pose the Guiding Questions below to verify student understanding. Have students work individually to complete Questions 6 and 7. Then call the class back together to have the students discuss and explain their work for Questions 6 and 7.

Guiding Questions

- What is a quadratic function?
- What does the *a* represent in the general form of a quadratic function? What does the *b* represent in the general form of a quadratic function? What does the *c* represent in the general form of a quadratic function?
- For the quadratic function $y = 2x^2 + 4x + 3$, what are the values of *a*, *b*, and *c*?
- In the quadratic function $y = 4x^2 + (-3x) + (-2)$, what are the values of *a*, *b*, and *c*?
- How can you simplify the equation $y = 4x^2 + (-3x) + (-2)$?

Investigate Problem 1

5. For each graph, write an equation that describes the problem situation. Be sure to define your variables. Write your answers using complete sentences.

The growth of the length can be described by $y = x$ where x is the frame number and y is the length in inches. The growth of the area can be described by $y = x^2$ where x is the frame number and y is the area in square inches.

6. Just the Math: Quadratic Function The equation that you wrote for the area, $y = x^2$, represents the simplest form of a *quadratic function*. A **quadratic function** is a function of the form $f(x) = ax^2 + bx + c$, where *a*, *b*, and *c* are constants with $a \neq 0$. The graph of a quadratic function is a U-shaped graph. What can you conclude about the rate of change of a quadratic function? Use a complete sentence in your answer.

Sample Answer: The rate of change of a quadratic function is not constant.

7. Identify the values of *a*, *b*, and *c* in each quadratic function below.

$f(x) = 2x^2 + 3x + 5$

$a = 2, b = 3, c = 5$

$h(x) = x^2 + 4x - 1$

$a = 1, b = 4, c = -1$

$g(x) = x^2 + 4$

$a = 1, b = 0, c = 4$

$f(x) = -x^2 - 2x + 8$

$a = -1, b = -2, c = 8$

$h(x) = x^2 - 3x$

$a = 1, b = -3, c = 0$

$g(x) = 10 - x^2$

$a = -1, b = 0, c = 10$

Problem 2 The Bouncing Ball

One of the programmers at your brother's company has the task of making the animations work on the web site. On one of the pages, he has to program an animation of a ball being thrown from one person to another person. The programmer uses a function to determine the path of the ball.

Notes Each term in a quadratic function is separated by an addition symbol when using the standard form of a quadratic function. If the quadratic function were $y = x^2 - 2x + 1$, then in standard form, the equation is $y = x^2 + (-2x) + 1$. So, in this instance, $a = 1$, $b = -2$, and $c = 1$.

Explore Together

Problem 2

Students will investigate another quadratic function.

Grouping

Ask for a student volunteer to read Problem 2 aloud. Have a student restate the problem. Pose the Guiding Questions below to verify student understanding. Have students work together in small groups to complete parts (A) through (E) of Problem 2.

Guiding Questions

- What is a pixel?
- What is meant by horizontal position? What is meant by vertical position?
- What happens to the values in the column for the horizontal position?
- What happens to the values of the vertical position as the horizontal position goes from 10 to 30 to 50 pixels?
- What happens to the values of the vertical position as the horizontal position goes from 50 to 70 to 90 pixels?
- How can the values for the vertical position increase, but then decrease later? How does that model the height of a ball that is thrown?
- Describe how you would predict what the graph of the data in the table in part (A) will look like.
- What 2 variable quantities are important in this situation?
- What bounds are appropriate for the horizontal position for the data in the table?
- What bounds are appropriate for the vertical position for the data in the table?
- What interval is appropriate for each of these variable quantities?

Problem 2 The Bouncing Ball

A. The table below shows some of the positions of the ball on the computer screen with respect to the origin as the animation plays. The origin represents the lower left-hand corner of the screen.

Quantity Name	**Horizontal position**	**Vertical position**
Unit	**pixels**	**pixels**
	10	20
	30	80
	50	100
	70	80
	90	20

Create a scatter plot of the path of the ball on the grid below. First, choose your bounds and intervals. Be sure to label your graph clearly.

Variable quantity	Lower bound	Upper bound	Interval
Horizontal position	**0**	**150**	**10**
Vertical position	**0**	**150**	**10**

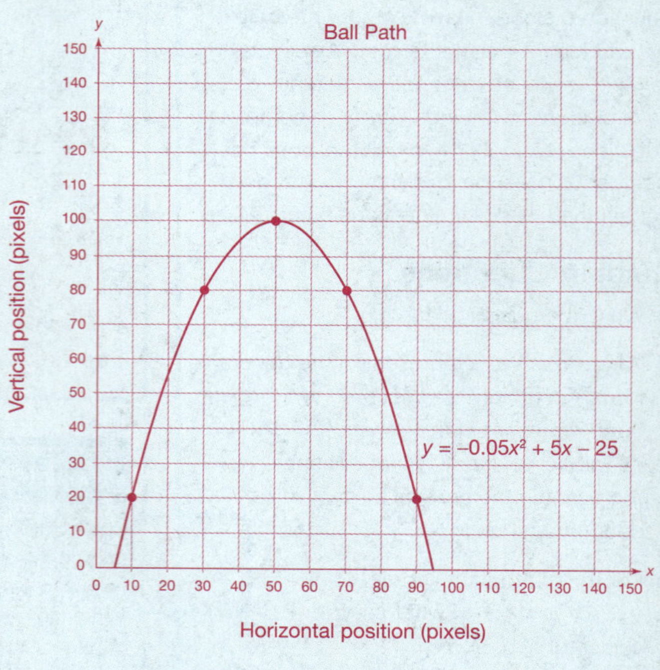

Notes The directions in part (B) ask the students to draw a smooth curve to fit the data. You may have to explain the meaning of the term *smooth curve* to some students.

Explore Together

Problem 2

Students will investigate the rates of change for the quadratic data that they graphed in part (A) of Problem 2.

Notes The students are asked to calculate rates of change in part (C). The Warm Up questions for this lesson allowed students to practice this skill. If the Warm Up questions were not used, you may want to review calculating the rate of change, or slope, with the students before they complete part (C) of Problem 2.

Key Formative Assessments

- Other than the height of a ball that is thrown into the air, what else might have a graph such as the one in part (A) of Problem 2?
- Compare the graphs of the quadratic functions in Question 3 of Problem 1 and in part (A) of Problem 2. What is similar in the two graphs? What is different?
- What does the graph of a quadratic function look like?
- How difficult was it to answer Question 5 of Problem 1 to identify the equation for the quadratic function in Problem 1?

Take Note

Order of Operations

1. Evaluate expressions inside grouping symbols such as ( ) or [ ].
2. Evaluate powers.
3. Multiply and divide from left to right.
4. Add and subtract from left to right.

Problem 2 The Bouncing Ball

B. Connect the points with a smooth curve.

C. Find the rates of change in the position of the ball as it moves from position to position. Record the results in the table.

Change in position	Rate of change
from (10, 20) to (30, 80)	$\frac{80 - 20}{30 - 10} = \frac{60}{20} = \frac{3}{1}$
from (30, 80) to (50, 100)	$\frac{100 - 80}{50 - 30} = \frac{20}{20} = \frac{1}{1}$
from (50, 100) to (70, 80)	$\frac{80 - 100}{70 - 50} = \frac{-20}{20} = \frac{-1}{1}$
from (70, 80) to (90, 20)	$\frac{20 - 80}{90 - 70} = \frac{-60}{20} = \frac{-3}{1}$

D. What do you notice about the rates of change? Use a complete sentence in your answer.

Sample Answer: The rates of change are all different.

E. What kind of function is represented by your graph? Use a complete sentence in your answer.

A quadratic function is represented by the graph.

Investigate Problem 2

1. The sporting goods company has seen the animation of the ball and wants the two people to be closer together and the ball to be thrown higher. The programmer has come up with a new path that is represented by the function

$f(x) = -0.05x^2 + 4x + 40.$

Before you can graph this new path, you need to be able to evaluate this function. In the order of operations, you should evaluate any powers first. So, to find the value of $f(10)$, substitute 10 for x and evaluate 10^2 first:

$$\begin{aligned} f(10) &= -0.05(10^2) + 4(10) + 40 \\ &= -0.05(100) + 4(10) + 40 \end{aligned}$$

Then multiply and finally add and subtract from left to right. Show all your work.

$$\begin{aligned} f(10) &= -0.05(100) + 4(10) + 40 \\ &= -5 + 40 + 40 \\ &= 75 \end{aligned}$$

Investigate Problem 2

Grouping

Ask for a student volunteer to read Question 1 aloud. Have a student restate the problem. Have the students complete Question 1 individually.

Explore Together

Investigate Problem 2

Students will create a table of values for a quadratic function. Then, they will graph the quadratic function.

Grouping

Call the class back together to discuss Question 1. Have students work together in small groups to complete Questions 2 and 3. Pose the Guiding Questions below to verify student understanding.

Call the class back together to have the students discuss and present their work for Questions 2 and 3.

Guiding Questions

- What information is already given in the table in Question 2?
- What information is not given in the table in Question 2?
- What is needed to calculate the values for the vertical positions in Question 2? Where can you find that information?
- What is the quadratic function that will be used to find the values for the vertical positions?
- How can you evaluate the equation to find the value for the vertical position for any given horizontal position?
- What two variable quantities are important in this problem situation?
- Which quantity will you represent on the horizontal axis in Question 3? Which quantity will you represent on the vertical axis in Question 3?
- What bounds and intervals are appropriate for the graph in Question 3?
- What do you predict will be the shape of the graph in Question 3? Why?

Investigate Problem 2

2. Complete the table below to show some of the new positions of the path of the ball.

Quantity Name	Horizontal position	Vertical position
Unit	pixels	pixels
	10	75
	20	100
	40	120
	60	100
	70	75

3. Create a graph of the path of the ball on the grid below. First, choose your bounds and intervals. Be sure to label your graph clearly.

Variable quantity	Lower bound	Upper bound	Interval
Horizontal postion	0	150	10
Vertical position	0	150	10

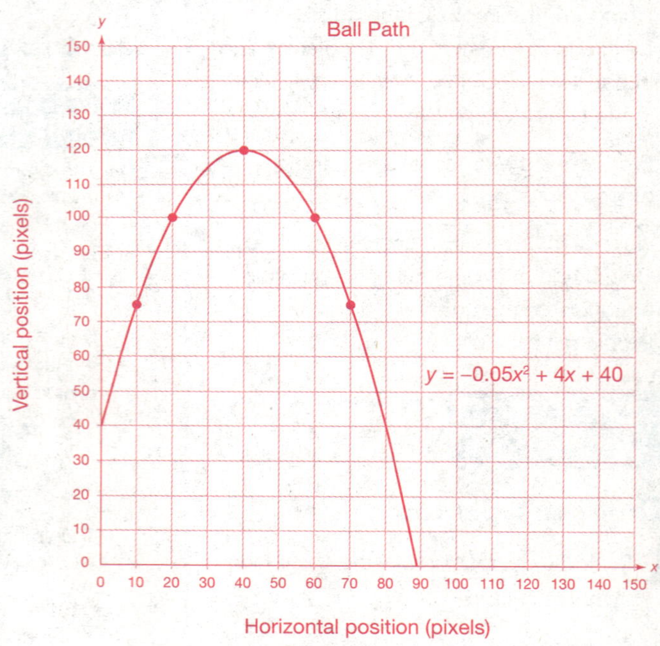

Explore Together

Investigate Problem 2

Students will analyze their graphs and then evaluate quadratic functions.

Grouping

Ask for a student volunteer to read Questions 4 and 5 aloud. Pose the Guiding Questions below to verify student understanding. Have students work together in small groups to complete Questions 4 and 5. Then, call the class back together to discuss and present their work.

Guiding Questions

- What does it mean for one graph to be higher than another graph?
- What is the same in the table of values in part (A) of Problem 1 and Question 2 of Problem 2? What is different?
- Why is there a value for 90 pixels for the horizontal position for the first situation but not the second situation?
- Is the vertical position higher for the graph in Question 3 of Problem 2 than for the graph in part (A) of Problem 1 at the 30 pixel horizontal position?
- Is the vertical position higher for the graph in Question 3 of Problem 2 than for the graph in part (A) of Problem 1 at the 40 pixel horizontal position?
- Is the vertical position higher for the graph in Question 3 of Problem 2 than for the graph in part (A) of Problem 1 at the 70 pixel horizontal position?
- What is the highest, or peak, vertical position for the graph in part (A) of Problem 1? What is the highest, or peak, vertical position for the graph in Question 3 of Problem 2?

Take Note

Whenever you see the share with the class icon, your group should prepare a short presentation to share with the class that shows how you solved the problem. Be prepared to ask questions during other groups' presentations and to answer questions during your presentation.

Investigate Problem 2

8

4. Is the highest point in this graph higher than the highest point in the graph in part (A)? If so, what is the difference in the heights? Write your answer using a complete sentence.

Yes; the difference in the heights is 20 pixels.

5. Evaluate each of the following quadratic functions for the given value of x. Show all your work.

$f(x) = x^2 + 5x + 7$; $f(4)$

$$\begin{aligned} f(4) &= 4^2 + 5(4) + 7 \\ &= 16 + 20 + 7 \\ &= 43 \end{aligned}$$

$g(x) = 25 - x^2$; $g(-5)$

$$\begin{aligned} g(-5) &= 25 - (-5)^2 \\ &= 25 - 25 \\ &= 0 \end{aligned}$$

$h(x) = -x^2 + 4x - 1$; $h(-2)$

$$\begin{aligned} h(-2) &= -(-2)^2 + 4(-2) - 1 \\ &= -4 - 8 - 1 \\ &= -13 \end{aligned}$$

$f(x) = 5x^2 - 18$; $f(0)$

$$\begin{aligned} f(0) &= 5(0)^2 - 18 \\ &= 0 - 18 \\ &= -18 \end{aligned}$$

$g(x) = 8x^2 - 3x + 10$; $g(2)$

$$\begin{aligned} g(2) &= 8(2)^2 - 3(2) + 10 \\ &= 8(4) - 6 + 10 \\ &= 32 - 6 + 10 \\ &= 36 \end{aligned}$$

$h(x) = -3x^2 + 15$; $h(10)$

$$\begin{aligned} h(10) &= -3(10)^2 + 15 \\ &= -3(100) + 15 \\ &= -300 + 15 \\ &= -285 \end{aligned}$$

6. In Lesson 5.1, you worked with linear functions. What are the similarities between linear functions and quadratic functions? What are the differences? Write your answers using complete sentences.

Sample Answer: Linear functions and quadratic functions are similar in that for each input value there is exactly one output value. Linear functions and quadratic functions are different in that linear functions are represented by straight lines and quadratic functions are represented by U-shaped curves. In linear functions, the highest power of any variable is the first power, and in quadratic functions, the highest power of any variable is the second power.

Grouping

Assign Question 6 individually as a writing project. Have the students write at least one paragraph. Then have several students read their responses during the next class session. Finally, discuss and summarize the responses with the class.

Wrap Up

Close

- Review all key terms and their definitions. Include the terms *rate of change, quadratic function,* and *evaluate*.
- You may also want to review any other vocabulary terms that were discussed during the lesson, which may include *area, scatter plot, dependent variable, independent variable, horizontal axis, vertical axis,* and *line of best fit.*
- Remind the students to write the key terms and their definitions in the notes section of their notebooks. You may also want the students to include examples.
- Ask the students to explain how they found rates of change in this lesson.
- Ask the students to summarize the characteristics of quadratic functions.
- Have the students create a chart comparing the characteristics of linear functions and quadratic functions. A Venn diagram could also be used for this comparison.
- Ask the students to explain what is meant by "evaluating a quadratic function."

Ties to the Cognitive Tutor Software

Many of the quadratic word problems in the Cognitive Tutor software involve area. Students are generally familiar with the idea that length times width equals area and that area is measured in square units, so using area is a comfortable context to introduce quadratics. However, they may not have associated the word *square* in square units with the meaning of square as power of two. Making this connection can help students remember the relationship between area and quadratic functions.

Follow Up

Assignment

Use the Assignment for Lesson 8.1 in the Student Assignments book. See the Teacher's Resources and Assessments book for answers.

Assessment

See the Assessments provided in the Teacher's Resources and Assessments book for Chapter 8.

Open-Ended Writing Task

Ask the students to write a paragraph explaining what they learned in this lesson. Have them write the information as if they were writing a letter to a friend who had been sick and missed the entire lesson. They will try to teach the main concepts to their friend in their letter.

Reflections

Insert your reflections on the lesson as it played out in class today.

What went well?

____

____

What did not go as well as you would have liked?

____

____

How would you like to change the lesson in order to improve the things that did not go well and capitalize on the things that did go well?

____

____

____

____

____

Notes

8

8.2 Satellite Dish

Parabolas

Learning By Doing Lesson Map

Get Ready

Objectives

In this lesson, you will:

- Graph quadratic functions.
- Find the line of symmetry of a parabola.
- Find the vertex of a parabola.
- Identify the maximum or minimum value of a function.

Key Terms

- parabola
- line of symmetry
- vertical line
- vertex
- minimum
- maximum

NCTM Content Standards

Grades 9–12 Expectations

Algebra Standards

- Generalize patterns using explicitly defined and recursively defined functions.
- Analyze functions of one variable by investigating rates of change, intercepts, zeros, asymptotes, and local and global behavior.
- Understand and compare the properties of classes of functions, including exponential, polynomial, rational, logarithmic, and periodic functions.
- Interpret representations of functions of two variables.
- Use symbolic algebra to represent and explain mathematical relationships.
- Identify essential quantitative relationships in a situation and determine the class or classes of functions that might model the relationships.
- Draw reasonable conclusions about a situation being modeled.

Lesson Overview

Within the context of this lesson, students will be asked to:

- Graph quadratic functions.
- Find the line of symmetry of a parabola graphically and algebraically.
- Find the vertex of a parabola graphically and algebraically.
- Identify the maximum or minimum value of a quadratic function.

Essential Questions

The following key questions are addressed in this lesson:

1. What is a satellite dish?
2. What is a parabola? What does a parabola represent?
3. What is a vertex, a line of symmetry, a maximum, and a minimum value?

Show The Way

Warm Up

Place the following questions or an applicable subset of these questions on the board before students enter class. Students should begin working as soon as they are seated.

Draw a line that divides each shape into 2 identical halves. Note that there is more than one possible answer in some cases.

8

1.

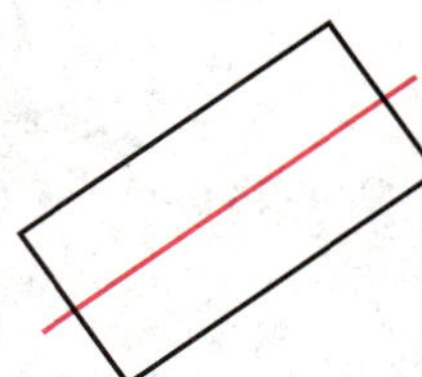

2.

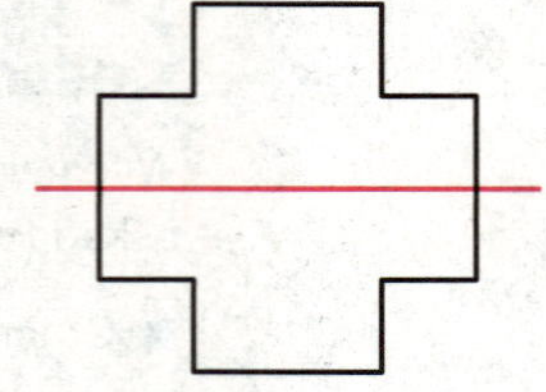

3.

4.

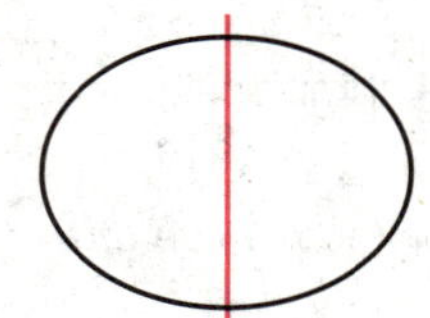

5.

6.

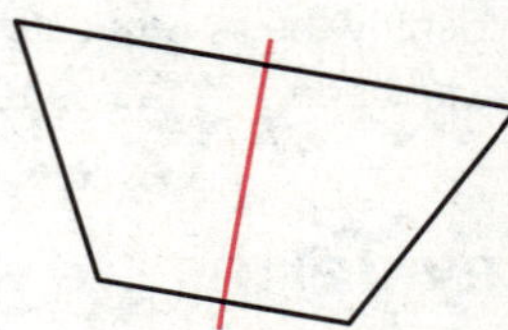

7.

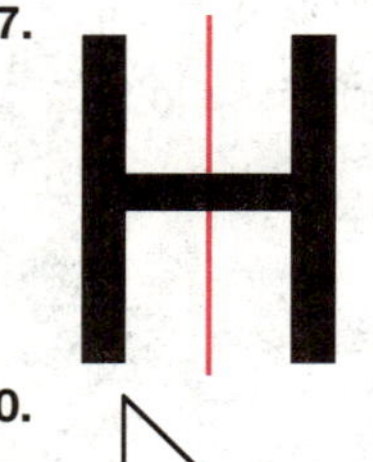

8.

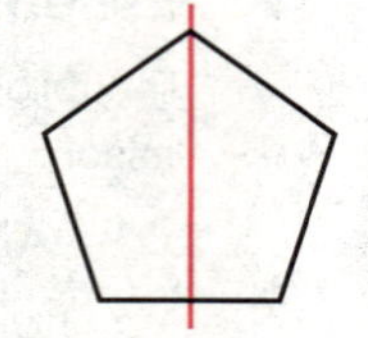

9.

10.

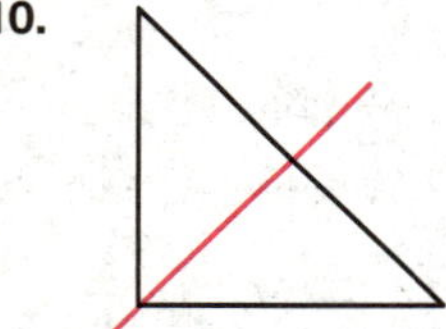

11.

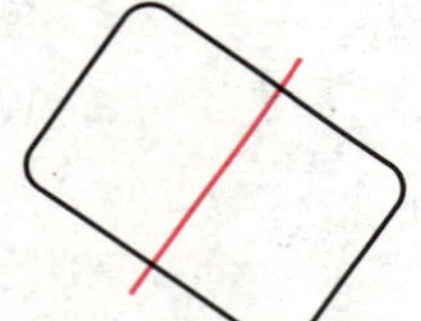

12.

Motivator

Begin the lesson with the motivator to get students thinking about the topic of the upcoming problem. This lesson is about the shape of a satellite dish. The motivating questions are about satellites and satellite dishes.

Ask the students the following questions to get them interested in the lesson.

- What is a satellite?
- How are satellites put into space?
- How are satellites kept in space?
- What is a satellite dish?
- What is a common use for a satellite dish?
- What is the shape of a satellite dish?

Explore Together

Problem 1

Students will graph and evaluate quadratic functions.

Grouping

Ask for a student volunteer to read the Scenario and Problem 1 aloud. Have a student restate the problem. Pose the Guiding Questions below to verify students' understanding. Have students work individually to complete part (A) of Problem 1.

Guiding Questions

- What is a satellite dish?
- What is the shape of a satellite dish?
- What is meant by the term *profile* of a satellite dish?
- If a marble were held on the edge of a satellite dish and released, would it follow a path along the profile of the dish?
- When the distance from the very center of the dish is 0 inches horizontally, what will the distance from the center be vertically?
- How can you represent the point that is 0 inches away from the center of the dish both horizontally and vertically?
- What does the solution (0, 0) for the function $y = \frac{1}{36}x^2$ represent in this situation?
- How can you solve for the vertical component of the profile if given the horizontal component?

Common Student Errors

Students will frequently forget that squaring a negative value will yield a positive result.

Students may be confused initially by the values that they calculate for the vertical component and why they are all non-negative, even if the horizontal component is negative.

SCENARIO A satellite dish is a type of antenna that transmits signals to and receives signals from satellites. Satellite dishes are most commonly used by people to receive satellite television transmissions. You can use a quadratic equation to model the profile, or outline, of a satellite dish.

Problem 1 Dish Design

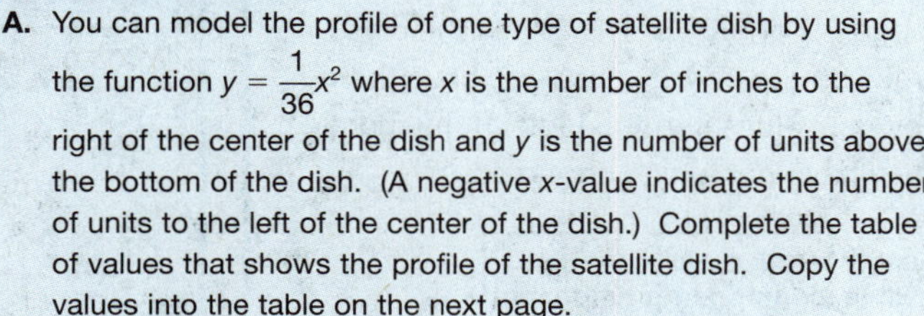

A. You can model the profile of one type of satellite dish by using the function $y = \frac{1}{36}x^2$ where x is the number of inches to the right of the center of the dish and y is the number of units above the bottom of the dish. (A negative x-value indicates the number of units to the left of the center of the dish.) Complete the table of values that shows the profile of the satellite dish. Copy the values into the table on the next page.

Quantity Name	Horizontal component	Vertical component
Unit	inches	inches
Expression	x	$\frac{1}{36}x^2$
	−24	16
	−18	9
	−12	4
	−6	1
	0	0
	6	1
	12	4
	18	9
	24	16

Explore Together

Problem 1

Students will graph the parabolic data from the table that they created in part (A).

Call the class back together to have the students discuss and explain their work for part (A) of Problem 1. When the class agrees on the correct answers, have them copy the values into the table in the margin if they have not already done so.

Quantity Name	Horizontal component	Vertical component
Unit	inches	inches
Expression	x	$\frac{1}{36}x^2$
	−24	16
	−18	9
	−12	4
	−6	1
	0	0
	6	1
	12	4
	18	9
	24	16

Grouping

Have the students complete part (B) of Problem 1 individually. Call the class back together to have the students discuss and present their work for part (B) of Problem 1.

Investigate Problem 1

Just the Math

Students will investigate parabolas in Question 1.

Grouping

Ask for a student volunteer to read Question 1 aloud. Have a student restate the problem. Pose the Guiding Questions to verify student understanding. Have students work together in small groups to complete Questions 1 and 2.

Problem 1 Dish Design

B. Create a graph of the quadratic function on the grid below. First, choose your bounds and intervals. Be sure to label your graph clearly.

Variable quantity	Lower bound	Upper bound	Interval
Horizontal component	−21	21	3
Vertical component	0	30	2

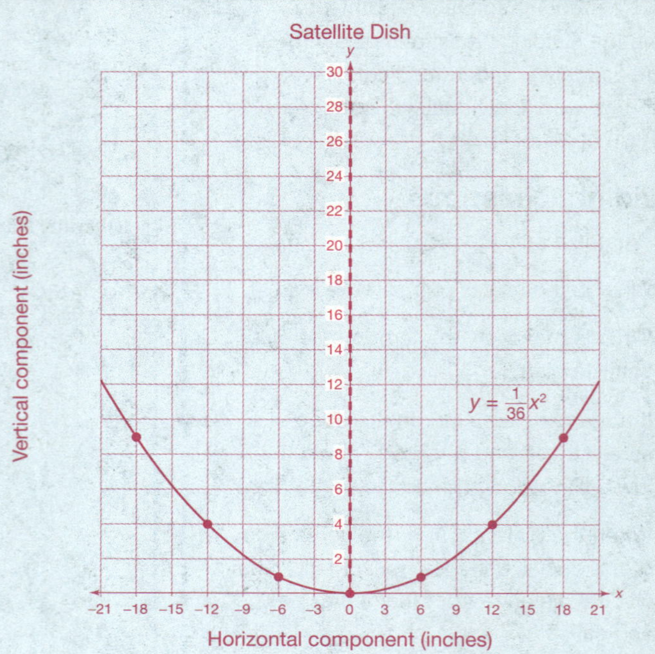

Investigate Problem 1

1. **Just the Math: Parabolas** Describe the shape of the graph. Use a complete sentence in your answer.

   Sample Answer: The graph is U-shaped.

   The graph of any quadratic function is called a **parabola.** How is the graph of a quadratic function different from the graph of a linear function? Use complete sentences in your answer.

   Sample Answer: The graph of a quadratic function is curved. The graph of a linear function is a straight line.

Guiding Questions

- What is a linear function?
- What is a quadratic function?
- What does the graph of a linear function look like?
- What does the graph of a quadratic function look like?

Explore Together

Investigate Problem 1

Students will investigate parabolic behavior of quadratic functions.

Call the class back together to have the students discuss and present their work for Questions 1 and 2.

Key Formative Assessments

- For each x value in a quadratic function, how many y values are possible?
- Are there 2 different y-values for each x-value for this quadratic function?
- What x-value has only one y-value for this quadratic function?
- Do you think that there will always be one x-value for a parabola that will have only one y-value when all of the other x-values will have 2 different y-values? Is this true for all parabolas?

The equation of a **vertical line** is $x = a$, where a is a real number.

Just the Math

Students will investigate the line of symmetry in Question 3.

Grouping

Ask for a student volunteer to read Question 3 aloud. Have a student restate the problem. Pose the Guiding Questions below to verify student understanding. Complete Question 3 together as a whole class.

Guiding Questions

- What is the meaning of the word *symmetric*?
- What is a line of symmetry?

Investigate Problem 1

2. Choose a positive height above the bottom of the dish. What are the corresponding x-values on the graph? Use a complete sentence in your answer.

   Answers will vary. Sample Answer: When $y = 1$, $x = -6$ and $x = 6$.

   Choose another positive height above the bottom of the dish. What are the corresponding x-values on the graph? Use a complete sentence in your answer.

   Answers will vary. Sample Answer: When $y = 4$, $x = -12$ and $x = 12$.

   Choose one more positive height above the bottom of the dish. What are the corresponding x-values on the graph? Use a complete sentence in your answer.

   Answers will vary. Sample Answer: When $y = 9$, $x = -18$ and $x = 18$.

   What do you notice about these x-values? Use a complete sentence in your answer.

   Sample Answer: For each positive y-value, the corresponding x-values are opposites.

3. **Just the Math: Line of Symmetry** Draw a dashed line on your graph in part (B) so that the graph on one side of the line is a mirror image of the graph on the other side of the line. Your line is called the **line of symmetry.** What is the equation of this line?

   $x = 0$

   For any quadratic function of the form $y = ax^2 + bx + c$, the equation of the line of symmetry is given by $x = -\frac{b}{2a}$.

   Think about the function $y = \frac{1}{36}x^2$. What is the value of a for this function? What is the value of b for this function? Write your answers using complete sentences.

   In the function for the satellite dish, $a = \frac{1}{36}$ and $b = 0$.

   Use the values of a and b to write the equation of the line of symmetry. Write your answer using a complete sentence.

   The equation of the line of symmetry is

   $$x = -\frac{b}{2a} = -\frac{0}{2\left(\frac{1}{36}\right)} = 0.$$

Common Student Errors

The algebraic formula $x = -\frac{b}{2a}$ that is used to locate the line of symmetry is confusing and out of context for many students. It may help to have the students graph some of the functions in Question 3 to verify the line of symmetry that they calculate.

Explore Together

Investigate Problem 1

Students will continue to investigate the behavior of parabolic graphs of quadratic functions.

Key Formative Assessments

- How can you find the location of a line of symmetry for any parabola by looking at the graph of the parabola?
- When would a parabola have a line of symmetry that is not located at $x = 0$?
- What would the graph of such a parabola look like?

Just the Math

Students will investigate the vertex of a parabola in Question 4.

Grouping

Ask for a student volunteer to read Question 4 aloud. Have a student restate the problem. Pose the Guiding Questions below to verify student understanding. Have students work individually to complete Question 4.

Guiding Questions

- What is a vertex?
- What is important about the vertex for a parabola?
- What do you think is the relationship between the vertex and the line of symmetry?

Call the class back together to have the students discuss and present their work for Question 4.

Key Formative Assessments

- How can you find the vertex of a parabola by looking at the graph of a quadratic function?
- In the function $y = \frac{1}{36}x^2$ for the profile of the satellite dish, you found the vertex to be the lowest point, or (0, 0), on your graph. Will the vertex always be the lowest point on the graph of a parabola?
- What other location is possible for the vertex of a parabola?

Investigate Problem 1

Find the equation of the line of symmetry for each quadratic function. Show all your work.

$y = x^2 + 2x + 1$

$x = -\frac{2}{2(1)} = -1$

$y = -3x^2 + 6x + 3$

$x = -\frac{6}{2(-3)} = -\frac{6}{(-6)} = 1$

$y = x^2 - 2x + 2$

$x = -\frac{(-2)}{2(1)} = -(-1) = 1$

$y = -x^2 - 4x - 1$

$x = -\frac{(-4)}{2(-1)} = -\left(\frac{-4}{-2}\right) = -2$

$y = 2x^2 - 20x + 54$

$x = -\frac{(-20)}{2(2)} = -\frac{(-20)}{4} = 5$

$y = -5x^2 + 8x - 10$

$x = -\frac{8}{2(-5)} = -\frac{8}{(-10)} = \frac{4}{5}$

4. Just the Math: Vertex What is the lowest point on your graph in part (B)? This point is called the **vertex.** What is different about this point from the other points on the graph? Use a complete sentence in your answer.

The lowest point is (0, 0). Sample Answer: The line of symmetry passes through the point.

The x-coordinate of the vertex is given by $x = -\frac{b}{2a}$. Does this make sense to you? Why? Use a complete sentence in your answer.

Sample Answer: Yes, because the vertex is on the line of symmetry.

Find the vertex of the graph of each quadratic function. Show all your work.

$y = x^2 - 8$

$x = -\frac{0}{2(1)} = 0$

$y = 0^2 - 8 = -8$

(0, –8)

$y = -2x^2 + 4x - 5$

$x = -\frac{4}{2(-2)} = -\frac{4}{(-4)} = 1$

$y = -2(1)^2 + 4(1) - 5 = -2 + 4 - 5 = -3$

(1, –3)

Explore Together

Investigate Problem 1

Students will consider maximum and minimum values and how they relate to the vertex point of a parabola as well as the domain and range for quadratic functions.

Just the Math

Maximum and minimum values will be formally introduced in Question 5.

Ask for a student volunteer to read Question 5 aloud. Have a student restate the problem. Pose the Guiding Questions below to verify student understanding. Complete Question 5 together as a whole class.

Guiding Questions

- What is a maximum value? What is a minimum value?
- How is the maximum value or the minimum value related to the vertex of a parabola?

Grouping

Ask for a student volunteer to read Question 6 aloud. Have a student restate the problem. Pose the Guiding Questions below to verify student understanding. Have students work individually to complete Questions 6 through 8.

Guiding Questions

- What is a domain? What is a range?
- How can you find the domain and the range of a function?

Key Formative Assessments

- How can you find a maximum or minimum value by looking at the graph of a parabola? How can you find a maximum or minimum value algebraically?
- Was the sign of the coefficient of x^2 positive or negative for the quadratic function that you graphed in part (B)? Was the graph of the function a parabola that opened upward or downward? Did it have a minimum or maximum value?
- Was the sign of the coefficient of x^2 positive or negative for the quadratic function that you graphed in Lesson 8.1? Was the graph of the function a parabola that opened upward or downward? Did it have a minimum or maximum value?
- What do you predict is the relationship between the sign of the coefficient of x^2 in a quadratic function and the graph of the function having a maximum or minimum value?

Investigate Problem 1

5. **Just the Math: Maximum and Minimum** When the parabola opens upward, such as your graph in part (B), the *y*-coordinate of the vertex is called the **minimum,** or lowest value, of the function. When the parabola opens downward, such as the path of the ball in Lesson 8.1, the *y*-coordinate of the vertex is called the **maximum,** or highest value, of the function. Do the graphs of linear functions have maximum or minimum values? Use complete sentences to explain your reasoning.

   Sample Answer: No, because a line extends in either direction infinitely.

6. What is the domain of the graph of your function in part (B)? What is the range of the graph of your function in part (B)? Do not consider the problem situation to answer the question. Use complete sentences in your answer.

   The domain is all real numbers. The range is all numbers greater than or equal to 0.

7. The satellite dish is four inches tall. Use this information to find the domain and range of the function in the problem situation. Show all your work and use a complete sentence in your answer.

   The domain is all real numbers from –12 to 12. The range is all real numbers from 0 to 4.

   How wide is the dish at its widest point? Use a complete sentence to explain your reasoning.

   24 inches; Sample Answer: When $y = 4$, $x = -12$ and $x = 12$, so the distance is 2(12 inches) = 24 inches wide.

8. How is the domain of a linear function the same as or different from the domain of a quadratic function? Use a complete sentence in your answer.

   Sample Answer: They are both all real numbers.

   How is the range of a linear function the same as or different from the range of a quadratic function? Use complete sentences in your answer.

   Sample Answer: The range of a linear function is all real numbers. The range of a quadratic function is not all real numbers.

Explore Together

Investigate Problem 1

Students will continue to investigate parabolic graphs of quadratic functions.

Call the class back together to have the students discuss and present their work for Questions 6 through 8.

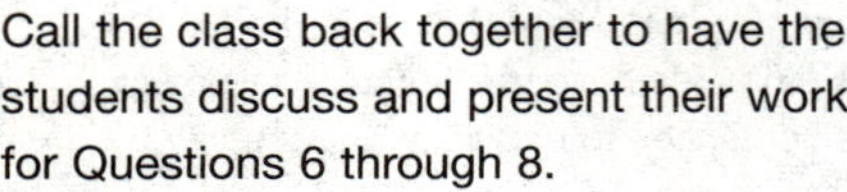

8

Grouping

Ask for a student volunteer to read Question 9 aloud. Have a student restate the situation. Pose the Guiding Questions below to verify student understanding. Have students work together in small groups to complete Questions 9 through 11.

Guiding Questions

- What is being asked in Question 9?
- How can you find the vertex of the function algebraically?
- How can you find the line of symmetry of the function algebraically?
- How can you graph a quadratic function?
- How can you find the domain and the range of a quadratic function?

Common Student Errors

Students may have difficulty finding the domain and range for these functions. Be ready to refocus students by asking them probing questions to have them remember what the domain and range for a function are and how to find them.

Note Many people will call a maximum value the max and a minimum value the min.

Investigate Problem 1

9. For each function, algebraically determine the vertex and the line of symmetry of the graph. Then draw the graph of the function. Identify the domain and range of the function. Use a complete sentence to tell whether the function has a maximum or minimum value.

$y = -x^2 + 4x$

$x = -\frac{4}{2(-1)} = -\frac{4}{(-2)} = 2; y = -(2)^2 + 4(2) = 4$

Vertex: (2, 4);

Line of symmetry: $x = 2$

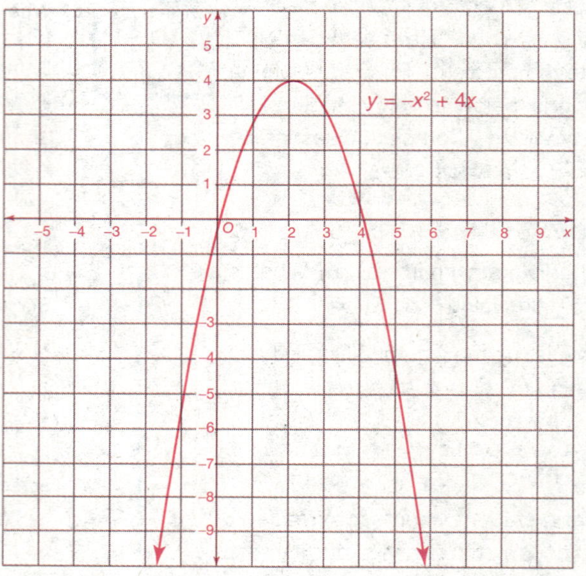

Domain: All real numbers; Range: All real numbers less than or equal to 4. The graph opens downward; therefore, the function has a maximum value.

Explore Together

Investigate Problem 1

Students will continue to investigate parabolic graphs of quadratic functions. These questions require the students to consolidate all components of their understanding of quadratic functions.

Common Student Errors

In Question 10, students may not easily recognize the relationship between the coefficient of the x^2 term and the quadratic function having a maximum or minimum value. This concept was discussed in the Key Formative Assessments earlier in this lesson. Even after discussing this idea, it still may be difficult for some students. You may want to have the students draw the graphs of several quadratic functions and identify whether they have a maximum or minimum value. Then have the students identify the sign of the coefficient of x^2 in the function to make the relationship more clear.

Call the class back together to have the students discuss and present their work for Questions 9 through 11.

Key Formative Assessments

- What is the name for the graph of a quadratic function?
- What is the general shape of a parabola?
- What is a vertex? How can you find a vertex algebraically? How can you find a vertex graphically?
- What is a line of symmetry? How is the line of symmetry related to the vertex?
- When does a parabola have a maximum value? When does a parabola have a minimum value?
- How can you find the maximum or minimum value?
- Can a quadratic function have both a maximum and a minimum value?

8

Investigate Problem 1

$y = 2x^2 - 4x + 1$

$x = -\frac{(-4)}{2(2)} = -\frac{(-4)}{4} = 1;$

$y = 2(1)^2 - 4(1) + 1 = 2 - 4 + 1 = -1$

Vertex: (1, –1);

Line of symmetry: $x = 1$

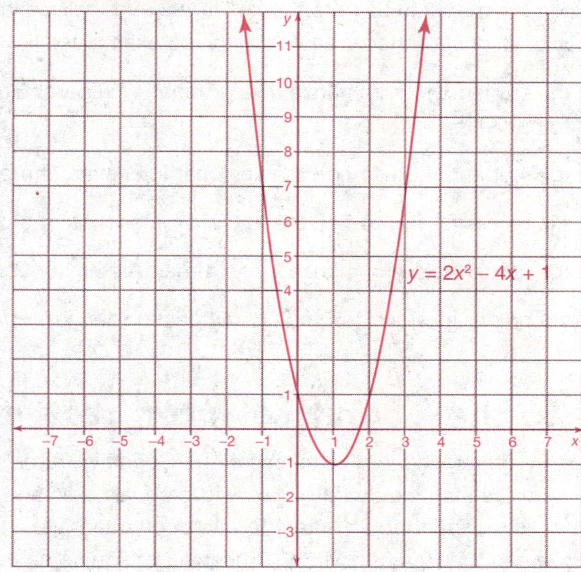

Domain: All real numbers; Range: All real numbers greater than or equal to –1. The graph opens upward; therefore, the function has a minimum value.

10. How can you tell from the equation for the parabola whether the parabola opens upward or downward? Use complete sentences in your answer.

Sample Answer: If *a* is positive, the parabola opens upward. If *a* is negative, the parabola opens downward.

11. Find the vertex of the graph of the quadratic function. Then tell whether the *y*-coordinate of the vertex is a minimum or a maximum.

$y = 2x^2 + 8x - 3$

$x = -\frac{8}{2(2)} = -\frac{8}{4} = -2$

$y = 2(-2)^2 + 8(-2) - 3$
$= -11$

(–2, –11); Minimum

$y = -x^2 + 10x + 3$

$x = -\frac{10}{2(-1)} = -(-5) = 5$

$y = -(5)^2 + 10(5) + 3$
$= 28$

(5, 28); Maximum

Wrap Up

Close

- Review all key terms and their definitions. Include the words and terms *parabola, line of symmetry, vertical line, vertex, maximum,* and *minimum.*
- You may also want to review any other vocabulary terms that were discussed during the lesson, which may include *linear function, quadratic function, exponent, mirror image, origin, domain,* and *range.*
- Remind the students to write the key terms and their definitions in the notes section of their notebooks. You may also want the students to include examples.
- Ask the students to summarize the definition of a parabola and the important characteristics of parabolas.
- Ask the students how to find the key characteristics of a parabola from a graph.
- Ask the students how to find the key characteristics of a parabola algebraically.
- Ask the students to summarize all patterns that they noticed or discussed during this lesson.
- Make a poster-size chart of the characteristics and patterns that they noticed.

8

Ties to the Cognitive Tutor Software

Many of the quadratic word problems in the Cognitive Tutor software involve area. Students are generally familiar with the idea that length times width equals area and that area is measured in square units, so using area is a comfortable context to introduce quadratics. However, they may not have associated the word *square* in square units with the meaning of square as the power of two. Making this connection can help students remember the relationship between area and quadratic functions.

Follow Up

Assignment

Use the Assignment for Lesson 8.2 in the Student Assignments book. See the Teacher's Resources and Assessments book for answers.

Assessment

See the Assessments provided in the Teacher's Resources and Assessments book for Chapter 8.

Open-Ended Writing Task

Ask the students to write a paragraph summarizing the following questions. What is a parabola? What items or motions in the real world can be modeled with parabolas?

Reflections

Insert your reflections on the lesson as it played out in class today.

What went well?

What did not go as well as you would have liked?

How would you like to change the lesson in order to improve the things that did not go well and capitalize on the things that did go well?

Notes

8

8.3

Dog Run

Comparing Linear and Quadratic Functions

Learning By Doing Lesson Map

Get Ready

8

Objectives

In this lesson, you will:

- Use linear and quadratic functions to model a situation.
- Determine the effect on the area of a rectangle when its length or width doubles.

Key Terms

- linear function
- quadratic function

NCTM Content Standards

Grades 9–12 Expectations

Number and Operations Standards

- Judge the effects of such operations as multiplication, division, and computing powers and roots on the magnitude of quantities.

Algebra Standards

- Generalize patterns using explicitly defined and recursively defined functions.
- Analyze functions of one variable by investigating rates of change, intercepts, zeros, asymptotes, and local and global behavior.
- Understand and compare the properties of classes of functions, including exponential, polynomial, rational, logarithmic, and periodic functions.
- Interpret representations of functions of two variables.
- Use symbolic algebra to represent and explain mathematical relationships.
- Judge the meaning, utility, and reasonableness of the results of symbol manipulations, including those carried out by technology.
- Identify essential quantitative relationships in a situation and determine the class or classes of functions that might model the relationships.
- Draw reasonable conclusions about a situation being modeled.
- Approximate and interpret rates of change from graphical and numerical data.

Geometry Standards

- Use geometric models to gain insights into, and answer questions in, other areas of mathematics.

Measurement Standards

- Understand and use formulas for the area, surface area, and volume of geometric figures, including cones, spheres, and cylinders.
- Apply informal concepts of successive approximation, upper and lower bounds, and limit in measurement situations.

Lesson Overview

Within the context of this lesson, students will be asked to:

- Create tables of values and graph linear and quadratic functions to model situations.
- Identify the effects on the area of a rectangle when the length or width is doubled.

Essential Questions

The following key questions are addressed in this lesson:

1. What is a dog run?
2. What shape is the dog run for this situation?
3. What is a linear function? What does the graph of a linear function look like?
4. What is a quadratic function? What does the graph of a quadratic function look like?
5. What is meant by the "maximum possible area"?

Show The Way

Warm Up

Place the following questions or an applicable subset of these questions on the board before students enter class. Students should begin working as soon as they are seated.

Calculate the line of symmetry and the vertex for each quadratic function. You may refer to your notes from the previous lesson.

8

1. $y = 3x^2 + 12x - 2$

   Line of symmetry: $x = -2$
   Vertex: $(-2, -14)$

2. $y = -8x^2 + 4x + 1$

   Line of symmetry: $x = \frac{1}{4}$

   Vertex: $\left(\frac{1}{4}, \frac{3}{2}\right)$

3. $y = -3x^2 + 24x - 12$

   Line of symmetry: $x = 4$
   Vertex: $(4, 36)$

4. $y = 2x^2 + 5x + 3.125$

   Line of symmetry: $x = -1.25$

   Vertex: $(-1.25, 0)$

5. $y = x^2 - 8x - 3$

   Line of symmetry: $x = 4$
   Vertex: $(4, -19)$

6. $y = 5x^2 + 1$

   Line of symmetry: $x = 0$
   Vertex: $(0, 1)$

Motivator

Begin the lesson with the motivator to get students thinking about the topic of the upcoming problem. This lesson is about creating an area for a dog to run. The motivating questions are about dog runs.

Ask the students the following questions to get them interested in the lesson.

- What is a dog run?
- What are possible ways to construct a dog run?
- How could you decide what type of dog run to make if you were to make one yourself?

Explore Together

Problem 1

Students will algebraically maximize the area for a rectangular dog run.

Grouping

Ask for a student volunteer to read the Scenario and Problem 1 aloud. Have a student restate the problem. Pose the Guiding Questions below to verify student understanding. Have students work together in small groups to complete parts (A) through (E) of Problem 1.

Guiding Questions

- What is a dog run?
- How many sides of the rectangular dog run will need to be fenced?
- What will be used for the fourth side of the dog run?
- How much fencing do the dog owners have to use?
- If we know the width of the dog run, and the total amount of fencing available, how can we find the length of the dog run?
- How can you find the area of the rectangle?
- Will we have to do anything differently to find the area of the rectangular dog run because one of the sides is not made of fencing?
- Which do you think the dog owners would prefer to make the largest possible: the width, the length, or the area of the dog run?

Note Finding the largest possible area is also called maximizing the area.

Common Student Errors

Students may incorrectly calculate the amount of fencing based on 2 widths rather than just 1.

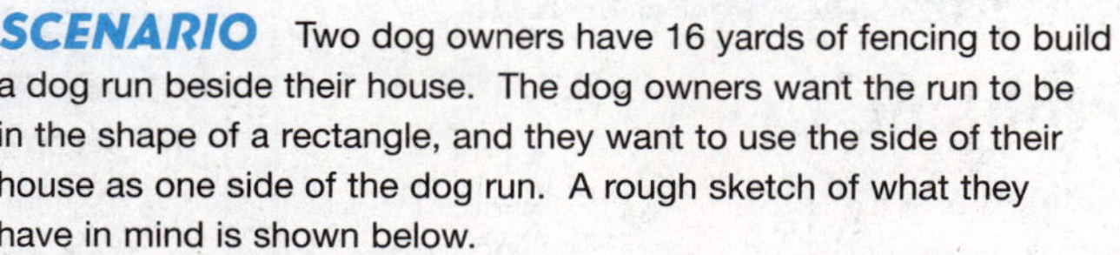

SCENARIO Two dog owners have 16 yards of fencing to build a dog run beside their house. The dog owners want the run to be in the shape of a rectangle, and they want to use the side of their house as one side of the dog run. A rough sketch of what they have in mind is shown below.

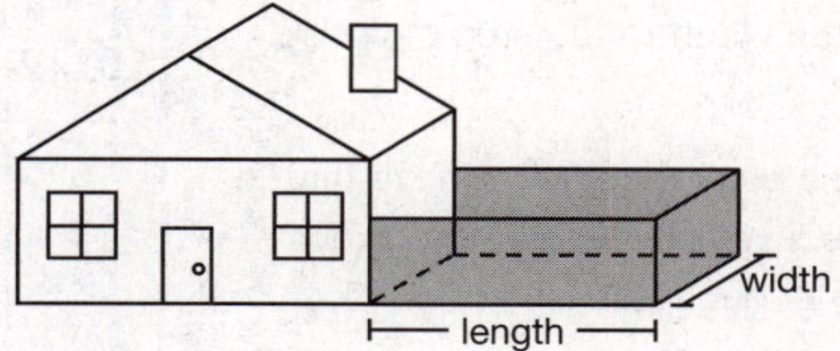

8

Problem 1 Deciding on the Dimensions

A. Suppose that the width of the dog run is 2 yards. Find the length of the dog run and the area of the dog run. Show all your work and use a complete sentence in your answer.

$16 - 2 = 14$; Length: $14 \div 2 = 7$; Area: $2(7) = 14$;

The length is 7 yards and the area is 14 square yards.

B. Suppose that the width of the dog run is 4 yards. Find the length of the dog run and the area of the dog run. Show all your work and use a complete sentence in your answer.

$16 - 4 = 12$; Length: $12 \div 2 = 6$; Area: $4(6) = 24$;

The length is 6 yards and the area is 24 square yards.

C. Suppose that the width of the dog run is 7 yards. Find the length of the dog run and the area of the dog run. Show all your work and use a complete sentence in your answer.

$16 - 7 = 9$; Length: $9 \div 2 = 4.5$; Area: $7(4.5) = 31.5$;

The length is 4.5 yards and the area is 31.5 square yards.

D. Suppose that the width of the dog run is 8 yards. Find the length of the dog run and the area of the dog run. Show all your work and use a complete sentence in your answer.

$16 - 8 = 8$; Length: $8 \div 2 = 4$; Area: $8(4) = 32$;

The length is 4 yards and the area is 32 square yards.

Explore Together

Problem 1

Students will investigate the relationships between the width and length of the dog run and of the width to the area of the dog run.

Call the class back together to have the students discuss and present their work for parts (A) through (E).

Investigate Problem 1

Grouping

Ask for a student volunteer to read Question 1 aloud. Have a student restate the problem. Complete Question 1 together as a whole class. Have students work together in small groups to complete Questions 2 through 7.

Common Student Errors

Initially, as the width increases, the area also increases, so students may incorrectly assume that this will always be the case.

Students are sometimes confused when they see the area increasing and then decreasing as the width increases. Ask students to predict what the graph of the area function will look like. If the students correctly realize that the function will be a parabola, ask them what the vertex represents in this situation. Students who are visual learners may better understand the situation if they imagine a parabolic graph.

Problem 1 Deciding on the Dimensions

E. Complete the table below to show different widths, lengths, and areas that can occur with sixteen yards of fencing. Copy the Width and Area columns of the table into the correct columns in the margin of page 380.

Width	Length	Area
yards	yards	square yards
0	8	0
2	7	14
4	6	24
6	5	30
8	4	32
10	3	30
12	2	24
14	1	14
16	0	0

Investigate Problem 1

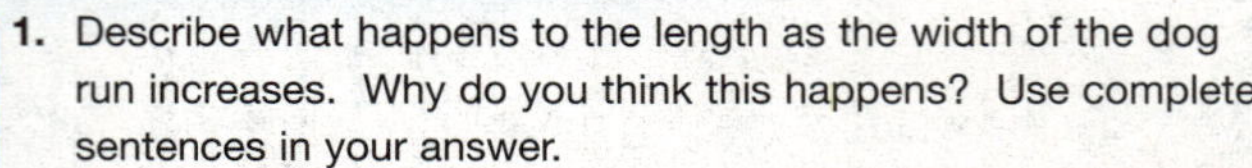

1. Describe what happens to the length as the width of the dog run increases. Why do you think this happens? Use complete sentences in your answer.
   Sample Answer: The length decreases as the width increases because the total amount of fencing available is constant.
2. Describe what happens to the area as the width of the dog run increases. Use a complete sentence in your answer.
   Sample Answer: The area increases and then decreases as the width increases.
3. Describe what happens to the length and area as the width of the dog run decreases. Use complete sentences in your answer.
   Sample Answer: The length increases as the width decreases. The area increases and then decreases as the width decreases.
4. Describe what happens to the width and area as the length of the dog run *increases*. Describe what happens to the width and area as the length of the dog run *decreases*. Use complete sentences in your answer.
   Sample Answer: The width decreases as the length increases. The area increases and then decreases as the length increases. The width increases as the length decreases. The area increases and then decreases as the length decreases.

Explore Together

Investigate Problem 1

Students will create a graph of the linear relationship between the width and the length of the dog run.

Common Student Errors

Some students have difficulty identifying which quantity to graph on the horizontal axis and which quantity to graph on the vertical axis when neither of the variables is clearly the independent variable.
In Question 6, the students are asked to graph the length as a function of the width. If students have difficulty, pose the Guiding Questions below to guide the students through the process and to deepen their understanding of functions.

Guiding Questions

- The cost for bowling is a function of the number of games played. Which variable is the independent variable? Which variable is the dependent variable?
- The distance driven is a function of the amount of time driven. Which variable is the independent variable? Which variable is the dependent variable?
- When we graph a function, which axis do we use to graph the independent variable? Which axis do we use to graph the dependent variable?

Call the class back together to have the students discuss and present their work for Questions 2 through 7.

Key Formative Assessments

- Why does a change in the length of 1 yard have the same effect as a change in the width of 2 yards?
- What is the minimum possible width? Would it make sense for a dog owner to use that width?
- What is the maximum possible width? Would it make sense for a dog owner to use that width?
- What is the domain for the width in this situation?

Investigate Problem 1

5. Compare how the area changes as the width changes to how the area changes as the length changes. Use complete sentences to explain your reasoning.

   Sample Answer: The changes are the same in that the area increases and then decreases when either measurement increases or decreases. Given a particular length and width, an increase of the width by two yards has the same effect on the area as an increase in the corresponding length by one yard.

6. Create a graph that shows the length as a function of the width on the grid below. First, choose your bounds and intervals. Be sure to label your graph clearly.

Variable quantity	Lower bound	Upper bound	Interval
Width	0	30	2
Length	0	15	1

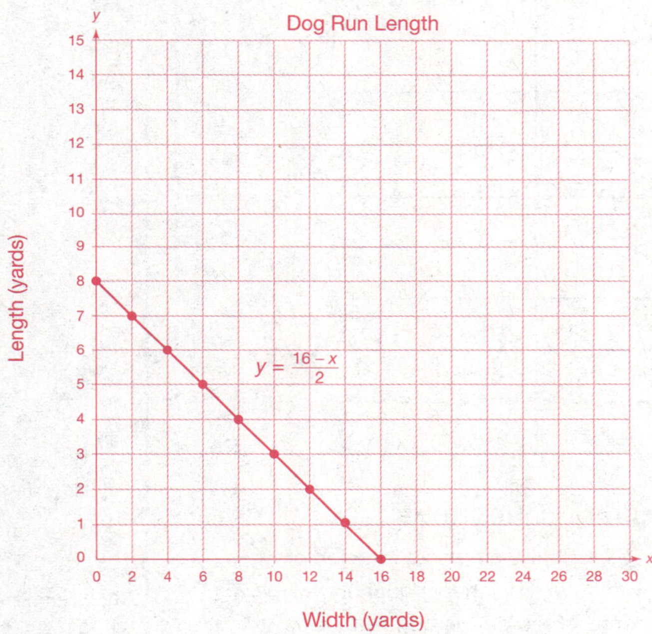

7. What kind of function is represented by the graph in Question 6? How do you know? Use a complete sentence in your answer.

   Sample Answer: The function is linear because the graph is a straight line.

- What is the minimum possible length? What is the maximum possible length?
- What is the range of possible values for the length?
- Why are the domain for the widths and the range for the lengths not equal?

8

Explore Together

Investigate Problem 1

Students will create a graph of the area as a function of the width of the dog run.

Grouping

Ask for a student volunteer to read Question 8 aloud. Have a student restate the problem. Pose the Guiding Questions below to verify student understanding. Have students work in small groups to complete Questions 8 through 15.

Width	Area
yards	square yards
0	0
2	14
4	24
6	30
8	32
10	30
12	24
14	14
16	0

Guiding Questions

- To graph the area as a function of the width, which variable quantity will be represented on the horizontal axis? Which variable quantity will be represented on the vertical axis?
- What is the domain that you found for the width?
- Based on the table of values in part (E) of Problem 1, what is the range of values for the area?
- What do you predict will be the shape of the graph of the area as a function of the width?
- What information did you use to make that prediction?

Investigate Problem 1

8. Create a graph that shows the area as a function of the width on the grid below. First, choose your bounds and intervals. Be sure to label your graph clearly.

Variable quantity	Lower bound	Upper bound	Interval
Width	0	30	2
Area	0	45	3

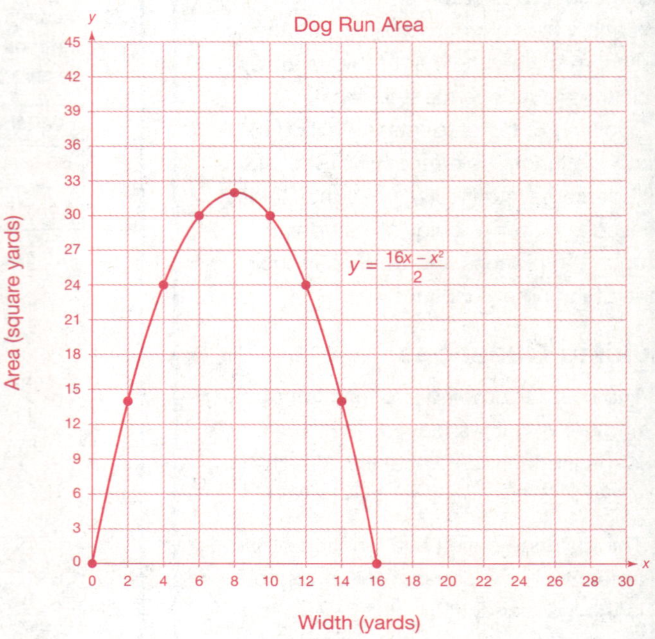

9. What kind of function is represented by the graph in Question 8? How do you know? Use a complete sentence in your answer.

Sample Answer: The function is quadratic because the graph is a parabola.

10. Determine the x-intercepts of each graph. What is the meaning of each x-intercept in the problem situation? Use complete sentences in your answer.

Sample Answer: The x-intercept of the graph of the linear function is 16. It indicates the width when the length is 0, which is not appropriate for the dog run. The x-intercepts of the graph of the quadratic function are 0 and 16. The x-intercepts indicate the width when the area is 0, which is not appropriate for the dog run.

- Do you think that the graph of the parabola will have a minimum or maximum value at the vertex? Why?
- Based on your table of values in part (E) of Problem 1, does the data support your prediction?
- What is an x-intercept for a graph? How can you calculate an x-intercept algebraically?
- What is a y-intercept for a graph? How can you calculate a y-intercept algebraically?

Explore Together

Investigate Problem 1

Students will answer questions based on their graph and interpret the meaning of the graph in the problem situation.

Common Student Errors

Students may not remember the meaning of rate of change. Pose the Guiding Questions below to guide students through Question 14 if needed.

Guiding Questions

- What is a rate of change?
- What is the rate of change of a linear function called?
- Is the rate of change constant for a linear function?
- Is the rate of change constant for a quadratic function?
- For widths between 0 yards and 8 yards, is the area increasing or decreasing as the width increases?
- Is the area increasing faster or slower as the width changes from 2 to 4 yards or from 6 to 8 yards? How do you know?
- For the widths between 8 yards and 16 yards, is the area increasing or decreasing as the width increases?
- Is the area decreasing faster or slower as the width changes from 8 to 10 yards or from 12 to 14 yards? How do you know?

Notes The field of calculus concentrates significantly on the rate of change of curves and functions. Developing an intuitive sense of varying rates of change in algebra will help students as they progress to more advanced studies of mathematics.

Call the class back together to have the students discuss and present their work for Questions 8 through 15.

Investigate Problem 1

11. How many *x*-intercepts can the graph of a linear function have? Use complete sentences to explain your reasoning.

Sample Answer: A line can have no *x*-intercept or one *x*-intercept because a line is constantly increasing, constantly decreasing, or is constantly horizontal.

12. How many *x*-intercepts can the graph of a quadratic function have? Use complete sentences to explain your reasoning.

Sample Answer: A parabola can have no *x*-intercept or one or two *x*-intercepts. If the parabola opens upward and its vertex is above the *x*-axis, it will never cross the *x*-axis. If the parabola has a vertex on the *x*-axis, it crosses the *x*-axis once. If the parabola opens upward and its vertex is below the *x*-axis, it crosses the *x*-axis twice.

13. Determine the *y*-intercepts of each graph. What is the meaning of each *y*-intercept in the problem situation? Use complete sentences in your answer.

Sample Answer: The *y*-intercept of the graph of the linear function is 8. It indicates the length when the width is 0, which is not appropriate for the dog run. The *y*-intercept of the graph of the quadratic function is 0. The *y*-intercept indicates the area when the width is 0, which is not appropriate for the dog run.

14. Describe the rates of change for each graph. Use complete sentences in your answer.

Sample Answer: The linear function has a constant rate of change and the quadratic function has a varying rate of change.

15. What is the greatest possible area? What are the length and width of the dog run with the greatest possible area? Use complete sentences to explain how you found your answer.

The largest possible area is 32 square yards. The width is 8 yards, and the length is 4 yards. Sample Answer: First, find the maximum of the quadratic function to find the greatest area. Then find the *x*-value that corresponds to the maximum to get the width. Then subtract this value from 16 and divide the result by 2 to get the length.

8

Problem 2 A Change in Plans

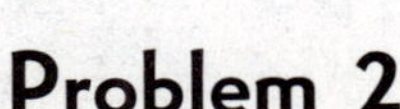

The owners read about a sale on the same exact fencing that they already have and decide to buy an additional 16 yards of fencing.

A. How many yards of fencing do they have now? Use a complete sentence in your answer.

They now have 32 yards of fencing.

Problem 2

Ask for a student volunteer to read part (A) of Problem 2 aloud. Have a student restate the problem.

Explore Together

Problem 2

Students will investigate another quadratic function with a similar scenario but a greater amount of fencing available.

Grouping

Have students work together in small groups to complete parts (A) through (D) of Problem 2. Pose the Guiding Questions below to verify student understanding.

Guiding Questions

- What information is given in the problem?
- How is this problem different from Problem 1?
- How is this problem the same as Problem 1?
- What effect will having twice the amount of fencing available have on the domain of the width?
- What effect will having twice the amount of fencing available have on the range of the width?
- With twice the amount of fencing available, will the maximum area also double?
- What information did you use to find your answer? How can you check to see whether your prediction is correct?

Common Student Errors

Many students will incorrectly assume that all values will simply double once the amount of available fencing is doubled. Pose the Guiding Questions above to have the students develop a hypothesis about how doubling the amount of available fencing will affect the width, length, and area of the dog run. Then allow the students to test their predictions in the questions for Problem 2.

Problem 2 A Change in Plans

B. Complete the table below to show different widths, lengths, and areas that can be made with the new amount of fencing.

Width	Length	Area
yards	yards	square yards
0	16	0
8	12	96
16	8	128
24	4	96
32	0	0

C. Create a graph that shows the length as a function of the width on the grid below. First, determine your bounds and intervals. Be sure to label your graph clearly.

Variable quantity	Lower bound	Upper bound	Interval
Width	0	45	3
Length	0	30	2

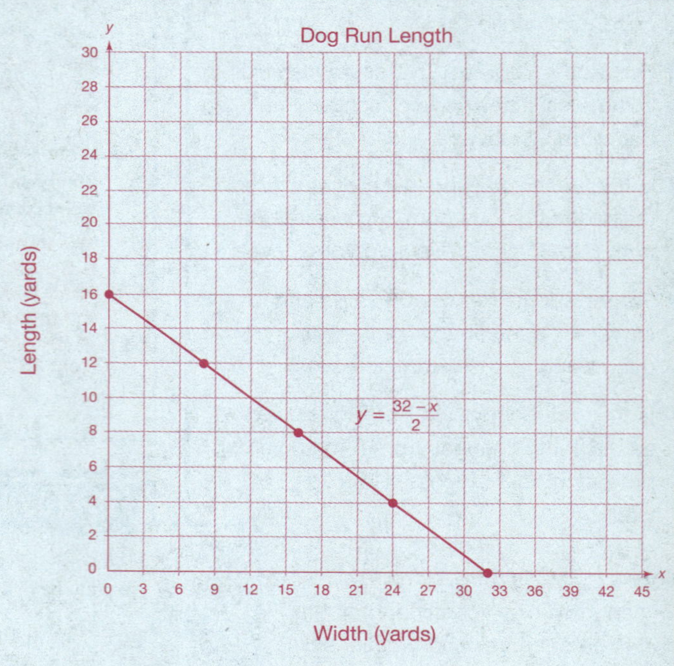

Explore Together

Investigate Problem 2

Students will investigate the effect on the range of a quadratic function when the values in the domain have been doubled.

Grouping

Call the class back together to have the students discuss and present their work for parts (A) through (D) of Problem 2.

Ask for a student volunteer to read Question 1 aloud. Have a student restate the problem. Pose the Guiding Questions below to verify student understanding. Have students work together in small groups to complete Questions 1 through 8.

Guiding Questions

- What are you being asked to find in Question 1?
- What information will you need to answer Question 1?
- How can you find the information you need to answer Question 1?
- Based on your work in parts (A) through (D) of Problem 2, how do you think the rates of change for the length as a function of the width will compare to that of Problem 1?
- Based on your work in parts (A) through (D) of Problem 2, how do you think the rates of change for the area as a function of the width will compare to that of Problem 1?

Common Student Errors

Some students may not answer Question 2 based on the context of the situation, but rather answer based on the mathematical definition of an x-intercept and of a y-intercept. Refocus any students who give answers such as "the x-intercept is the value for y when x is zero" by having them explain what an x-value of zero means in this situation. Then ask the students what the rectangle would look like if $x = 0$. Would a dog be able to get exercise in a dog run with a width of zero yards?

Problem 2 A Change in Plans

D. Create a graph that shows the area as a function of the width on the grid below. First, choose your bounds and intervals. Be sure to label your graph clearly.

Variable quantity	Lower bound	Upper bound	Interval
Width	0	45	3
Area	0	150	10

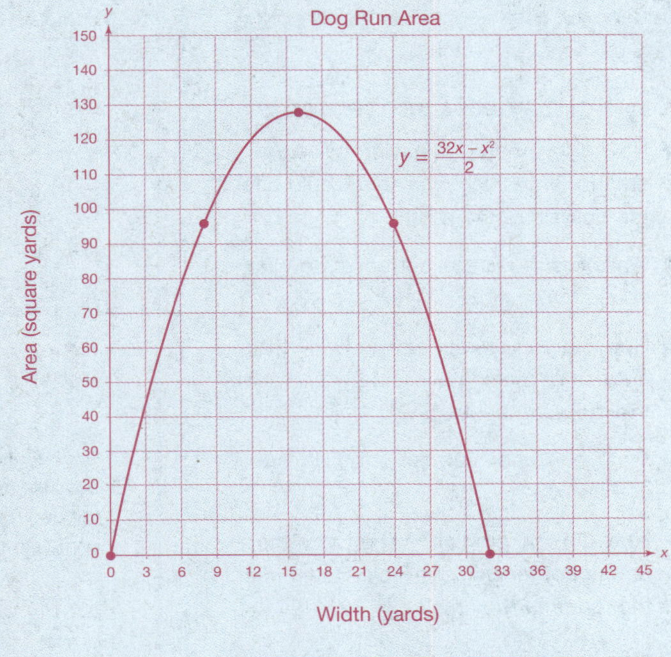

8

Investigate Problem 2

1. Describe the rates of change for each of the graphs. Use complete sentences in your answer.
   Sample Answer: The linear function has a constant rate of change, and the quadratic function has a changing rate of change.
2. What are the x- and y-intercepts of the graph of the linear function? What is their meaning in this problem situation? Use complete sentences in your answer.
   The x-intercept is 32, and the y-intercept is 16. These values indicate lengths and widths that do not make sense in the problem situation.

Explore Together

Investigate Problem 2

Students will continue to investigate the effect on the range of a quadratic function when the values in the domain have been doubled.

Grouping

Call the class back together to have the students discuss and present their work for Questions 1 through 8.

Key Formative Assessments

- How did you predict having twice the amount of fencing available would affect the domain of the width?
- Was your prediction correct? How do you know?
- How did you predict having twice the amount of fencing available would affect the range of the length?
- Was your prediction correct? How do you know?
- How did you predict having twice the amount of fencing available would affect the range of the area?
- Was your prediction correct? How do you know?
- How can you explain these changes algebraically?

Note You may want to use the term *maximize the area* in this situation to talk about finding the greatest possible area.

Investigate Problem 2

3. What are the *x*- and *y*-intercepts of the graph of the quadratic function? What is their meaning in the problem situation? Use complete sentences in your answer.

   The *x*-intercepts are 0 and 32, and the *y*-intercept is 0. These values indicate widths and areas that do not make sense in the problem situation.

4. What is the greatest possible area? What are the length and width of the dog run with the greatest possible area? Use a complete sentence to explain how you found your answer.

   The greatest possible area is 128 square yards. The width is 16 yards, and the length is 8 yards. Sample Answer: First, find the maximum of the quadratic function to find the greatest area. Then find the *x*-value that corresponds to the maximum to get the width. Then subtract this value from 32 and divide the result by 2 to get the length.

5. How does the amount of fencing the owners have now compare to the amount of fencing the owners had in Problem 1? Use a complete sentence in your answer.

   Sample Answer: The amount of fencing they have now is twice the amount of fencing they had in Problem 1.

6. How do the length and width of the dog run with the greatest possible area in this problem compare to the length and width of the dog run with the greatest possible area in Problem 1? Use a complete sentence in your answer.

   Sample Answer: Both the length and width in this problem are two times the length and width in Problem 1.

7. How do the greatest possible areas in this problem and Problem 1 compare?

   Sample Answer: This area, 128 square yards, is four times the area in Problem 1, 32 square yards.

8. Use complete sentences to explain why the difference in the areas is more than the differences in the lengths and widths.

   Sample Answer: The area is a square measure so it grows faster than the length and width, which are linear measures.

   Because the width is multiplied by a factor of 2 and the length is also multiplied by a factor of 2, the area is multiplied by a factor of 2 times 2, or 4 according to the equation for area.

Wrap Up

Close

- Review all key terms and their definitions. Include the terms *linear function* and *quadratic function.*
- You may also want to review any other vocabulary terms that were discussed during the lesson, which may include *vertex, line of symmetry, dimensions, area, maximize the area, x-intercept,* and *y-intercept.*
- Remind the students to write the key terms and their definitions in the notes section of their notebooks. You may also want the students to include examples.
- Ask the students to explain the mathematics that they learned and that they reviewed in the lesson.
- Ask the students whether their prediction for the effect of doubling the width was correct.
- Ask the students whether their prediction for the effect of doubling the area was correct.
- Ask the students how they think tripling the amount of available fencing would affect the domain of the width, the range of the length, and the range of the area. Have the students test their conjectures as a class.

8

Ties to the Cognitive Tutor Software

When students use the Cognitive Tutor software for quadratic word problems, one of the difficult concepts for them to grasp is that the same dependent (y) value may be associated with two different independent (x) values. They experience this differently more in the table than in the graph. Graphically, students should be encouraged to use symmetry to understand the relationship between two points with the same y-value.

Follow Up

Assignment

Use the Assignment for Lesson 8.3 in the Student Assignments book. See the Teacher's Resources and Assessments book for answers.

Assessment

See the Assessments provided in the Teacher's Resources and Assessments book for Chapter 8.

Open-Ended Writing Task

Ask the students to write a paragraph discussing how they predict having only 8 yards of fencing available would affect the maximum possible area for the dog run. Then have them test their prediction by creating a table of values and a graph of the situation. The students should identify the maximum possible area of a dog run using only 8 yards of fencing. Finally, students should compare the actual value to their predicted value.

Reflections

Insert your reflections on the lesson as it played out in class today.

What went well?

8

What did not go as well as you would have liked?

How would you like to change the lesson in order to improve the things that did not go well and capitalize on the things that did go well?

Notes

8.4 Guitar Strings and Other Things

Square Roots and Radicals

Learning By Doing Lesson Map

Get Ready

Objectives

In this lesson, you will:

- Evaluate the square root of a perfect square.
- Approximate a square root.

Key Terms

- square root
- positive square root
- negative square root
- principal square root
- radical symbol
- radicand
- perfect square

NCTM Content Standards

Grades 9–12 Expectations

Number and Operation Standards

- Use number-theory arguments to justify relationships involving whole numbers.
- Develop fluency in operations with real numbers, vectors, and matrices, using mental computation or paper-and-pencil calculations for simple cases and technology for more complicated cases.

Algebra Standards

- Generalize patterns using explicitly defined and recursively defined functions.
- Use symbolic algebra to represent and explain mathematical relationships.
- Judge the meaning, utility, and reasonableness of the results of symbol manipulations, including those carried out by technology.
- Draw reasonable conclusions about a situation being modeled.

Lesson Overview

Within the context of this lesson, students will be asked to:

- Evaluate the square root of a perfect square.
- Approximate the square root of values that are not perfect squares.

Essential Questions

The following key questions are addressed in this lesson:

1. What is a square root?
2. What is a perfect square?
3. How can you approximate a square root?
4. How can you identify a radicand?
5. How can you evaluate a square root?

Show The Way

8

Warm Up

Place the following questions or an applicable subset of these questions on the board before students enter class. Students should begin working as soon as they are seated.

Evaluate each expression.

1.	3^2	9	**2.**	$(-3)^2$	9	**3.**	-3^2	−9
4.	$(-10)^2$	100	**5.**	-10^2	−100	**6.**	10^2	100
7.	$(-5)^3$	−125	**8.**	-5^3	−125	**9.**	5^3	125
10.	$(2)^5$	32	**11.**	$(-2)^5$	−32	**12.**	$-(2)^5$	−32

Motivator

Begin the lesson with the motivator to get students thinking about the topic of the upcoming problem. This lesson is about sound waves from a guitar string. The motivating questions are about sound waves from musical instruments.

Ask the students the following questions to get them interested in the lesson.

- What is a sound wave?
- How can a sound wave be created by a vibration of a guitar string?
- What other musical instruments make sound by the vibration of strings?
- What does it mean to tune a guitar?

Explore Together

Problem 1

Students will investigate square roots of numbers by first calculating the square of numbers then reversing the process.

Grouping

Ask for a student volunteer to read the Scenario and Problem 1 aloud. Have a student restate the problem. Pose the Guiding Questions below to verify student understanding. Have students work together in small groups to complete parts (A) through (D) of Problem 1.

Take Note

A *cycle* of a wave is the motion of the string up and then down one time.

Guiding Questions

- What information is given in this problem?
- What does it mean to pluck a guitar string?
- Why might the variable v represent the wave speed?
- If you are given a wave speed, how can you find the tension?

Common Student Errors

Students may need to be reminded of the steps to take to find the square of a number. The Warm Up questions for this section gave the students additional practice with this skill.

Notes Students will solve the questions on this page fairly quickly, especially if they are using calculators. If you have not already done so, you may want to take this opportunity to show the students how to find the square of a number using their calculators as well as showing them how to find other powers of numbers using the ^ key or other appropriate keys on their calculators.

SCENARIO When you pluck a string on a guitar, the string vibrates and produces sound. When the string vibrates, the vibrations are repeating waves of movement up and down, as shown below.

If the guitar is not tuned properly, the correct notes will not be played, and the result may not sound musical. To tune a guitar properly requires a change in the *tension* of the strings. The tension can be thought of as the amount of stretch on the string between two fixed points. A string with the correct tension produces the correct wave speed, which in turn produces the correct sounds.

Problem 1 Good Vibrations

Consider a string that weighs approximately 0.0026 pound per inch and is 34 inches long. An equation that relates the wave speed v in cycles per second and tension t in pounds is $v^2 = t$.

A. Find the tension of the string if the wave speed is 9.5 cycles per second. Show all your work and use a complete sentence in your answer.

$t = 9.5^2 = 90.25$
The tension is 90.25 pounds.

B. Find the tension of the string if the wave speed is 8.5 cycles per second. Show all your work and use a complete sentence in your answer.

$t = 8.5^2 = 72.25$
The tension is 72.25 pounds.

C. Find the tension of the string if the wave speed is 7.6 cycles per second. Show all your work and use a complete sentence in your answer.

$t = 7.6^2 = 57.76$
The tension is 57.76 pounds.

D. What happens to the tension as the wave speed increases? Use a complete sentence in your answer.

The tension increases as the wave speed increases.

Call the class back together to have the students discuss and present their work for parts (A) through (D) of Problem 1.

Explore Together

Investigate Problem 1

Students will find the square root of perfect square numbers intuitively.

Grouping

Ask for a student volunteer to read Question 1 aloud. Pose the Guiding Questions below to verify student understanding. Have students work together in small groups to complete Questions 1 and 2.

Guiding Questions

- What information was given in Problem 1 parts (A) through (E)?
- What information is given in Question 1?
- How is the given information different?
- How can you find the answer to Question 1?

Take Note

Finding the square root of a number is the *inverse operation* of finding the square of a number.

Call the class back together to have the students discuss and present their work for Questions 1 and 2.

Take Note

The symbol, $\sqrt{\ }$, is called the **radical symbol.** The number underneath a radical symbol is called the **radicand.**

Just the Math

The terminology for square roots will be formally introduced in Question 3. Ask for a student volunteer to read Question 3 aloud. Have a student restate the problem. Pose the Guiding Questions at the right to verify student understanding.

Grouping

Complete Question 3 together as a whole class. Have students complete Questions 4 and 5 individually.

Investigate Problem 1

1. Write an equation that you can use to find the wave speed of a string when the tension on the string is 81 pounds.

$v^2 = 81$

What must the wave speed be? How do you know? Use a complete sentence in your answer.

The wave speed must be 9 cycles per second because the square of 9 is 81.

2. Write an equation that you can use to find the wave speed of a string when the tension on the string is 36 pounds.

$v^2 = 36$

What must the wave speed be? How do you know? Use a complete sentence in your answer.

The wave speed must be 6 cycles per second because the square of 6 is 36.

3. Just the Math: Square Root Your answers to Questions 1 and 2 are *square roots* of 81 and 36, respectively. Formally, you can say that a number b is a **square root** of a if $b^2 = a$.

So, 9 is a square root of 81 because $9^2 = 81$ and 6 is a square root of 36 because $6^2 = 36$.

Is there another number whose square is 81? If so, name the number.

Yes. –9.

Is there another number whose square is 36? If so, name the number.

Yes. –6.

Every positive number has two square roots: a **positive square root** and a **negative square root.** So, you can see that the square roots of 81 are 9 and –9, and the square roots of 36 are 6 and –6. The positive square root is called the **principal square root.** An expression such as $\sqrt{36}$ indicates that you should find the principal, or positive, square root of 36.

4. Complete each statement below.

$\sqrt{4} = \boxed{2}$ $\qquad -\sqrt{25} = \boxed{-5}$

$-\sqrt{100} = \boxed{-10}$ $\qquad \sqrt{49} = \boxed{7}$

Guiding Questions

- What is a square root? What is a positive square root? What is a negative square root? What is a principal square root?
- How can you algebraically write the statement that the negative square root of 9 is –3?

Explore Together

Investigate Problem 1

Students will investigate the square roots of numbers that are not perfect squares. Students will approximate the values.

Call the class back together to have the students discuss and present their work for Questions 4 and 5.

Grouping

Ask for a student volunteer to read Question 6 aloud. Have a student restate the problem. Pose the Guiding Questions below to verify student understanding. Have students work together in small groups to complete Questions 6 through 8.

Guiding Questions

- What is the principal square root of 25?
- What is the principal square root of 36?
- Is the following statement true or false? $25 < 30 < 36$
- If you agree that $25 < 30 < 36$ is true, what can you say about the relationship of the following principal roots: $\sqrt{25}$, $\sqrt{30}$, and $\sqrt{36}$?
- If $\sqrt{25} < \sqrt{30} < \sqrt{36}$, and $\sqrt{25}$ is 5 and $\sqrt{36}$ is 6, what can you conclude about $\sqrt{30}$?

Take Note

Remember that the symbol $\approx$ means "is approximately equal to."

Notes It is important for students to complete Questions 6 through 12 by hand without the help of a calculator. These problems are teaching the concept of a square root and so, the use of a calculator would inhibit student understanding at this point.

Another common symbol for an approximation is an equal sign with a dot above, $\doteq$.

Call the class back together to have the students discuss and present their work for Questions 6 through 8.

Investigate Problem 1

5. Each of the radicands in Question 4 is a **perfect square**. Can you explain why these numbers are called perfect squares? Use a complete sentence in your answer.

Sample Answer: Each number can be written as a square of a positive integer.

6. Write an equation that you can use to find the wave speed of a string when the tension on the string is 42 pounds.

$v^2 = 42$

What number represents the wave speed? Write your answer as a radical.

$\sqrt{42}$

Can you write this number as a positive integer? Why or why not? Use a complete sentence in your answer.

Sample Answer: No, because the radicand is not a perfect square.

7. Because 42 is not a perfect square, we have to approximate the value of $\sqrt{42}$. To do this, we will use perfect squares. Complete the statements below.

The perfect square that is closest to 42 and is less than 42 is 36.

The perfect square that is closest to 42 and is greater than 42 is 49.

So, 42 is between 36 and 49 and $\sqrt{42}$ is between $\sqrt{36} = 6$ and $\sqrt{49} = 7$.

Estimate $\sqrt{42}$ by choosing numbers between 6 and 7. Test each number by finding its square and seeing how close it is to 42.

$6.4^2 =$ 40.96 $\qquad 6.5^2 =$ 42.25

Which number is closer to 42?

42.25

So, $\sqrt{42} \approx$ 6.5.

8. What is the wave speed of a string if the tension is 42 pounds? Use a complete sentence in your answer.

The wave speed is approximately 6.5 cycles per second.

9. What happens to the wave speed as the tension increases? Use a complete sentence in your answer.

The wave speed increases as the tension increases.

8

Explore Together

Investigate Problem 1

Students will approximate principal square roots for non-perfect-square numbers.

Grouping

Have students work in small groups to complete Questions 9 through 12.

8

Common Student Errors

Students may not understand the value of approximating principal roots manually. They may feel that because the technology exists for calculator approximations that there is no purpose in manual approximation. Talk to them about their frustration, but reassure them that the purpose of the manual approximation is to create a deeper understanding of the concept of square roots. The concept of a square root is vital to their understanding of more advanced mathematics topics in the future.

Call the class back together to have the students discuss and present their work for Questions 9 through 12.

Key Formative Assessments

- How can you find the square of a number?
- What is a perfect square?
- List 5 examples of perfect squares.
- How can you find the square root of a perfect square?
- What signs can a square root have?
- What number has only one possible square root? What is the square root of zero?
- What is a positive square root? Why do we often refer to a positive square root as the principal square root?
- What is the radicand in the expression $\sqrt{100}$?
- If you were to approximate $\sqrt{51}$, what two whole numbers would you know it is between?

Investigate Problem 1

10. Approximate $\sqrt{13}$ to the nearest tenth. First complete each statement below. Show all your work.

$\boxed{9} < 13 < \boxed{16}$

$\sqrt{\boxed{9}} < \sqrt{13} < \sqrt{\boxed{16}}$

$\boxed{3} < \sqrt{13} < \boxed{4}$

Sample Answer: $3.5^2 = 12.25$

$3.6^2 = 12.96$

$3.7^2 = 13.69$

$\sqrt{13} \approx 3.6$

11. Approximate $\sqrt{30}$ to the nearest tenth. First complete each statement below. Show all your work.

$\boxed{25} < 30 < \boxed{36}$

$\sqrt{\boxed{25}} < \sqrt{30} < \sqrt{\boxed{36}}$

$\boxed{5} < \sqrt{30} < \boxed{6}$

Sample Answer: $5.4^2 = 29.16$

$5.5^2 = 30.25$

$\sqrt{30} \approx 5.5$

12. Approximate $\sqrt{75}$ to the nearest tenth. First complete each statement below. Show all your work.

$\boxed{64} < 75 < \boxed{81}$

$\sqrt{\boxed{64}} < \sqrt{75} < \sqrt{\boxed{81}}$

$\boxed{8} < \sqrt{75} < \boxed{9}$

Sample Answer: $8.5^2 = 72.25$

$8.6^2 = 73.96$

$8.7^2 = 75.69$

$\sqrt{75} \approx 8.7$

Wrap Up

Close

- Review all key terms and their definitions. Include the terms *square root, positive square root, negative square root, principal square root, radical symbol, radicand,* and *perfect square.*
- You may also want to review any other vocabulary terms that were discussed during the lesson, which may include *cycle, vibration, increase, decrease, square of a number, powers,* and *exponents.*
- Remind the students to write the key terms and their definitions in the notes section of their notebooks. You may also want the students to include examples.
- Ask the students to explain the meaning of the square of a number.
- Ask the students to explain the meaning of a square root.
- Have the students explain the relationship between the square of a number and the square root of a number.
- Ask the students to predict what the square of the square root of a number is. Have them test their prediction by finding the square of the square root of 9.
- Ask the students to explain why we very often are only interested in the principal square root of a value and not the negative square root of a number. When in life would we find situations in which only the positive square root would be a valid solution to a quadratic function?

Ties to the Cognitive Tutor Software

Students practice using perfect squares to approximate square roots in the Cognitive Tutor software. The general technique of memorizing familiar benchmark numbers in order to approximate an unknown value is an important skill that can help students make sense of contexts involving square roots.

Follow Up

Assignment

Use the Assignment for Lesson 8.4 in the Student Assignments book. See the Teacher's Resources and Assessments book for answers.

Assessment

See the Assessments provided in the Teacher's Resources and Assessments book for Chapter 8.

Open-Ended Writing Task

Ask the students to write a short paragraph responding to the following statement: "The square root of a number has to be positive or negative."

8

Reflections

Insert your reflections on the lesson as it played out in class today.

What went well?

What did not go as well as you would have liked?

How would you like to change the lesson in order to improve the things that did not go well and capitalize on the things that did go well?

Notes

8.5

Tent Designing Competition

Solving by Factoring and Extracting Square Roots

Learning By Doing Lesson Map

Get Ready

8

Objectives

In this lesson, you will:

- Solve a quadratic equation by factoring.
- Solve a quadratic equation by extracting square roots.

Key Terms

- parabola
- intercepts
- pi

NCTM Content Standards

Grades 9–12 Expectations

Number and Operations Standards

- Judge the effects of such operations as multiplication, division, and computing powers and roots on the magnitude of quantities.

Algebra Standards

- Generalize patterns using explicitly defined and recursively defined functions.
- Interpret representations of functions of two variables.
- Use symbolic algebra to represent and explain mathematical relationships.
- Judge the meaning, utility, and reasonableness of the results of symbol manipulations, including those carried out by technology.
- Draw reasonable conclusions about a situation being modeled.

Geometry Standards

- Draw and construct representations of two- and three-dimensional geometric objects using a variety of tools.

Measurement Standards

- Understand and use formulas for the area, surface area, and volume of geometric figures, including cones, spheres, and cylinders.

Lesson Overview

Within the context of this lesson, students will be asked to:

- Solve a quadratic function by factoring and setting the factors of the function equal to zero.
- Solve a quadratic function by extracting square roots.

Essential Questions

The following key questions are addressed in this lesson:

1. What is a tent design?
2. What is a parabolic shape for a tent?
3. How can you calculate the intercepts for a quadratic equation?
4. What is pi?
5. How can you solve a quadratic function?

Show The Way

Warm Up

Place the following questions or an applicable subset of these questions on the board before students enter class. Students should begin working as soon as they are seated.

Simplify each expression using the distributive property.

8

1. $2(x - 3)$

$2x - 6$

2. $-\frac{1}{2}(8x + 12)$

$-4x - 6$

3. $\frac{4}{5}\left(\frac{15}{8}x^2 - 25x - \frac{30}{4}\right)$

$\frac{3}{2}x^2 - 20x - 6$

4. $-0.25(12x^3 - x^2 + 2x + 7)$

$-3x^3 + 0.25x^2 - 0.5x - 1.75$

5. $\frac{1}{2}(5x^2 + 4x + 2)(5 - 3)$

$5x^2 + 4x + 2$

6. $-(x - 2)(3 + 1)$

$-4x + 8$

Motivator

Begin the lesson with the motivator to get students thinking about the topic of the upcoming problem. This lesson is about creating a design for a tent. The motivating questions are about a tent design.

Ask the students the following questions to get them interested in the lesson.

- What is a tent design?
- What shapes for tents have you seen?
- What do you think are important characteristics of a tent design that will make a consumer choose your tent?

Explore Together

Problem 1

Students will evaluate quadratic functions in factored form to find the height of a parabolic tent at various horizontal distances from the center of the tent base.

Grouping

Ask for a student volunteer to read the Scenario and Problem 1 aloud. Have a student restate the problem. Pose the Guiding Questions below to verify student understanding. Have students work together in small groups to complete parts (A) through (E) of Problem 1.

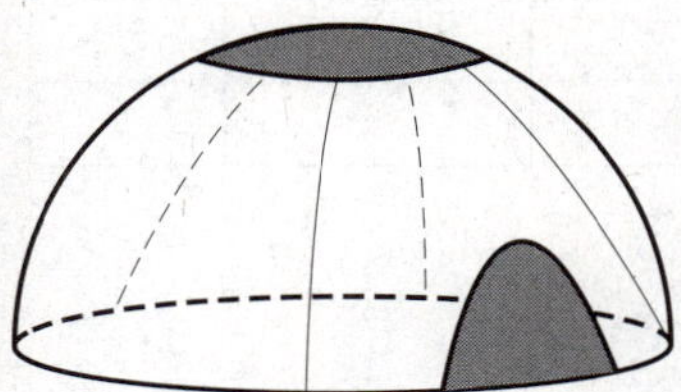

Take Note

The equation of the tent is a quadratic equation in *factored* form. You will learn more about factoring in Chapter 10.

Guiding Questions

- What information is given in this problem?
- What is the shape of the tent?
- Where do you think the tent will be tallest?
- Where do you think the tent will be shortest?
- What does the variable x represent in this situation?
- What does the variable y represent in this situation?
- What will an x-value of zero represent in this situation?
- How can you find the value of y for a given value of x?

SCENARIO You and your friend are working together in a competition to design a camping tent. Your design is based on a *parabola*. Your idea is to take a part of a parabola that opens downward and rotate it around to create the shape of the tent as shown at the left.

Problem 1 Planning the Tent Shape

You are testing out different parabolic shapes for the tent. The first shape can be modeled by the equation $y = -\frac{1}{2}(x - 4)(x + 4)$, where x is the number of feet to the right of the center and y is the height of the tent in feet.

A. What is the height of the tent two feet to the right of the center? Show all your work and use a complete sentence in your answer.

$$y = -\frac{1}{2}(2 - 4)(2 + 4)$$
$$= -\frac{1}{2}(-2)(6)$$
$$= 6$$

The height is six feet.

B. What is the height of the tent two feet to the left of the center? Show all your work and use a complete sentence in your answer.

$$y = -\frac{1}{2}(-2 - 4)(-2 + 4)$$
$$= -\frac{1}{2}(-6)(2)$$
$$= 6$$

The height is six feet.

C. What is the height of the tent four feet to the right of the center? Show all your work and use a complete sentence in your answer.

$$y = -\frac{1}{2}(4 - 4)(4 + 4)$$
$$= -\frac{1}{2}(0)(8)$$
$$= 0$$

The height is zero feet.

D. What is the height of the tent four feet to the left of the center? Show all your work and use a complete sentence in your answer.

$$y = -\frac{1}{2}(-4 - 4)(-4 + 4)$$
$$= -\frac{1}{2}(-8)(0)$$
$$= 0$$

The height is zero feet.

Common Student Errors

Students may forget to use the order of operations to evaluate the expressions. You may need to remind the students of the correct order of operations.

Explore Together

Problem 1

Students will continue to evaluate quadratic functions in factored form to find the height of a parabolic tent at various horizontal distances from the center of the tent base.

Call the class back together to have the students discuss and present their work for parts (A) through (E) of Problem 1.

8

Investigate Problem 1

Students will graph their equation to model the tent height as a function of the horizontal distance from the center.

Grouping

Ask for a student volunteer to read Question 1 aloud. Have a student restate the problem. Pose the Guiding Questions below to verify student understanding.

Have students work together in small groups to complete Questions 1 through 4.

Guiding Questions

- What two quantities are you being asked to graph in this situation?
- Which quantity is represented by the x-value? Will the x-value be graphed on the horizontal or vertical axis?
- What is the smallest possible distance from the center point on the base? What is the largest reasonable distance from the center point on the base?
- What is the smallest possible height of the tent at any point? What is the largest possible height?

Common Student Errors

Students may have difficulty understanding that the x-value represents the relative position compared to the center of the base of the tent. The negative values for x may pose the most difficulty for the students. If this is an issue, you can model the situation for the students pointing to a spot on the floor to be considered the center of the base of the tent. Show that if you move one foot to the students' right, that it would be an x-value of positive one. If you move one foot to the students' left, that it would be an x-value of negative one. The y-value would be the height of the tent at that point.

Problem 1 Planning the Tent Shape

E. What is the height of the tent at the center? What does this height represent? Show all your work and use a complete sentence in your answer.

$$y = -\frac{1}{2}(0 - 4)(0 + 4)$$

$$= -\frac{1}{2}(-4)(4)$$

$$= 8$$

The height, eight feet, is the maximum height.

Investigate Problem 1

1. Create a graph of the tent shape on the grid below. First, choose your bounds and intervals. Be sure to label your graph clearly.

Variable quantity	Lower bound	Upper bound	Interval
Horizontal position	−7	8	1
Height	0	15	1

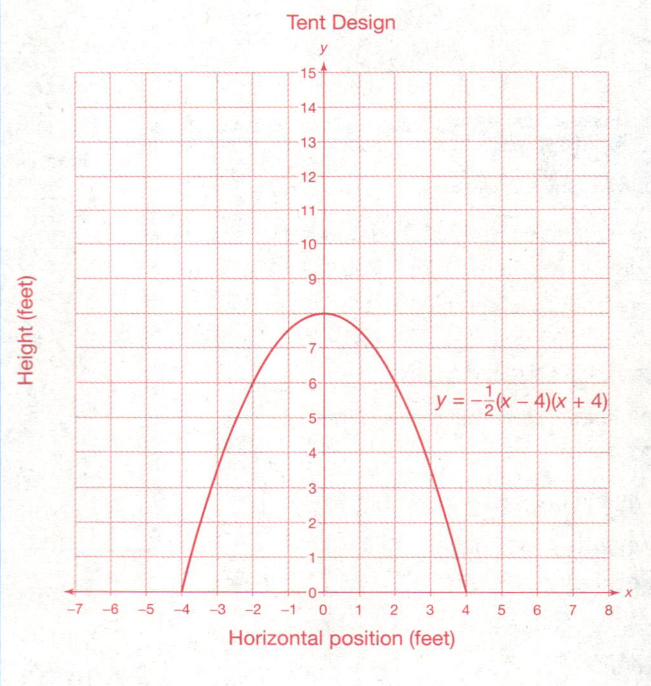

Explore Together

Investigate Problem 1

Students will identify the *intercepts* and explain their meaning in this situation.

Call the class back together to have the students discuss and present their work for Questions 1 through 4.

Key Formative Assessments

- What values did you find for the x-intercepts in this situation?
- Do the values make sense?
- Which x-intercept value did you use for the x-value in Question 4?
- Would you have found the same result if you used the other x-intercept? Why is that true?
- If a product of factors is zero, is it always true that at least one of the factors must be equal to zero? Is there any other possible way to multiply factors and get a product of zero?

Grouping

Ask for a student volunteer to read Question 5 aloud. Have a student restate the problem. Have the students work together in small groups to complete Question 5.

Common Student Errors

The concept that a product of zero is only possible if at least one of the factors is equal to zero is a vital concept to the study of advanced mathematics. This is well worth any time that must be spent in class to verify student understanding. Some students will not understand this concept without further explanation.

Investigate Problem 1

2. What is the y-intercept of the graph? What does it represent in the problem situation? Use complete sentences in your answer.

   Sample Answer: The y-intercept, 8, represents the maximum height of the tent.

3. What are the x-intercepts of the graph? What do they represent in the problem situation? Use complete sentences in your answer.

   Sample Answer: The x-intercepts, –4 and 4, represent the left and right edges of the tent.

   How wide is your tent? Use a complete sentence in your answer.

   The tent is eight feet wide.

4. Consider the equation below that you could use to *algebraically* find the x-intercepts of the graph. To do this, substitute 0 for y.

   $0 = -\frac{1}{2}(x - 4)(x + 4)$

   Substitute one of your x-intercepts into this equation. Then simplify.

   $0 = -\frac{1}{2}(\boxed{4} - 4)(\boxed{4} + 4)$

   $0 = -\frac{1}{2}(\boxed{0})(\boxed{8})$

   $0 = \boxed{0}$

   Why is the product equal to zero? Use a complete sentence in your answer.

   Sample Answer: Because one of the factors is zero.

5. You are also considering a different parabolic tent design that is modeled by the equation $y = -0.24(x - 5)(x + 5)$. Write an equation that you can use to find the x-intercepts of the parabola.

   $\mathbf{0 = -0.24(x - 5)(x + 5)}$

   What x-values do you think will be solutions of this equation? Use complete sentences to explain your reasoning.

   Sample Answer: The solutions are 5 and –5 because when $x = 5$, then $(x - 5) = 0$ and zero multiplied by any number is zero. Likewise, when $x = -5$, then $(x + 5) = 0$ and zero multiplied by any number is zero.

Explore Together

Investigate Problem 1

Students will solve a quadratic equation in factored form by substitution.

Call the class back together to have the students discuss and present their work for Question 5.

8

Key Formative Assessments

- What is an x-intercept?
- How many x-intercepts did you find for this situation?
- How did you find the x-intercepts?
- What do the x-intercepts represent in this situation?

Grouping

Ask for a student volunteer to read Question 6 aloud. Have a student restate the problem. Pose the Guiding Questions below to verify student understanding. Have students work in small groups to complete Questions 6 and 7.

Guiding Questions

- What is a y-intercept?
- Do you expect more than one y-intercept for this situation? Why or why not?
- How can you find the y-intercept?
- How can you choose which quantity to represent on the x-axis and which to represent on the y-axis?
- How can you choose the bounds to use for a graph in this situation?
- How do you think the graph of this situation will compare to the graph of the original situation?

Call the class back together to have the students discuss and present their work for Questions 6 and 7.

Investigate Problem 1

Check your answers by substituting them into the equation that you wrote. Show all your work.

$0 \stackrel{?}{=} -0.24(5 - 5)(5 + 5)$

$0 \stackrel{?}{=} -0.24(0)(10)$

$0 = 0$

$0 \stackrel{?}{=} -0.24(-5 - 5)(-5 + 5)$

$0 \stackrel{?}{=} -0.24(-10)(0)$

$0 = 0$

What do the x-intercepts mean in the problem situation? Use a complete sentence in your answer.

Sample Answer: The x-intercepts indicate the bottom left and right edges of the tent.

6. Algebraically find the y-intercept. What does it mean in the problem situation? Use a complete sentence to explain.

$y = -0.24(0 - 5)(0 + 5) = -0.24(-5)(5) = 6$

The y-intercept, 6, represents the maximum height of the tent.

7. Create a graph of the new tent shape on the grid below. First, choose your bounds and intervals.

Variable quantity	Lower bound	Upper bound	Interval
Horizontal position	−7	8	1
Height	0	15	1

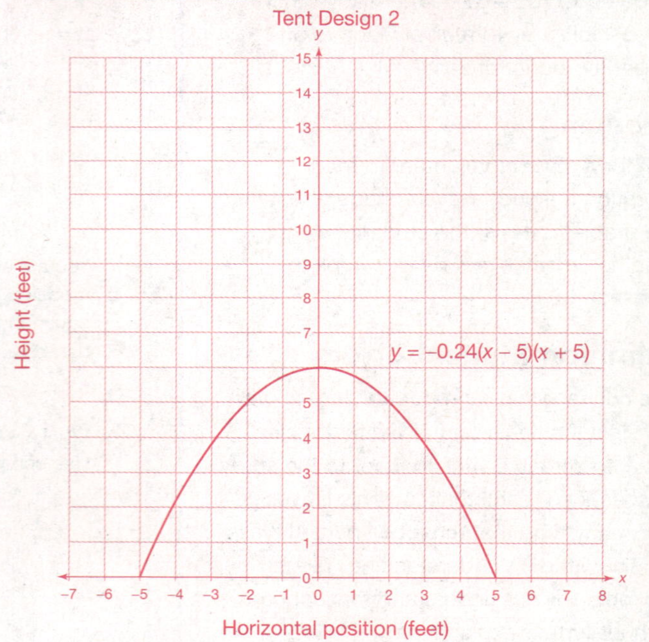

Explore Together

Problem 2

Take Note

Pi, written by using the Greek letter π, is a constant that is the ratio of a circle's circumference to its diameter.

Pi is an irrational number whose value is approximately 3.14.

Students will investigate the volume for their tent designs.

Grouping

Ask for a student volunteer to read Problem 2 aloud. Have a student restate the problem. Have the students complete parts (A) through (C) of Problem 2 individually.

Call the class back together to have the students discuss and explain their results for parts (A) through (C) of Problem 2.

Take Note

Volume is measured in *cubic* units. Because the dimensions of the tent are measured in feet, the volume of the tent is measured in cubic feet.

Grouping

Ask for a student volunteer to read part (D) aloud. Have a student restate the problem. Pose the Guiding Questions below to verify student understanding. Have students work in small groups to complete part (D) of Problem 2.

Guiding Questions

- What is pi? Is pi a constant or a variable?
- Why do you think this value is represented by a Greek symbol rather than by a letter of the American alphabet?
- How can you find the volume of each tent design?

Problem 2 Tent Volume

Another consideration in your tent design is the tent's volume, or the amount of space inside the tent. The volume of your tent is related to the maximum tent width and maximum tent height by the equation $V = \frac{\pi}{8}w^2h$ where V is the volume, w is the maximum tent width, and h is the maximum tent height.

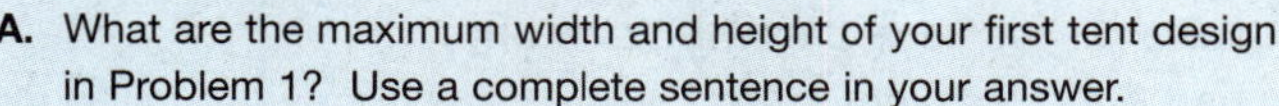

A. What are the maximum width and height of your first tent design in Problem 1? Use a complete sentence in your answer.

The width is eight feet, and the height is eight feet.

B. What are the maximum width and height of your second tent design in Problem 1? Use a complete sentence in your answer.

The width is ten feet, and the height is six feet.

C. Which tent do you expect to have more volume? Why? Use complete sentences in your answer.

Answers will vary.

D. Find the volume of each tent design in Problem 1. Use 3.14 for π. Show all your work and use complete sentences in your answer.

Tent 1: $\frac{\pi}{8}(8^2)(8) = 64\pi \approx 200.96$

Tent 2: $\frac{\pi}{8}(10^2)(6) = 75\pi \approx 235.5$

The volume of the first tent is approximately 201 cubic feet, and the volume of the second tent is approximately 236 cubic feet.

Which tent has the greater volume?

The second tent has the greater volume.

Which measurement, the width or the height, has more of an effect on the volume? Use complete sentences to explain your reasoning.

Sample Answer: The width has more of an effect because it is being squared in the equation for the volume.

8

Explore Together

Investigate Problem 2

Students will continue to investigate the volume of the tent designs in the form of quadratic functions.

Call the class back together to have the students discuss and present their work for part (D) of Problem 2.

8

Grouping

Ask for a student volunteer to read Question 1 aloud. Have a student restate the problem. Pose the Guiding Questions below to verify student understanding. Have students work together in small groups to complete Questions 1 and 2 of Problem 2.

Guiding Questions

- What information is given in this situation?
- How is the information given in Problem 2 different from the information given in Problem 1?
- What is the same in both Problem 1 and Problem 2?
- How can you solve to find the maximum height of the tent if you are given the volume and the maximum width?
- Will finding the height be easier or more difficult than finding the volume in Problem 1? Why?
- What will you have to do differently in solving for the height than when you solved for the volume of the tent?

Common Student Errors

These questions may be more difficult for some students. You may need to call the class back together to work through Question 1 as a whole class if needed. If so, have the students complete Question 2 in small groups or individually to check for understanding.

Call the class back together to have the students discuss and present their work for Questions 1 and 2.

Investigate Problem 2

1. You and your friend decide to create more tent designs based on the volume of the tent. Your first design based on the volume will have a volume of 392.5 cubic feet and a width of 10 feet. Write an equation that you can use to find the height of the tent.

$392.5 = \frac{\pi}{8}(10^2)h$

Find the height of the tent. Use 3.14 for π. Show all your work and use a complete sentence in your answer.

$392.5 = \frac{\pi}{8}(10^2)h$

$392.5 = 12.5\pi h$

$\frac{392.5}{12.5\pi} = h$

$10 \approx h$

The height is about 10 feet.

2. Your second design based on the volume will have a volume of 314 cubic feet and a height of 8 feet. Write an equation that you can use to find the width of the tent.

$314 = \frac{\pi}{8}w^2(8)$

Get the variable by itself on one side of the equation and simplify. Use 3.14 for π and show all your work.

$314 = \frac{\pi}{8}w^2(8)$

$\frac{314}{\pi} = w^2$

$100 \approx w^2$

Can you visually tell which two numbers are the solutions of this equation? If so, what are the numbers? Use a complete sentence in your answer.

Yes. The numbers are 10 and –10.

Does each number represent the tent width? Use a complete sentence to explain your reasoning.

No. Sample Answer: The solution –10 does not make sense in the problem situation because width cannot be negative.

What is the tent width? Use a complete sentence in your answer.

The tent width is 10 feet.

Explore Together

Investigate Problem 2

Students will solve a quadratic equation by extracting square roots.

Grouping

Ask for a student volunteer to read Question 3 aloud. Have a student restate the problem. Have students work in small groups to complete Questions 3 through 5.

Notes If class time is limited, you can assign Questions 3 through 5 as part of a homework assignment. When the students return for the next class session, have them work together in their small groups to discuss and consider their responses to these questions. Then, call the class back together to have the students discuss and present their work for these questions.

Key Formative Assessments

- What are factors of a product?
- What does it mean to represent an equation in factored form?
- How can you solve a quadratic function by representing the original equation in factored form?
- Why do we find the values where each factor of a product is equal to zero to find the solution to a quadratic equation? What does the zero represent in such an equation?
- How can you evaluate a quadratic function for y if the value for x is given?
- Are there any other ways that you can solve a quadratic function for the x-value if the y-value is given?
- Why can't we solve the equation $x^2 = -8$?
- What was the most important mathematical concept that you learned in this lesson?

Investigate Problem 2

3. Your third design based on the volume will have a volume of 400 cubic feet and a height of 6 feet. Write an equation that you can use to find the width of the tent.

$400 = \frac{\pi}{8}w^2(6)$

Isolate the variable on one side of the equation and simplify. Use 3.14 for π and show all your work. If necessary, round to the nearest whole number.

$400 = \frac{\pi}{8}w^2(6)$

$\frac{400}{0.75\pi} = w^2$

$170 \approx w^2$

Can you visually tell which two numbers are the solutions of this equation?

No.

A solution to your equation will be the number whose square is 170. So, one solution is the square root of 170. What is the other solution of this equation? Use a complete sentence in your answer.

The other solution is $-\sqrt{170}$.

Use a calculator to approximate the solutions of the equation to the nearest tenth.

$\sqrt{170} \approx 13.0; -\sqrt{170} \approx -13.0$

What is the width of your tent? Use a complete sentence in your answer.

The tent's width is about 13 feet.

4. What is the solution of the equation $x^2 = 0$? Use a complete sentence to explain your reasoning.

Sample Answer: The solution is zero because zero is the only number whose square is zero.

5. What is the solution of the equation $x^2 = -5$? Use a complete sentence to explain your reasoning.

Sample Answer: There is no solution because the square of a number is always positive.

8

Wrap Up

Close

- Review all key terms and their definitions. Include the words *parabola, intercepts*, and *pi.*
- You may also want to review any other vocabulary terms that were discussed during the lesson, which may include *relative position, product, horizontal axis,* and *vertical axis.*
- Remind the students to write the key terms and their definitions in the notes section of their notebooks. You may also want the students to include examples.
- Ask the students to explain how they solved for the missing values in this lesson.
- Ask the students to summarize how to solve quadratic functions using the methods discussed in this lesson.
- Ask the students if they think there might be any other possible ways to solve quadratic functions and to explain their ideas.
- Pull together all of the closing information suggested by the students. Give a brief summary of solving quadratic equations by using the methods presented in this lesson. Be sure to point out any misconceptions or over-generalizations that the students suggested during the lesson to make sure they don't have false understanding.

Ties to the Cognitive Tutor Software

Most students need to be told that objects accelerate as they fall and that the height of the object can be modeled with a quadratic function. When working with vertical motion problems in the Cognitive Tutor software, students are practicing using a known formula to model the situation, rather than constructing the formula from an understanding of the situation.

Follow Up

Assignment

Use the Assignment for Lesson 8.5 in the Student Assignments book. See the Teacher's Resources and Assessments book for answers.

Assessment

See the Assessments provided in the Teacher's Resources and Assessments book for Chapter 8.

Open-Ended Writing Task

Ask the students to write an exit slip explaining what mathematical ideas or concepts about which they are still confused after working through this lesson. For what do they want further explanation? Collect the exit slips as the students leave the room. Address their concerns at the beginning of the next class session.

Reflections

Insert your reflections on the lesson as it played out in class today.

What went well?

___

___

What did not go as well as you would have liked?

___

___

How would you like to change the lesson in order to improve the things that did not go well and capitalize on the things that did go well?

___

___

___

___

___

Notes

8

8

8.6 Kicking a Soccer Ball

Using the Quadratic Formula to Solve Quadratic Equations

Learning By Doing Lesson Map

Get Ready

8

Objectives

In this lesson, you will:

- Solve a quadratic equation by using the quadratic formula.
- Find the value of the discriminant.

Key Terms

- quadratic formula
- discriminant

NCTM Content Standards

Grades 9–12 Expectations

Number and Operations Standards

- Develop fluency in operations with real numbers, vectors, and matrices, using mental computation or paper-and-pencil calculations for simple cases and technology for more complicated cases.
- Judge the reasonableness of numerical computations and their results.

Algebra Standards

- Interpret representations of functions of two variables.
- Use symbolic algebra to represent and explain mathematical relationships.
- Use a variety of symbolic representations, including recursive and parametric equations, for functions and relations.
- Identify essential quantitative relationships in a situation and determine the class or classes of functions that might model the relationships.
- Draw reasonable conclusions about a situation being modeled.

Lesson Overview

Within the context of this lesson, students will be asked to:

- Solve a quadratic function by using the quadratic formula.
- Find the value of the discriminant.
- Identify the number of roots for a quadratic equation based on the sign of the discriminant.

Essential Questions

The following key questions are addressed in this lesson:

1. What is the quadratic formula?
2. What is a solution to a quadratic equation?
3. When can you use the quadratic formula?
4. How can you use the quadratic formula to find solutions for quadratic equations?
5. What is a discriminant and why is it important?

Show The Way

8

Warm Up

Place the following questions or an applicable subset of these questions on the board before students enter class. Students should begin working as soon as they are seated.

Write each equation in the form $y = ax^2 + bx + c$. State the value of a, b, and c. Finally, calculate the line of symmetry and the vertex.

1. $y + 2 = 2x^2 - 12x$

$y = 2x^2 - 12x - 2$; $a = 2$, $b = -12$; $c = -2$

Line of symmetry: $x = 3$; Vertex: $(3, -20)$

2. $8 = y - 4x^2 - 16x + 1$

$y = 4x^2 + 16x + 7$; $a = 4$, $b = 16$; $c = 7$

Line of symmetry: $x = -2$; Vertex: $(-2, -9)$

3. $2y + 10x = 6x^2 + 22x - 18$

$y = 3x^2 + 6x - 9$; $a = 3$, $b = 6$; $c = -9$

Line of symmetry: $x = -1$; Vertex: $(-1, -12)$

4. $3 = y - x^2$

$y = x^2 + 3$; $a = 1$, $b = 0$; $c = 3$

Line of symmetry: $x = 0$; Vertex: $(0, 3)$

Motivator

Begin the lesson with the motivator to get students thinking about the topic of the upcoming problem. This lesson is about the path of a kicked soccer ball. The motivating questions are about students' experiences in kicking a soccer ball.

Ask the students the following questions to get them interested in the lesson.

- What is the best approach to kicking a soccer ball?
- What path does a kicked soccer ball usually take?
- Is it better to be able to kick the ball up higher in the air or a longer distance?

Explore Together

Problem 1

Students will evaluate a quadratic function to find the height of a soccer ball at various horizontal distances away from the kick.

Grouping

Ask for a student volunteer to read the Scenario and Problem 1 aloud. Have a student restate the problem. Pose the Guiding Questions below to verify student understanding. Have students work together in small groups to complete parts (A) and (B) of Problem 1.

Guiding Questions

- What information is given in this problem?
- What does it mean to kick a soccer ball in a controlled environment? Why is that information included in the problem?
- What is the shape of the path traveled by the soccer ball?
- What is the meaning of the y-value in this situation?
- What is the meaning of the x-value in this situation?
- Which variable is the independent variable? Which is the dependent variable?
- Which is easier to find, a y-value if given an x-value or an x-value if given a y-value for this situation? Why?

Common Student Errors

Students may forget to use the order of operations to evaluate the expressions. You may need to remind the students of the correct order of operations.

Students may interpret the word approximate differently in part (B) giving different possible correct answers.

SCENARIO A friend of yours is working on a project that involves the path of a soccer ball. She tells you that she has collected data for several similar soccer kicks in a controlled environment (with no wind and minimum spin on the ball). She has modeled the general path of the ball using a quadratic function. You are interested in her model because you are studying quadratic functions in your math class.

Problem 1 The Path of a Soccer Ball

Your friend's model is $y = -0.01x^2 + 0.6x$ where x is the horizontal distance that the ball has traveled in meters and y is the vertical distance that the ball has traveled in meters.

A. Complete the table of values that shows the vertical and horizontal distances that the ball has traveled. Copy the values into the table on the next page.

Quantity Name	Horizontal distance	Vertical distance
Unit	meters	meters
Expression	x	$-0.01x^2 + 0.6x$
	0	0
	5	2.75
	10	5.00
	25	8.75
	50	5.00
	75	−11.25
	100	−40.00
	150	−135.00

B. Can you approximate from your table how far the ball traveled before it hit the ground? If so, describe the distance. Use a complete sentence in your answer.

Answers will vary.

8

Explore Together

Problem 1

Quantity Name	Horizontal distance	Vertical distance
Unit	meters	meters
Expression	x	$-0.01x^2 + 0.6x$
	0	0
	5	2.75
	10	5.00
	25	8.75
	50	5.00
	75	−11.25
	100	−40.00
	150	−135.00

Students will continue to evaluate quadratic functions to find the height of a soccer ball at various horizontal distances from the kick.

Call the class back together to have the students discuss and explain their work for parts (A) and (B) of Problem 1. When you have verified the answers to the table in part (A) as a whole class, have the students copy their answers into the table on this page if they have not already done so.

Students will graph their equation to model the height of the soccer ball as a function of the horizontal distance from the point that the ball was kicked.

Grouping

Have students work together in small groups to compete parts (C) through (F) of Problem 1.

Call the class back together to have the students discuss and present their work for parts (C) through (F) of Problem 1.

Problem 1 The Path of a Soccer Ball

C. Create a graph of the path of the ball on the grid below. First, choose your bounds and intervals. Be sure to label your graph clearly.

Variable quantity	Lower bound	Upper bound	Interval
Horizontal distance	0	75	5
Vertical distance	0	15	1

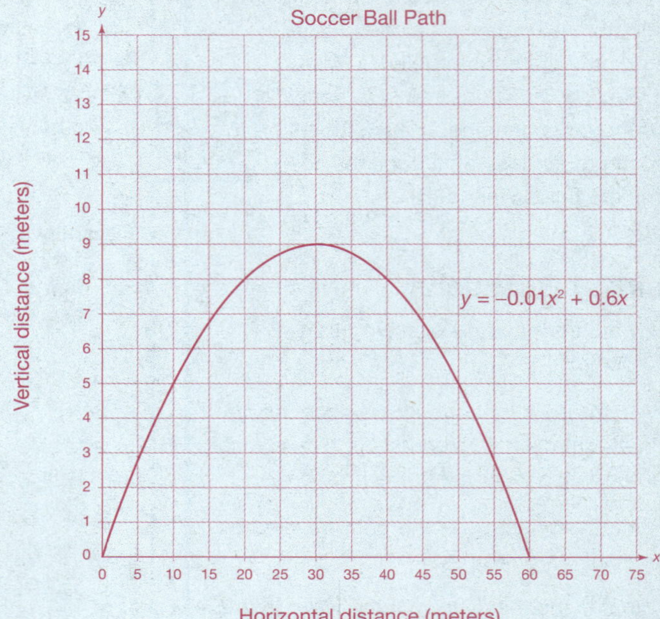

D. Use your graph to determine the maximum height of the ball. Use a complete sentence in your answer.

Sample Answer: The ball reaches a height of 9 meters.

E. What is the y-intercept of the graph? What does it represent in this problem situation? Use a complete sentence in your answer.

Sample Answer: The y-intercept, 0, represents the starting height of the ball in meters.

F. How far does the ball travel horizontally before it hits the ground? Use a complete sentence in your answer.

Sample Answer: The ball travels 60 meters horizontally.

Explore Together

Investigate Problem 1

Students will solve a quadratic equation for a non-whole number value.

Grouping

Ask for a student volunteer to read Question 1 aloud. Have a student restate the problem. Have the students complete Questions 1 and 2 individually. Then call the class together to discuss the questions.

Take Note

The symbol $\pm$ means "plus or minus" and is a compact way to write a solution. For instance, $x = m \pm n$ is the compact notation for $x = m + n$ and $x = m - n$.

Just the Math

The students will be introduced to the quadratic formula to solve quadratic equations in Question 3.

Grouping

Ask for a student volunteer to read Question 3 aloud. Have a student restate the problem. Pose the Guiding Questions below to verify student understanding. Complete Question 3 together as a whole class.

Guiding Questions

- What is the quadratic formula?
- What does x represent in the quadratic formula?
- What does a solution to an equation represent?
- How can an equation have more than one possible solution for a given value of y?
- What is the standard form of a quadratic equation?

Investigate Problem 1

1. In terms of the graph of the function, how can you interpret your answer to part (F)? Use a complete sentence in your answer.

   The distance 60 meters is one of the *x*-intercepts of the graph.

2. Write an equation that you can use to algebraically find the answer to the question in part (F).

   $0 = -0.01x^2 + 0.6x$

   Can you visually determine the solutions to this equation?

   Sample Answer: Yes.

   Can you solve this equation by using the methods that you learned in the previous lesson?

   No.

3. **Just the Math: Quadratic Formula** To solve the equation in Question 2, we can use the *quadratic formula*. The **quadratic formula** states that the solutions to the equation $ax^2 + bx + c = 0$ when $a \neq 0$ are given by

   $$x = \frac{-b \pm \sqrt{b^2 - 4ac}}{2a}.$$

   You will see where this formula comes from in Chapter 13.

   For instance, consider the equation $2x^2 - 3x + 1 = 0$.

   What are the values of a, b, and c in this equation? Write your answer using a complete sentence.

   In this equation, $a = 2$, $b = -3$, and $c = 1$.

   To find the solutions of the equation, substitute the values for a, b, and c into the quadratic formula and simplify:

   $$x = \frac{-(\boxed{-3}) \pm \sqrt{(\boxed{-3})^2 - 4(\boxed{2})(\boxed{1})}}{2(\boxed{2})}$$

   $$= \frac{\boxed{3} \pm \sqrt{\boxed{9} - \boxed{8}}}{\boxed{4}}$$

   $$= \frac{3 \pm \sqrt{\boxed{1}}}{4}$$

   $$= \frac{3 \pm \boxed{1}}{4}$$

   So, the solutions are $x = \frac{3+1}{4} = \frac{4}{4} = 1$ and $x = \frac{3-1}{4} = \frac{2}{4} = \frac{1}{2}$.

- In the quadratic equation $y = -3x^2 - 2x + 4$, what is the value of a? What is the value of b? What is the value of c? What is the value of $-b$? What is the value of $2a$? What is the value of b^2? What is the value of $b^2 - 4ac$?

Common Student Errors

Students may need to be reminded of how to identify a, b, and c from a quadratic equation.

Explore Together

Investigate Problem 1

Students will solve quadratic equations by using the quadratic formula.

Take Note

Remember that you must have zero on one side of the equation in order to use the quadratic formula.

8

Grouping

Ask for a student volunteer to read Questions 4 and 5 aloud. Have a student restate the problem. Pose the Guiding Questions below to verify student understanding. Have students work in small groups to complete Questions 4 and 5.

Guiding Questions

- What is the equation for the vertical height of the soccer ball as a function of the horizontal height of the soccer ball?
- What will you have to set your equation equal to in order to use the quadratic formula to solve your equation?
- What will the value(s) that you find using the quadratic formula represent in this situation?
- How can you find the values of *a*, *b*, and *c* for your equation?
- What is the quadratic formula?
- Question 5 asks you to find the horizontal distance when the height is 5 meters. How can you use the quadratic formula to solve for this situation? In the second part of Question 5, what do you have to do to each side of the equation before you can use the quadratic formula?

Call the class back together to have the students discuss and present their work for Questions 4 and 5.

Investigate Problem 1

For each quadratic equation below, find the values of *a*, *b*, and *c*.

$5x^2 + 6x + 1 = 0$

$a = 5, b = 6, c = 1$

$8x^2 + 4x - 6 = 0$

$a = 8, b = 4, c = -6$

$10x^2 - 1 = 0$

$a = 10, b = 0, c = -1$

$-x^2 + 8x = 2$

$a = -1, b = 8, c = -2$

4. Use the quadratic formula to find the horizontal distance that the ball travels before it hits the ground. Show all your work. Use a complete sentence to describe the answers that you find.

$0 = -0.01x^2 + 0.6x$; $a = -0.01, b = 0.6, c = 0$

$$x = \frac{-0.6 \pm \sqrt{0.6^2 - 4(-0.01)(0)}}{2(-0.01)}$$

$$= \frac{-0.6 \pm \sqrt{0.36}}{-0.02}$$

$$= \frac{-0.6 \pm 0.6}{-0.02}$$

$$x = \frac{-0.6 - 0.6}{-0.02} = \frac{-1.2}{-0.02} = 60; \quad x = \frac{-0.6 + 0.6}{-0.02} = 0$$

The value of *x* represents the horizontal distance that the ball traveled. So, the ball travels 60 meters horizontally.

5. Write an equation that you can use to find the horizontal distance the ball has traveled when it reaches a height of five meters.

$5 = -0.01x^2 + 0.6x$

Solve the equation. Show all your work and use a complete sentence in your answer.

$0 = -0.01x^2 + 0.6x - 5$; $a = -0.01, b = 0.6, c = -5$

$$x = \frac{-0.6 \pm \sqrt{0.6^2 - 4(-0.01)(-5)}}{2(-0.01)}$$

$$= \frac{-0.6 \pm \sqrt{0.36 - 0.20}}{-0.02}$$

$$= \frac{-0.6 \pm \sqrt{0.16}}{-0.02}$$

$$= \frac{-0.6 \pm 0.4}{-0.02}$$

$$x = \frac{-0.6 - 0.4}{-0.02} = \frac{-1}{-0.02} = 50; \quad x = \frac{-0.6 + 0.4}{-0.02} = \frac{-0.2}{-0.02} = 10$$

The ball is 5 meters high when it has traveled 10 meters and when it has traveled 50 meters.

Explore Together

Investigate Problem 1

Students will practice the skill of solving quadratic equations using the quadratic formula.

Grouping

Ask for a student volunteer to read Question 6 aloud. Have a student restate the problem. Have students complete Question 6 individually.

Call the class back together to have the students discuss and explain their results for Question 6.

After the students explain their conclusions for Question 6, ask for a student volunteer to read the sentence between Question 6 and the Summary Box about the discriminant.

Ask the students to develop a rule for the number of solutions based on the sign of the discriminant. Their rule should include all the information given in the sample answer to Question 6.

Key Formative Assessments

- What is a quadratic equation?
- What is the meaning of a solution to a quadratic equation?
- What 3 ways have we learned to solve quadratic equations?
- How can you solve a quadratic equation by representing the original equation in factored form?
- How can you solve a quadratic equation by extracting square roots?
- How can you solve a quadratic equation using the quadratic formula?
- What is a discriminant and what can it tell us?

Investigate Problem 1

Does your answer make sense? Use complete sentences to explain your reasoning.

Sample Answer: Yes, because the ball rises and then it falls, so there are two horizontal distances for which the height is five meters.

6. For each quadratic equation below, find the value of $\sqrt{b^2 - 4ac}$. Show all your work. Write your answer as a radical.

$x^2 + 7x - 2 = 0$

$\sqrt{7^2 - 4(1)(-2)} = \sqrt{49 + 8} = \sqrt{57}$

$-3x^2 + 4x - 8 = 0$

$\sqrt{4^2 - 4(-3)(-8)} = \sqrt{16 - 96} = \sqrt{-80}$

$8x^2 + 8x + 2 = 0$

$\sqrt{8^2 - 4(8)(2)} = \sqrt{64 - 64} = \sqrt{0}$

$3x^2 - 5x + 2 = 0$

$\sqrt{(-5)^2 - 4(3)(2)} = \sqrt{25 - 24} = \sqrt{1}$

What can you conclude about the number of solutions of each of the quadratic equations above? Use complete sentences in your answer.

Sample Answer: When the radicand is positive, there are two solutions. When the radicand is zero, there is one solution. When the radicand is negative, there is no solution.

The expression $b^2 - 4ac$ is called the **discriminant** of the quadratic formula.

Summary — Solving Quadratic Equations

In this lesson and the previous lesson, you explored three different methods for solving a quadratic equation, depending on its form:

- An equation in the form $(x - a)(x - b) = 0$ is solved by determining the value of x that makes each factor zero. This method is called **factoring.**
- An equation in the form $x^2 = b$ is solved by recognizing that $x = \sqrt{b}$ and $x = -\sqrt{b}$ satisfy the equation $x^2 = b$. This method is called **extracting square roots.**
- An equation in the form $ax^2 + bx + c = 0$ is solved by using the **quadratic formula:** $x = \dfrac{-b \pm \sqrt{b^2 - 4ac}}{2a}$.

8

Wrap Up

Close

- Review all key terms and their definitions including *quadratic formula* and *discriminant.*
- You may also want to review any other vocabulary terms that were discussed during the lesson, which may include *solution, approximate, square root, factoring,* and *extracting square roots.*
- Remind the students to write the key terms and their definitions in the notes section of their notebooks. You may also want the students to include examples.
- Ask the students to explain how to solve quadratic equations using the quadratic formula.
- Ask the students to summarize how to solve quadratic equations using the methods discussed in this lesson and the previous lesson. Ask the students why they think we need all 3 ways to solve quadratic equations.
- Pull together all of the closing information suggested by the students and give a brief summary of solving quadratic equations using each of the 3 methods presented in this lesson and the previous lesson. Be sure to point out any misconceptions or overgeneralizations that the students suggested during the lesson to make sure that they don't have false understanding.

Ties to the Cognitive Tutor Software

Most students need to be told that objects accelerate as they fall and that the height of the object can be modeled with a quadratic function. When working with vertical motion problems in the Cognitive Tutor software, students are practicing using a known formula to model the situation, rather than constructing the formula from an understanding of the situation.

Follow Up

Assignment

Use the Assignment for Lesson 8.6 in the Student Assignments book. See the Teacher's Resources and Assessments book for answers.

Assessment

See the Assessments provided in the Teacher's Resources and Assessments book for Chapter 8.

Open-Ended Writing Task

Ask the students to write a short paragraph explaining when they would use each of the 3 ways to solve quadratic equations.

Reflections

Insert your reflections on the lesson as it played out in class today.

What went well?

What did not go as well as you would have liked?

How would you like to change the lesson in order to improve the things that did not go well and capitalize on the things that did go well?

Notes

8

8

8.7 Pumpkin Catapult

Using a Vertical Motion Model

Learning By Doing Lesson Map

Get Ready

8

Objective

In this lesson, you will:

- Write and use a vertical motion model.

Key Term

- vertical motion model

NCTM Content Standards

Grades 9–12 Expectations

Number and Operation Standards

- Develop fluency in operations with real numbers, vectors, and matrices, using mental computation or paper-and-pencil calculations for simple cases and technology for more complicated cases.
- Judge the reasonableness of numerical computations and their results.

Algebra Standards

- Interpret representations of functions of two variables.
- Use symbolic algebra to represent and explain mathematical relationships.
- Judge the meaning, utility, and reasonableness of the results of symbol manipulations, including those carried out by technology.
- Draw reasonable conclusions about a situation being modeled.

Measurement Standards

- Apply informal concepts of successive approximation, upper and lower bounds, and limit in measurement situations.

Lesson Overview

Within the context of this lesson, students will be asked to:

- Write and use a vertical motion model to represent the height of a pumpkin as a function of time.
- Write a quadratic equation to model the vertical motion to represent the height of a pumpkin as a function of horizontal distance.
- Graph a vertical motion model that represents a real-life situation.

Essential Questions

The following key questions are addressed in this lesson:

1. What is a pumpkin catapult?
2. What is a vertical motion model?
3. How can you calculate vertical height if given time or horizontal distance?
4. How can you calculate time or horizontal distance if given vertical height?

Show The Way

Warm Up

Place the following questions or an applicable subset of these questions on the board before students enter class. Students should begin working as soon as they are seated.

Solve each quadratic equation by using the quadratic formula. Show the equation that you used; the values for *a*, *b*, *c*, and the discriminant; and the solutions.

8

1. $0 = x^2 - 12x + 2$

$0 = x^2 - 12x + 2$; $a = 1$, $b = -12$; $c = 2$

Discriminant: 136;

$x = \frac{-(-12) \pm \sqrt{(-12)^2 - 4(1)(2)}}{2(1)}$;

$x \approx 0.169$; $x \approx 11.831$

2. $8 = -4x^2 - 16x + 18$

$0 = -4x^2 - 16x + 10$; $a = -4$, $b = -16$; $c = 10$

Discriminant: 416;

$x = \frac{-(-16) \pm \sqrt{(-16)^2 - 4(-4)(10)}}{2(-4)}$;

$x \approx -4.550$; $x \approx 0.550$

3. $5x = 3x^2 + 11x - 9$

$0 = 3x^2 + 6x - 9$; $a = 3$, $b = 6$; $c = -9$

Discriminant: 144; $x = \frac{-6 \pm \sqrt{6^2 - 4(3)(-9)}}{2(3)}$;

$x = -3$; $x = 1$

4. $8 - x = x^2$

$0 = x^2 + x - 8$; $a = 1$, $b = 1$; $c = -8$

Discriminant: 33; $x = \frac{-1 \pm \sqrt{1^2 - 4(1)(-8)}}{2(1)}$;

$x \approx -3.372$; $x \approx 2.372$

Motivator

Begin the lesson with the motivator to get students thinking about the topic of the upcoming problem. This lesson is about a pumpkin catapult contest. The motivating questions are about unusual contests.

Ask the students the following questions to get them interested in the lesson.

- What is a pumpkin catapult?
- This lesson is based on an unusual contest in which pumpkins are catapulted through the air. What other unusual contests are you aware of?
- Have you participated in any unusual contests?

Explore Together

Problem 1

Students will use the vertical motion model to represent the parabolic flight of a pumpkin hurled through the air by a catapult.

Take Note

Often, in a model, when the independent variable represents time, the variable t is used instead of x.

Grouping

Ask for a student volunteer to read the Scenario and part (A) of Problem 1 aloud. Have a student restate the problem. Pose the Guiding Questions below to verify student understanding. Work together as a whole class to complete parts (A) and (B) of Problem 1.

Guiding Questions

- What information is given in this problem?
- What is a catapult?
- In what shape is the path that the pumpkin will fly?
- What is the equation used to model vertical motion?
- What is the value given for the initial height in Problem 1? What variable represents the initial height in the vertical motion model?
- What is the name used to represent the rate that the pumpkin is traveling? (*Hint:* Similar to speed.)
- What is the value given for the initial velocity in Problem 1? What variable represents the initial velocity in the vertical motion model?
- What two values will vary in this situation? What variable letters are used to represent these varying quantities?

SCENARIO Every year, the city of Millsboro, Delaware, holds a competition called the World Championship Punkin Chunkin, which is a pumpkin throwing competition. Participants build a pumpkin catapult that hurls a pumpkin. The catapult that hurls the pumpkin the farthest is the winner.

Problem 1 A Pumpkin Catapult

You can model the motion of a pumpkin that is released by a catapult by using the **vertical motion model** $y = -16t^2 + vt + h$, where t is the time that the object has been moving in seconds, v is the initial velocity (speed) of the object in feet per second, h is the initial height of the object in feet, and y is the height of the object in feet at time t seconds.

A. Suppose that a catapult is designed to hurl a pumpkin from a height of 30 feet at an initial velocity of 212 feet per second. Write a quadratic function that models the height of the pumpkin in terms of time.

$y = -16t^2 + 212t + 30$

B. Write an equation that you can use to determine when the pumpkin will hit the ground. Then solve the equation. Show all your work.

$0 = -16t^2 + 212t + 30$; $a = -16$, $b = 212$, $c = 30$

$$t = \frac{-212 \pm \sqrt{212^2 - 4(-16)(30)}}{2(-16)}$$

$$= \frac{-212 \pm \sqrt{44,944 + 1920}}{-32}$$

$$= \frac{-212 \pm \sqrt{46,864}}{-32}$$

$$t = \frac{-212 + \sqrt{46,864}}{-32} \approx -0.14 \qquad t = \frac{-212 - \sqrt{46,864}}{-32} \approx 13.39$$

Do both solutions have meaning in the problem situation? Use complete sentences to explain your reasoning.

Sample Answer: No, the negative solution does not make sense because *t* represents time and time is positive.

Notes Velocity can be positive or negative. Velocity is a measure of both speed and direction.

You should show your students how to use their calculators to limit the number of approximations that they make. They will get the most accurate answers by only rounding once. So, the value for t should be entered into the calculator in a single entry as $(-212 + \sqrt{46,864})/(-32)$.

Explore Together

Problem 1

Students will create a table of values and draw a graph to model the vertical motion of the pumpkin.

Grouping

Ask for a student volunteer to read part (C) of Problem 1 aloud. Have a student restate the problem. Pose the Guiding Questions below to verify student understanding. Have students work together in small groups to complete parts (C) and (D) of Problem 1.

Guiding Questions

- What two quantities will you represent in the table of values?
- Which is the independent variable, the time or the height?
- What is the unit for the time?
- What do you think are reasonable values for the time?
- What is the dependent variable?
- What is the unit for the height?
- How can you determine the range of reasonable values for the height of the pumpkin after you complete your table?
- What would a positive height represent in this situation?
- What would a height of zero represent in this situation?
- What would a negative height represent in this situation?
- Are negative values for the height reasonable in this situation?
- If we calculate a negative height at a given time, what can we conclude about that time?

8

Problem 1 A Pumpkin Catapult

When does the pumpkin hit the ground? Use a complete sentence in your answer.

The pumpkin hits the ground after about 13.39 seconds.

C. Complete the table of values that shows the height of the pumpkin in terms of time.

Quantity Name	Time	Height
Unit	seconds	feet
Expression	t	$-16t^2 + 212t + 30$
	0	30
	2	390
	5	690
	10	550
	12	270
	13	82
	15	–390

D. Create a graph of the model to see the path of the pumpkin on the grid on the next page. First, choose your bounds and intervals. Be sure to label your graph clearly.

Variable quantity	Lower bound	Upper bound	Interval
Time	0	15	1
Height	0	900	60

Common Student Errors

Students may incorrectly try to include the ordered pair of 15 seconds and –390 feet on their graph. They need to discount this ordered pair as impossible for this situation.

You may need to talk to the students about the advantages and disadvantages of algebraic models.

Explore Together

Problem 1

Call the class back together to have the students discuss and present their work for parts (C) and (D).

Investigate Problem 1

Students will use the vertical motion model, the table of values, the graph, and algebraic solutions to answer questions about the situation being modeled.

Grouping

Ask for a student volunteer to read Question 1 aloud. Have a student restate the problem. Have the students complete Question 1 individually. Then discuss their responses as a whole class.

Ask for a student volunteer to read Question 2 aloud. Have a student restate the problem. Pose the Guiding Questions below to verify student understanding. Have students work together in small groups to complete Questions 2 through 4.

Guiding Questions

- What are you asked to do in Question 2?
- How can you calculate the answer to Question 2?
- What numbers would make sense in this situation?

Common Student Errors

Students may not immediately know how to complete Question 3. Be ready to refocus students by posing the Guiding Questions and asking them what usually occurs at the maximum or minimum points in a parabola. When they realize that they need to solve for the vertex, ask them how they have solved for the vertex value in previous lessons.

Problem 1 A Pumpkin Catapult

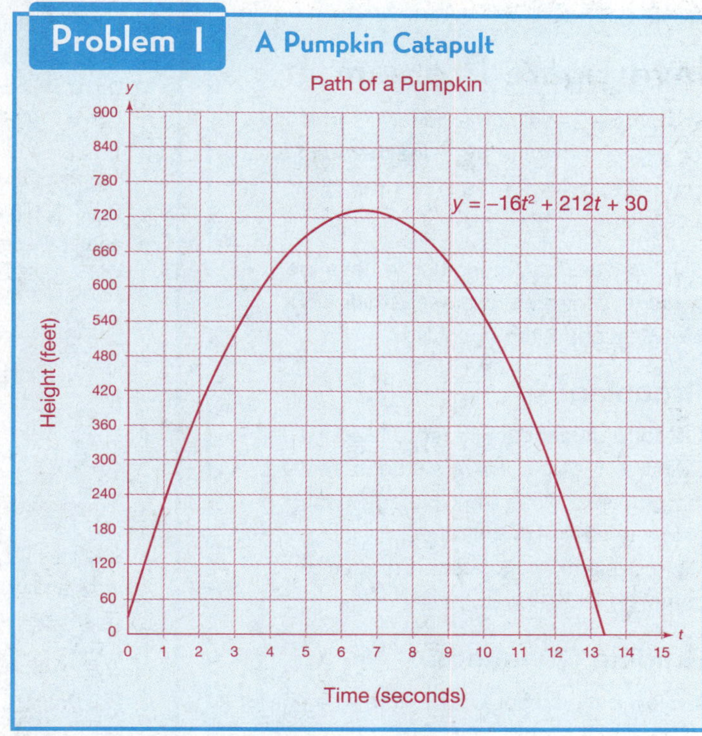

Investigate Problem 1

1. Does your answer to part (B) make sense in terms of the graph? Write your answer using a complete sentence.

   Yes, the time of about 13.39 seconds is one of the *x*-intercepts of the graph, which corresponds to when the height is zero, or the ground.

2. What is the height of the pumpkin three seconds after it is launched from the catapult? Show all your work and use a complete sentence in your answer.

   $y = -16(3)^2 + 212(3) + 30 = -16(9) + 636 + 30 = 522$
   The height of the pumpkin is 522 feet.

   What is the height of the pumpkin eight seconds after it is launched from the catapult? Show all your work and use a complete sentence in your answer.

   $y = -16(8)^2 + 212(8) + 30 = -16(64) + 1696 + 30 = 702$
   The height of the pumpkin is 702 feet.

   What is the height of the pumpkin 20 seconds after it is launched from the catapult? Show all your work and use a complete sentence in your answer.

   $y = -16(20)^2 + 212(20) + 30 = -16(400) + 4240 + 30 = -2130$
   The height of the pumpkin is –2130 feet.

8

Explore Together

Investigate Problem 1

Students will evaluate a quadratic function for a given time and then use the quadratic formula to solve for the time when given a height.

Call the class back together to have the students discuss and present their work for Questions 2 through 4.

8

Grouping

Ask for a student volunteer to read Question 5 aloud. Have a student restate the problem. Pose the Guiding Questions below to verify student understanding. Have students work in small groups to complete Question 5.

Guiding Questions

- How can you solve for a time when given a value for the height of the pumpkin?
- What is the quadratic formula?
- What is the equation required to be set equal to in order to use the quadratic formula?
- How can you set the equation equal to zero if the height is given to be a value different from zero?
- How can you find the values of *a*, *b*, and *c* to substitute in the quadratic formula?

Call the class back together to have the students discuss and present their work for Question 5.

Investigate Problem 1

Do all of your answers to Question 2 make sense? Use a complete sentence to explain your reasoning.

Sample Answer: No, the last answer does not make sense because the pumpkin would be beneath the ground.

3. When is the pumpkin at its highest point? Show all your work and use a complete sentence in your answer.

$a = -16, b = 212$

$$t = -\frac{b}{2a} = -\frac{212}{2(-16)} = 6.625$$

The pumpkin is at its highest point after 6.625 seconds.

4. What is the maximum height of the pumpkin? Show all your work and use a complete sentence in your answer.

$$y = -16(6.625)^2 + 212(6.625) + 30 = -702.25 + 1404.5 + 30 = 732.25$$

The pumpkin reaches a height of 732.25 feet.

5. When is the pumpkin at a height of 500 feet? Show all your work and use a complete sentence in your answer.

$$500 = -16t^2 + 212t + 30$$

$$0 = -16t^2 + 212t - 470 \qquad a = -16, b = 212, c = -470$$

$$t = \frac{-212 \pm \sqrt{212^2 - 4(-16)(-470)}}{2(-16)} = \frac{-212 \pm \sqrt{44{,}944 - 30{,}080}}{-32} = \frac{-212 \pm \sqrt{14{,}864}}{-32}$$

$$t = \frac{-212 + \sqrt{14{,}864}}{-32} \approx 2.82 \qquad t = \frac{-212 - \sqrt{14{,}864}}{-32} \approx 10.43$$

The pumpkin will reach a height of 500 feet at about 2.82 seconds and 10.43 seconds.

Is your answer confirmed by your graph?

Sample Answer: Yes.

Explore Together

Problem 2

Students will consider a different quadratic model for the same catapult as discussed in Problem 1.

Grouping

Ask for a student volunteer to read Problem 2 aloud. Have a student restate the problem. Have students work in small groups to complete parts (A) through (E) of Problem 2.

Notes The answer given in part (E) is a sample answer only. Student answers will depend on how they approximate the maximum distance for the catapult.

Common Student Errors

Some students may not recognize that in Problem 2 they are being asked to use a new model that is representing the vertical distance as a function of the horizontal distance. Pose the Guiding Questions below to verify student understanding, if necessary.

Guiding Questions

- The equation that you wrote in Problem 1 represented the vertical height of the pumpkin as a function of time. In Problem 2, the vertical height is represented as a function of what variable quantity?
- Why might we want to consider the vertical height as a function of the horizontal distance rather than as a function of time?
- What was the meaning of the y-value for the original function in Problem 1 when the time was equal to zero?
- What is the meaning of the y-value for the new function in Problem 2 when the horizontal distance is equal to zero?
- What was the meaning of the t-values when the y-value was equal to zero in the original function in Problem 1?
- What is the meaning of the x-values when the y-value is equal to zero in the new function in Problem 2?

Problem 2 How Far Can the Pumpkin Go?

In 2005, the winner in the catapult division of the World Championship Punkin Chunkin hurled a pumpkin 2862.28 feet.

A. A model for the path of a pumpkin being launched from the catapult described in Problem 1 is $y = -0.00036x^2 + x + 30$, where x is the horizontal distance of the pumpkin in feet and y is the vertical distance of the pumpkin in feet. According to the model, what is the pumpkin's height when it has traveled 500 feet horizontally? Show all your work and use a complete sentence in your answer.

$y = -0.00036(500)^2 + 500 + 30 = -90 + 500 + 30 = 440$

The pumpkin's height is 440 feet.

B. What is the pumpkin's height when it has traveled 1000 feet horizontally? Show all your work and use a complete sentence in your answer.

$y = -0.00036(1000)^2 + 1000 + 30 = -360 + 1000 + 30 = 670$

The pumpkin's height is 670 feet.

C. What is the pumpkin's height when it has traveled 2500 feet horizontally? Show all your work and use a complete sentence in your answer.

$y = -0.00036(2500)^2 + 2500 + 30 = -2250 + 2500 + 30 = 280$

The pumpkin's height is 280 feet.

D. What is the pumpkin's height when it has traveled 3000 feet horizontally? Show all your work and use a complete sentence in your answer.

$y = -0.00036(3000)^2 + 3000 + 30 = -3240 + 3000 + 30 = -210$

The pumpkin's height is –210 feet.

E. Do you think that this catapult has a chance of beating the 2005 catapult winner? Use complete sentences to explain your reasoning.

Sample Answer: Yes. The height will be 0 when the pumpkin hits the ground. Because the height at 2500 feet is positive and the height at 3000 feet is negative, the height will be zero between 2500 feet and 3000 feet.

Call the class back together to have the students discuss and present their work for parts (A) through (E) of Problem 2.

8

Explore Together

Investigate Problem 2

Students will create a graph to model the pumpkin's vertical height as a function of the horizontal distance. Then, students will solve their new quadratic equation using the quadratic formula.

Grouping

Have students work together in small groups to complete Questions 1 through 3.

Common Student Errors

Some students may still struggle to choose appropriate bounds and intervals for their graphs. Be ready to redirect students making such errors. Remind the students that their graph should represent the entire situation and should not have too many values beyond the domain and range represented.

8

Investigate Problem 2

1. Algebraically determine the horizontal distance the pumpkin travels before it hits the ground. Does it beat the winner? Show all your work and use a complete sentence in your answer.

   $0 = -0.00036x^2 + x + 30 \quad a = -0.00036, b = 1, c = 30$

   $$x = \frac{-1 \pm \sqrt{1^2 - 4(-0.00036)(30)}}{2(-0.00036)}$$

   $$= \frac{-1 \pm \sqrt{1 + 0.0432}}{-0.00072}$$

   $$= \frac{-1 \pm \sqrt{1.0432}}{-0.00072}$$

   $$x = \frac{-1 + \sqrt{1.0432}}{-0.00072} \approx -30$$

   $$x = \frac{-1 - \sqrt{1.0432}}{-0.00072} \approx 2807$$

   The pumpkin travels about 2807 feet. It does not beat the winner.

2. Create a graph of the model to see the path of the pumpkin on the grid below. First, choose your bounds and intervals.

Variable quantity	Lower bound	Upper bound	Interval
Horizontal distance	0	3000	200
Height	0	900	60

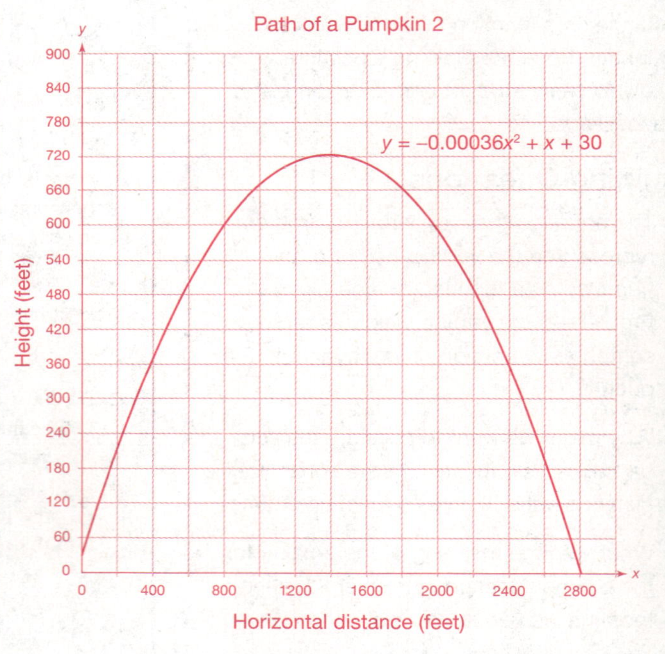

Explore Together

Investigate Problem 2

Common Student Errors

Even after solving several quadratic equations using the quadratic formula, some students will forget to subtract the value of 550 from both sides of the equation before finding the values for a, b, and c. Some students will make careless arithmetic errors when using the quadratic formula.

Call the class back together to have the students discuss and present their work for Questions 1 through 3.

Key Formative Assessments

- What is a vertical motion model?
- What variables did you compare in Problem 1?
- What variables did you compare in Problem 2?
- What information were you able to find from your original equation in Problem 1?
- What information were you able to find from your new equation in Problem 2?
- How were you able to use the quadratic formula in this lesson?

Investigate Problem 2

3. Use the information from Problem 1 and this problem to find the *horizontal* distance the pumpkin travels after 10 seconds. Show all your work and use a complete sentence in your answer.

Height: $y = 550$ (from table in Problem 1 part (C))

$$550 = -0.00036x^2 + x + 30$$

$$0 = -0.00036x^2 + x + -520$$

$$a = -0.00036,\ b = 1,\ c = -520$$

$$x = \frac{-1 \pm \sqrt{1^2 - 4(-0.00036)(-520)}}{2(-0.00036)}$$

$$= \frac{-1 \pm \sqrt{1 - 0.7488}}{-0.00072}$$

$$= \frac{-1 \pm \sqrt{0.2512}}{-0.00072}$$

$$x = \frac{-1 + \sqrt{0.2512}}{-0.00072} \approx 693 \qquad x = \frac{-1 - \sqrt{0.2512}}{-0.00072} \approx 2085$$

The pumpkin travels about 2085 feet horizontally.

8

Wrap Up

Close

- Review all key terms and their definitions. Include the term *vertical motion model.*
- You may also want to review any other vocabulary terms that were discussed during the lesson, which may include *initial velocity, speed, rounding, model, vertex, catapult, the Quadratic Formula, independent variable,* and *dependent variable.*
- Remind the students to write the key terms and their definitions in the notes section of their notebooks. You may also want the students to include examples.
- Ask the students to explain how they were able to model vertical motion.
- Ask the students to summarize each component of the vertical motion model presented in Problem 1.
- Have the students state what a solution to a quadratic equation represents and what is meant by a solution or root of a function.

Ties to the Cognitive Tutor Software

Most students need to be told that objects accelerate as they fall and that the height of the object can be modeled with a quadratic function. When working with vertical motion problems in the Cognitive Tutor software, students are practicing using a known formula to model the situation, rather than constructing the formula from an understanding of the situation.

Follow Up

Assignment

Use the Assignment for Lesson 8.7 in the Student Assignments book. See the Teacher's Resources and Assessments book for answers.

Assessment

See the Assessments provided in the Teacher's Resources and Assessments book for Chapter 8.

Open-Ended Writing Task

Ask the students to write a short paragraph comparing their work in Problem 1 and in Problem 2 of this lesson. How were the problems similar and how were they different?

Reflections

Insert your reflections on the lesson as it played out in class today.

What went well?

What did not go as well as you would have liked?

How would you like to change the lesson in order to improve the things that did not go well and capitalize on the things that did go well?

Notes

8.8 Viewing the Night Sky

Using Quadratic Functions

Learning By Doing Lesson Map

Get Ready

Objective

In this lesson, you will:

- Analyze a quadratic function that models the shape of an object.

Key Terms

- axis of symmetry
- vertex
- domain
- range

NCTM Content Standards

Grades 9–12 Expectations

Algebra Standards

- Understand relations and functions and select, convert flexibly among, and use various representations for them.
- Analyze functions of one variable by investigating rates of change, intercepts, zeros, asymptotes, and local and global behavior.
- Interpret representations of functions of two variables.
- Draw reasonable conclusions about a situation being modeled.

Lesson Overview

Within the context of this lesson, students will be asked to:

- Graph quadratic functions.
- Analyze two quadratic functions that model the shape of a telescope lens.
- Evaluate quadratic functions for given values.

Essential Questions

The following key questions are addressed in this lesson:

1. What is a telescope?
2. What is the vertex for a telescope lens?
3. How can you find the domain and the range for a telescope lens?
4. What is the quadratic function used to find the lens height?
5. How can you evaluate a quadratic function for given values?

Show The Way

Warm Up

Place the following questions or an applicable subset of these questions on the board before students enter class. Students should begin working as soon as they are seated.

Solve each quadratic equation by using any method. Show your work.

8

1. $0 = 3(x + 4)(x - 5)$

$x = -4; x = 5$

2. $3x^2 = 27$

$x = -3; x = 3$

3. $0 = 7x^2 + 2x - 5$

$x = -1; x = \frac{5}{7}$

4. $-1 = -2x^2 - 6x$

$x \approx -3.158; x \approx 0.158$

Motivator

Begin the lesson with the motivator to get students thinking about the topic of the upcoming problem. This lesson is about modeling the shape of a telescope. The motivating questions are about telescopes and their lenses.

Ask the students the following questions to get them interested in the lesson.

- What is a telescope?
- What is a lens?
- What are telescopes used for?
- What is the largest telescope you have heard of? (Possibly the Hubble Space Telescope.)

You may want to reference the Hubble Space Telescope. A web site with information about the Hubble Space Telescope is: hubble.nasa.gov.

Explore Together

Problem 1

Students will investigate 2 quadratic functions that model the parabolic shape of a lens of a telescope.

Problems 1 and 2 are similar mathematically.

Grouping

Ask for a student volunteer to read the Scenario and Problem 1 aloud. Have a student restate the problem. Ask for another student volunteer to read Problem 2 aloud. Have a student restate the problem. Pose the Guiding Questions below to verify student understanding. Because the mathematics in Problems 1 and 2 are similar, assign Problem 1 parts (A) through (D) and Questions 1 through 4 of Problem 1 to half of the students. Assign Problem 2 parts (A) through (D) and Questions 1 through 4 of Problem 2 to the other half of the students. Have the students work in small groups to complete the questions assigned to them.

Guiding Questions

- What information is given in this problem?
- What is a telescope?
- What is the equation given to model the shape of a lens?
- What is the dependent variable for the equation?
- What does the variable y represent for this situation?
- On what quantities does y depend?
- What measurement for the lens is not used to determine the value of the height of the lens?
- If the aperture is not used to find the height of the lens for any given value of x, why is the aperture important in this situation? What information about x can be found from the aperture?

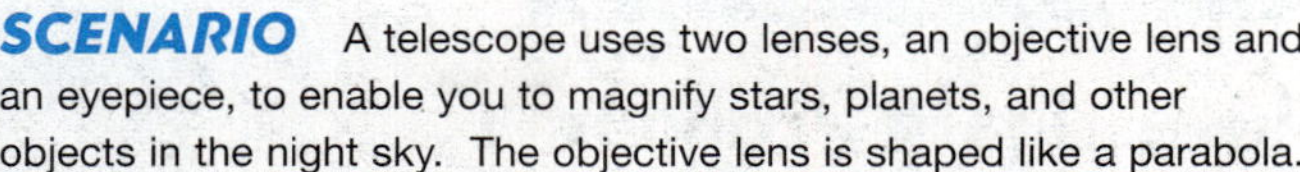

SCENARIO A telescope uses two lenses, an objective lens and an eyepiece, to enable you to magnify stars, planets, and other objects in the night sky. The objective lens is shaped like a parabola.

Telescopes are described by the *aperture* (pronounced ăpər-chər) and the *focal length*. The aperture is the width of the objective lens. The focal length is a bit more complicated. It is the distance from the *vertex* (the lowest point) of the lens to a point, called the focal point, on the *axis of symmetry*. The focal point is the point at which light rays coming into the telescope meet after they bounce off the lens.

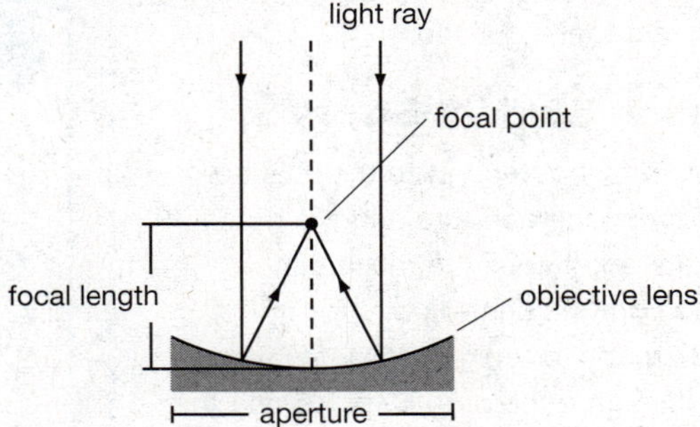

The shape of the lens can be described by the quadratic function $y = \frac{1}{4p}x^2$, where x is the number of units to the right of the axis of symmetry, y is the height of the lens, and p is the focal length.

Problem 1 Size of the Lens

A. The aperture of one telescope model is 6 inches and the focal length of the objective lens is 48 inches. Write the function that represents the shape of the lens.

$y = \frac{1}{192}x^2$

B. What is the **axis of symmetry** of the graph of the lens? Use a complete sentence in your answer.

The axis of symmetry is the y-axis, or the line $x = 0$.

C. What point is the *vertex* of the lens? Use a complete sentence in your answer.

The vertex is (0, 0).

Explore Together

Investigate Problem 1

Students will continue to investigate quadratic functions that model the parabolic shape of a lens of a telescope.

Grouping

Students will be working in small groups to complete these questions.

8

Common Student Errors

Students may not recognize the relationship between the aperture and the domain. Half of the aperture is the maximum distance from the center point for the lens. Be ready to redirect students who do not think through part (D) by posing the Guiding Questions below.

Guiding Questions

- What does x represent in this situation?
- What measurement of the lens relates to the x value?
- What is the distance between an x-value of –1 and an x-value of 1?
- What is the distance between an x-value of –2 and an x-value of 2?
- What two x-values would have a total distance equal to the aperture length?
- How do those values relate to the domain of the possible x-values?

Common Student Errors

Students may struggle to remember how to round values to the nearest hundredth. They may also forget the meaning of the term *range*. Pose the Guiding Questions below to verify student understanding.

Guiding Questions

- How can you round an answer to the nearest hundredth?
- When do you round a value up to the next highest number?
- What does the range represent?
- How can you determine the range for this situation?

Problem 1 Size of the Lens

D. What is the *domain* of this function in the problem situation? Use a complete sentence in your explanation.

Sample Answer: Because the axis of symmetry is $x = 0$ and the lens is 6 inches wide, the domain is all real numbers from –3 to 3.

Investigate Problem 1

1. Complete the table of values that shows the shape of the lens. Use your domain as a guide for choosing the x-values. If necessary, round your answers to the nearest hundredth.

Quantity Name	Horizontal position	Height
Unit	inches	inches
Expression	x	$\frac{1}{192}x^2$
	–3	0.05
	–2	0.02
	–1	0.01
	0	0
	1	0.01
	2	0.02
	3	0.05

2. What is the *range* of the function? Show all your work and use a complete sentence in your answer.

At $x = 3$: $y = \frac{1}{192}(3^2) = 0.046875^{-}\ 0.05$

The range is all real numbers from 0 to about 0.05.

Use complete sentences to explain how you found your answer to Question 2.

Sample Answer: The parabola opens upward. This means that the vertex (0, 0), is the lowest point on the graph, so the range starts at 0. It also means that the greatest value for y will occur when $x = 3$ and $x = -3$.

3. What is the maximum height of this lens? If necessary, round your answer to the nearest thousandth. Use a complete sentence to explain how you found your answer.

The maximum height of the lens is about 0.05 inch, which is the greatest value of y that occurs when $x = 3$ and $x = -3$.

Explore Together

Investigate Problem 1

Students will graph the first quadratic function for this lesson.

Grouping

If the students are completing Problem 1 at the same time, call the class back together to have the students discuss and present their work for parts (A) through (D) and Questions 1 through 4 of Problem 1.

Problem 2

Grouping

Pose the Guiding Questions below to verify student understanding. Have the students work together in small groups to complete parts (A) through (D) and Questions 1 through 4.

Guiding Questions

- What information is given in this problem?
- How does the situation in Problem 2 differ from the situation in Problem 1?
- How is the situation in Problem 2 the same as the situation in Problem 1?
- What information will be used to find the equation for y?

Notes If the students all completed Problem 1 and time does not allow you to complete both problems in one class period, you may want to assign any remaining work in Problem 2 as part of a homework assignment. At the beginning of the next class session, have the students meet together to discuss the work that they completed as a homework assignment.

Investigate Problem 1

4. Create a graph of the quadratic function on the grid below. First, choose your bounds and intervals. Be sure to label your graph clearly.

Variable quantity	Lower bound	Upper bound	Interval
Horizontal position	–3.5	3.5	0.5
Height	0	0.15	0.01

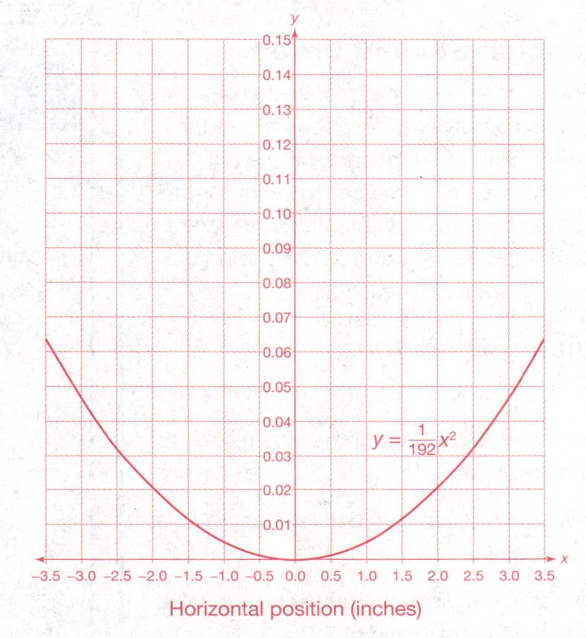

Problem 2 Increase the Focal Length

A. Now consider a different telescope. This telescope has the same aperture as the one in part (A), but its focal length is 56 inches. Write the function that represents the lens shape.

$$y = \frac{1}{224}x^2$$

B. What is the axis of symmetry of the graph of the lens? Use a complete sentence in your answer.

The axis of symmetry is the y-axis, or the line $x = 0$.

Explore Together

Investigate Problem 2

Students will continue to investigate quadratic functions that model the parabolic shape of a lens of a telescope.

Grouping

Students will be working in small groups to complete these questions.

8

Common Student Errors

Students may again not recognize the relationship between the aperture and the domain. Half of the aperture is the maximum distance from the center point for the lens. Be ready to redirect students who do not think through part (D) by posing the following Guiding Questions.

Guiding Questions

- What does x represent in this situation?
- What measurement of the lens relates to the x value?
- What is the distance between an x-value of –1 and an x-value of 1?
- What is the distance between an x-value of –2 and an x-value of 2?
- What two x-values would have a total distance equal to the aperture length?
- How do those values relate to the domain of the possible x-values?

Common Student Errors

Students may struggle to remember how to round values to the nearest thousandth. They may also forget the meaning of the term *range*. Pose the Guiding Questions below to verify student understanding.

Guiding Questions

- How can you round an answer to the nearest thousandth?
- When do you round a value up to the next higher number?
- What does the range represent?
- How can you determine the range for this situation?

Problem 2 Increase the Focal Length

C. What is the vertex of the lens? Use a complete sentence in your answer.

The vertex is (0, 0).

D. What is the domain of this function in the problem situation? Use a complete sentence in your explanation.

Sample Answer: Because the axis of symmetry is $x = 0$ and the lens is 6 inches wide, the domain is all real numbers from –3 to 3.

Investigate Problem 2

1. Complete the table of values that shows the shape of the lens. Use your domain as a guide for choosing the x-values. If necessary, round your answers to the nearest thousandth.

Quantity Name	Horizontal position	Height
Unit	inches	inches
Expression	x	$\frac{1}{224}x^2$
	–3	0.040
	–2	0.018
	–1	0.004
	0	0
	1	0.004
	2	0.018
	3	0.040

2. What is the range of the function? Show all your work and use a complete sentence in your answer.

At $x = 3$: $y = \frac{1}{224}(3^2) \approx 0.04018$

The range is all real numbers from 0 to about 0.040.

Use complete sentences to explain how you found your answer to Question 2.

Sample Answer: The parabola opens upward. This means that the vertex (0, 0), is the lowest point on the graph, so the range starts at 0. It also means that the greatest value for y will occur when $x = 3$ and $x = -3$.

Explore Together

Investigate Problem 2

Students will graph the quadratic function representing the shape of the lens in Problem 2.

Call the class back together.

- If half of the class completed Problem 1 and the other half completed Problem 2, form new groups with 1 or 2 students who completed Problem 1 and 1 or 2 students who completed Problem 2 together in each new group. Have them explain Problems 1 and 2 through Question 4 to each other. When every group member understands and completes the problem that they were not originally assigned, have students work in their new groups to complete Questions 5 and 6.
- If the students completed both Problems 1 and 2, have them debrief and explain their work through Question 4 of Problem 2. Then assign Questions 5 and 6 to be completed by working together in their small groups or as part of a homework assignment.

Investigate Problem 2

3. What is the maximum height of this lens? If necessary, round your answer to the nearest thousandth. Use a complete sentence to explain how you found your answer.

The maximum height of the lens is 0.040 inch, which is the greatest value of y that occurs when $x = 3$ and $x = -3$.

4. Create a graph of the quadratic function on the grid below. First, choose your bounds and intervals. Be sure to label your graph clearly.

Variable quantity	Lower bound	Upper bound	Interval
Horizontal position	**−3.5**	**3.5**	**0.5**
Height	**0**	**0.15**	**0.01**

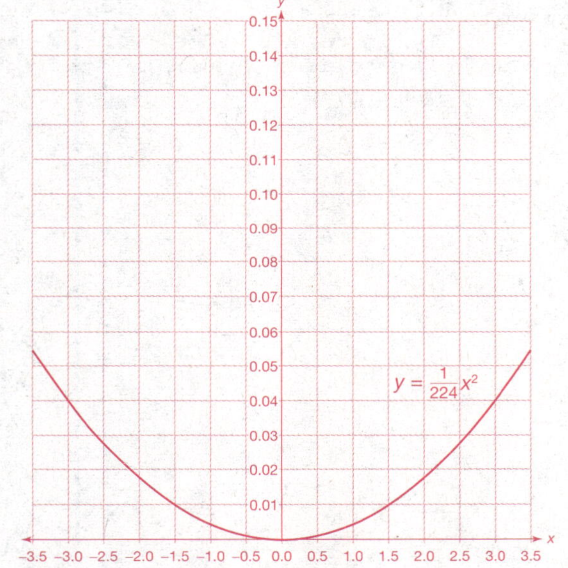

Explore Together

Investigate Problem 2

Call the class back together to have the students discuss and present their work for Questions 5 and 6.

Key Formative Assessments

- How did you find the domain of the functions?
- How did you find the range of the functions?
- How did you find the vertex for each of the functions?
- What was the significance of the vertex for each of the functions?
- How did you find the axis of symmetry for each of the functions?

8

Investigate Problem 2

5. What is the difference in the widths of the lenses? What is the difference in the heights of the lenses? Use complete sentences in your answer.

The lenses have the same width and height depending on how the answers are rounded.

6. How does the focal length affect the shape of the lens? Use a complete sentence in your answer.

Sample Answer: The greater the focal length, the flatter the lens.

Use complete sentences to explain how you found your answer to Question 6.

Sample Answer: The domains of both functions are the same, but the greatest value in the range of the first lens is greater than the greatest value in the range of the second lens. So, the focal length of the second lens is greater than the focal length of the first lens.

Wrap Up

Close

- Review all key terms and their definitions. Include the terms *axis of symmetry, vertex, domain,* and *range.*
- You may also want to review any other vocabulary terms that were discussed during the lesson, which may include *telescope, lens, aperture, focal length,* and *light ray.*
- Remind the students to write the key terms and their definitions in the notes section of their notebooks. You may also want the students to include examples.
- Ask the students to explain the following:
  - What is a quadratic function?
  - How are quadratic functions used to model situations?
  - What types of situations are modeled well with quadratic functions?
  - What are the important concepts for quadratic functions that have been discussed in this chapter?
  - What do you still need to know more about before you can take a test on quadratic functions?
- Have the students work in small groups to create a review guide for the chapter and also a matching worksheet with the key terms that were discussed in the chapter.

Ties to the Cognitive Tutor Software

Most students need to be told that objects accelerate as they fall and that the height of the object can be modeled with a quadratic function. When working with vertical motion problems in the Cognitive Tutor software, students are practicing using a known formula to model the situation, rather than constructing the formula from an understanding of the situation.

Follow Up

Assignment

Use the Assignment for Lesson 8.8 in the Student Assignments book. See the Teacher's Resources and Assessments book for answers.

Assessment

See the Assessments provided in the Teacher's Resources and Assessments book for Chapter 8.

Open-Ended Writing Task

Ask the students to write a paragraph summarizing what they know about quadratic functions. They should also state any concepts for quadratic functions about which they are still confused.

Reflections

Insert your reflections on the lesson as it played out in class today.

What went well?

8

What did not go as well as you would have liked?

How would you like to change the lesson in order to improve the things that did not go well and capitalize on the things that did go well?

Notes

8

Looking Ahead to Chapter 9

Focus

In Chapter 9, you will learn how to identify prime and composite numbers, factor composite numbers by using prime factorizations, and write numbers in scientific notation. You will also work with powers and learn properties of powers, as well as different forms of *n*th roots of numbers.

Chapter Warm-up

Answer these questions to help you review skills that you will need in Chapter 9.

Use mental math to find the value of *x*.

1. $3x = 81$ **27**

2. $5x = 105$ **21**

3. $17x = 85$ **5**

Write each expression as a power.

4. (8)(8)(8)(8)

8^4

5. 11 • 11 • 11 • 11 • 11 • 11

11^6

6. (–13)(–13)(–13)

$(-13)^3$

9

Evaluate each power.

7. 2^6 **64**

8. 5^2 **25**

9. 10^4 **10,000**

Read the problem scenario below.

You work for a cooking supply manufacturer and you are in charge of making circular pizza pans. Your boss would like to know the area of each circular pan to get an idea of the amount of material that will be needed for each of the pizza pans. Note that the area of a circle is $A = \pi r^2$, where r is the radius of the circle. Use 3.14 for π.

10. What is the area of a pizza pan that has a radius of 3 inches?

28.26 square inches

11. What is the area of a pizza pan that has a radius of 6 inches?

113.04 square inches

12. One pizza pan has a radius of 7 inches. Another pizza pan has a radius of 5 inches. How much more area does the first pizza pan have than the second pizza pan?

The first pizza pan has 75.36 more square inches than the second pizza pan.

Key Terms

factors ■ p. 422
prime number ■ p. 423
composite number ■ p. 423
prime factorization ■ p. 423
power ■ p. 425
exponent ■ p. 425
product ■ p. 425
quotient ■ p. 425
positive exponent ■ p. 429
negative exponent ■ p. 430
zero exponent ■ p. 430
standard form ■ p. 433
scientific notation ■ p. 433
cube root ■ p. 444
index ■ p. 444
*n*th root ■ p. 445
radicand ■ p. 445
rational exponent ■ p. 447
radical ■ p. 447

CHAPTER

9

Properties of Exponents

In 2004, Steve Fossett set a world record for the fastest trip around the world in a sailboat. He and his crew of thirteen made the journey in 58 days and 9 hours. In Lesson 9.6, you will examine the capsize factor of a boat by using its weight and width.

9

9.1 The Museum of Natural History

Powers and Prime Factorization

Learning By Doing Lesson Map

Get Ready

Objectives

In this lesson, you will:

- List the factors of numbers.
- Identify prime and composite numbers.
- Write the prime factorizations of numbers.

Key Terms

- factors
- prime number
- composite number
- prime factorization

NCTM Content Standards

Grades 9-12 Expectations

Number and Operations Standards

- Use number-theory arguments to justify relationships involving whole numbers.

Algebra Standards

- Use symbolic algebra to represent and explain mathematical relationships.
- Identify essential quantitative relationships in a situation and determine the class or classes of functions that might model the relationships.
- Use symbolic expressions, including iterative and recursive forms, to represent relationships arising from various contexts.
- Draw reasonable conclusions about a situation being modeled.

Lesson Overview

Within the context of this lesson, students will be asked to:

- Find and list the whole number factors of given numbers.
- Classify whole numbers as either prime, composite, or neither.
- Find and write the prime factorization of given numbers.
- Find and list any common whole number factors for given pairs of numbers.

Essential Questions

The following key questions are addressed in this lesson:

1. What is a museum of natural history?
2. What is a prime number?
3. What is a composite number?
4. What is a factor?
5. How can you write the prime factorization for a given number?

Show The Way

Warm Up

Place the following questions or an applicable subset of these questions on the board before students enter class. Students should begin working as soon as they are seated.

Simplify each expression.

1. $-3(2x - 4) + 1$

   $-6x + 13$

2. $\frac{2}{3}(9x^2 - 18x + 12)$

   $6x^2 - 12x + 8$

3. $\frac{3}{7}\left(\frac{14}{9}x^2 - 21x - \frac{35}{12}\right)$

   $\frac{2}{3}x^2 - 9x - \frac{5}{4}$

4. $-0.75(12x^3 - x^2 + 3x + 5)$

   $-9x^3 + 0.75x^2 - 2.25x - 3.75$

9

5. $\frac{1}{8}(4^2)(7x^2 - 6x + 5)(3 - 4)$

   $-14x^2 + 12x - 10$

6. $-(3x - 4)(5 + 3) - \frac{14}{2}$

   $-24x + 25$

Motivator

Begin the lesson with the motivator to get students thinking about the topic of the upcoming problem. This lesson is about grouping adults and students for a school trip to a museum. The motivating questions are about a school trip.

Ask the students the following questions to get them interested in the lesson.

- What is a Museum of Natural History?
- Have you ever gone to a Museum of Natural History?
- Where is the nearest Museum of Natural History?
- What is the school trip that you most enjoyed and why?

Explore Together

Problem 1

Students will investigate whole number factors of given values.

Grouping

Ask for a student volunteer to read the Scenario and Problem 1 aloud. Have a student restate the problem. Pose the Guiding Questions below to verify student understanding. Have students work together in small groups to complete parts (A) through (E) of Problem 1.

Guiding Questions

- What information is given in this problem?
- What are the requirements to have an appropriate number of groups?
- How can you decide if a given number of groups is possible?

Common Student Errors

Students may incorrectly ignore the restriction of having the same number of adults in each group and the same number of students in each group.

Another common error is for the students to misinterpret the problem to think that they need the same number of students as the number of adults in each group. In that case, they would answer no to every question.

Call the class back together to have the students discuss and explain their work for parts (A) through (E) of Problem 1.

SCENARIO Some of the students at an elementary school are going on a field trip to a museum of natural history. Some adults are also going on the trip to serve as guides. The school principal wants to divide the students and adults into groups so that each group has the same number of students and the same number of adults.

Problem 1 Splitting Up

Suppose that there are 12 students and 4 adults going on the trip.

A. Can there be two groups of students and adults? Use complete sentences to explain your reasoning.

Yes. If there are two groups, then there are exactly 6 students in each group because $12 \div 2 = 6$ and there are exactly 2 adults in each group because $4 \div 2 = 2$.

B. Can there be three groups of students and adults? Use complete sentences to explain your reasoning.

No. There can be three groups of students, but there cannot be three groups of adults because four adults cannot be divided evenly into three groups.

C. Can there be four groups of students and adults? Use complete sentences to explain your reasoning.

Yes. If there are four groups, then there are exactly 3 students in each group because $12 \div 4 = 3$ and there is exactly 1 adult in each group because $4 \div 4 = 1$.

D. Can there be five groups of students and adults? Use complete sentences to explain your reasoning.

No, because 12 students cannot be divided evenly into 5 groups and 4 adults cannot be divided evenly into 5 groups.

E. What are the different-sized groups that are possible for this number of students and adults? Describe the number of groups, and the numbers of students and adults in each group. Use complete sentences in your answer.

There can be one group of twelve students and four adults, two groups, each with six students and two adults, or four groups, each with three students and one adult.

9

Explore Together

Investigate Problem 1

Students will continue to investigate whole number factors of given values by determining possible group sizes with given restrictions.

Grouping

Ask for a student volunteer to read Question 1 aloud. Have a student restate the problem. Pose the Guiding Questions below to verify student understanding. Have students work together in small groups to complete Questions 1 through 4.

Guiding Questions

- How is the information in Question 1 different from the information in parts (A) through (E) of this problem? How is the information the same?
- How can you determine the number of possible groups for any given number of adults and number of students?

Call the class back together to have the students discuss and present their work for Questions 1 through 4.

Just the Math

Students will be formally introduced to the concept of the factors of numbers. The factors discussed in this lesson are the whole number factors. Ask for a student volunteer to read Question 5 aloud. Have a student restate the problem. Have the students complete Question 5 individually.

Common Student Errors

Some students have difficulty understanding the concept of factors. It may help students if you connect the term *factors* with the idea that factories make products and factors in math multiply to give a **product**. [Thanks to Debra Walters, third grade teacher in Gibsonia, PA, for this learning aide.]

9

Investigate Problem 1

1. Now, suppose that there are 30 students and 6 adults going on the trip. What are the different-sized groups that are possible for this number of students and adults? Describe the number of groups, and the numbers of students and adults in each group. Use complete sentences to explain your reasoning.

   There can be one group of 30 students and 6 adults. There can be two groups of 15 students and 3 adults because 30 ÷ 2 = 15 and 6 ÷ 2 = 3. There can be three groups of 10 students and 2 adults because 30 ÷ 3 = 10 and 6 ÷ 3 = 2. There can be six groups of 5 students and 1 adult because 30 ÷ 6 = 5 and 6 ÷ 6 = 1.

2. Now suppose that there are 29 students and 5 adults going on the trip. What are the different-sized groups that are possible for this number of students and adults? Describe the number of groups, and the number of students and adults in each group. Use complete sentences to explain your reasoning.

   There can be one group of 29 students and 5 adults because 29 ÷ 1 = 29 and 5 ÷ 1 = 5.

3. How does the number of groups relate to the number of students and the number of adults going on the trip? Use complete sentences in your answer.

   Sample Answer: In each case, the number of groups evenly divides both the total number of students and the total number of adults.

4. How did you determine the answers to part (E) and Questions 1 and 2? Use complete sentences to explain your method.

   Sample Answer: In each case, the number of groups is the number that evenly divides both the total number of students and the total number of adults. To find the number of students and the number of adults in each group, divide the total number of students and the total number of adults by the number of groups.

5. **Just the Math: Factors of a Number** The **factors** of a given number are the numbers that evenly divide the given number. For each pair of numbers, list all the factors.

4 and 12	6 and 30	5 and 29
4: 1, 2, 4	**6: 1, 2, 3, 6**	**5: 1, 5**
12: 1, 2, 3, 4, 6, 12	**30: 1, 2, 3, 5, 6, 10, 15, 30**	**29: 1, 29**

Explore Together

Investigate Problem 1

Students will continue to develop an understanding of whole number factors. They will then extend their understanding to write prime factorizations of numbers.

Call the class back together to have the students discuss and present their work for Question 5.

Just the Math

Students will classify whole numbers as prime numbers or composite numbers.

Grouping

Ask for a student volunteer to read Question 6 aloud. Have a student restate the problem. Pose the Guiding Questions below to verify student understanding. Have students work individually to complete Question 6.

Guiding Questions

- What is a prime number?
- How can you decide whether a number is a prime number?
- What is an example of another prime number?
- What is a composite number?
- How can you decide whether a number is a composite number?
- What is an example of another composite number?

Call the class back together to have the students discuss and present their work for Question 6.

Just the Math

Students will write prime factorizations of numbers.

Grouping

Ask for a student volunteer to read Question 7 aloud. Have a student restate the problem. Complete Questions 7 and 8 together as a class.

Investigate Problem 1

In part (E) and in Questions 1 and 2, you were finding all the *common* factors of the numbers of students and adults. List the common factors of each pair of numbers in Question 5.

4 and 12: 1, 2, 4; 6 and 30: 1, 2, 3, 6; 5 and 29: 1

In part (E) and in Questions 1 and 2, did you find all the different-sized groups? How do you know? Use complete sentences in your answer.

Sample Answer: Yes. In each case, the numbers of groups match the common factors.

6. **Just the Math: Prime Numbers** What do you notice about the factors of 5 and 29?

   Sample Answer: The factors of 5 and 29 are 1 and the number itself.

   The numbers 5 and 29 are called *prime numbers*. A **prime number** is a number greater than 1 that has exactly two whole number factors, 1 and itself. The numbers 4, 12, 6, and 30 are **composite numbers,** or numbers greater than 1 that have more than two whole number factors. Make a list of the first 10 prime numbers after the prime number 2.

   3, 5, 7, 11, 13, 17, 19, 23, 29, 31

7. **Just the Math: Prime Factorization** Every composite number can be written as a product of prime numbers. This is called the **prime factorization** of a number. For instance, write 36 as the product of prime factors.

   $36 = \underline{2} \cdot \underline{2} \cdot \underline{3} \cdot \underline{3}$

   You can use your knowledge of powers from Lesson 1.2 to write this prime factorization as products of powers: $2^2 \cdot 3^2$. Write the prime factorization of each of the numbers in Question 5. Write your answers as products of powers.

   $4 = 2^2$; $12 = 2^2 \cdot 3$;

   $6 = 2 \cdot 3$; $30 = 2 \cdot 3 \cdot 5$;

   5 and 29 are prime numbers.

8. Use complete sentences to explain how you can use the prime factorization of a number to help you determine the common factors of a pair of numbers.

   Sample Answer: You can find the common factors of a pair of numbers by first determining the primes and the number of each prime that are common to both numbers in the pair. The common factors are all possible combinations of these primes.

Note While 1 is a common whole number factor for all pairs of numbers, it is not included in Question 7 because it is not prime.

9

Explore Together

Investigate Problem 1

Students will use prime factorizations to find the common factors for given pairs of numbers.

Grouping

Ask for a student volunteer to read Question 9 aloud. Have a student restate the problem. Pose the Guiding Questions below to verify student understanding. Have students work in small groups to complete Questions 9 through 11.

Guiding Questions

- What is a common factor?
- How can you find common factors?

Call the class back together to have the students discuss and present their work for Questions 9 through 11.

Key Formative Assessments

- How can you determine whether a value is a factor of a number?
- What is a prime number?
- What is a composite number?
- Is zero a prime number?
- Is zero a composite number?
- Is one a prime number?
- Is one a composite number?
- How many whole number factors can you divide evenly into the number one?
- Is there any other number that you can think of that is not prime or composite but is still positive?
- How can you find and write the prime factorization of a number?
- When might we use a prime factorization of a number?

Investigate Problem 1

9. Use prime factorizations to find the common factors of each pair of numbers. Show all your work and use a complete sentence in your answer.

9 and 12

$9 = 3^2$; $12 = 2^2 \cdot 3$

3; The common factor is 3.

21 and 42

$21 = 3 \cdot 7$; $42 = 2 \cdot 3 \cdot 7$

3, 7, $3 \cdot 7 = 21$; The common factors are 3, 7, and 21.

36 and 40

$36 = 2^2 \cdot 3^2$; $40 = 2^3 \cdot 5$

2, $2^2 = 4$; The common factors are 2 and 4.

54 and 24

$54 = 2 \cdot 3^3$; $24 = 2^3 \cdot 3$

2, 3, $2 \cdot 3 = 6$; The common factors are 2, 3, and 6.

31 and 50

31 is a prime number; $50 = 2 \cdot 5^2$

There are no common factors other than 1.

10. Suppose that 126 students and 48 adults are going on the trip. What are the different-sized groups that are possible for this number of students and adults? Describe the number of groups, and the numbers of students and adults in each group. Show all your work and use a complete sentence in your answer.

$126 = 2 \cdot 3^2 \cdot 7$; $48 = 2^4 \cdot 3$

Common factors: 2, 3, 6

There can be one group, with 126 students and 48 adults. There can be two groups, each with 63 students and 24 adults because $126 \div 2 = 63$ and $48 \div 2 = 24$. There can be three groups, each with 42 students and 16 adults because $126 \div 3 = 42$ and $48 \div 3 = 16$. There can be six groups, each with 21 students and 8 adults because $126 \div 6 = 21$ and $48 \div 6 = 8$.

11. Which of the groupings would you choose from Question 10? Use complete sentences to explain your answer.

Sample Answer: The six groups seem more manageable for a museum tour.

Take Note

Whenever you see the share with the class icon, your group should prepare a short presentation to share with the class that describes how you solved the problem. Be prepared to ask questions during other groups' presentations and to answer questions during your presentation.

Wrap Up

Close

- Review all key terms and their definitions. Include the terms *factor, prime number, composite number,* and *prime factorization.*
- You may also want to review any other vocabulary terms that were discussed during the lesson which may include *whole numbers, products,* and *common factors*.
- Remind the students to write the key terms and their definitions in the notes section of their notebooks. You may also want the students to include examples.
- Ask the students to compare and contrast each of the key terms.
- Have each student write a whole number on a small piece of paper. Collect the numbers and put them into a small container. Make a 2-column chart on the front board or a piece of poster paper that has the first column labeled as prime numbers and the second column labeled as composite numbers. Have students randomly choose numbers on the small papers one at a time. Have the students work together as a class to classify each number as prime or composite. Choose some of the numbers from each list and have the students write a prime factorization of each. Finally, choose one pair of numbers at a time and ask the students to find the common factors for each pair.

Ties to the Cognitive Tutor Software

In the Cognitive Tutor software, students work on scientific notation problems involving related bases (e.g. 6×10^7 and 6×10^8) and related exponents (e.g. 2×10^8 and 5×10^8). By grouping related problems together, students come to understand the meaning of the different parts of the notation.

Follow Up

Assignment

Use the Assignment for Lesson 9.1 in the Student Assignments book. See the Teacher's Resources and Assessments book for answers.

Assessment

See the Assessments provided in the Teacher's Resources and Assessments book for Chapter 9.

Open-Ended Writing Task

Ask the students to write a pair of whole numbers on a piece of paper. Have them classify each number as prime or composite. Have them write the prime factorization of each number. Finally, have the students find and list any common factors for their pair of numbers. Have them give you their work for this activity as an exit pass when they are leaving class.

Reflections

Insert your reflections on the lesson as it played out in class today.

What went well?

What did not go as well as you would have liked?

How would you like to change the lesson in order to improve the things that did not go well and capitalize on the things that did go well?

9

Notes

9.2

Bits and Bytes

Multiplying and Dividing Powers

Learning By Doing Lesson Map

Get Ready

Objectives

In this lesson, you will:

- Write numbers as powers.
- Multiply powers.
- Divide powers.

Key Terms

- power
- exponent
- product
- quotient

Materials

- calculator

NCTM Content Standards

Grades 9-12 Expectations

Number and Operations Standards

- Develop a deeper understanding of very large and very small numbers and of various representations of them.
- Use number-theory arguments to justify relationships involving whole numbers.
- Judge the effects of such operations as multiplication, division, and computing powers and roots on the magnitude of quantities.
- Develop fluency in operations with real numbers, vectors, and matrices, using mental computation or paper-and-pencil calculations for simple cases and technology for more complicated cases.

Algebra Standards

- Generalize patterns using explicitly defined and recursively defined functions.

Lesson Overview

Within the context of this lesson, students will be asked to:

- Work with whole numbers of large magnitude.
- Write large numbers as powers of the same base.
- Multiply and divide large numbers.
- Develop a rule to multiply large numbers by finding the product of powers of the same base.
- Develop a rule to divide large numbers by finding the quotient of powers of the same base.

Essential Questions

The following key questions are addressed in this lesson:

1. What is computer memory space?
2. What is a quotient?
3. What is a product?
4. What is a power?
5. What is an exponent?

Show The Way

Warm Up

Place the following questions or an applicable subset of these questions on the board before students enter class. Students should begin working as soon as they are seated.

Write the prime factorization for each number.

1. 156 $2^2 \cdot 3 \cdot 13$
2. 18 $2 \cdot 3^2$
3. 25 5^2
4. 181 **181 is prime.**
5. 14 $2 \cdot 7$
6. 625 5^4
7. 1275 $3 \cdot 5^2 \cdot 17$
8. 2340 $2^2 \cdot 3^2 \cdot 5 \cdot 13$

9

List the common factors for each pair of numbers.

9. 18 and 156 **2, 3, 6**
10. 14 and 18 **2**
11. 25 and 625 **5, 25**
12. 156 and 1275 **3**
13. 156 and 2340 **2, 3, 4, 6, 12, 13, 26, 39, 52, 78, 156**
14. 625 and 2340 **5**

Motivator

Begin the lesson with the motivator to get students thinking about the topic of the upcoming problem. This lesson is about the amount of space available in different forms of technology. The motivating questions are about memory space for technology.

Ask the students the following questions to get them interested in the lesson.

- What is memory space for technology?
- What types of technology use memory space?
- What units of measurement do you know of for memory?
- What else do you know about memory capacity for computers and technology?

Explore Together

Problem 1

Students will investigate very large numbers by writing them as powers.

Grouping

Ask for a student volunteer to read the Scenario and Problem 1 aloud. Have a student restate the problem. Pose the Guiding Questions below to verify student understanding. Have students work together in small groups to complete parts (A) through (D) of Problem 1.

Guiding Questions

- What types of devices use memory?
- Which is a larger unit of memory, a bit or a byte?
- How many bits of memory are equal to 2 bytes?
- How many bytes of memory are equal to 80 bits?
- What is a prime number? What is a prime factorization?
- What is the prime factorization of 27?
- How can you find and write the prime factorization of 32?

Note Because the numbers in this problem are very large, it is especially important to have the students use their calculators and work in groups to complete this lesson.

Common Student Errors

Students will often feel overwhelmed with the magnitude of the numbers in this lesson. Reassure them and remind them of the strategy to find prime factorizations one factor at a time.

Call the class back together and have the students discuss and explain their work for parts (A) through (D) of Problem 1.

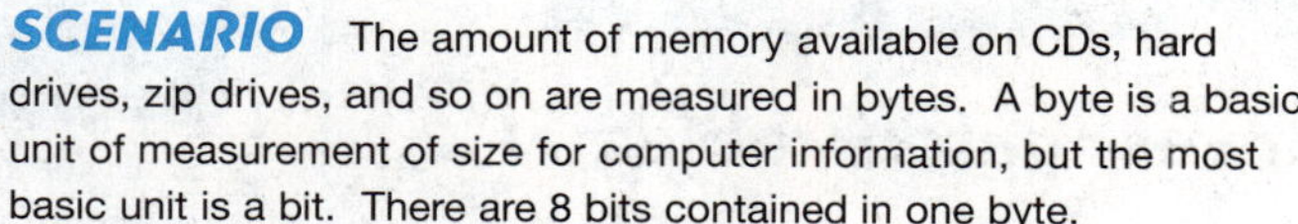

SCENARIO The amount of memory available on CDs, hard drives, zip drives, and so on are measured in bytes. A byte is a basic unit of measurement of size for computer information, but the most basic unit is a bit. There are 8 bits contained in one byte.

Problem 1 Kilobytes, Megabytes, and Gigabytes

The most common units of measurement for computer data storage are kilobytes, megabytes, and gigabytes.

A. One kilobyte is the same as 1024 bytes. Write the prime factorization of the number of bytes in one kilobyte.

$1024 = 2^{10}$

B. One megabyte is the same as 1,048,576 bytes. Write the prime factorization of the number of bytes in one megabyte.

$1{,}048{,}576 = 2^{20}$

C. One gigabyte is the same as 1,073,741,824 bytes. Write the prime factorization of the number of bytes in one gigabyte.

$1{,}073{,}741{,}824 = 2^{30}$

D. What do the prime factorizations in parts (A) through (C) have in common? Use a complete sentence in your answer.

Sample Answer: The prime factorizations are all powers of 2.

9

Investigate Problem 1

1. Divide the number of bytes in one megabyte by the number of bytes in one kilobyte. Show all your work.

$\frac{1{,}048{,}576}{1024} = 1024$

What does this number represent? Use a complete sentence in your answer.

Sample Answer: The number represents the number of times greater one megabyte is than one kilobyte.

2. Write the numerator, denominator, and quotient from Question 1 as powers of 2.

$\frac{2^{20}}{2^{10}} = 2^{10}$

Investigate Problem 1

Grouping

Ask for a student volunteer to read Question 1 aloud. Have a student restate the problem. Pose the Guiding Questions on the next page to verify student understanding. Have students work together in small groups to complete Question 1 through Question 7.

Explore Together

Investigate Problem 1

Students will develop a rule in Question 6 to divide numbers with large values by using their prime factorizations.

Guiding Questions

- How can you divide large numbers?
- What makes division of large numbers more difficult than division of small numbers if the process is the same?
- What is a quotient?

Call the class back together to have the students discuss and present their work for Questions 1 through 7.

Common Student Errors

Some students may incorrectly generalize the results in Questions 2 and 4 to think that the result for a division of powers will be the base to ten less than the original exponent in the numerator.

Key Formative Assessments

- Which is easier, long division of the original numbers or using the rule that you developed in Question 6?
- Why would you want to use the rule that you developed?
- Would your rule work for numbers written as powers of different bases? Why or why not?
- How can you write your rule as an algebraic formula?

9

Investigate Problem 1

3. Divide the number of bytes in one gigabyte by the number of bytes in one kilobyte. Show all your work and use a complete sentence in your answer.

$\frac{1,073,741,824}{1024} = 1,048,576$

What does this number represent? Use a complete sentence in your answer.

Sample Answer: The number represents the number of times greater one gigabyte is than one kilobyte.

4. Write the numerator, denominator, and quotient from Question 3 as powers of 2.

$\frac{2^{30}}{2^{10}} = 2^{20}$

5. In Question 2, how do the bases of all the powers compare? In Question 4, how do the bases of all the powers compare? In each division problem, how does the exponent in the quotient relate to the exponents in the numerator and the denominator? Use complete sentences in your answer.

Sample Answer: In Questions 2 and 4, all the bases are the same. The exponent in the quotient is the difference of the exponent in the numerator and the exponent in the denominator.

6. Write a rule that you can use to find the quotient of two powers that have the same base. Use a complete sentence in your answer.

Sample Answer: The quotient is a power that has the same base as the powers in the numerator and denominator, and the exponent is found by subtracting the exponent in the denominator from the exponent in the numerator.

7. Use your rule from Question 6 to simplify each expression. Show all your work and write your answer as a power.

$\frac{4^9}{4^8}$ $\frac{5^6}{5^3}$ $\frac{10^{12}}{10^8}$

$\frac{4^9}{4^8} = 4^{9-8}$ $= 4^1$

$\frac{5^6}{5^3} = 5^{6-3}$ $= 5^3$

$\frac{10^{12}}{10^8} = 10^{12-8}$ $= 10^4$

Explore Together

Problem 2

Students will develop a rule to multiply numbers with large values using their prime factorizations.

Grouping

Ask for a student volunteer to read Problem 2 aloud. Have a student restate the problem. Pose the Guiding Questions below to verify student understanding. Have students work together in small groups to complete parts (A) through (C) of Problem 2.

Guiding Questions

- What is meant by the term *external*? What is meant by the term *internal*?
- What information is given in this problem?
- How does this problem differ from Problem 1?
- How are the problems similar?

Call the class back together to have the students discuss and present their work for parts (A) through (C) of Problem 2.

Investigate Problem 2

Grouping

Ask for a student volunteer to read Question 1 of Problem 2 aloud. Have a student restate the problem. Pose the Guiding Questions below to verify student understanding. Have students work together in small groups to complete Questions 1 through 4 of Problem 2.

Guiding Questions

- What information is given in this problem?
- What is a factor?
- What is a product?

Problem 2 Storage Options

External (outside) storage for a computer comes in different forms: CD, DVD, external hard drive, flash drive, and so on. The sizes for these different options vary and get larger as new technologies develop.

A. One model of hard drive can store up to 4 gigabytes of data. Find the number of bytes that this hard drive can store. Remember that 1 gigabyte = 1,073,741,824 bytes. Show all your work and use a complete sentence in your answer.

1,073,741,824(4) = 4,294,967,296;

The hard drive can store 4,294,967,296 bytes.

B. One model of flash drive can store up to 128 megabytes of data. Find the number of bytes that this flash drive can store. Remember that 1 megabyte = 1,048,576 bytes. Show all your work and use a complete sentence in your answer.

1,048,576(128) = 134,217,728;

The flash drive can store 134,217,728 bytes.

C. Use complete sentences to explain how you found your answers to parts (A) and (B).

Sample Answer: To find the number of bytes stored by the hard drive, multiply the number of bytes in one gigabyte by the number of gigabytes available on the hard drive. To find the number of bytes stored by the flash drive, multiply the number of bytes in one megabyte by the number of megabytes available on the flash drive.

Investigate Problem 2

1. Write the multiplication problem that you used in part (A) to find the number of bytes of storage for the hard drive.

   1,073,741,824(4) = 4,294,967,296

   Write each factor in the product as a power of 2. Then write the result as a power of 2. What do you notice about the exponents? Use a complete sentence in your answer.

   $2^{30}(2^2) = 2^{32}$; Sample Answer: The exponent in the product is the sum of the exponents in the factors.

2. Write the multiplication problem that you used in part (B) to find the number of bytes of storage for the hard drive.

   1,048,576(128) = 134,217,728

9

Explore Together

Investigate Problem 2

Students will use the rule in Question 5 that they developed in Question 4 for multiplying large numbers that are written as prime factorizations.

Grouping

Students will be working together in small groups to complete Questions 1 through 4 of Problem 2.

Call the class back together to have the students discuss and present their work for Questions 1 through 4 of Problem 2.

Grouping

Ask for a student volunteer to read Question 5 aloud. Have a student restate the problem. Have students work together in small groups to complete Questions 5 through 7 of Problem 2.

Call the class back together to have the students discuss and present their work for Questions 5 through 7.

Key Formative Assessments

- Which is easier, multiplication of the original numbers or using the rule that you developed in Question 4 of Problem 2?
- Why would you want to use the rule that you developed in Question 4?
- Would your rule work for numbers written as powers of different bases? Why or why not?
- How can you write your rule as an algebraic formula?
- How do the rules that you developed for division and for multiplication differ? How are the rules similar?
- How can you find the product of $3^2 \cdot 3^3 \cdot 3^5$? Does that follow with the rule that you developed for a product in Question 4 of Problem 2?

9

Investigate Problem 2

Write each factor in the product as a power of 2. Then write the result as a power of 2. What do you notice about the exponents? Use a complete sentence in your answer.

$2^{20}(2^7) = 2^{27}$; Sample Answer: The exponent in the product is the sum of the exponents in the factors.

3. How do the bases of the powers in Question 1 compare? How do the bases of the powers in Question 2 compare? Use a complete sentence in your answer.

The bases in Question 1 are the same and the bases in Question 2 are the same.

4. Write a rule that you can use to find the product of two powers that have the same base. Use a complete sentence in your answer.

Sample Answer: The product is a power that has the same base as the powers in the factors and the exponent is found by adding the exponents in the factors.

5. Use your rule from Question 4 to simplify each expression. Show all your work and write your answer as a power.

$5^6 \cdot 5^8$

$5^6 \cdot 5^8 = 5^{6+8}$
$= 5^{14}$

$2^3 \cdot 2^5$

$2^3 \cdot 2^5 = 2^{3+5}$
$= 2^8$

$1^9 \cdot 1^7$

$1^9 \cdot 1^7 = 1^{9+7}$
$= 1^{16}$

6. Simplify each expression, if possible. Show all your work and write your answer as a power.

$4^5 \cdot 4^5$

$4^5 \cdot 4^5 = 4^{5+5}$
$= 4^{10}$

$\frac{8^{10}}{8^3}$

$\frac{8^{10}}{8^3} = 8^{10-3}$
$= 8^7$

$2^3 \cdot 3^2$

$2^3 \cdot 3^2$

$\frac{6^4}{5^4}$

$\frac{6^4}{5^4}$

$10^5 \cdot 10^1$

$10^5 \cdot 10^1 = 10^{5+1}$
$= 10^6$

$8^4 \cdot 3^4$

$8^4 \cdot 3^4$

7. Describe the situations in which you would use powers to multiply and divide. Use complete sentences in your answer.

Sample Answer: Use powers to multiply and divide when both numbers in a product and quotient are very large and both numbers can be written as powers with the same base.

Wrap Up

Close

- Review all key terms and their definitions. Include the terms *power, exponent, product,* and *quotient.*
- You may also want to review any other vocabulary terms that were discussed during the lesson which may include *large numbers, prime numbers, prime factorizations, internal,* and *external.*
- Remind the students to write the key terms and their definitions in the notes section of their notebooks. You may also want the students to include examples.
- Ask the students to restate, then compare and contrast the two rules that they developed in this lesson.
- Have the students write two powers with the same base. Then have them find the product of the powers and the quotient of the larger power divided by the smaller power.
- Have the students trade their numbers with another student and have their partner find the product of their numbers and the quotient of the larger power divided by the smaller power.

Ties to the Cognitive Tutor Software

In the Cognitive Tutor software, students work on scientific notation problems involving related bases (e.g., 6×10^7 and 6×10^8) and related exponents (e.g., 2×10^8 and 5×10^8). By grouping related problems together, students come to understand the meaning of the different parts of the notation.

9

Follow Up

Assignment

Use the Assignment for Lesson 9.2 in the Student Assignments book. See the Teacher's Resources and Assessments book for answers.

Assessment

See the Assessments provided in the Teacher's Resources and Assessments book for Chapter 9.

Open-Ended Writing Task

Ask the students to restate the rule that they developed in Question 6 of Problem 1. Then ask them to write a response to the following statement, "The quotient of $\frac{2^4}{2^6} = 2^{-2}$ therefore, 2^{-2} must be equal to $\frac{1}{2^2} = \frac{1}{4}$." Next, have the students predict the value of 3^{-2} and explain their prediction.

Reflections

Insert your reflections on the lesson as it played out in class today.

What went well?

What did not go as well as you would have liked?

How would you like to change the lesson in order to improve the things that did not go well and capitalize on the things that did go well?

9

Notes

9.3

As Time Goes By

Zero and Negative Exponents

Learning By Doing Lesson Map

Get Ready

Objectives

In this lesson, you will:

- Write a number as a power.
- Evaluate powers with positive, negative, and zero exponents.

Key Terms

- positive exponent
- negative exponent
- zero exponent

NCTM Content Standards

Grades 9-12 Expectations

Number and Operations Standards

- Develop a deeper understanding of very large and very small numbers and of various representations of them.
- Use number-theory arguments to justify relationships involving whole numbers.
- Judge the effects of such operations as multiplication, division, and computing powers and roots on the magnitude of quantities.
- Develop fluency in operations with real numbers, vectors, and matrices, using mental computation or paper-and-pencil calculations for simple cases and technology for more complicated cases.

Algebra Standards

- Generalize patterns using explicitly defined and recursively defined functions.
- Understand the meaning of equivalent forms of expressions, equations, inequalities, and relations.

Measurement Standards

- Use unit analysis to check measurement computations.

Lesson Overview

Within the context of this lesson, students will be asked to:

- Use unit analysis to find equivalent times.
- Write numbers as powers.
- Evaluate powers that have positive exponents, powers that have negative exponents, and powers that have an exponent of zero.
- Evaluate the products and the quotients of powers with various exponents.

Essential Questions

The following key questions are addressed in this lesson:

1. What is a computer operation or communication?
2. What is a positive exponent?
3. What is a negative exponent?
4. What is a power with an exponent of zero?
5. How can you evaluate powers with various exponents?

Show The Way

Warm Up

Place the following questions or an applicable subset of these questions on the board before students enter class. Students should begin working as soon as they are seated.

Simplify each expression, if possible. Write your answer as a power.

1. $3^4 \cdot 3^8$ 3^{12}

2. $7^6 \cdot 7^3 \cdot 7$ 7^{10}

3. $5^4 \cdot 5^7 \cdot 6^2 \cdot 6^3$ $5^{11} \cdot 6^5$

4. $\frac{3^8}{3^3}$ 3^5

5. $\frac{3^3}{3^8}$ $\frac{1}{3^5}$

6. $\frac{8^6}{3^7}$ $\frac{8^6}{3^7}$

9

Motivator

Begin the lesson with the motivator to get students thinking about the topic of the upcoming problem. This lesson is about the short amount of time for computer operations. The motivating questions are about computer speeds.

Ask the students the following questions to get them interested in the lesson.

- How fast do computers seem to process information?
- When you type information into a computer, how much time do you estimate passes between when you press the key on the keyboard and when the corresponding letter or value appears on the computer monitor?
- Do most computer operations or communications seem to take more or less than one second?

Explore Together

Problem 1

Students will calculate the number of seconds that are equivalent to a given number of milliseconds using unit analysis and then using powers.

Grouping

Ask for a student volunteer to read the Scenario and Problem 1 aloud. Have a student restate the problem. Pose the Guiding Questions below to verify student understanding. Have students work together in small groups to complete parts (A) through (D) of Problem 1.

Guiding Questions

- What information is given in this problem?
- What is a millisecond?
- How can you determine the number of seconds that are equal to a given number of milliseconds using unit analysis?
- How many seconds are equal to 5000 milliseconds?

Common Student Errors

Students often make careless errors when working with numbers that are so small. Remind them to be careful and to check their work for reasonableness.

Call the class back together to have the students discuss and explain their work for parts (A) through (D) of Problem 1.

SCENARIO Computer operations and communications between electronic equipment happen so fast that they need to be measured in units of time that are smaller than seconds. Three units of time that are smaller than seconds are milliseconds, microseconds, and nanoseconds.

Problem 1 Faster and Faster

A. One millisecond is the same as $\frac{1}{1000}$ second. Complete the statement below to determine the number of seconds that there are in 10,000 milliseconds.

$$10{,}000 \text{ \sout{milliseconds}}\left(\frac{1 \text{ second}}{1000 \text{ \sout{milliseconds}}}\right) = \frac{\boxed{10{,}000} \text{ seconds}}{\boxed{1000}}$$

$$= \boxed{10} \text{ seconds}$$

There are __10__ seconds in 10,000 milliseconds.

B. Complete the statement below to determine the number of seconds there are in 1000 milliseconds.

$$1000 \text{ \sout{milliseconds}}\left(\frac{1 \text{ second}}{1000 \text{ \sout{milliseconds}}}\right) = \frac{\boxed{1000} \text{ seconds}}{\boxed{1000}}$$

$$= \boxed{1} \text{ second}$$

There is __1__ second in 1000 milliseconds.

C. Complete the statement below to determine the number of seconds there are in 100 milliseconds.

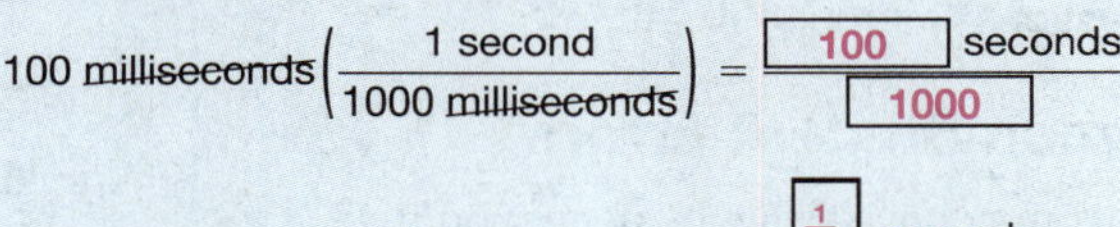

$$100 \text{ \sout{milliseconds}}\left(\frac{1 \text{ second}}{1000 \text{ \sout{milliseconds}}}\right) = \frac{\boxed{100} \text{ seconds}}{\boxed{1000}}$$

$$= \boxed{\tfrac{1}{10}} \text{ second}$$

There is __$\frac{1}{10}$__ second in 1000 milliseconds.

D. How does the numerator compare to the denominator in each problem in parts (A) through (C)? Use complete sentences in your answer.

In part (A), the numerator is greater than the denominator.
In part (B), the numerator and denominator are equal.
In part (C), the denominator is greater than the numerator.

Explore Together

Investigate Problem 1

Students will investigate bases with an exponent of zero as well as bases with negative exponents.

Grouping

Ask for a student volunteer to read Question 1 aloud. Have a student restate the problem. Pose the Guiding Questions below to verify student understanding. Complete Question 1 together as a class. Have students work together in small groups to complete Questions 2 through 4.

Guiding Questions

- What is the quotient of any number divided by itself?
- Why do we define a number raised to an exponent of zero to be 1?

Call the class back together to have the students discuss and present their work for Questions 2 through 4.

Common Student Errors

Students may have difficulty accepting the definition that a base to a zero exponent is defined to be 1. Refocus the students by having them summarize the information in Questions 2 through 4 to help them understand this definition. Also tell them that the rules of mathematics have to be defined in a way that they never contradict other rules.

Grouping

Ask for a student volunteer to read Question 5 aloud. Have a student restate the problem. Have students work together in small groups to complete Question 5.

Call the class back together to have the students discuss and present their work for Question 5. Have the students summarize their findings from Question 5 as a rule for negative exponents of bases.

Investigate Problem 1

1. Show how you could use powers to find the answer to part (A). Show all your work.

$\frac{10^4}{10^3} = 10^{4-3} = 10^1$

2. Write the numerator and denominator from the statement in part (B) as a quotient of powers.

$\frac{10^3}{10^3}$

According to your answer in part (B), what is the value of this quotient?

1

Use the rule for finding a quotient of powers to simplify the quotient of powers above. Show your work and write your answer as a power.

$\frac{10^3}{10^3} = 10^{3-3} = 10^0$

3. Consider the quotient $\frac{2^4}{2^4}$. What is the value of this quotient?

Use a complete sentence to explain your reasoning.

Sample Answer: Because the numerator is equal to the denominator, the value must be one.

Write your answer as a power of 2. Show your work.

$\frac{2^4}{2^4} = 2^{4-4} = 2^0$

4. What is the value of any number raised to the power of zero? Use a complete sentence in your answer.

The value of any number raised to the power of zero is one.

5. Write the numerator and denominator from the statement in part (C) as a quotient of powers.

$\frac{10^2}{10^3}$

Use the rule for finding a quotient of powers to write the quotient as a power of 10.

$\frac{10^2}{10^3} = 10^{2-3} = 10^{-1}$

According to your answer to part (C), what is the value of this power of 10?

$\frac{1}{10}$

So, $10^{-1} = \frac{1}{10^1}$.

Explore Together

Investigate Problem 1

Students will investigate powers involving negative exponents.

Grouping

Ask for a student volunteer to read Question 6 aloud. Have a student restate the problem. Pose the Guiding Questions below to verify student understanding. Have the students complete Question 6 and 7 individually.

Guiding Questions

- What are you asked to do in these questions?
- How can you complete Question 6?

Call the class back together to have the students discuss and present their work for Question 6 and 7.

Grouping

Ask for a student volunteer to read Question 8 aloud. Have a student restate the problem. Pose the Guiding Questions below to verify student understanding. Have students work together in small groups to complete Question 8 and 9.

Guiding Questions

- What is being asked in Question 8?
- Are 2^3 and 2^{-3} equivalent powers?
- How can you rewrite a power that has a negative exponent so that the exponent is positive?

Call the class back together to have the students discuss and present their work for Question 8 and 9.

Grouping

Have the students complete Question 10 individually.

Call the class back together to have the students discuss and present their work for Question 10.

Investigate Problem 1

6. Complete the table below that shows small units of time.

Unit	Number of seconds	Number of seconds as a power with a positive exponent	Number of seconds as a power with a negative exponent
Millisecond	$\frac{1}{1000}$	$\frac{1}{10^{3}}$	10^{-3}
Microsecond	$\frac{1}{1,000,000}$	$\frac{1}{10^{6}}$	10^{-6}
Nanosecond	$\frac{1}{1,000,000,000}$	$\frac{1}{10^{9}}$	10^{-9}

7. Use complete sentences to explain how you can write a power with a negative exponent as a power with a positive exponent.

Sample Answer: Write the power in the denominator of a fraction whose numerator is one and change the sign of the exponent in the power from negative to positive.

8. Rewrite the power so that the exponent is positive.

5^{-3} $\qquad$ 4^{-6} $\qquad$ 3^{-10}

$\frac{1}{5^3}$ $\qquad$ $\frac{1}{4^6}$ $\qquad$ $\frac{1}{3^{10}}$

9. Suppose that you are given a fraction that has a power in the denominator. How can you rewrite the value so that the power is no longer in the denominator? Use complete sentences in your answer.

Sample Answer: Write the power in the numerator and change the sign of the exponent in the power from negative to positive (or from positive to negative).

10. Rewrite the fraction so that there is no power in the denominator.

$\frac{1}{8^3}$ $\qquad$ $\frac{1}{1^5}$ $\qquad$ $\frac{1}{6^7}$

8^{-3} $\qquad$ 1^{-5} $\qquad$ 6^{-7}

9

Explore Together

Investigate Problem 1

Students will summarize their findings about powers and use their summaries to simplify expressions involving operations of powers.

Grouping

Ask for a student volunteer to read Question 11 aloud. Have a student restate the problem. Pose the Guiding Questions below to verify student understanding. Have the students complete Question 11 individually.

Guiding Questions

- What is being asked in Question 11?
- What is the difference between a power and an exponent?

Call the class back together to have the students discuss and explain their work for Question 11.

Grouping

Ask for a student volunteer to read Question 12 aloud. Have a student restate the problem. Have students work together in small groups to complete Question 12.

Call the class back together to have the students discuss and present their work for Question 12.

Key Formative Assessments

- What is the value of any power with an exponent of zero?
- What do you know about any power with a positive exponent?
- What do you know about any power with a negative exponent?
- How can you evaluate powers with negative exponents? How can you evaluate powers with positive exponents? How can you evaluate powers with zero as an exponent?

9

Investigate Problem 1

11. What can you conclude about the value of a power when the exponent is greater than zero? Use a complete sentence in your answer.

The value of the power is greater than one.

What can you conclude about the value of a power when the exponent is less than zero? Use a complete sentence in your answer.

The value of the power is less than one.

What can you conclude about the value of a power when the exponent is zero? Use a complete sentence in your answer.

The value of the power is equal to one.

12. Use what you have learned so far in this chapter to simplify each expression completely. Show all your work.

$\frac{8^3}{8^5}$

$\frac{8^3}{8^5} = 8^{3-5} = 8^{-2} = \frac{1}{8^2} = \frac{1}{64}$

$6^{-3} \cdot 6^2$

$6^{-3} \cdot 6^2 = 6^{-3+2} = 6^{-1} = \frac{1}{6}$

$10^0 \cdot 10^{-3}$

$10^0 \cdot 10^{-3} = 10^{0-3} = 10^{-3} = \frac{1}{10^3} = \frac{1}{1000}$

$\frac{2^0}{2^4}$

$\frac{2^0}{2^4} = 2^{0-4} = 2^{-4} = \frac{1}{2^4} = \frac{1}{16}$

$\frac{2^{-3}}{2^2}$

$\frac{2^{-3}}{2^2} = 2^{-3-2} = 2^{-5} = \frac{1}{2^5} = \frac{1}{32}$

$5^4 \cdot 5^{-2}$

$5^4 \cdot 5^{-2} = 5^{4+(-2)} = 5^2 = 25$

Wrap Up

Close

- Review all key terms and their definitions. Include the terms *positive exponent, negative exponent,* and *zero exponent.*
- You may also want to review any other vocabulary terms that were discussed during the lesson which may include *power, exponent, unit analysis, product,* and *quotient.*
- Remind the students to write the key terms and their definitions in the notes section of their notebooks. You may also want the students to include examples.
- Ask the students to summarize what they discovered in this lesson.
- Ask the students to explain why we define any power with an exponent of zero to be 1.
- Have the students summarize how to evaluate powers with positive exponents.
- Have the students summarize how to evaluate powers with negative exponents.

Ties to the Cognitive Tutor Software

Students often think that negative exponents will result in negative numbers. In the Cognitive Tutor software, students work on contextual problems involving negative exponents, helping to counter this misconception.

9

Follow Up

Assignment

Use the Assignment for Lesson 9.3 in the Student Assignments book. See the Teacher's Resources and Assessments book for answers.

Assessment

See the Assessments provided in the Teacher's Resources and Assessments book for Chapter 9.

Open-Ended Writing Task

Ask the students to conjecture what the value of $\frac{1}{2^{-3}}$ would be. They should explain their theory in a short paragraph and also think of a way that they could test their conjecture to determine whether they are correct.

Reflections

Insert your reflections on the lesson as it played out in class today.

What went well?

____

____

What did not go as well as you would have liked?

____

____

How would you like to change the lesson in order to improve the things that did not go well and capitalize on the things that did go well?

____

____

____

____

____

Notes

9.4 Large and Small Measurements

Scientific Notation

Learning By Doing Lesson Map

Get Ready

Objectives

In this lesson, you will:

- Write numbers in scientific notation.
- Write numbers in standard form.

Key Terms

- standard form
- scientific notation

NCTM Content Standards

Grades 9-12 Expectations

Number and Operations Standards

- Develop a deeper understanding of very large and very small numbers and of various representations of them.
- Use number-theory arguments to justify relationships involving whole numbers.
- Judge the effects of operations such as multiplication, division, and computing powers and roots on the magnitude of quantities.
- Develop fluency in operations with real numbers, vectors, and matrices, using mental computation or paper-and-pencil calculations for simple cases and technology for more complicated cases.

Algebra Standards

- Understand the meaning of equivalent forms of expressions, equations, inequalities, and relations.

Lesson Overview

Within the context of this lesson, students will be asked to:

- Write very large and very small numbers that are represented in standard form by using scientific notation.
- Write numbers that are represented in scientific notation by using standard form.

Essential Questions

The following key questions are addressed in this lesson:

1. What is scientific notation?
2. When does it help to write numbers in scientific notation?
3. How can you write a number in scientific notation?
4. What is standard form?
5. How can you write a number in standard form?

Show The Way

Warm Up

Place the following questions or an applicable subset of these questions on the board before students enter class. Students should begin working as soon as they are seated.

Tell which of the numbers in each group are greater than one or equal to one, but less than ten.

1. 1; 14; 4.2; 8.9349; 10.0; and 12.04 **1; 4.2; 8.9349**

2. 6.328; 5343.1; 0.35; 4.367543; and 8.0054 **6.328; 4.367543; and 8.0054**

Complete each statement to make it true.

3. 15 • __1000__ = 15,000

4. 2.13 • __10,000,000__ = 21,300,000

9

5. 1.435 • 0.1 = __0.1435__

6. 1.435 • 0.01 = __0.01435__

7. 1.435 • 0.00001 = __0.00001435__

8. __5.76__ • 0.001 = 0.00576

Motivator

Begin the lesson with the motivator to get students thinking about the topic of the upcoming problem. This lesson is about representing very large and very small numbers. The motivating questions are about very large and very small numbers.

Ask the students the following questions to get them interested in the lesson.

- What quantities in our universe can you think of that are represented with very large numbers?
- What quantities in our universe can you think of that are represented with very small numbers
- What is the largest number that you can think of?
- What makes working with very large and very small numbers difficult?

Explore Together

Problem 1

Students will write large numbers using scientific notation and will use standard form to write numbers that are in scientific notation.

Take Note

To write a number in *standard form* means to write the number as a numeral.

Grouping

Ask for a student volunteer to read the Scenario and Problem 1 aloud. Have a student restate the problem. Pose the Guiding Questions below to verify student understanding. Have students work together in small groups to complete parts (A) through (E) of Problem 1.

Guiding Questions

- Is it difficult or tedious to read the number 152 out loud?
- Is it difficult or tedious to read the number 152,129,573,289,341 out loud?
- Are there any numbers larger than 152 but smaller than 152,129,573,289,341 that are easy to read out loud?
- What is the smallest number that you feel is large enough to be difficult or tedious to read out loud?
- What are the requirements to write a number in scientific notation?
- What are some numbers that you can think of that are greater than or equal to one, but less than ten?

Common Student Errors

Students may have difficulty reading the numbers in this problem out loud. However, reading very large or very small numbers is an important skill. Encourage the students to read them correctly.

SCENARIO In our universe, we encounter very large numbers, such as the diameter of Earth, the distances to Mars, other planets and stars, and so on. We also encounter very small numbers, such as the weight of a butterfly, the length of a blood cell, the width of a grain of sand, and so on. **Scientific notation** is a shorthand notation that is used to write these numbers so that they can be more easily used in computations.

Problem 1 The Big Stuff

Scientific notation uses powers of ten to represent large and small numbers. A number is written in scientific notation if it is the product of a number greater than or equal to one and less than 10 and a power of 10. You can use scientific notation to describe the distances from the sun to planets in the solar system.

A. Earth's average distance from the sun is 1.496×10^8 kilometers. Complete the statement below to write this distance in standard form.

1.496×10^8 kilometers = **149,600,000** kilometers

B. Mercury's average distance from the sun is 5.8×10^7 kilometers. Complete the statement below to write this distance in standard form.

5.8×10^7 kilometers = **58,000,000** kilometers

C. What happens to the value of a power of ten as the exponent gets larger? Use a complete sentence in your answer.

Sample Answer: The value gets larger.

D. Mars' average distance from the sun is 227,900,000 kilometers. Complete the statement below to write this distance in scientific notation.

227,900,000 kilometers = $2.279 \times$ **10^8** kilometers

E. Saturn's average distance from the sun is 1,433,000,000 kilometers. Complete the statement below to write the diameter in scientific notation.

1,433,000,000 kilometers = **1.433** $\times 10^9$ kilometers

Call the class back together to have the students discuss and explain their work for parts (A) through (E) of Problem 1.

Explore Together

Investigate Problem 1

Students will continue to investigate numbers in scientific notation.

Grouping

Ask for a student volunteer to read Question 1 aloud. Have a student restate the problem. Pose the Guiding Questions below to verify student understanding. Have students work together in small groups to complete Question 1 through 5.

Take Note

Note that when you compare two numbers written in scientific notation that have the same exponent, the number that is larger is the number with the larger value between 1 and 10. For instance, the number 3.4×10^5 is greater than the number 3.2×10^5.

9

Guiding Questions

- How can you determine whether numbers are written in scientific notation?
- What do numbers written in scientific notation look like?

Call the class back together to have the students discuss and present their work for Questions 1 through 5.

Problem 2

Ask for a student volunteer to read Problem 2 aloud. Have a student restate the problem. After students understand the situation, complete part (A) together as a class.

Guiding Questions

- What is the value of the number 10^{-1} when written as a fraction? What is the value of the number 10^{-1} when written as a decimal?
- What is the value of the number 10^{-5} when written as a fraction? What is the value of the number 10^{-5} when written as a decimal?

Investigate Problem 1

1. The following numbers represent the closest distance of a comet to the sun during the comet's orbit. Which distances are *not* written in scientific notation? How do you know? Use complete sentences in your explanation.

   Comet Halley: 8.78×10^7 kilometers

   Comet Encke: 50×10^6 kilometers

   Comet Wild 2: 2.368×10^8 kilometers

   Comet Hale-Bopp: 136×10^6 kilometers

   Comet Encke's distance and Comet Hale-Bopp's distance are not written in scientific notation because the numbers being multiplied by the power of 10 are not between 1 and 10.

2. Write the distances that you identified in Question 1 in scientific notation.

   Comet Encke: 5.0×10^7 kilometers;

   Comet Hale-Bopp: 1.36×10^8 kilometers

3. Without writing the distances in standard form, list the comets in order from shortest distance from the sun to longest distance from the sun.

   Comet Encke, Comet Halley, Comet Hale-Bopp, Comet Wild 2

4. Use a complete sentence to explain how to compare two large numbers that are in scientific notation.

   The number with the greater exponent is the larger number.

5. Write each of the distances from Question 1 in standard form.

   Comet Halley: **87,800,000 kilometers**

   Comet Encke: **50,000,000 kilometers**

   Comet Wild 2: **236,800,000 kilometers**

   Comet Hale-Bopp: **136,000,000 kilometers**

Problem 2 The Small Stuff

You can also use scientific notation to write very small numbers.

A. The diameter of a large grain of sand is 2.0×10^{-1} centimeter. Complete the statement below to write the diameter in standard form.

   2.0×10^{-1} centimeter = **0.2** centimeter

Explore Together

Problem 2

Students will continue to investigate using scientific notation to represent numbers that have a value between zero and one.

Grouping

Have students work together in small groups to complete parts (B) through (E) of Problem 2.

Common Student Errors

Students often will write a value like 8.4×10^{-4} incorrectly by one factor of 10. Most frequently, students will think the value is 0.000084. Some students make this mistake by thinking that the number of zeros to the right of the decimal point is the same as the exponent after the negative sign. Other students will incorrectly write 0.0084. When errors like these occur, refocus the students by having them write the power of 10 as a decimal value and then read the number out loud. If they can correctly identify the number 10^{-4} as 0.0001, then they are more likely to understand what it means to multiply a number by 10^{-4}.

Call the class back together to have the students discuss and present their work for parts (B) through (E) of Problem 2.

Investigate Problem 2

Grouping

Have students work together in small groups to complete Question 1 through Question 4 of Investigate Problem 2.

Problem 2 The Small Stuff

B. The diameter of a red blood cell is 8.4×10^{-4} centimeter. Complete the statement below to write this diameter in standard form.

8.4×10^{-4} centimeter = **0.00084** centimeter

C. What happens to the value of a number as the absolute value of the exponent of a power of ten gets larger? Use a complete sentence in your answer.

Sample Answer: The value of the number gets smaller.

D. The diameter of a bacterium is 0.00002 centimeter. Complete the statement below to write this diameter in scientific notation.

0.00002 centimeter = $2.0 \times$ **10^{-5}** centimeter

E. The diameter of a human hair is 0.0025 centimeter. Complete the statement below to write the diameter in scientific notation.

0.0025 centimeter = **2.5** $\times 10^{-3}$ centimeter

Investigate Problem 2

1. The following numbers represent the diameters of the nucleus (inside) of different kinds of atoms. Which diameters are *not* written in scientific notation? How do you know? Use complete sentences in your explanation.

   Aluminum atom nucleus: 7.2×10^{-13} centimeter

   Cobalt atom nucleus: 0.93×10^{-15} centimeter

   Gold atom nucleus: 14×10^{-11} centimeter

   Sodium atom nucleus: 6.8×10^{-13} centimeter

   The diameters of the cobalt and gold atom nuclei are not written in scientific notation because the numbers being multiplied by the power of 10 are not between 1 and 10.

2. Write the diameters that you identified in Question 1 in scientific notation.

   Cobalt atom nucleus: 9.3×10^{-16} centimeter

   Gold atom nucleus: 1.4×10^{-10} centimeter

3. Without writing the diameters in standard form, list the atoms by diameter from least to greatest.

   cobalt, sodium, aluminum, gold

9

Explore Together

Investigate Problem 2

Students will continue to use scientific notation to represent numbers with values that are between zero and one.

Call the class back together to have the students discuss and present their work for Question 1 through Question 4 of Problem 2.

Grouping

Ask for a student volunteer to read Question 5 aloud. Have a student restate the problem. Pose the Guiding Questions below to verify student understanding. Have students work in small groups to complete Questions 5 through 7.

Guiding Questions

- On what page will you find the values for the diameters to answer Question 5?
- How can you write the diameters in scientific notation?

Call the class back together to have the students discuss and present their work for Questions 5 through 7.

Key Formative Assessments

- What type of exponent is written in a power of 10 to represent a very large number in scientific notation?
- What type of exponent is written in a power of 10 to represent a very small number in scientific notation?
- How can you write a number in scientific notation?
- Give 5 examples of numbers that are written in standard form.
- How can you find the value of a number that is written in scientific notation?
- How can you write a number in standard form if it is written in scientific notation?

9

Investigate Problem 2

4. Use a complete sentence to explain how to compare two small numbers that are in scientific notation.

The number with the greater exponent is the larger number.

5. Write each of the diameters from Question 1 in standard form.

Aluminum atom nucleus: **0.0000000000072 centimeter**

Cobalt atom nucleus: **0.00000000000000093 centimeter**

Gold atom nucleus: **0.00000000014 centimeter**

Sodium atom nucleus: **0.0000000000068 centimeter**

6. Complete the table that shows measurements for different objects.

Object	Measurement	Measurement in standard form	Measurement in scientific notation
Jupiter	average distance from sun (kilometers)	778,600,000	**7.786×10^8**
Water molecule	diameter (centimeters)	**0.0000000282**	2.82×10^{-8}
Comet Hyakutake	nearest distance to the sun (kilometers)	**34,000,000**	3.4×10^7
Atom	diameter (centimeters)	0.00000005	**5.0×10^{-8}**
Dust speck	diameter (centimeters)	**0.00003**	3.0×10^{-5}
Saturn's E ring	inner radius (kilometers)	300,000	**3.0×10^5**

7. What happens to the decimal point in a number after you multiply the number by a power of ten with an exponent that is an integer greater than zero?

Sample Answer: The decimal point moves to the right.

What happens to the decimal point in a number after you multiply the number by a power of ten with an exponent that is an integer less than zero?

Sample Answer: The decimal point moves to the left.

Wrap Up

Close

- Review all key terms and their definitions. Include the terms *standard form* and *scientific notation.*
- You may also want to review any other vocabulary terms that were discussed during the lesson which may include *computation, diameter, decimal, power, exponent,* and *negative exponent.*
- Remind the students to write the key terms and their definitions in the notes section of their notebooks. You may also want the students to include examples.
- Ask the students to explain the meaning of the standard form of a number.
- Ask the students to explain the meaning of scientific notation.
- Hand each student a small piece of paper. Have the students write two very small numbers and two very large numbers in standard form. Next, have the students trade their papers with someone else in the class. (If there are an odd number of students three students can trade theirs so that each has a different paper than their own.) Then have the students write the numbers that they were given in scientific notation. Finally, collect the papers. Then, choose some of the papers and write the numbers in standard form and in scientific notation on the board exactly as they are written on the papers. Have the students work together as a class to agree or disagree with each pair.

Ties to the Cognitive Tutor Software

In the Cognitive Tutor software, students convert numbers to and from scientific notation so that they understand that scientific notation can be used as an equivalent numeric representation. They also compare numbers in scientific notation because, in many cases, scientific notation is used when an approximation to an exact number is all that is needed. In such cases, getting a sense of the order of magnitude of the number (the exponent) is more important than its exact equivalent.

Follow Up

Assignment

Use the Assignment for Lesson 9.4 in the Student Assignments book. See the Teacher's Resources and Assessments book for answers.

Assessment

See the Assessments provided in the Teacher's Resources and Assessments book for Chapter 9.

Open-Ended Writing Task

Ask the students to write a summary of how to write numbers in scientific notation. Have them include at least one example of a large number in standard form rewritten in scientific notation and at least one example of the same for a very small number.

Reflections

Insert your reflections on the lesson as it played out in class today.

What went well?

What did not go as well as you would have liked?

How would you like to change the lesson in order to improve the things that did not go well and capitalize on the things that did go well?

9

Notes

9.5 The Beat Goes On

Properties of Powers

Learning By Doing Lesson Map

Get Ready

Objectives

In this lesson, you will:

- Use the power of a power property.
- Use the power of a product property.
- Use the power of a quotient property.

Key Terms

- power
- product
- quotient

Materials

- Poster paper for the closure activity

NCTM Content Standards

Grades 9-12 Expectations

Number and Operations Standards

- Compare and contrast the properties of numbers and number systems, including the rational and real numbers, and understand complex numbers as solutions to quadratic equations that do not have real solutions.
- Use number-theory arguments to justify relationships involving whole numbers.
- Judge the effects of such operations as multiplication, division, and computing powers and roots on the magnitude of quantities.
- Develop fluency in operations with real numbers, vectors, and matrices, using mental computation or paper-and-pencil calculations for simple cases and technology for more complicated cases.

Measurement Standards

- Understand and use formulas for the area, surface area, and volume of geometric figures, including cones, spheres, and cylinders.

Lesson Overview

Within the context of this lesson, students will be asked to:

- Calculate the lengths of radii when given diameters.
- Calculate the area of circular drumheads.
- Develop and apply the power of a power property.
- Develop and apply the power of a product property.
- Develop and apply the power of a quotient property.

Essential Questions

The following key questions are addressed in this lesson:

1. What is a drumhead?
2. How can you calculate the area of a circle?
3. How can you calculate the length of a radius if given the length of the diameter of a circle?
4. What is the power of a power property?
5. What is the power of a product property?
6. What is the power of a quotient property?

Show The Way

Warm Up

Place the following questions or an applicable subset of these questions on the board before students enter class. Students should begin working as soon as they are seated.

Simplify each expression. Where applicable, $x \neq 0$.

1. 5^0 1
2. $(-8)^0$ 1
3. $-(8)^0$ −1
4. -8^0 −1
5. π^0 1
6. 10^2 100
7. $\frac{x^3}{x^3}$ 1
8. -4^3 −64
9. $\frac{12x}{3x}$ 4
10. $(-1)^5$ −1
11. $x^2(x^5)$ x^7
12. $\frac{1}{\left(\frac{1}{3}\right)}$ 3
13. $\frac{x^2}{x^5} = x^{\boxed{-3}} = \frac{\boxed{1}}{\boxed{x^3}}$
14. $2x^{-5} = \frac{\boxed{2}}{\boxed{x^5}}$
15. $4x^5(x^{-7}) = 4\boxed{x}^{\boxed{-2}} = \frac{4}{\boxed{x}^{\boxed{2}}}$

9

Motivator

Begin the lesson with the motivator to get students thinking about the topic of the upcoming problem. This lesson is about the sound produced by differently sized drums. The motivating questions are about drums.

Ask the students the following questions to get them interested in the lesson.

- Do you play the drums?
- How are drums made?
- What geometric shape are most drums?

Explore Together

Problem 1

Students will investigate powers by first calculating the area of circular drumheads.

Take Note

In a circle, the radius is half the length of the diameter.

Take Note

Recall that π is an irrational number whose value is approximately 3.14.

Grouping

Ask for a student volunteer to read the Scenario and Problem 1 aloud. Have a student restate the problem. Pose the Guiding Questions below to verify student understanding. Have students work together in small groups to complete part (A) through (C) of Problem 1.

Guiding Questions

- What information is given in this problem? What are you asked to find?
- What measurement is referred to as the drumhead?
- How can you calculate the length of the radius of the drumhead?
- What is area?
- How can you calculate the area of a drumhead?

Call the class back together to have the students discuss and present their work for part (A) through (C) of Problem 1.

Investigate Problem 1

Grouping

Ask for a student volunteer to read Question 1 aloud. Have a student restate the problem. Complete Question 1 together as a class.

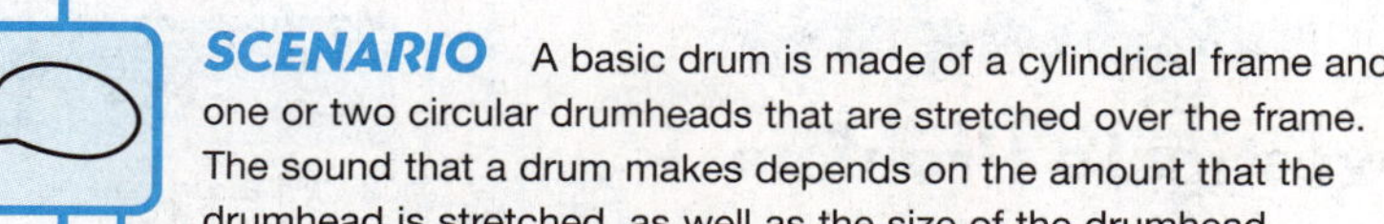

SCENARIO A basic drum is made of a cylindrical frame and one or two circular drumheads that are stretched over the frame. The sound that a drum makes depends on the amount that the drumhead is stretched, as well as the size of the drumhead.

Problem 1 Area of a Drumhead

A. A snare drum's drumhead is 8 inches. This measurement describes the diameter of the drumhead. Find the area of the drumhead by using the formula for the area of a circle: $A = \pi r^2$, where r is the radius and A is the area. Show all your work and use a complete sentence in your answer. Leave your answer in terms of π.

$r = 8 \div 2 = 4$

$A = \pi(4)^2 = 16\pi$

The area is 16π square inches.

B. A bass drum's drumhead is 18 inches. Find the area of the drumhead. Show all your work and use a complete sentence in your answer. Leave your answer in terms of π.

$r = 18 \div 2 = 9$

$A = \pi(9)^2 = 81\pi$

The area is 81π square inches.

C. A tom-tom's drumhead is 16 inches. Find the area of the drumhead. Show all your work and use a complete sentence in your answer. Leave your answer in terms of π.

$r = 16 \div 2 = 8$

$A = \pi(8)^2 = 64\pi$

The area is 64π square inches.

Investigate Problem 1

1. For each answer in parts (A) through (C), write the prime factorization of the number that is multiplied by π.

   Area of 8-inch drumhead: $A = \boxed{2^4}\pi$

   Area of 18-inch drumhead: $A = \boxed{3^4}\pi$

   Area of 16-inch drumhead: $A = \boxed{2^6}\pi$

9

Explore Together

Investigate Problem 1

Take Note

The plural of radius is *radii.*

Students will develop the concept and also a rule to simplify an expression that is a power of a power.

Take Note

A *conjecture* is a possible explanation that can be tested by further investigation.

Ask for a student volunteer to read the Take Note boxes on this page aloud. Have a student restate each of the notes. Pose the Guiding Questions below to verify student understanding.

Guiding Questions

- What does the word singular mean? What does the word plural mean?
- What is the plural form of the term diameter?
- Explain the term conjecture in your own words.

Grouping

After students understand the terms radii and conjecture, have the students work together in small groups to complete Question 2 through 6.

Call the class back together to have the students discuss and present their work for Questions 1 through 6.

Problem 2

Grouping

Ask for a student volunteer to read part (A) Problem 2 aloud. Have a student restate the problem. Pose the Guiding Questions on the next page to verify student understanding. Have students work together in small groups to complete part (A) through (D) of Problem 2.

Investigate Problem 1

2. What do you notice about the radii of the drumheads in parts (A) through (C)? Use a complete sentence in your answer.

Sample Answer: Each radius can be written as a power.

3. Write each radius from the problem situation as a power.

Radius of 8-inch drumhead: $r =$ 2^2

Radius of 18-inch drumhead: $r =$ 3^2

Radius of 16-inch drumhead: $r =$ 2^3

4. In finding each drumhead area, you squared the radius of the drumhead. Review your work in Questions 1 and 3. What is the result of squaring a power? Use a complete sentence in your answer.

Sample Answer: The exponent of the power is multiplied by two.

5. What do you think is the result of cubing a power? Write an example that demonstrates your conjecture. Use a complete sentence in your answer.

Sample Answer: Consider $(2^2)^3$.

$(2^2)^3 = 4^3 = 64$; $64 = 2^6 = 2^{2(3)}$

The exponent of the power is multiplied by three.

6. Write a rule using a complete sentence that explains how to simplify a power that has a base that is a power, such as:

$(2^4)^5$. ← exponent

base is a power

Sample Answer: To simplify a power of a power, multiply the exponents.

Problem 2 More Bass Drum Drumheads

A. Bass drum drumheads come in a multitude of sizes. Another size of a bass drum drumhead is 24 inches. Find the area of the drumhead. Show all your work and use a complete sentence in your answer. Leave your answer in terms of π.

$r = 24 \div 2 = 12$

$A = \pi(12)^2 = 144\pi$

The area is 144π square inches.

B. Write the prime factorization of the radius of the drumhead in part (A).

$r =$ $2^2(3)$

Now write the prime factorization of the area.

$A =$ $2^4 \cdot 3^2$ π

Explore Together

Problem 2

Students will develop the concept and a rule to simplify an expression that is a power of a product.

Guiding Questions

- What dimension of the circular drumhead is given to be 24 inches?
- How can you find the length of the radius of the drumhead?

Call the class back together to have the students discuss and present their work for parts (A) through (D) of Problem 2.

Investigate Problem 2

Grouping

Ask for a student volunteer to read Question 1 aloud. Have a student restate the problem. Pose the Guiding Questions below to verify student understanding. Have students work together in small groups to complete Question 1 and 2.

Guiding Questions

- What is a power? What is an exponent?
- What is the base of a power?
- What is a product? What is a quotient?

Call the class back together to have the students discuss and present their work for Question 1 and 2.

Key Formative Assessments

- What rule or property did you develop in Problem 1 to simplify the power of a power?
- Have a student give an example of a power of a power. Have the class simplify it using their property. Repeat this a few times to verify student understanding.
- What rule or property did you develop in Problem 2 to simplify the power of a product?

Problem 2 More Bass Drumheads

How are the prime factorizations related? Use a complete sentence in your answer.

Sample Answer: In the area, the prime factors of the radius have been squared.

C. Another bass drum drumhead size is 36 inches. Find the area of the drumhead. Show all your work and use a complete sentence in your answer. Leave your answer in terms of π.

$r = 36 \div 2 = 18$

$A = \pi(18)^2 = 324\pi$

The area is 324π square inches.

D. Write the prime factorization of the radius of the drumhead in part (C).

$r = \boxed{2(3^2)}$

Now write the prime factorization of the area.

$A = \boxed{2^2 \cdot 3^4}\pi$

How are the prime factorizations related? Use a complete sentence in your answer.

Sample Answer: In the area, the prime factors of the radius have been squared.

9

Investigate Problem 2

1. Use a complete sentence to explain how you can simplify a power that has a product as its base without first simplifying the base.

   Sample Answer: To simplify a power whose base is a product, apply the exponent to each factor.

2. Now consider a power that has a quotient as its base, such as: $\left(\frac{3}{4}\right)^{10}$.

   Write a conjecture for a method of how to simplify this kind of power. Write an example that demonstrates your conjecture. Use a complete sentence in your answer.

   Sample Answer:

   $$\left(\frac{1}{2}\right)^3 = \frac{1}{2}\left(\frac{1}{2}\right)\left(\frac{1}{2}\right) = \frac{1(1)(1)}{2(2)(2)} = \frac{1^3}{2^3}$$

   To simplify a power that has a quotient as its base, apply the exponent to both the numerator and denominator of the quotient.

- Have a student give an example of a power of a product. Have the class simplify it using their property. Repeat this a few times to verify student understanding.
- What rule or property did you develop in Problem 2 to simplify the power of a quotient?
- Have a student give an example of a power of a quotient. Have the class simplify it using their property. Repeat this a few times to verify student understanding.

Explore Together

Investigate Problem 2

Students will apply the properties of powers that they have developed in this lesson.

Grouping

Ask for a student volunteer to read Question 3 aloud. Have a student restate the problem. Pose the Guiding Questions below to verify student understanding. Have students work in small groups to complete Questions 3 through 5.

Guiding Questions

- What properties will you use in Question 3?
- What other questions do you still have that you want to have answered before you practice simplifying powers using these properties?

9

Common Student Errors

Students may have difficulty using the properties for simplifying powers of powers, products of powers, and quotients of powers if they only memorize the rules. Emphasize that the students need to understand the properties and why they are true. Then, the students will always be able to apply these rules in appropriate situations.

Students may also have difficulty remembering the properties from previous lessons. Reinforce the importance of these properties as lifelong skills that will be needed often. If the students have been careful to write down all the properties and notes in a notes section of their notebooks, they will be able to refer to that section for reference for these questions as well as to study for tests and quizzes.

Call the class back together to have the students discuss and present their work for Questions 3 through 5.

Investigate Problem 2

3. Use your properties of powers to simplify each expression. Show all your work.

$(3^2)^3$

$(3^2)^3 = 3^{2(3)}$

$= 3^6$

$= 729$

$(15 \cdot 12)^2$

$(15 \cdot 12)^2 = 15^2 \cdot 12^2$

$= 225(144)$

$= 32{,}400$

$\left(\frac{2}{3}\right)^4$

$\left(\frac{2}{3}\right)^4 = \frac{2^4}{3^4}$

$= \frac{16}{81}$

$(4^5)^0$

$(4^5)^0 = 4^{5(0)}$

$= 4^0$

$= 1$

4. A summary of the rules for powers that you have learned so far in this chapter is shown below. Complete each property.

Product rule of powers: $a^b a^c = a^{\boxed{b+c}}$

Quotient rule of powers: $\frac{a^b}{a^c} = a^{\boxed{b-c}}$

Negative exponents: $\frac{1}{a^b} = a^{\boxed{-b}}$

Zero exponents: $a^0 = \boxed{1}$

Power of a power rule: $(a^b)^c = a^{\boxed{bc}}$

Power of a product rule: $(ab)^c = a^{\boxed{c}}b^{\boxed{c}}$

Power of a quotient rule: $\left(\frac{a}{b}\right)^c = \frac{a^{\boxed{c}}}{b^{\boxed{c}}}$

5. All the properties that you have used so far apply to algebraic expressions too. Simplify each expression. Show all your work.

$(x^3)^3$

$(x^3)^3 = x^{3(3)}$

$= x^9$

$(2p)^2$

$(2p)^2 = 2^2p^2$

$= 4p^2$

y^3y^7

$y^3y^7 = y^{3+7}$

$= y^{10}$

$\left(\frac{x}{4}\right)^3$

$\left(\frac{x}{4}\right)^3 = \frac{x^3}{4^3}$

$= \frac{x^3}{64}$

Explore Together

Investigate Problem 2

Students will identify the property that justifies each step in simplifying expressions involving powers.

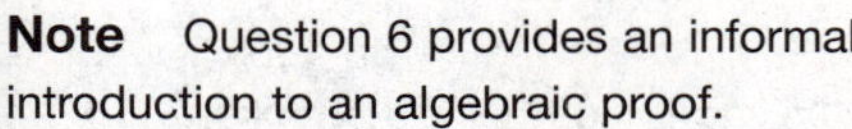

Note Question 6 provides an informal introduction to an algebraic proof.

Grouping

Ask for a student volunteer to read Question 6 aloud. Have a student restate the problem. Pose the Guiding Questions below to verify student understanding. Complete the first part of Question 6 together as a class. Have students work together in small groups to complete the remaining part of Question 6.

Take Note

Recall that in Lesson 4.5, you learned other properties of real numbers.

Guiding Questions

- What is a property?
- How can you determine which property is used for each step in simplifying an expression?

Call the class back together to have the students discuss and present their work for Question 6.

Key Formative Assessments

- How can you determine whether a value is a factor?
- How can you simplify the power of a power?
- How can you simplify the power of a product?
- How can you simplify the power of a quotient?

Investigate Problem 2

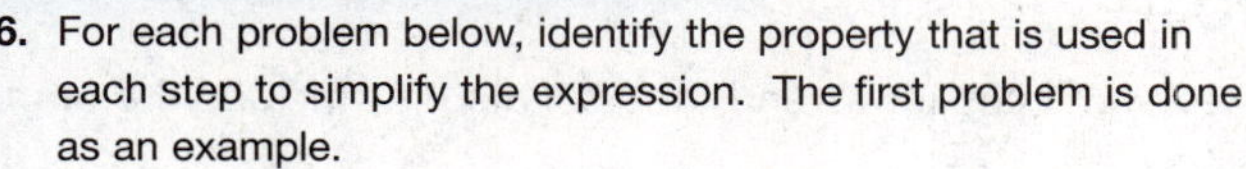

6. For each problem below, identify the property that is used in each step to simplify the expression. The first problem is done as an example.

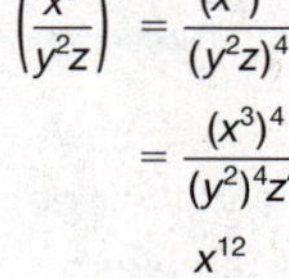

$\left(\frac{x^3}{y^2z}\right)^4 = \frac{(x^3)^4}{(y^2z)^4}$	Power of a quotient rule
$= \frac{(x^3)^4}{(y^2)^4z^4}$	Power of a product rule
$= \frac{x^{12}}{y^8z^4}$	Power of a power rule

$(2ab^2)^3(-3b)^2 = 2^3a^3(b^2)^3(-3)^2b^2$	Power of a product rule
$= 8a^3(b^2)^3(9)b^2$	Evaluate powers.
$= 8a^3b^6(9)b^2$	Power of a power rule
$= 8(9)a^3b^6b^2$	Commutative prop. of multi.
$= 72a^3b^6b^2$	Multiply.
$= 72a^3b^8$	Product rule of powers

$\left(\frac{x^2}{2y^3}\right)^4\left(\frac{xy^2}{3}\right)^3 = \frac{(x^2)^4}{(2y^3)^4} \cdot \frac{(xy^2)^3}{3^3}$	Power of a quotient rule
$= \frac{(x^2)^4}{2^4(y^3)^4} \cdot \frac{x^3(y^2)^3}{3^3}$	Power of a product rule
$= \frac{x^8}{2^4y^{12}} \cdot \frac{x^3y^6}{3^3}$	Power of a power rule
$= \frac{x^8x^3y^6}{2^4y^{12}(3^3)}$	Mult. numerators and denominators
$= \frac{x^{11}y^6}{2^4y^{12}(3^3)}$	Product rule of powers
$= \frac{x^{11}}{2^4y^6(3^3)}$	Quotient rule of powers
$= \frac{x^{11}}{16y^6(27)}$	Evaluate powers
$= \frac{x^{11}}{432y^6}$	Multiply

9

Wrap Up

Close

- Review all key terms and their definitions. Include the terms *power, product,* and *quotient.*
- You may also want to review any other vocabulary terms that were discussed during the lesson which may include *drumhead, area, circular, singular, plural, radius, radii, diameter, conjecture, hypothesis, theory, factor, squared,* and *cubed.*
- Remind the students to write the key terms and their definitions in the notes section of their notebooks. You may also want the students to include examples.
- Ask the students to work together in small groups. Have the groups write each algebraic property that they used in this lesson.
- Prepare a piece of poster paper (multiple sheets of it) while the students are working in groups. Label the papers with the title "Algebraic Properties." Have each group take turns adding 2 properties to the papers until no groups have any more properties to add.
- Have the students work in small groups to develop an example for each property that was listed on the poster papers. Choose 2 or 3 examples of each property to be written beside each property on the poster paper.
- Have the students copy the list of properties and examples into their notes section of their notebooks or you can make a handout for them of these to save class time. Keep the posters on the walls for several days to reinforce these properties with the students.

Ties to the Cognitive Tutor Software

In the Cognitive Tutor software, students convert numbers to and from scientific notation so that they understand that scientific notation can be used as an equivalent numeric representation. They also compare numbers in scientific notation because, in many cases, scientific notation is used when an approximation to an exact number is all that is needed. In such cases, getting a sense of the order of magnitude of the number (the exponent) is more important than its exact equivalent.

Follow Up

Assignment

Use the Assignment for Lesson 9.5 in the Student Assignments book. See the Teacher's Resources and Assessments book for answers.

Assessment

See the Assessments provided in the Teacher's Resources and Assessments book for Chapter 9.

Open-Ended Writing Task

Ask the students to write a short essay in response to the following questions. Why do we refer to a number raised to an exponent of 2 as the number "squared"? Why do we refer to a number raised to an exponent of 3 as the number "cubed"?

Reflections

Insert your reflections on the lesson as it played out in class today.

What went well?

What did not go as well as you would have liked?

How would you like to change the lesson in order to improve the things that did not go well and capitalize on the things that did go well?

9

Notes

9

9.6 Sailing Away

Radicals and Rational Exponents

Learning By Doing Lesson Map

Get Ready

Objectives

In this lesson, you will:

- Find the *n*th root of a number.
- Write an expression in radical form.
- Write an expression in rational exponent form.

Key Terms

- cube root
- index
- *n*th root
- radicand
- rational exponent
- radical

NCTM Content Standards

Grades 9-12 Expectations

Number and Operations Standards

- Compare and contrast the properties of numbers and number systems, including the rational and real numbers, and understand complex numbers as solutions to quadratic equations that do not have real solutions.
- Use number-theory arguments to justify relationships involving whole numbers.
- Judge the effects of such operations as multiplication, division, and computing powers and roots on the magnitude of quantities.

Algebra Standards

- Understand the meaning of equivalent forms of expressions, equations, inequalities, and relations.
- Identify essential quantitative relationships in a situation and determine the class or classes of functions that might model the relationships.

Lesson Overview

Within the context of this lesson, students will be asked to:

- Evaluate expressions to solve equations.
- Solve for the *n*th root of numbers.
- Develop and apply a rule for writing expressions in radical form.
- Develop and apply a rule for writing expressions in rational exponent form.

Essential Questions

The following key questions are addressed in this lesson:

1. What is a root?
2. What is a radical?
3. What is a rational exponent?
4. How can you write a radical as a power?
5. How can you write a power as a radical?

Show The Way

Warm Up

Place the following questions or an applicable subset of these questions on the board before students enter class. Students should begin working as soon as they are seated.

Simplify each expression. Where applicable, $x \neq 0$.

1. $(x^2)^3$ x^6
2. $(2^2)^3$ $2^6 = 64$
3. $(-2^2)^3$ $(-4)^3 = -64$
4. $-(2^2)^3$ $-(4)^3 = -64$
5. $(3x)^2$ $9x^2$
6. $(-5x^3)^2$ $25x^6$
7. $\left(\frac{1}{5}\right)^3$ $\frac{1}{125}$
8. $\left(\frac{x}{2y}\right)^3$ $\frac{x^3}{8y^3}$
9. $\left(\frac{x}{3}\right)^4$ $\frac{x^4}{81}$
10. $(-4a^7b^8c^9)^4$ $256a^{28}b^{32}c^{36}$
11. $(x^{-2})^4$ $x^{-8} = \frac{1}{x^8}$
12. $(x^{-2})^{-3}$ x^6
13. $\left(\frac{1}{x}\right)^{-1}$ x
14. $(-y)^{-1} = \frac{\boxed{-1}}{\boxed{y}}$
15. $\left(\frac{-y}{x}\right)^{-1} = \frac{\boxed{-x}}{\boxed{y}}$

9

Motivator

Begin the lesson with the motivator to get students thinking about the topic of the upcoming problem. This lesson is about different types of boats. The motivating questions are about boating.

Ask the students the following questions to get them interested in the lesson.

- Have you ever had the opportunity to take a trip on a boat or ride in a boat for any reason?
- Did you enjoy the experience?
- In what type of boat did you travel?
- What surprised you about being on a boat that you didn't expect before you were on it?

Explore Together

Problem 1

Students will evaluate expressions involving a base with an exponent of 3.

Grouping

Ask for a student volunteer to read the Scenario and Problem 1 aloud. Have a student restate the problem. Pose the Guiding Questions below to verify student understanding. Have students work together in small groups to complete parts (A) through (C) of Problem 1.

Take Note

When a power has an exponent of 3, it is called the third power. When you evaluate a third power, you are *cubing* the base.

Guiding Questions

- What information is given in this problem?
- What is a beam width?
- What does it mean for a boat to capsize?
- What is the equation to calculate the weight of a boat?
- What unit is used for the boat weight?
- How can you calculate the weight of a boat?
- What do we do when we square a number?
- What does it mean to cube a number?
- What is a cube?

Common Student Errors

Students may incorrectly multiply 64 and the ratio of b and c before cubing the ratio of b and c.

SCENARIO Some boats move by skimming across the surface of the water, while other boats move by pushing through the water. The boats that push through the water have a body, or hull, that is called a displacement hull. Sailboats are boats with displacement hulls. Important measurements of a boat that can be used to determine the boat's speed and stability are labeled on the figure below.

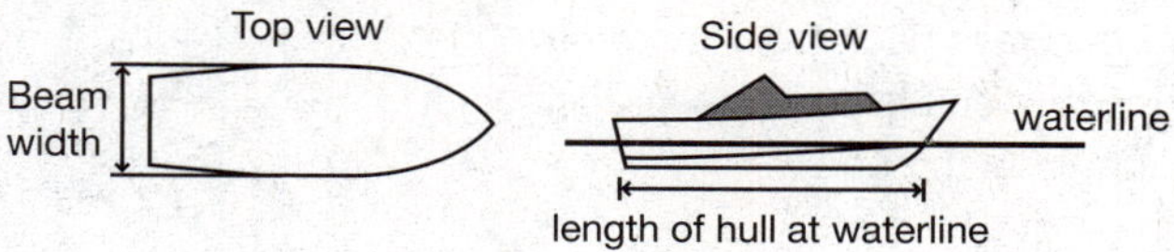

Problem 1 Boat Stability

The weight of a boat depends on its beam width. An equation that relates a boat's weight to its beam width is $w = 64\left(\frac{b}{c}\right)^3$, where w is the boat's weight in pounds, b is the beam width in feet, and c is the capsize factor. When the capsize factor c is less than 2, the boat is less likely to capsize, or turn over.

A. Suppose that a boat has a capsize factor of three and a beam width of six feet. What is the weight of the boat? Show your work and use a complete sentence in your answer.

$w = 64\left(\frac{6}{3}\right)^3$

$w = 64(2)^3$

$w = 64(8)$

$w = 512$

The boat weighs 512 pounds.

B. Suppose that a boat has a capsize factor of three and a beam width of nine feet. What is the weight of the boat? Show your work and use a complete sentence in your answer.

$w = 64\left(\frac{9}{3}\right)^3$

$w = 64(3)^3$

$w = 64(27)$

$w = 1728$

The boat weighs 1728 pounds.

9

Explore Together

Problem 1

Students will solve the equation for the weight of a boat.

Call the class back together to have the students discuss and present their work for parts (A) through (C).

Grouping

Ask for a student volunteer to read part (D) aloud. Have a student restate the problem. Pose the Guiding Questions below to verify student understanding. Have students work together in small groups to complete parts (D) through (F) of Problem 1.

Guiding Questions

- How is the information given in part (D) different from the information in parts (A) through (C)? What is the same?
- How can you determine the beam width with the given information?

Common Student Errors

Some students will incorrectly cube the value of 125 in parts (D) and (E) rather than finding the cube root of 125.

Call the class back together to have the students discuss and present their work for parts (D) through (F) of Problem 1.

Take Note

Finding the cube root of a number is the *inverse operation* of finding the cube of a number.

Investigate Problem 1

Just the Math

Students will be formally introduced to the concept of the cube root of a number.

Problem 1 Boat Stability

C. Use complete sentences to explain how you found your answers to parts (A) and (B).

Sample Answer: First, find the quotient of the values for *b* and *c*. Then, evaluate the power and multiply the result by 64.

D. Suppose that a boat has a capsize factor of 2 and weighs 1000 pounds. Write an equation that you can use to find the beam width of the boat.

$$1000 = 64\left(\frac{b}{2}\right)^3$$

In the equation that you wrote, get the power by itself on one side of the equation. Show your work.

$$1000 = 64\left(\frac{b}{2}\right)^3$$

$$1000 = 64\left(\frac{b^3}{8}\right)$$

$$1000 = 8b^3$$

$$125 = b^3$$

E. To find the value of b, you need to find the number that when multiplied by itself three times is equal to 125. What is this number? Use a complete sentence in your answer.

The value of *b* is 5.

F. What is the boat's beam width? Use a complete sentence in your answer.

The boat's beam width is five feet.

Investigate Problem 1

1. Just the Math: Cube Root Your answer to part (E) is the *cube root* of 125. You can say that a number b is a **cube root** of a number a if $b^3 = a$.

So, 5 is a cube root of 125 because $5^3 = 125$.

Is there another number whose cube is 125?

No.

A cube root is designated by using the symbol $\sqrt[3]{\ }$. The number 3 is called the **index** of the radical. So, $\sqrt[3]{125} = 5$.

Complete each statement below.

$\sqrt[3]{8} = \boxed{2}$ $\sqrt[3]{-27} = \boxed{-3}$

Grouping

Ask for a student volunteer to read Question 1 aloud. Have a student restate the problem. Pose the Guiding Questions on the next page to verify student understanding. Complete Question 1 together as a class.

Explore Together

Investigate Problem 1

Students will investigate the *n*th root of given values.

Guiding Questions

- What is a cube root?
- What is 3^3?
- What is the cube root of 27?
- What is the cube root of –8?

Take Note

It is understood that $\sqrt{\ }$ and $\sqrt[2]{\ }$ indicate a square root.

Just the Math

Students will be introduced to the terminology for the *n*th root.

Grouping

Ask for a student volunteer to read Question 2 aloud. Have a student restate the problem. Pose the Guiding Questions below to verify student understanding. Have students work together in small groups to complete Questions 2 through 5.

Guiding Questions

- What do we call the number 3 in the expression $\sqrt[3]{27}$?
- What do we call the number 27 in the expression $\sqrt[3]{27}$?
- What number is equal to $\sqrt[3]{27}$? Is there any other number that is equal to $\sqrt[3]{27}$?
- What is a square root? What is a cube root?
- In the statement $b^n = a$, what does the b represent? What does the n represent? What does the a represent?
- If we know $b^n = a$, then what is the *n*th root of a? How would we write that using a radical?
- In the expression $b = \sqrt[n]{a}$, what is the index? What is the radicand?

Investigate Problem 1

2. **Just the Math: *n*th Roots** You can extend the idea of square roots and cube roots. Use n to represent a positive number. Then a number b is the ***n*th root of *a*** if $b^n = a$. For instance, 2 is the fourth root of 16 because $2^4 = 16$.

   Complete each statement below.

   3 is the fifth root of 243 because $\boxed{3^5} = 243$.

   -2 is the cube root of -8 because $(-2)^3 = \boxed{-8}$.

   4 is the $\boxed{\text{sixth}}$ root of 4096 because $4^6 = 4096$.

   The *n*th root of a number a is designated as $\sqrt[n]{a}$, where n is the index of the radical and a is the *radicand*.

3. Complete each statement below.

   $\sqrt{100} = \boxed{10}$ because $\boxed{10}^2 = 100$.

   $\sqrt[3]{216} = \boxed{6}$ because $\boxed{6}^3 = 216$.

   $\sqrt[4]{81} = \boxed{3}$ because $\boxed{3}^4 = 81$.

   $\sqrt[5]{-32} = \boxed{-2}$ because $(\boxed{-2})^5 = -32$.

4. Notice that a power can be positive or negative, depending on the base and the exponent. When the exponent of a power is an even number and the base is a positive number, is the value of the power a positive number or a negative number? How do you know? Use a complete sentence to explain.

   Sample Answer: The value is always positive because the product of positive numbers is always positive.

   When the exponent of a power is an even number and the base is negative, is the value of the power a positive number or a negative number? Use a complete sentence to explain.

   Sample Answer: The value is always positive because the exponent is even and the product of an even number of negative numbers is positive.

5. When the exponent of a power is an odd number and the base is a positive number, is the value of the power a positive number or negative number? Use a complete sentence to explain.

   Sample Answer: The value is always positive because the product of positive numbers is always positive.

   When the exponent of a power is an odd number and the base is negative, is the value of the power a positive number or negative number? Use a complete sentence to explain.

   Sample Answer: The value is always negative because the exponent is odd and the product of an odd number of negative numbers is negative.

Call the class back together to have the students discuss and present their work for Questions 2 through 5.

Explore Together

Problem 2

Side view

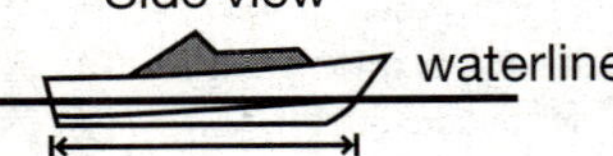

length of hull
at waterline

Students will work with an equation to calculate the speed of a boat and the length of a boat using powers and roots.

Grouping

Ask for a student volunteer to read Problem 2 aloud. Have a student restate the problem. Pose the Guiding Questions below to verify student understanding. Have students work in small groups to complete parts (A) through (D) of Problem 2.

Guiding Questions

- What information is given in this problem?
- What does s represent?
- In what unit is the speed measured?
- What does ℓ represent?
- In what units are the length measured?
- How can calculate the speed if you are given a length for a boat?
- How can you calculate the length if you are given the speed for a boat?

Call the class back together to have the students discuss and present their work for parts (A) through (D) of Problem 2.

Investigate Problem 2

Grouping

Ask for a student volunteer to read Question 1 aloud. Have the students complete Question 1 individually. When they are finished, discuss Question 1 together as a class.

9

Problem 2 Boat Speed

The speed of a boat depends on the length of the hull at the waterline. An equation that relates the speed s in knots and the length ℓ in feet of the hull at the waterline is $s = 1.34\sqrt{\ell}$.

A. What is the speed of a boat that has a length of 16 feet at the waterline? Show your work and use a complete sentence in your answer.

$s = 1.34\sqrt{16}$

$s = 1.34(4)$

$s = 5.36$

The speed of the boat is 5.36 knots.

B. Write an equation that you can use to find the length of a boat at its waterline if the boat's speed is 6.7 knots.

$6.7 = 1.34\sqrt{\ell}$

C. In the equation that you wrote, get the radical by itself on one side of the equation. Show your work.

$6.7 = 1.34\sqrt{\ell}$

$5 = \sqrt{\ell}$

D. What is the length of the boat at the waterline? How do you know? Use complete sentences in your answer.

The length of the boat at the waterline is 25 feet because the square root of 25 is 5.

Your answer is the value of which variable? Use a complete sentence in your answer.

The answer is the value of the variable ℓ.

Investigate Problem 2

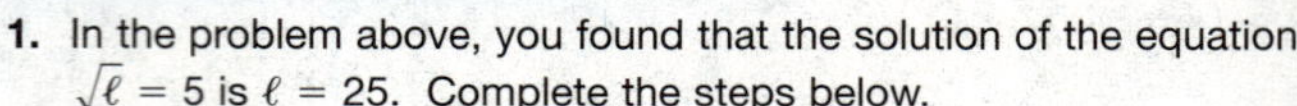

1. In the problem above, you found that the solution of the equation $\sqrt{\ell} = 5$ is $\ell = 25$. Complete the steps below.

$\ell = 25$ Given

$\ell = \boxed{5^2}$ Write 25 as a power.

$\ell = \boxed{(\sqrt{\ell})^2}$ Replace 5 by $\sqrt{\ell}$.

Explore Together

Investigate Problem 2

Students will develop a rule for fractional exponents to represent roots.

Grouping

Ask for a student volunteer to read Question 2 aloud. Have a student restate the problem. Pose the Guiding Questions below to verify student understanding. Have students work individually to complete Question 2.

Guiding Questions

- If we find the square root of a number and then square the result, will we always get the same number that we started with?
- If we square a number and then take the square root of that number, would we always get the same number? (Could we also get the opposite of the original number for this situation?)
- What is a root?

Take Note

Remember that a *rational number* is a number of the form $\frac{a}{b}$ where $b \neq 0$.

Call the class back together to have the students discuss and present their work for Question 2.

Take Note

A boat's *displacement* is the volume of water that is moved by the weight of the boat in the water.

Just the Math

The students will use rational exponents to represent radicals as powers. Ask for a student volunteer to read Question 3 aloud. Have a student restate the problem. Pose the Guiding Questions on the next page to verify student understanding. Complete Question 3 together as a class. Have students work together in small groups to complete Questions 4 through 6.

Investigate Problem 2

2. Because the definition of the *n*th root involves powers, it would be nice if we could write an *n*th root of a number as a power of the number. Consider your equation from Question 1:

$\ell^1 = (\sqrt{\ell})^2.$

If we could write the square root of ℓ as a power of ℓ, we would have $\ell^1 = (\ell^?)^2$. You know that when you find the power of a power, you multiply the exponents: $(a^b)^c = a^{\boxed{bc}}$.

Represent the square root as an exponent so the exponent times 2 is equal to 1. Use complete sentences to explain.

Sample Answer: I would use $\frac{1}{2}$ as the exponent because $(\ell^{1/2})^2 = \ell^{1/2(2)} = \ell^1$

3. Just the Math: Radicals and Rational Exponents
A **rational exponent** is an exponent that is a rational number. We can write each *n*th root as a rational exponent in the following way: If *n* is an integer greater than 1, then $\sqrt[n]{a} = a^{1/n}$. Write each radical as a power.

$\sqrt[3]{7}$ $\quad$ $\sqrt[5]{x}$ $\quad$ $\sqrt{y}$

$7^{1/3}$ $\quad$ $x^{1/5}$ $\quad$ $y^{1/2}$

Write each power as a *radical*.

$8^{1/4}$ $\quad$ $z^{1/5}$ $\quad$ $m^{1/3}$

$\sqrt[4]{8}$ $\quad$ $\sqrt[5]{z}$ $\quad$ $\sqrt[3]{m}$

4. Write the equation in Problem 2 in rational exponent form.

$s = 1.34\ell^{1/2}$

5. Another nautical equation can be used to determine how fast a sailboat can travel in light wind. This equation is $s = \frac{a}{d^{2/3}}$ where *a* is the sail area and *d* is the boat's displacement. You can write this equation in radical form by using the properties you know about powers. Use the properties of powers to write the equation in radical form.

$s = \frac{a}{d^{2/3}}$ Given equation

$s = \frac{a}{\boxed{d^{(1/3)2}}}$ Write power as a product.

$s = \frac{a}{\boxed{(d^{1/3})^2}}$ Power of a power property

$s = \frac{a}{\boxed{(\sqrt[3]{d})^2}}$ Definition of rational exponent

9

Explore Together

Investigate Problem 2

Students will simplify expressions involving rational exponents, radicals, products, and quotients.

Guiding Questions

- What is a rational number?
- How can you write a radical as a power using rational exponents?
- What does s represent in this equation? In what unit is the speed measured?
- What does d represent in this equation? What is displacement?
- What does a represent in this equation?

Call the class back together to have the students discuss and present their work for Questions 4 through 6.

9

Grouping

Ask for a student volunteer to read Question 7 aloud. Have a student restate the problem. Have students work in small groups to complete Questions 7 and 8.

Call the class back together to have the students discuss and present their work for Questions 7 and 8.

Key Formative Assessments

- What is a radical?
- What is a root?
- What is a cube root?
- How can you find the nth root of a number?
- What is an index?
- What is a radicand?
- How can you find the value of an expression with a rational exponent?
- How can you write a radical using a rational exponent?
- How can you find the value of a radical?
- How can you write a power with a rational exponent as a radical?

Investigate Problem 2

6. The equation in Question 5 gives you s, the *sail area to displacement ratio.* Use the equation to find s for a boat that has a sail area of 600 square feet and a displacement of 125 cubic feet. Show your work and use a complete sentence in your answer.

$s = \frac{600}{(\sqrt[3]{125})^2}$

$s = \frac{600}{5^2}$

$s = \frac{600}{25}$

$s = 24$ The sail area to displacement ratio is 24.

7. Write each expression in radical form. Show your work and simplify your answer, if possible.

$4^{3/2}$

$4^{3/2} = 4^{(1/2)3} = (4^{1/2})^3 = (\sqrt{4})^3 = 2^3 = 8$

$5^{3/4}$

$5^{3/4} = 5^{(1/4)3} = (5^{1/4})^3 = (\sqrt[4]{5})^3$

$x^{4/5}$

$x^{4/5} = x^{(1/5)4} = (x^{1/5})^4 = (\sqrt[5]{x})^4$

$y^{2/3}$

$y^{2/3} = y^{(1/3)2} = (y^{1/3})^2 = (\sqrt[3]{y})^2$

8. Write each expression in rational exponent form. Show your work and simplify your answer, if possible.

$(\sqrt[4]{2})^3$

$(\sqrt[4]{2})^3 = (2^{1/4})^3 = 2^{(1/4)3} = 2^{3/4}$

$(\sqrt{5})^4$

$(\sqrt{5})^4 = (5^{1/2})^4 = 5^{(1/2)4} = 5^{4/2} = 5^2 = 25$

$(\sqrt[5]{x})^8$

$(\sqrt[5]{x})^8 = (x^{1/5})^8 = x^{(1/5)8} = x^{8/5}$

$(\sqrt[5]{y})^{10}$

$(\sqrt[5]{y})^{10} = (y^{1/5})^{10} = y^{(1/5)10} = y^{10/5} = y^2$

- What does squaring a number mean?
- What does taking the square root of a number mean?
- What does cubing a number mean?
- What types of situations can you think of that can be modeled using radicals?

Wrap Up

Close

- Review all key terms and their definitions. Include the terms *cube root, nth root, index, radical, radicand,* and *rational exponent.*
- You may also want to review any other vocabulary terms that were discussed during the lesson which may include *ratio, rational number, displacement, powers,* and *exponents.*
- Remind the students to write the key terms and their definitions in the notes section of their notebooks. You may also want the students to include examples.
- Ask the students to compare and contrast each of the key terms.
- Have the students summarize the mathematics that they learned in this lesson.
- Have each student write an expression as a radical on a small piece of paper. Have the students trade papers with a partner or among 3 students so that everyone has a different paper than their own. Have the students write the expression as a power using a rational exponent and simplify it if possible. Have the students trade papers again and ask the students to check the work on the paper that they received. (It can be their original paper or a new one to them.)

Follow Up

Assignment

Use the Assignment for Lesson 9.6 in the Student Assignments book. See the Teacher's Resources and Assessments book for answers.

Assessment

See the Assessments provided in the Teacher's Resources and Assessments book for Chapter 9.

Open-Ended Writing Task

Ask the students to write a short essay explaining the steps to take to evaluate a cube root of a number compared to the steps to cube a number. What is the same, what is different, and how are they related?

Reflections

Insert your reflections on the lesson as it played out in class today.

What went well?

____

____

What did not go as well as you would have liked?

____

____

How would you like to change the lesson in order to improve the things that did not go well and capitalize on the things that did go well?

____

____

____

____

____

9

Notes

Looking Ahead to Chapter 10

Focus

In Chapter 10, you will learn about polynomials, including how to add, subtract, multiply, and divide polynomials. You will also learn about polynomial and rational functions.

Chapter Warm-up

Answer these questions to help you review skills that you will need in Chapter 10.

Solve each two-step equation.

1. $5x + 3 = 2x - 12$

$x = -5$

2. $4x = 7 - 3x$

$x = 1$

3. $\frac{x}{5} + 3 = \frac{x}{10}$

$x = -30$

Simplify each expression.

4. $(3^2)^5$

3^{10}

5. $x^7 \cdot x^{-5}(x^2)^{-3}$

$\frac{1}{x^4}$

6. $\left(\frac{xy^5}{y^{-2}}\right)^{-1}$

$\frac{1}{xy^7}$

Read the problem scenario below.

You have just planted a flower bed in a 3-foot wide by 5-foot long rectangular section of your yard. After planting some flowers, you decide that you would like to add to the width of your garden so that there is more space to plant flowers. Let x represent the amount added to the length of your garden.

10

7. Represent the area of your expanded garden with an expression in two ways by using the distributive property.

$5(3 + x) = 15 + 5x$

8. What is the area of your garden if you expand the width by 4 feet?

$5(3 + 4) = 5(7) = 35$ square feet

Key Terms

polynomial ■ p. 454
term ■ p. 454
coefficient ■ p. 454
degree of a term ■ p. 454
degree of the polynomial ■ p. 454
monomial ■ p. 455
binomial ■ p. 455
trinomial ■ p. 455
standard form ■ p. 455
Vertical Line Test ■ p. 457
combine like terms ■ p. 460
distributive property ■ p. 460
area model ■ p. 464
divisor ■ p. 468
dividend ■ p. 468
remainder ■ p. 468
FOIL pattern ■ p. 472
square of a binomial sum ■ p. 473
square of a binomial difference ■ p. 475
rational expression ■ p. 483
domain ■ p. 483
restrict the domain ■ p. 485

CHAPTER

10

Polynomial Functions and Rational Expressions

Stained glass windows have been used as decoration in homes, religious buildings, and other buildings since the 7th century. In Lesson 10.4, you will find the area of a stained glass window.

10.1

Water Balloons

Polynomials and Polynomial Functions

Learning By Doing Lesson Map

Get Ready

Objectives

In this lesson, you will:

- Identify terms and coefficients of polynomials.
- Classify polynomials by the number of terms.
- Classify polynomials by degree.
- Write polynomials in standard form.
- Use the Vertical Line Test to determine whether equations are functions.
- Identify essential quantitative relationships in a situation and determine the class or classes of functions that might model the relationships.
- Draw reasonable conclusions about a situation being modeled.

Key Terms

- polynomial
- term
- coefficient
- degree
- monomial
- binomial
- trinomial
- standard from
- Vertical Line Test

NCTM Content Standards

Grades 9–12 Expectations

Number and Operations Standards

- Judge the reasonableness of numerical computations and their results.

Algebra Standards

- Generalize patterns using explicitly defined and recursively defined functions.
- Understand relations and functions and select, convert flexibly among, and use various representations for them.
- Understand and compare the properties of classes of functions, including exponential, polynomial, rational, logarithmic, and periodic functions.
- Interpret representations of functions of two variables.
- Use symbolic algebra to represent and explain mathematical relationships.

Lesson Overview

Within the context of this lesson, students will be asked to:

- Identify if an expression is a polynomial.
- Identify terms and coefficients of polynomials.
- Classify polynomials by the number of terms as well as by the highest degree term.
- Write polynomials in standard form.
- Apply the Vertical Line Test to graphs of equations to determine whether the graphed equations are functions.

Essential Questions

The following key questions are addressed in this section:

1. What is a polynomial?
2. What is vertical motion?
3. How can you determine the coefficient of a term?
4. How can you determine the degree of a term? How can you determine the degree of a polynomial?
5. How can you determine whether a graphed equation is a function?
6. How can you write a function in standard form?
7. What is a monomial? What is a binomial? What is a trinomial?

Show The Way

Warm Up

Place the following questions or an applicable subset of these questions on the board before students enter class. Students should begin working as soon as they are seated.

The vertical motion model for a ball hit by a baseball bat is $y = -16t^2 + 40t + 2.5$, where y is the height in feet and t is the time in seconds. Answer each question. Recall that the general equation for a vertical motion model is $y = -16t^2 + vt + h$, where v is the initial velocity in feet per second and h is the initial height in feet.

1. What is velocity? **Velocity is the speed and direction in which an object is moving.**
2. What is the initial velocity? **40 feet per second.**
3. What does the constant 2.5 represent in the equation? **The initial height from where the object was thrown was 2.5 feet off of the ground.**
4. What is the height of the ball 1 second after the ball was hit? **26.5 feet**
5. What shape will the graph of this equation look like? **A parabola that opens down.**
6. What is the vertex point of the parabola? **The vertex point is at (1.25, 27.5).**
7. What is the meaning of the vertex point for the parabola? **This means that the ball is at the highest point of 27.5 feet, 1.25 seconds after the ball is hit.**
8. What is the axis of symmetry for this parabola? **The axis of symmetry of this parabola is $x = 1.25$.**
9. When would the ball hit the ground? **The ball will hit the ground about 2.56 seconds after it is hit. (Assuming no one else catches the ball before it hits the ground.)**

10

Motivator

Begin the lesson with the motivator to get students thinking about the topic of the upcoming problem. This lesson is about the path of a tossed water balloon. The motivating questions are about tossing water balloons.

Ask the students the following questions to get them interested in the lesson.

- What is a water balloon?
- Why would someone fill a balloon with water?
- When have you tossed water balloons?
- What did the path of the balloon look like when you threw it?

Explore Together

Problem 1

Students will investigate vertical motion for a thrown water balloon.

Grouping

Ask for a student volunteer to read the Scenario and Problem 1 aloud. Have a student restate the problem. Pose the Guiding Questions below to verify student understanding. Complete part (A) of Problem 1 together as a whole class. When the students understand the situation, have them work together in small groups to complete parts (B) through (F) of Problem 1. Then call the class back together to discuss and present their work for parts (B) through (F) of Problem 1.

Take Note

The *vertical motion model* is $y = -16t^2 + vt + h$, where t is the time in seconds that the object has been moving, v is the initial velocity (speed) in feet per second of the object, h is the initial height in feet of the object, and y is the height in feet of the object at time t.

Guiding Questions

- What information is given in this problem?
- What is the meaning of velocity?
- What is vertical motion?
- What formula did you find in Chapter 8 to model vertical motion?
- In the formula $y = -16t^2 + vt + h$, what is the meaning of the variable t? What is the meaning of the variable y? What is the meaning of the variable v? What is the meaning of the variable h?
- Will the water balloon travel in a horizontal line across the field? What will the path of the balloon look like?

SCENARIO On a calm day, you and a friend are tossing water balloons in a field trying to hit a boulder in the field. The balloons travel in a path that is in the shape of a parabola.

Problem 1 Ready, Set, Launch

A. On your first throw, the balloon leaves your hand 3 feet above the ground at a velocity of 20 feet per second. Use what you learned in Chapter 8 about vertical motion models to write an equation that gives the height of the balloon in terms of time.

$y = -16t^2 + 20t + 3$

B. What is the height of the balloon after one second? Show your work and use a complete sentence to explain your answer.

$y = -16(1)^2 + 20(1) + 3 = -16 + 20 + 3 = 7$

The balloon is at a height of seven feet after one second.

C. What is the height of the balloon after two seconds? Show your work and use a complete sentence in your answer.

$y = -16(2)^2 + 20(2) + 3 = -64 + 40 + 3 = -21$

The balloon is on the ground after two seconds.

D. On your second throw, the balloon leaves your hand at ground level at a velocity of 30 feet per second. Write an equation that gives the height of the balloon in terms of time.

$y = -16t^2 + 30t$

E. What is the height of the balloon after one second? Show your work and use a complete sentence in your answer.

$y = -16(1)^2 + 30(1) = -16 + 30 = 14$

The balloon is at a height of 14 feet after one second.

F. What is the height of the balloon after two seconds? Show your work and use a complete sentence in your answer.

$y = -16(2)^2 + 30(2) = -64 + 60 = -4$

The balloon is on the ground after two seconds.

- What will a positive value for the height represent in this situation?
- Is it possible for the water balloon to have a negative height? What would a negative height represent?
- What would a height of zero represent?

Explore Together

Investigate Problem 1

Students will compare two polynomials used to model vertical motion.

Take Note

Whole numbers are the numbers 0, 1, 2, 3, and so on.

Grouping

Ask for a student volunteer to read Question 1 aloud. Have a student restate the problem. Have students work together in small groups to complete Question 1. Then call the class back together to have the students discuss and explain their work for Question 1.

Just the Math

Students will be formally introduced to terminology for polynomials in Questions 2 and 3. It is important for students to recognize that terms in a polynomial are separated by addition. This is true because a polynomial such as $x^2 - 2$ can be written as $x^2 + (-2)$. So, x^2 and -2 are each terms of this polynomial.

Grouping

Ask for a student volunteer to read Question 2 aloud. Have a student restate the problem. Pose the Guiding Questions below to verify student understanding. Complete Questions 2 and 3 together as a whole class. Then have students work together in small groups to complete Questions 4 and 5.

Guiding Questions

- What is a polynomial? What are the important characteristics of a polynomial?
- What is a term? What is a coefficient?
- In the polynomial $-3x^2 + 4x + (-5)$ list the terms? What is the coefficient of x^2? What is the coefficient of x?
- What is a degree of a term? What is a degree of a polynomial?
- How can you find the degree of a term? How can you find the degree of a polynomial?

Investigate Problem 1

1. How are your models the same? Use a complete sentence in your answer.

Sample Answer: They are both quadratic equations.

How are your models different? Use a complete sentence in your answer.

Sample Answer: The *y*-intercepts of the graphs of the equations are different and the numbers multiplied by *t* are different.

2. Just the Math: Polynomials The expressions $-16t^2 + 20t + 3$ and $-16t^2 + 30t$ that model the heights of the balloons are *polynomials*. Each expression is a **polynomial** because it is a sum of products of the form ax^k, where a is a real number and k is a whole number. Each product is a **term** and the number being multiplied by a power is a **coefficient.** For each of the polynomials above, name the terms and coefficients. Use complete sentences in your answer.

The terms of $-16t^2 + 20t + 3$ are $-16t^2$, $20t$, and 3.
The coefficients of $-16t^2 + 20t + 3$ are -16, 20, and 3.

The terms of $-16t^2 + 30t$ are $-16t^2$ and $30t$.
The coefficients of $-16t^2 + 30t$ are -16 and 30.

3. For each of your polynomial models, what is the greatest exponent? Use a complete sentence in your answer.

The greatest exponent in each model is two.

The **degree of a term** in a polynomial is the exponent of the term. The greatest exponent in a polynomial determines the **degree of the polynomial.** For instance, in the polynomial $4x + 3$, the greatest exponent is 1, so the degree of the polynomial is 1. What is the degree of each of your polynomial models? Use a complete sentence in your answer.

Each polynomial model has degree two.

4. What kind of expression is a polynomial of degree 0? Give an example and use a complete sentence to explain your reasoning.

Sample Answer: $4x^0 = 4$; A polynomial of degree 0 is a constant expression.

What kind of expression is a polynomial of degree 1? Give an example and use a complete sentence to explain.

Sample Answer: $2x + 1$; A polynomial of degree 1 is a linear expression.

Explore Together

Investigate Problem 1

Students will classify polynomials according to the number of terms in each polynomial.

Call the class back together to have the students discuss and present their work for Question 4.

Grouping

Ask for a student volunteer to read Question 5 aloud. Have a student restate the problem. Pose the Guiding Questions below to verify student understanding. Have students work together in small groups to complete Questions 5 and 6.

Guiding Questions

- What is a term for a polynomial?
- How can you find the number of terms?
- What is a monomial? What is a binomial? What is a trinomial?
- Why do you think that we only have special names for polynomials with 1, 2, or 3 terms?

Call the class back together to have the students discuss and present their work for Questions 5 and 6.

Just the Math

The concept of the standard form of a polynomial will be introduced in Question 7.

Grouping

Ask for a student volunteer to read Question 7 aloud. Have a student restate the problem. Pose the Guiding Questions below to verify student understanding. Have students work together in small groups to complete Questions 7 and 8.

Guiding Questions

- What does it mean for something to have a standard or to be standardized?
- Why is it important for us to be able to write polynomials in a standard form? What would happen if several people were considering a polynomial to represent the motion of the space shuttle, but each had the polynomial represented in a different way?
- How can you write a polynomial in standard form?

Notes Only the basic approach to standard form is presented in this problem. Students will not yet be asked to write a polynomial that has more than one variable in standard form.

Investigate Problem 1

What kind of expression is a polynomial of degree 2? Give an example and use a complete sentence to explain your reasoning.

Sample Answer: x^2; A polynomial of degree 2 is a quadratic expression.

A polynomial of degree 3 is a called a cubic polynomial. Write an example of a cubic polynomial.

Sample Answer: x^3

5. For each of your polynomial models in parts (A) and (D), find the number of terms in the model. Use a complete sentence in your answer.

The model for the first throw has three terms and the model for the second throw has two terms.

Polynomials with only one term are **monomials.** Polynomials with exactly two terms are **binomials.** Polynomials with exactly three terms are **trinomials.** Classify each polynomial model in parts (A) and (D) by its number of terms. Use a complete sentence in your answer.

The polynomial model for the first throw is a trinomial and the polynomial model for the second throw is a binomial.

6. Give an example of a monomial of degree 3.

Sample Answer: $4x^3$

Give an example of a trinomial of degree 5.

Sample Answer: $x^5 + 2x - 3$

7. Just the Math: Standard Form of a Polynomial
Later in this chapter, we will be adding, subtracting, multiplying, and dividing polynomials. To make this process easier, it is helpful to write polynomials in *standard form*. A polynomial is written in **standard form** by writing the terms in *descending* order, starting with the term with the greatest degree and ending with the term with the least degree. Write each polynomial in standard form.

$6 + 5x$
$5x + 6$

$7 - x^2$
$-x^2 + 7$

$4 + 3x + 4x^2$
$4x^2 + 3x + 4$

$4 + 3x^2 + 9x - x^3$
$-x^3 + 3x^2 + 9x + 4$

$5 - 6x^4$
$-6x^4 + 5$

$x^6 - 4x^3 + 16$
$x^6 - 4x^3 + 16$

10

Explore Together

Investigate Problem 1

Students will complete a summary question to review the words polynomials, terms, and degrees.

Call the class back together to have the students discuss and present their work for Questions 7 and 8.

Key Formative Assessments

- What are the requirements for an expression to be considered a polynomial?
- How can you determine whether an expression is a polynomial?
- How can you determine where a term begins and where the term ends?
- How can you determine the number of terms in a polynomial?
- What are the requirements for a polynomial to be in standard form?

Problem 2

Grouping

Ask for a student volunteer to read Problem 2 aloud. Have a student restate the problem. Pose the Guiding Questions below to verify student understanding. Have students work together in small groups to complete parts (A) through (C) of Problem 2.

Guiding Questions

- What information is given in this problem?
- How is this problem similar to Problem 1? How is this problem different from Problem 1?
- How can you write an equation to model the motion for the balloon thrown by your friend?

If the students finish Problem 1 during the class period and do not have enough time to complete Problem 2 during class, parts (A) through (C) of Problem 2 can be assigned as part of a homework assignment. At the beginning of the next class session, call the students together to discuss and present parts (A) through (C) of Problem 2 and then continue with the problem as suggested on the next few pages.

10

Investigate Problem 1

8. Determine whether each algebraic expression is a polynomial. If the expression is a polynomial, classify it by its degree and number of terms. If the expression is not a polynomial, use a complete sentence to explain why it is not a polynomial.

$4x^2 + 3x - 1$

polynomial of degree 2; trinomial

$3x^{-2} + 4x - 1$

Not a polynomial because the power on the first term is not a whole number.

$4x^6 + 1$

polynomial of degree 6; binomial

$10 - x^4$

polynomial of degree 4; binomial

$2\sqrt{x} + 3x - 4$

Not a polynomial because a polynomial does not contain square roots.

25

polynomial of degree 0; monomial

Problem 2 The Balloon's Path

It is now your friend's turn to throw water balloons at the boulder.

A. In your friend's first throw, the balloon leaves her hand 2 feet above the ground at a velocity of 36 feet per second. Write an equation that gives the height of the balloon in terms of time.

$y = -16t^2 + 36t + 2$

B. Complete the table of values that shows the height of the balloon in terms of time.

Quantity Name	Time	Height
Unit	seconds	feet
Expression	t	$-16t^2 + 36t + 2$
	0.0	2
	0.5	16
	1.0	22
	2.0	10
	2.5	–8
	3.0	–34

Explore Together

Problem 2

Students will graph a parabolic function representing the height of a water balloon as a function of the time after the balloon was thrown.

Call the class back together to have the students discuss and present their work for parts (A) through (C) of Problem 2.

Key Formative Assessments

- What two quantities were compared for your graph in part (C)?
- What is the unit for time? What is the unit for height?
- What is the shape of the graph?
- Does this shape make sense for this problem situation? Why or why not?
- Why does the parabola open downward rather than upward?
- How many *y*-intercepts are there?
- What is the meaning of the *y*-intercept for your graph?
- How many *x*-intercepts are there?
- Why is there only one *x*-intercept?
- What is the meaning of the *x*-intercept?

Investigate Problem 2

Grouping

Ask for a student volunteer to read Questions 1 and 2 aloud. Have a student restate the problem. Pose the Guiding Questions below to verify student understanding. Complete Questions 1 and 2 together as a whole class.

Guiding Questions

- What is a function? What does it mean for every input value to have exactly one output value?

Problem 2 The Balloon's Path

C. Create a graph of the model to see the path of the balloon on the grid below. First, choose your bounds and intervals. Be sure to label your graph clearly.

Variable quantity	Lower bound	Upper bound	Interval
Time	0	3.75	0.25
Height	0	30	2

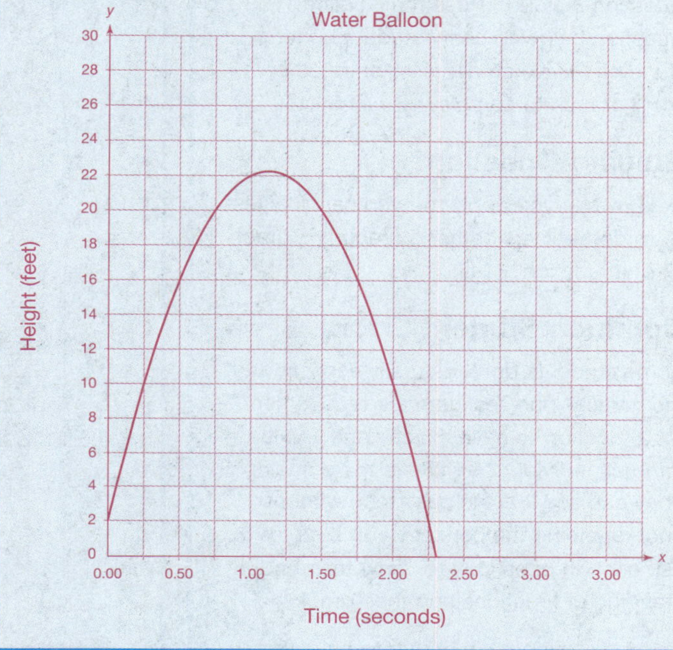

Investigate Problem 2

1. Is the equation that you wrote in part (A) a function? How do you know? Use a complete sentence in your answer.

   Sample Answer: Yes. For every input, there is only one output.

2. **Just the Math: Vertical Line Test** You can use a graph to determine whether an equation is a function. The **Vertical Line Test** states that an equation is a function if you can pass a vertical line through any part of the graph of the equation and the line intersects the graph, at most, one time. Consider your graph in part (C). Does your graph pass the Vertical Line Test?

   Yes.

- What is the Vertical Line Test and how can you use it? Does the vertical line have to pass through just one point at a selected *x*-value, or does it have to pass through exactly one point at each *x*-value?

10

Explore Together

Investigate Problem 2

Students will determine whether a given graph is the graph of a function by applying the Vertical Line Test.

Grouping

Ask for a student volunteer to read Question 3 aloud. Pose the Guiding Question below to verify student understanding. Have students work together in small groups to complete Question 3.

Guiding Question

- How can you apply the Vertical Line Test to determine whether a graph is a graph of a function?

Common Student Errors

Some students don't naturally associate the Vertical Line Test with the requirement for a function to have exactly one y-value for each individual x-value in the domain. They can see it as an easy trick without understanding the concept. Be sure to ask enough questions to verify their understanding of functions and their graphs.

Call the class back together to have the students discuss and present their work for Question 3.

10

Take Note

Whenever you see the share with the class icon, your group should prepare a short presentation to share with the class that describes how you solved the problem. Be prepared to ask questions during other groups' presentations and to answer questions during your presentation.

Investigate Problem 2

3. Consider each equation and its graph. Use the Vertical Line Test to determine whether the equation is a function.

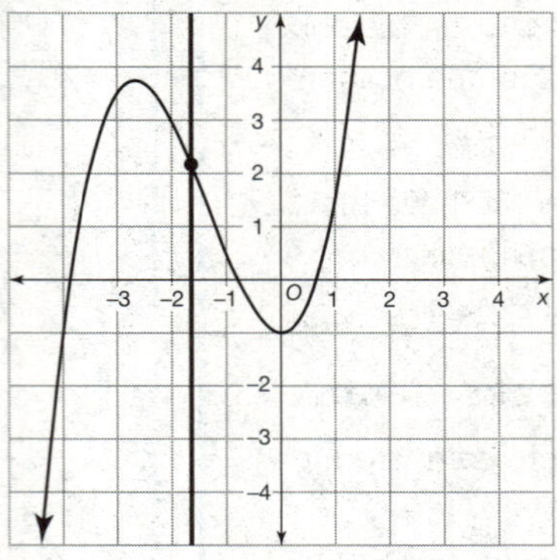

The equation $y = \frac{1}{2}x^3 + 2x^2 - 1$ is a function.

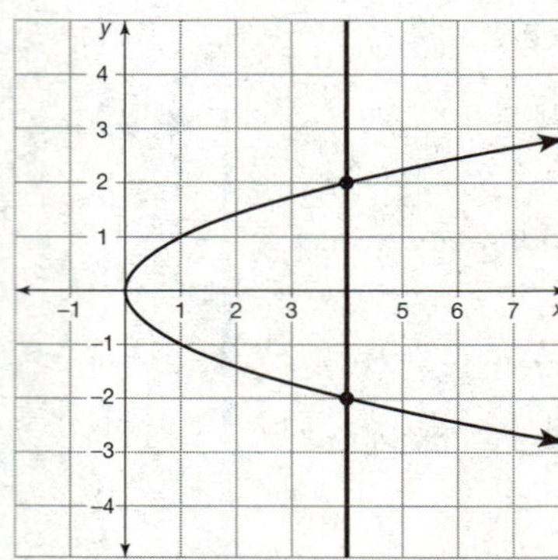

The equation $x = y^2$ is not a function.

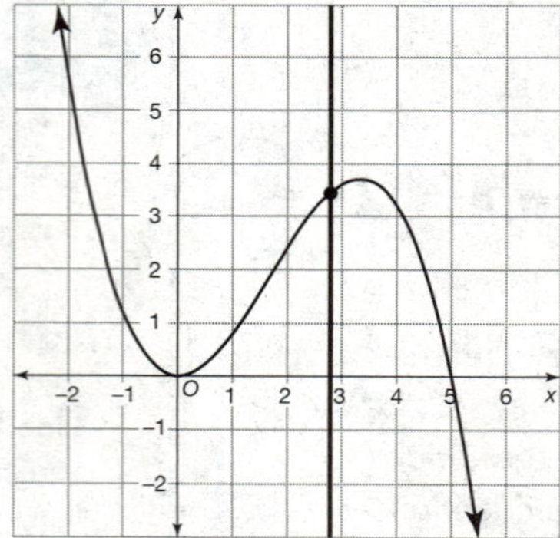

The equation $y = -0.2x^3 + x^2$ is a function.

Key Formative Assessments

- What is a coefficient in a polynomial?
- Are coefficients always whole numbers? Are they always positive?
- How can you determine the degree of a polynomial? Is it possible to have a degree of polynomial that is $\frac{1}{2}$?
- What is a monomial? What is a binomial? What is a trinomial?
- How can you write a polynomial in standard form?
- How can you determine whether a graph is the graph of a function?
- Are all linear graphs functions? What type of line is not a function? Is a horizontal line a function?

Wrap Up

Close

- Review all key terms and their definitions. Include the terms *polynomial, term, coefficient, degree, monomial, binomial, trinomial, standard form,* and *Vertical Line Test.*
- You may also want to review any other vocabulary terms that were discussed during the lesson which may include *velocity, speed, vertical motion, x-value, y-value, x-intercept, y-intercept, whole number, exponent, function, linear function, quadratic function, cubic function, domain, range, equation,* and *expression.*
- Remind the students to write the key terms and their definitions in the notes section of their notebooks. You may also want the students to include examples.
- Ask the students to each suggest a monomial, a binomial, a trinomial, and an expression or equation that is not a polynomial by writing them down on paper. While they are doing that, divide the front board into 4 sections. Label the first with the name monomial, the second with binomial, the third with trinomial, and the fourth with not a polynomial. Have the students then write their suggestions in the appropriate place on the front board. Begin a discussion by asking several students to choose one of the equations that is not a polynomial and explain what polynomial requirement it fails. Next, ask the students to verify the classifications of monomial, binomial, and trinomial. If they disagree with any of the suggestions, have them explain why. Then, ask the students to identify the polynomials that have a degree of 1, then repeat for a degree of 2, etc. Ask which polynomials are written in standard form. If any are not in standard form, ask volunteers to rewrite them in standard form. Finally, ask them to determine which polynomials are functions and which are not functions. Have the students explain their answers.

Ties to the Cognitive Tutor Software

The Cognitive Tutor software's skill model focuses on two basic tasks in adding and subtracting polynomials. First, students need to learn to apply the concept of "like terms" to expressions with any number of terms and any order. Second, students need to understand that, once these like terms are identified, you can combine them by focusing on the coefficient, not the exponents.

Follow Up

Assignment

Use the Assignment for Lesson 10.1 in the Student Assignments book. See the Teacher's Resources and Assessments book for answers.

Assessment

See the Assessments provided in the Teacher's Resources and Assessments book for Chapter 10.

Open-Ended Writing Task

Have the students each write 2 monomials, 2 binomials, and 2 trinomials. Have the students then write each of their polynomials in standard form, determine whether each is a function, identify the coefficients of each term, identify the degree of each term, and identify the degree of each polynomial. At the beginning of the next class session, have the students trade their work and verify the answers for their partners.

Reflections

Insert your reflections on the lesson as it played out in class today.

What went well?

What did not go as well as you would have liked?

How would you like to change the lesson in order to improve the things that did not go well and capitalize on the things that did go well?

10

Notes

10.2

Play Ball!

Adding and Subtracting Polynomials

Learning By Doing Lesson Map

Get Ready

Objectives

In this lesson, you will:

- Add polynomials.
- Subtract polynomials.

Key Terms

- combine like terms
- distributive property
- add
- subtract

NCTM Content Standards

Grades 9–12 Expectations

Number and Operations Standards

- Develop a deeper understanding of very large and very small numbers and of various representations of them.
- Judge the effects of operations such as multiplication, division, and computing powers and roots on the magnitude of quantities.
- Develop an understanding of properties of, and representations for, the addition and multiplication of vectors and matrices.
- Develop fluency in operations with real numbers, vectors, and matrices, using mental computation or paper-and-pencil calculations for simple cases and technology for more complicated cases.
- Judge the reasonableness of numerical computations and their results.

Algebra Standards

- Understand and perform transformations, such as arithmetically combining, composing, and inverting commonly used functions, using technology to perform such operations on more complicated symbolic expressions.
- Understand and compare the properties of classes of functions, including exponential, polynomial, rational, logarithmic, and periodic functions.
- Interpret representations of functions of two variables.
- Write equivalent forms of expressions, inequalities, and systems of equations and solve them with fluency—mentally or with paper and pencil in simple cases and using technology in all cases.
- Use symbolic algebra to represent and explain mathematical relationships.
- Use a variety of symbolic representations, including recursive and parametric equations, for functions and relations.
- Identify essential quantitative relationships in a situation and determine the class or classes of functions that might model the relationships.
- Draw reasonable conclusions about a situation being modeled.

Lesson Overview

Within the context of this lesson, students will be asked to:

- Evaluate two polynomials for the same x-value, then find the sum of the results.
- Add two polynomials, then evaluate the sum for a given value of x.
- Compare the results for the two sums.
- Add polynomials.
- Subtract polynomials.
- Simplify expressions requiring addition and subtraction of polynomials.

Essential Questions

The following key questions are addressed in this section:

1. What is the distributive property?
2. What are like terms?
3. How can you combine like terms?
4. How can you add polynomials?
5. How can you subtract polynomials?

Show The Way

Warm Up

Place the following questions or an applicable subset of these questions on the board before students enter class. Students should begin working as soon as they are seated.

Simplify each expression.

1. $10x - 4x$ **$6x$**

2. $15x - 2x + 3x$ **$16x$**

3. $\frac{(30 + 6)}{(2 \cdot 6)}$ **3**

4. $\frac{5(10 + 4) - 16}{6}$ **9**

5. $3(2 - x)$ **$6 - 3x$**

6. $3(x^2 + x)$ **$3x^2 + 3x$**

7. $\frac{-8}{6}\left(\frac{12}{16}\right)$ **-1**

8. $-5(7x^3 + 2x^2 + 3x)$

$-35x^3 - 10x^2 - 15x$

9. $3(x + 2) - 4$ **$3x^2 + 2$**

Motivator

Begin the lesson with the motivator to get students thinking about the topic of the upcoming problem. This lesson is about attendance at baseball games. The motivating questions are about attending baseball games.

Ask the students the following questions to get them interested in the lesson.

10

- Have you attended a major league baseball game?
- What stadium(s) have you attended?
- What is your favorite stadium? Why?
- What is the closest major league baseball stadium and team to our school?
- Is that team part of the National League or the American League?

Explore Together

Problem 1

Students will evaluate cubic functions for given values of x. Students will then find the sum of the evaluated functions for the same x-value.

Grouping

Ask for a student volunteer to read the Scenario and Problem 1 aloud. Have a student restate the problem. Pose the Guiding Questions below to verify student understanding. Have students work together in small groups to complete parts (A) and (B) of Problem 1. Then call the class back together to have the students discuss and present their work for parts (A) and (B).

Guiding Questions

- What information is given in this problem?
- What is the degree of each given polynomial representing attendance?
- What does x represent in each equation? What does y represent in each equation?
- How are the equations similar? How are the equations different?
- What are you asked to do in this problem?
- How can you determine the x-value for a given year?
- How can you evaluate a given function for a specified value of x?

Common Student Errors

Students will often multiply the coefficient of a variable term by the value for the variable before evaluating the exponent. Be ready to remind the students of the order of operations for evaluating an expression.

Students also often struggle with numbers of large magnitude. It is difficult for them to comprehend these numbers because they can't visualize them easily. Remind the students to check their answers for reasonableness.

SCENARIO A friend of yours loves baseball and plans to visit every major league baseball park in the country. The major league is made up of two divisions, the National League and the American League.

Problem 1 Batter Up

Your friend recently read an article that stated that the attendance at National League baseball games from 1990 through 2001 can be modeled by the function

$$y = -86{,}584x^3 + 1{,}592{,}363x^2 - 5{,}692{,}368x + 24{,}488{,}926$$

where x is the number of years since 1990 and y is the number of people who attended. The article also stated that the attendance at American League baseball games from 1990 through 2001 can be modeled by the function

$$y = -56{,}554x^3 + 1{,}075{,}426x^2 - 4{,}806{,}571x + 30{,}280{,}751$$

where x is the number of years since 1990 and y is the number of people who attended.

A. Find the attendance for each league in 1995. Show your work and use a complete sentence in your answer.

National League:
$y = -86{,}584(5)^3 + 1{,}592{,}363(5)^2 - 5{,}692{,}368(5) + 24{,}488{,}926$
$y = 25{,}013{,}161$

American League:
$y = -56{,}554(5)^3 + 1{,}075{,}426(5)^2 - 4{,}806{,}571(5) + 30{,}280{,}751$
$y = 26{,}064{,}296$

In 1995, the attendance for the National League was 25,013,161 people and the attendance for the American League was 26,064,296 people.

B. Find the total attendance at all games in major league baseball in 1995. Show your work and use a complete sentence in your answer.

25,013,161 + 26,064,296 = 51,077,457

The total attendance in 1995 was 51,077,457 people.

How did you find your answer to part (B)? Use a complete sentence in your answer.

Sample Answer: I added the 1995 National League attendance to the 1995 American League attendance.

10

Explore Together

Problem 1

Students will add two polynomials by combining the like terms.

Grouping

Ask for a student volunteer to read part (C) of Problem 1 aloud. Have a student restate the problem. Pose the Guiding Questions below to verify student understanding. Have students work together in small groups to complete parts (C) and (D) of Problem 1.

Guiding Questions

- How do parts (C) and (D) differ from parts (A) and (B)?
- How will you find the x-value for the year 2000?

Call the class back together to have the students discuss and present their work for parts (C) and (D) of Problem 1.

Key Formative Assessments

- How could you find the predicted attendance for this year for each of the two major leagues in baseball?
- How many years since 1990 is it now? What value will we use for x?
- What is the predicted attendance for the National League this year? What is the predicted attendance for the American League this year? What is the total predicted attendance for this year in major league baseball?
- Do you think there is a simpler way to get a predicted total attendance than finding the individual values separately and then adding them together?

Investigate Problem 1

Take Note

Remember that you use a **distributive property** to combine like terms:

$$4x + 10x = x(4 + 10)$$
$$= x(14)$$
$$= 14x$$

10

Problem 1 Batter Up

C. Find the attendance for each league in 2000. Show your work and use a complete sentence in your answer.

National League:
$y = -86{,}584(10)^3 + 1{,}592{,}363(10)^2 - 5{,}692{,}368(10) + 24{,}488{,}926$
$y = 40{,}217{,}546$

American League:
$y = -56{,}554(10)^3 + 1{,}075{,}426(10)^2 - 4{,}806{,}571(10) + 30{,}280{,}751$
$y = 33{,}203{,}641$

In 2000, the attendance for the National League was 40,217,546 people and the attendance for the American League was 33,203,641 people.

D. Find the total attendance at all games in major league baseball in 2000. Show your work and use a complete sentence in your answer.

40,217,546 + 33,203,641 = 73,421,187

The total attendance in 2000 was 73,421,187 people.

How did you find your answer to part (D)? Use a complete sentence in your answer.

Sample Answer: I added the 2000 National League attendance to the 2000 American League attendance.

Investigate Problem 1

1. How would you write a function that you could use to find the total attendance at all major league baseball games? Use a complete sentence in your answer.

   Sample Answer: Add the function for the National League attendance to the function for the American League attendance.

2. You can add or subtract two polynomials by **combining like terms**. What are the like terms in the attendance polynomial models?

   $-86{,}584x^3$ **and** $-56{,}554x^3$; $1{,}592{,}363x^2$ **and** $1{,}075{,}426x^2$; $-5{,}692{,}368x$ **and** $-4{,}806{,}571x$; **24,488,926 and 30,280,751**

   Find the sum of each group of like terms. What is the model for total attendance at all major league baseball games?

   $y = -143{,}138x^3 + 2{,}667{,}789x^2 - 10{,}498{,}939x + 54{,}769{,}677$

Grouping

Ask for a student volunteer to read Question 1 aloud. Have a student restate the problem. Have students work together in small groups to complete Questions 1 through 3.

Explore Together

Investigate Problem 1

Students will compare the result of finding the sum of two polynomials evaluated at the same x-value to the result of finding the sum of the polynomials and then evaluating at the same x-value.

Call the class back together to have the students discuss and present their work for Questions 1 through 3.

Key Formative Assessments

- How can you add two polynomials?
- What are like terms?
- Does the degree of each term have to be equal for the terms to be considered like terms?
- How can you combine like terms?
- How do you use the distributive property to combine like terms?
- Which method is easier, evaluating the individual attendance equations for a given year and then adding their values or adding the like terms first and then evaluating the new equation?
- What other situation can you think of in which it might be easier to add polynomials before evaluating?

Grouping

Ask for a student volunteer to read Question 4 aloud. Have a student restate the problem. Pose the Guiding Questions below to verify student understanding. Have the students complete Question 4 individually. Then have the students discuss and explain their work for Question 4 with the class.

Guiding Questions

- What are we asked to do in Question 4?
- How can you subtract polynomials?

Investigate Problem 1

3. Use your model to find the total attendance in 1995 and 2000. Show your work and use a complete sentence in your answer.

1995:
$y = -143{,}138(5)^3 + 2{,}667{,}789(5)^2 - 10{,}498{,}939(5) + 54{,}769{,}677$
$y = 51{,}077{,}457$

2000:
$y = -143{,}138(10)^3 + 2{,}667{,}789(10)^2 - 10{,}498{,}939(10) + 54{,}769{,}677$
$y = 73{,}421{,}187$

In 1995, the total attendance was 51,077,457 people and in 2000, the total attendance was 73,421,187 people.

How do these answers compare to those in Problem 1 parts (B) and (D)? Use a complete sentence in your answer.

Sample Answer: The answers are the same as those in Problem 1 parts (B) and (D).

When is it useful for you to find the sum of two functions first, and then evaluate the function rather than evaluating each function separately and then finding the sum? Use a complete sentence in your answer.

Sample Answer: It is useful when you have to find many sums.

4. You can also write a function that shows how many more people attended National League games each year than attended American League games. Describe how you can find this function. Use a complete sentence in your answer.

Sample Answer: You can find this function by subtracting the function for American League game attendance from the function for National League game attendance.

Complete the statement below that gives the function described above.

$y = (-86{,}584x^3 + 1{,}592{,}363x^2 - 5{,}692{,}368x + 24{,}488{,}926)$
$- (-56{,}554x^3 + 1{,}075{,}426x^2 - 4{,}806{,}571x + 30{,}280{,}751)$

Because you are subtracting one function from another function, you must subtract *each term* of the second function from the first function. To do this, use the distributive property to distribute the negative sign to each term in the second function.

$y = -86{,}584x^3 + 1{,}592{,}363x^2 - 5{,}692{,}368x + 24{,}488{,}926$
$+ 56{,}554x^3 - 1{,}075{,}426x^2 + 4{,}806{,}571x - 30{,}280{,}751$

Common Student Errors

Some students may have difficulty subtracting polynomials. Remind the students that subtraction is the same as adding the opposite of the second quantity.

Explore Together

Investigate Problem 1

Students will practice adding and subtracting polynomials.

Grouping

Ask for a student volunteer to read Question 5 aloud. Have a student restate the problem.

If time permits during the class period, have students work together in small groups to complete Questions 5 and 6.

If time is limited at the end of the class session, you can assign Questions 5 and 6 as part of a homework assignment.

Common Student Errors

Students will frequently confuse unlike terms to be like terms because they have the same base. You may need to stop and briefly review terminology of the base and the exponent for powers. Remind the students that terms are only considered to be like if both the base and exponent match exactly.

Conversely, other students will incorrectly assume that terms are like only if the coefficient, the base, and the exponent all match exactly. It may help these students to be reminded that the coefficient is a counter. It tells us the number of that variable term that exists.

Call the class back together to have the students discuss and present their work for Questions 5 and 6.

Key Formative Assessments

- What are like terms?
- How can you use the distributive property to combine like terms?
- How can you add polynomials? When is it useful to add polynomials?
- How can you subtract polynomials? When is it useful to subtract polynomials?

10

Now combine like terms and simplify.

$y = -30{,}030x^3 + 516{,}937x^2 - 885{,}797x - 5{,}791{,}825$

5. What was the difference in attendance between the leagues in 1998? Which league had more people attending games in 1998? How do you know? Show your work and use complete sentences in your answer.

$y = -30{,}030(8)^3 + 516{,}937(8)^2 - 885{,}797(8) - 5{,}791{,}825$
$y = 4{,}830{,}407$

Sample Answer: The difference was 4,830,407 people. The National League had more people attending because the difference is the attendance of the American League subtracted from the attendance of the National League and this difference is positive.

What was the difference in attendance between the leagues in 1992? Which league had more people attending games in 1992? How do you know? Show your work and use a complete sentence in your answer.

$y = -30{,}030(2)^3 + 516{,}937(2)^2 - 885{,}797(2) - 5{,}791{,}825$
$y = -5{,}735{,}911$

Sample Answer: The difference was –5,735,911 people. The American League had more people attending because the difference is the attendance of the American League subtracted from the attendance of the National League and this difference is negative.

6. Simplify each expression by finding the sum or difference. Show your work.

$(4x^2 + 3x - 5) + (x^2 - 8x + 4)$

$$(4x^2 + 3x - 5) + (x^2 - 8x + 4) = (4x^2 + x^2) + (3x - 8x) + (-5 + 4)$$
$$= 5x^2 - 5x - 1$$

$(2x^2 + 3x - 4) - (2x^2 + 5x - 6)$

$$(2x^2 + 3x - 4) - (2x^2 + 5x - 6) = (2x^2 - 2x^2) + (3x - 5x) + (-4 + 6)$$
$$= -2x + 2$$

$(4x^3 + 5x - 2) + (7x^2 - 8x + 9)$

$$(4x^3 + 5x - 2) + (7x^2 - 8x + 9) = 4x^3 + 7x^2 + (5x - 8x) + (-2 + 9)$$
$$= 4x^3 + 7x^2 - 3x + 7$$

$(9x^4 - 5) - (8x^4 - 2x^3 + x)$

$$(9x^4 - 5) - (8x^4 - 2x^3 + x) = (9x^4 - 8x^4) + 2x^3 - x - 5$$
$$= x^4 + 2x^3 - x - 5$$

Wrap Up

Close

- Review all key terms and their definitions. Include the terms *combine like terms, distributive property, add,* and *subtract.*
- You may also want to review any other vocabulary terms that were discussed during the lesson which may include *order of operations, coefficient, terms, like terms, polynomial, function, evaluate, magnitude, sum, difference, base, exponent,* and *powers.*
- Remind the students to write the key terms and their definitions in the notes section of their notebooks. You may also want the students to include examples.
- Ask the students to compare and contrast the process of adding polynomials and subtracting polynomials.
- Ask the students to write a polynomial with at least 4 terms. Ask several students to write their polynomial on the front board. Choose a pair of the polynomials and have the class find the sum and then the difference of those polynomials. Repeat this process with other pairs of polynomials on the board as many times as you think is needed to help the students fully understand adding and subtracting polynomials.

Ties to the Cognitive Tutor Software

The Cognitive Tutor software's skill model focuses on two basic tasks in adding and subtracting polynomials. First, students need to learn to apply the concept of "like terms" to expressions with any number of terms and any order. Second, students need to understand that, once these like terms are identified, you can combine them by focusing on the coefficient, not the exponents.

Follow Up

Assignment

Use the Assignment for Lesson 10.2 in the Student Assignments book. See the Teacher's Resources and Assessments book for answers.

Assessment

See the Assessments provided in the Teacher's Resources and Assessments book for Chapter 10.

Open-Ended Writing Task

Have the students create two polynomials with at least 3 terms each. Have the students write the polynomials in standard form and then find the sum and the difference of the two polynomials. Ask the students to write a few sentences to explain their opinion as to what effect the order of the two polynomials has for both the sum and for the difference. During the next class session, discuss their work and responses for this activity as a whole class.

Reflections

Insert your reflections on the lesson as it played out in class today.

What went well?

What did not go as well as you would have liked?

How would you like to change the lesson in order to improve the things that did not go well and capitalize on the things that did go well?

10

Notes

10.3

Se Habla Español

Multiplying and Dividing Polynomials

Learning By Doing Lesson Map

Get Ready

Objectives

In this lesson, you will:

- Use an area model to multiply polynomials.
- Use distributive properties to multiply polynomials.
- Use long division to divide polynomials.

Key Terms

- area model
- distributive property
- divisor
- dividend
- remainder

NCTM Content Standards

Grades 9–12 Expectations

Number and Operations Standards

- Develop a deeper understanding of very large and very small numbers and of various representations of them.
- Judge the effects of such operations as multiplication, division, and computing powers and roots on the magnitude of quantities.
- Develop fluency in operations with real numbers, vectors, and matrices, using mental computation or paper-and-pencil calculations for simple cases and technology for more complicated cases.

Algebra Standards

- Understand and perform transformations, such as arithmetically combining, composing, and inverting commonly used functions, using technology to perform such operations on more complicated symbolic expressions.
- Interpret representations of functions of two variables.

Geometry Standards

- Draw and construct representations of two- and three-dimensional geometric objects using a variety of tools.
- Use geometric models to gain insights and answer questions in other areas of mathematics.

Lesson Overview

Within the context of this lesson, students will be asked to:

- Evaluate polynomials for a given value of x and then multiply the results.
- Multiply the polynomials, then evaluate the product for the same given value of x.
- Compare the answer from each process and identify them as resulting in the same value.
- Follow a similar procedure for division of polynomials.
- Use long division to divide polynomials.
- Students will identify the divisor, dividend, and remainder when dividing polynomials.

Essential Questions

The following key questions are addressed in this section:

1. What is an area model?
2. What is the distributive property and how can it help you to multiply polynomials?
3. How can you multiply polynomials?
4. What is long division?
5. How can you use long division to divide polynomials?

Show The Way

Warm Up

Place the following questions or an applicable subset of these questions on the board before students enter class. Students should begin working as soon as they are seated.

Use the properties of exponents to simplify each expression. Write your answer as a power.

1. $2^3 \cdot 2^4$ 2^7

2. $7^2 \cdot 7^5 \cdot 7$ 7^8

3. $4^5 \cdot 4^5 \cdot 8^2 \cdot 8^3$ $4^{10} \cdot 8^5$

4. $\frac{5^9}{5^3}$ 5^6

5. $\frac{2}{2^6}$ $\frac{1}{2^5}$

6. $\frac{(-8)^7}{(-8)^6}$ -8

Write each percent as a decimal.

7. 8% 0.08

8. 41% 0.41

Write each decimal as a percent.

9. 0.16 16%

10. 0.0007 0.07%

10

Motivator

Begin the lesson with the motivator to get students thinking about the topic of the upcoming problem. This lesson is about the number of students who take foreign language classes. The motivating questions are about foreign language classes.

Ask the students the following questions to get them interested in the lesson.

- What foreign language classes are offered at our school?
- Have you ever taken a foreign language class?
- What language did you study or are you studying?
- How did you choose which language to study?
- What other foreign language would you like our school to offer?

Explore Together

Problem 1

Students will evaluate two separate polynomials for a given x-value and then multiply the results.

Grouping

Ask for a student volunteer to read the Scenario and Problem 1 aloud. Have a student restate the problem. Pose the Guiding Questions below to verify student understanding. Have students work together in small groups to complete parts (A) through (D) of Problem 1.

Guiding Questions

- What information is given in this problem?
- What is a foreign language?
- What does the equation $y = -0.0005t^2 + 0.02t + 0.23$ represent?
- What is the meaning of t in that equation?
- How would you interpret a y-value of 0.16?
- What x-value would represent the year 1982? How can you find the x-value for a given year?
- What does the equation $y = -3t^3 + 134t^2 - 1786t + 18{,}398$ represent?
- How would you interpret a y-value of 12? How would you interpret a y-value of 1356?
- What number is equivalent to 1356 thousand?

Take Note

In Lesson 3.3, you learned how to use the percent equation. The *percent equation* is an equation of the form $a = pb$ where p is the percent in decimal form and the numbers that are being compared are a and b.

SCENARIO Your high school is trying to decide whether to offer more foreign language classes. The guidance counselor finds a model for the percent of high school students that were in a foreign language class in high school during the years from 1985 to 2000, which is

$$y = -0.0005t^2 + 0.02t + 0.23$$

where t represents the number of years since 1980 and y is the percent (in decimal form) of all high school students who were in a foreign language class. The guidance counselor also finds that the total number of students in high school during the years from 1985 to 2000 can be modeled by

$$y = -3t^3 + 134t^2 - 1786t + 18{,}398$$

where t represents the number of years since 1980 and y is the number of students in thousands.

Problem 1 Learning a Foreign Language

A. What percent of high school students were in a foreign language class in 1990? Show your work and use a complete sentence in your answer.

$y = -0.0005(10)^2 + 0.02(10) + 0.23$
$y = 0.38$

In 1990, 38% of high school students were in a foreign language class.

B. Find the total number of students in high school in 1990. Show your work and use a complete sentence in your answer.

$y = -3(10)^3 + 134(10)^2 - 1786(10) + 18{,}398$
$y = 10{,}938$

In 1990, 10,938 thousand students were in high school.

C. In 1990, how many students were in a foreign language class? Show your work and use a complete sentence in your answer.

Sample Answer: 0.38(10,938) = 4156.44

In 1990, approximately 4156 thousand students were in a foreign language class.

D. Use complete sentences to explain how you found your answer to part (C).

Sample Answer: To find the number of students in a foreign language class, multiply the total number of students in high school by the percent that are in a foreign language class.

Notes Students may also correctly answer part (B) with 10,938,000 students.

Students can also correctly complete part (C) using a proportion.

Call the class back together to have the students discuss and present their work for parts (A) through (D).

10

Explore Together

Investigate Problem 1

Students will use an area model to multiply two polynomials together.

Grouping

Ask for a student volunteer to read Question 1 aloud. Have a student restate the problem. Complete Question 1 together as a class.

Just the Math

Question 2 will help visual learners to better comprehend multiplication of polynomials. You will need to explain to the students that the variable x represents an unknown value. We can represent x in the model by choosing any length. However, we must represent each length of x with the same length. The constant of 1 represents the length of 1 unit. For instance, we might want it to represent 1 inch or 1 centimeter. Question 3 will extend their understanding to multiply polynomials symbolically.

Grouping

Ask for a student volunteer to read Question 2 and Question 3 aloud. Have a student restate the problem. Pose the Guiding Questions below to verify student understanding. Complete Questions 2 and 3 together as a whole class.

Guiding Questions

- What is a geometric model?
- What is area?
- How can you find the area of a rectangle?
- What does x represent in this situation?
- If x is a positive number, which will be larger, $3x$ or $4x + 1$? Why?
- How can you use an area model to determine the product of two polynomials?
- How can you multiply polynomials symbolically?

10

Investigate Problem 1

1. Use a complete sentence to explain how you can use the functions given in Problem 1 to write a function that gives the number of students that were in a foreign language class during the years from 1985 to 2000.

   Sample Answer: Multiply the function for the total number of students by the function for the percent of the total number of students that are taking a foreign language class.

2. **Just the Math: Area Model** You can use an **area model** to multiply two polynomials. Suppose that you want to find the product of $3x$ and $4x + 1$. Consider these polynomials to be the length and width of a rectangle as shown below. The area of the rectangle is the product of the polynomials.

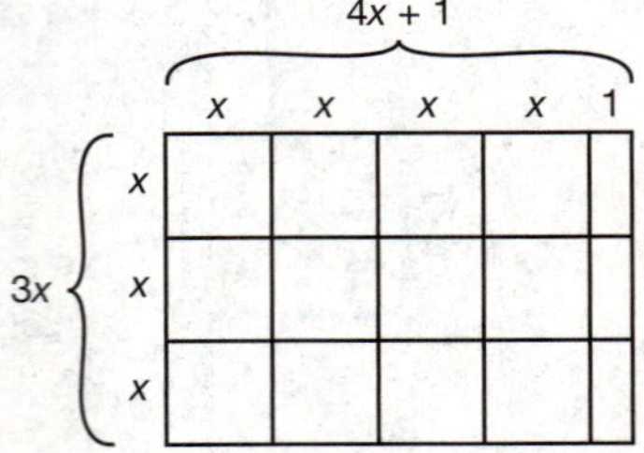

   What is the area of each square in the model? Use a complete sentence in your answer.

   The area of each square is (length)(width) = $x(x) = x^2$.

   What is the area of each small rectangle in the model? Use a complete sentence in your answer.

   The area of each small rectangle is (length)(width) = $x(1) = x$.

   What is the area of the entire rectangle? Show your work and use a complete sentence in your answer.

   $12(x^2) + 3(x) = 12x^2 + 3x$; The area of the rectangle is $12x^2 + 3x$ square units.

3. **Just the Math: Multiplying Polynomials** To multiply two polynomials, you need to use the distributive properties of multiplication and the properties of exponents. The number of times that you need to use a distributive property depends on the number of terms in the polynomials. For instance, to multiply the polynomials $3x$ and $4x + 1$ from Question 2 above, you only need to use the distributive property once. Use the distributive property to complete the first step below.

   $(3x)(4x + 1) = (\underline{3x})(\underline{4x}) + (\underline{3x})(\underline{1})$

Take Note

Recall the distributive properties of multiplication from Lesson 4.4:

$a(b + c) = ab + ac \quad a(b - c) = ab - ac$

Two other related distributive properties are:

$(b + c)a = ba + ca \quad (b - c)a = ba - ca$

Explore Together

Investigate Problem 1

Students will investigate multiplying polynomials.

Notes In Question 3, students could have started by multiplying each term of the polynomial $(x^2 - 3x + 2)$ by the binomial $(x + 1)$, and then simplifying. This will give the same product and may be easier for some students than the method suggested in Question 3.

Grouping

Ask for a student volunteer to read Question 4 aloud. Have a student restate the problem. Have the students complete Question 4 individually. Then discuss the answers together as a whole class.

Grouping

Ask for a student volunteer to read Question 5 aloud. Pose the Guiding Questions below to verify student understanding. Have students work together in small groups to complete Questions 5 through 7.

Guiding Questions

- How can you multiply polynomials?
- What types of errors do you think are common in these questions?
- After you multiply each term of one polynomial by each term of the other polynomial, what should you do next?
- When can you simplify the results from multiplying polynomials?

Common Student Errors

Multiplying polynomials is a difficult concept for students. Because it is abstract and out of context, sign errors, multiplication errors, and general arithmetic errors are common. Remind the students to check over their work carefully. It is often helpful for the students to rewrite the original polynomials in terms of addition only rather than subtraction. For instance, to multiply $(x^2 - 5)(-x^2 - 4x - 2)$, students could rewrite the problem as $(x^2 + (-5))((-x^2) + (-4x) + (-2))$, then multiply. This might help to prevent sign errors.

Investigate Problem 1

To simplify each product, remember that multiplication is commutative (you can multiply numbers in any order) and use the properties of exponents that you learned in Chapter 9. For instance, $(2x)(5x) = 2(5)(x)(x) = 10x^2$. Complete the product of $3x$ and $4x + 1$ below.

$(3x)(4x + 1) = (3x)(4x) + (3x)(1) = \underline{12x^2} + \underline{3x}$

To multiply the polynomials $x + 1$ and $x^2 - 3x + 2$, you need to use the distributive property twice. First, use a distributive property to multiply each term of $x + 1$ by the polynomial $x^2 - 3x + 2$. Complete the first step below.

$(x + 1)(x^2 - 3x + 2) = (x)(\underline{x^2 - 3x + 2}) + (1)(\underline{x^2 - 3x + 2})$

Now, distribute x to each term of $x^2 - 3x + 2$ and distribute 1 to each term of $x^2 - 3x + 2$. Complete the second step below:

$(x + 1)(x^2 - 3x + 2)$

$= (x)(x^2) + (\underline{x})(\underline{-3x}) + (\underline{x})(\underline{2}) + (\underline{1})(\underline{x^2})$

$+ (\underline{1})(\underline{-3x}) + (\underline{1})(\underline{2})$

Now multiply and collect like terms. Show your work and write your answer as a polynomial in standard form.

$(x + 1)(x^2 - 3x + 2) = x^3 - 3x^2 + 2x + x^2 - 3x + 2$

$= x^3 - 2x^2 - x + 2$

4. What do you notice about the products of the terms in Question 3 when you used a distributive property the second time? Use a complete sentence in your answer.

Sample Answer: The products are formed by multiplying each term in one polynomial by each term in the other polynomial.

5. Find each product. Show all your work.

$2x(x + 3)$

$2x(x) + 2x(3) = 2x^2 + 6x$

$5x^2(7x - 1)$

$(5x^2)(7x) - (5x^2)(1) = 35x^3 - 5x^2$

$(x + 1)(x + 3)$

$(x + 1)(x) + (x + 1)(3)$

$= x(x) + 1(x) + x(3) + 1(3)$

$= x^2 + x + 3x + 3$

$= x^2 + 4x + 3$

$(x^2 - 4)(2x + 3)$

$(x^2 - 4)(2x) + (x^2 - 4)(3)$

$= x^2(2x) - 4(2x) + x^2(3) - 4(3)$

$= 2x^3 - 8x + 3x^2 - 12$

$= 2x^3 + 3x^2 - 8x - 12$

$(x - 5)(x^2 + 3x + 1)$

$(x - 5)(x^2) + (x - 5)(3x) + (x - 5)(1)$

$= x(x^2) - 5(x^2) + x(3x) - 5(3x) + x(1) - 5(1)$

$= x^3 - 5x^2 + 3x^2 - 15x + x - 5$

$= x^3 - 2x^2 - 14x - 5$

10

Explore Together

Investigate Problem 1

Students will multiply the polynomials from the beginning of the lesson to determine the total number of students taking a foreign language class for any given year in the domain.

Grouping

Students will be working together in small groups to complete Questions 6 and 7.

Common Student Errors

Some students many not understand why they are multiplying the two polynomials in Question 6 to find the number of students who are taking a foreign language. Pose the Guiding Questions below to verify student understanding of this concept.

Guiding Questions

- What is the function that represents the percent of the total number of high school students enrolled in a foreign language?
- What is the function that represents the total number of high school students for any given year in the domain?
- If 20% of the students were enrolled in a foreign language class and there were 1000 total students, how would you find the number of students enrolled in a foreign language class?
- Why would you multiply those values?

Call the class back together to have the students discuss and present their work for Questions 5 through 7.

Notes Students may give the answer to Question 7 as 5,280,000. Be sure to have the students read this number correctly as five million two hundred eighty thousand students.

Investigate Problem 1

6. Use multiplication to find the function that gives the number of students in thousands that were in a foreign language class during the years from 1985 to 2000.

$(-0.0005t^2 + 0.02t + 0.23)(-3t^3 + 134t^2 - 1786t + 18{,}398)$

First, distribute each term of the polynomial for the percent to each term of the polynomial for the total number of students.

$-0.0005t^2(-3t^3 + 134t^2 - 1786t + 18{,}398) +$
$0.02t(-3t^3 + 134t^2 - 1786t + 18{,}398) +$
$0.23(-3t^3 + 134t^2 - 1786t + 18{,}398)$

Complete the steps below. Show your work.

$-0.0005t^2(-3t^3 + 134t^2 - 1786t + 18{,}398)$

$= (-0.0005t^2)(-3t^3) + (-0.0005t^2)(134t^2) + (-0.0005t^2)(-1786t) + (-0.0005t^2)(18{,}398)$

$= 0.0015t^5 - 0.067t^4 + 0.893t^3 - 9.199t^2$

$0.02t(-3t^3 + 134t^2 - 1786t + 18{,}398)$

$= (0.02t)(-3t^3) + (0.02t)(134t^2) + (0.02t^2)(-1786t) + (0.02t)(18{,}398)$

$= -0.06t^4 + 2.68t^3 - 35.72t^2 + 367.96t$

$0.23(-3t^3 + 134t^2 - 1786t + 18{,}398)$

$= (0.23)(-3t^3) + (0.23)(134t^2) + (0.23)(-1786t) + (0.23)(18{,}398)$

$= -0.69t^3 + 30.82t^2 - 410.78t + 4231.54$

Now combine like terms and simplify. Show your work.

$(0.0015t^5) + (-0.067t^4 - 0.06t^4) + (0.893t^3 + 2.68t^3 - 0.69t^3) + (-9.199t^2 - 35.72t^2 + 30.82t^2) + (367.96t - 410.78t) + 4231.54 = 0.0015t^5 - 0.127t^4 + 2.883t^3 - 14.099t^2 - 42.82t + 4231.54$

What is your function?

$y = 0.0015t^5 - 0.127t^4 + 2.883t^3 - 14.099t^2 - 42.82t + 4231.54$

7. How many students were in a foreign language class in 2000? Show all your work and use a complete sentence in your answer.

$y = 0.0015(20)^5 - 0.127(20)^4 + 2.883(20)^3 - 14.099(20)^2 - 42.82(20) + 4231.54$

$y = 5279.54$

In 2000, approximately 5280 thousand students were in a foreign language class.

Explore Together

Problem 2

Students will evaluate polynomial expressions and will calculate the percent of the total number of students who are taking a Spanish class.

Grouping

Ask for a student volunteer to read Problem 2 aloud. Have a student restate the problem. Pose the Guiding Questions below to verify student understanding. Have students work together in small groups to complete parts (A) through (D) of Problem 2.

Guiding Questions

- What information is given in this problem?
- How does this problem differ from Problem 1? How is this problem similar to Problem 1?
- What are you asked to find in part (A) of Problem 2? How can you find the number of students that were enrolled in a Spanish class in 1990?
- How many years after 1980 was 1990?

Note Students can also complete part (C) by writing and solving a proportion rather than the method suggested in the sample solution.

Call the class back together to have the students discuss and present their work for parts (A) through (D) of Problem 2.

Investigate Problem 2

Grouping

Ask for a student volunteer to read Question 1 aloud. Have a student restate the problem. Pose the Guiding Questions at the right to verify student understanding. Complete Question 1 together as a whole class.

Problem 2 Spanish, Anyone?

A model for the number of high school students that were in a Spanish class in high school during the years from 1985 to 2000 is $y = 4t^2 + 14t + 2137$, where t represents the number of years since 1980 and y is the number of high school students in thousands that were in a Spanish class.

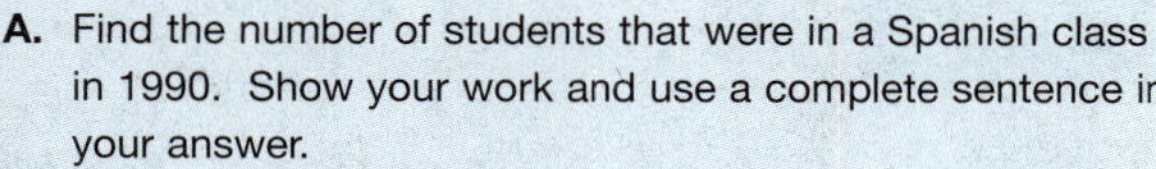

A. Find the number of students that were in a Spanish class in 1990. Show your work and use a complete sentence in your answer.

$y = 4(10)^2 + 14(10) + 2137 = 2677$

In 1990, 2677 thousand high school students were in a Spanish class.

B. Find the total number of students in high school in 1990. You can use the function $y = -3t^3 + 134t^2 - 1786t + 18{,}398$ from Problem 1. Show your work and use a complete sentence in your answer.

$y = -3(10)^3 + 134(10)^2 - 1786(10) + 18{,}398$

$y = 10{,}938$

In 1990, 10,938 thousand students were in high school.

C. In 1990, what percent of all high school students were in a Spanish class? Show your work and use a complete sentence in your answer.

Sample Answer: $2677 \div 10{,}938 \approx 0.24$

In 1990, approximately 24 percent of all students were in a Spanish class.

D. Use complete sentences to explain how you found your answer to part (C).

Sample Answer: To find the percent of students that were in a Spanish class, divide the number of students that were in a Spanish class by the total number of students in high school.

10

Investigate Problem 2

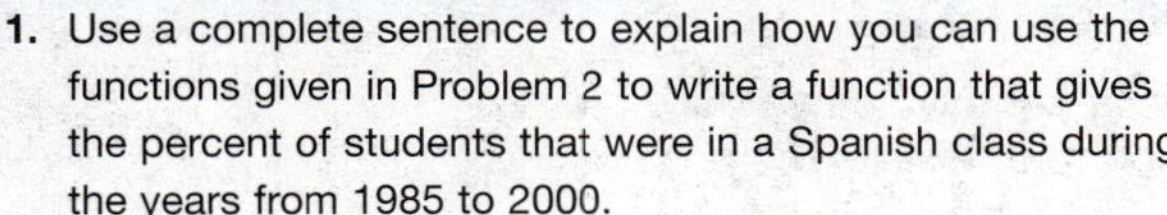

1. Use a complete sentence to explain how you can use the functions given in Problem 2 to write a function that gives the percent of students that were in a Spanish class during the years from 1985 to 2000.

Sample Answer: Divide the function for the number of students that are in a Spanish class by the function for the total number of students in high school.

Guiding Questions

- How did you find the percent of students that were enrolled in Spanish Classes in 1990?
- How would your equation change to allow you to find the percent of students enrolled in Spanish classes for any given year in the domain?

Explore Together

Investigate Problem 2

Students will divide one polynomial by another polynomial.

Take Note

In the *quotient* $a \div b$, a is the *dividend* and b is the *divisor*.

Grouping

Ask for a student volunteer to read Question 2 aloud. Have a student restate the problem. Pose the Guiding Questions below to verify student understanding. Complete Question 2 together as a whole class working slowly through each step.

Guiding Questions

- What information is given in this problem?
- When dividing 3240 by 24, what is the divisor? What is the dividend?
- To estimate the quotient 3240 ÷ 24, we would first think of how many times we could divide 3 by 24 getting 0. Then we would think of how many times we can divide 32 by 24 getting 1. How would you finish dividing 3240 by 24 using long division? Show all of your work.

Take Note

Whenever you divide two polynomials, it is important to write the polynomials in *standard form*.

Common Student Errors

The process of using long division to divide whole numbers is still confusing to some students. Extending that concept to divide polynomials is significantly more difficult for many students to understand. Try to relate the process to division of whole numbers as best as possible to help students follow the solution in Question 2. When you have finished Question 2, have students take turns explaining each step to verify student understanding.

Investigate Problem 2

2. You can divide two polynomials just like you divide numbers by using long division. For instance, consider the quotient of $10x^3 + 13x^2 + 2x - 4$ and $2x + 1$ shown near the bottom of the page. Begin by finding the quotient of the first terms. In this case, what is $10x^3$ divided by $2x$?

$5x^2$

This is the first term of the quotient. Write this term in the quotient below. Then multiply this expression by the polynomial $2x + 1$. Subtract the result from $10x^3 + 13x^2$ to get $8x^2$ (see Step 1).

Bring down the next term, $2x$, from the **dividend**. You need to do this because the product of $2x + 1$ and the second term of the quotient will have two terms. Now, we have to divide $8x^2$ by the $2x$ in the **divisor**. What is the result? This is the second term of the quotient.

$4x$

Write this term in the quotient below. Then multiply this expression by the polynomial $2x + 1$. Subtract the result from $8x^2 + 2x$ (see Step 2 below).

Bring down the last term, –4, from the dividend. Finally, we have to divide $-2x$ by $2x$. What is the result? This is the last term of the quotient.

-1

Write this term in the quotient below. Then multiply this expression by the polynomial $2x + 1$. Subtract the result from $-2x - 4$ (see Step 3 below).

Because the difference –3 is a term whose degree is less than the degree of the divisor ($2x$), we are done. This number –3 is the **remainder**. Write the remainder over the divisor below.

$$5x^2 + 4x + (-1) + \frac{-3}{2x+1}$$

$$
\begin{array}{rll}
2x + 1 \overline{)10x^3 + 13x^2 + 2x - 4} & & \\
-(10x^3 + 5x^2) & & \text{Step 1} \\
8x^2 + 2x & & \\
-(8x^2 + 4x) & & \text{Step 2} \\
-2x - 4 & & \\
-(-2x - 1) & & \text{Step 3} \\
-3 & &
\end{array}
$$

10

Explore Together

Investigate Problem 2

Students will continue to divide polynomials with long division and will check their results with multiplication.

Take Note

When the dividend and divisor polynomials are in standard form and the difference between consecutive exponents is greater than 1, you should write in a placeholder for the missing terms. For instance, you should write $x^3 + 4x - 1$ as $x^3 + 0x^2 + 4x - 1$.

Key Formative Assessments

- What are the basic steps for using long division to divide polynomials?
- Where do you put the polynomial that is the divisor?
- Where do you put the polynomial that is the dividend?
- How can you verify your answer after you have divided polynomials? Why does multiplying the quotient and remainder by the dividend equal the divisor? Will that always work for any division problem?

Grouping

Ask for a student volunteer to read Question 3 aloud. Have a student restate the problem. Have students work together in small groups to complete Questions 3 and 4.

If class time is limited when the students are ready to begin Questions 3 and 4, you can assign these questions as part of a homework assignment.

Investigate Problem 2

You can check your answer by multiplying the divisor and quotient and adding the remainder:

Divisor **Quotient** **Remainder** **Dividend**

$$(2x + 1)(\underline{5x^2 + 4x - 1}) + (\underline{-3}) \stackrel{?}{=} 10x^3 + 13x^2 + 2x - 4$$

$$\underline{10x^3 + 8x^2 - 2x + 5x^2 + 4x - 1 - 3} \stackrel{?}{=} 10x^3 + 13x^2 + 2x - 4$$

$$\underline{10x^3 + 13x^2 + 2x - 4} = 10x^3 + 13x^2 + 2x - 4$$

3. Find each quotient. Show all your work.

$(x^2 + 6x + 5) \div (x + 1) = \underline{x + 5}$

$$\begin{array}{r} x + 5 \\ x + 1 \overline{\smash{)}\, x^2 + 6x + 5} \\ \underline{-(x^2 + x)} \\ 5x + 5 \\ \underline{-(5x + 5)} \\ 0 \end{array}$$

$(2x^2 + 5x - 12) \div (2x - 3) = \underline{x + 4}$

$$\begin{array}{r} x + 4 \\ 2x - 3 \overline{\smash{)}\, 2x^2 + 5x - 12} \\ \underline{-(2x^2 - 3x)} \\ 8x - 12 \\ \underline{-(8x - 12)} \\ 0 \end{array}$$

10

Explore Together

Investigate Problem 2

Students will continue to divide polynomials using long division.

In Question 4 of Problem 2, students will perform the division for the situation in Problem 2. The students will find an equation to model the percentage of the total number of high school students that are enrolled in a Spanish class during any given year.

Grouping

Students will be working in small groups to complete these questions.

Common Student Errors

Some students incorrectly reverse the divisor and the dividend when using long division to divide whole numbers and will often do the same when dividing polynomials. Be ready to refocus the students and ask them guiding questions to help them recognize their error and divide appropriately.

Arithmetic errors will also be common. As with multiplying polynomials, it will help some students to rewrite the original polynomials in terms of addition rather than subtraction.

Call the class back together to have the students discuss and present their work for Questions 3 and 4.

Notes Synthetic division for polynomials is not discussed in this lesson. It is too early to present that topic to students who are first learning how to divide polynomials. Synthetic division should be saved until students are very proficient with long division and clearly understand the concepts at work in the process.

Key Formative Assessments

- How can you multiply polynomials?
- How can you use an area model to help multiply polynomials?

Investigate Problem 2

$(x^3 + 2x - 6) \div (x + 1) = x^2 - x + 3 - \frac{9}{x + 1}$

$x^2 - x + 3 - \frac{9}{x+1}$

$x + 1\overline{)x^3 + 0x^2 + 2x - 6}$

$-(x^3 + x^2)$

$-x^2 + 2x$

$-(-x^2 - x)$

$3x - 6$

$-(3x + 3)$

-9

4. Complete the division problem below to find the function that represents the percentage of the total number of high school students that are enrolled in a Spanish class.

$(-3t^3 + 134t^2 - 1786.00t + 18{,}398) \div (4t^2 + 14t + 2137)$

$= \underline{-0.75t} + \underline{36.125} + \frac{-689t - 58{,}801.125}{4t^2 + 14t + 2137}$

$4t^2 + 14t + 2137\overline{)-3t^3 + 134t^2 - 1786.00t + 18{,}398}$

$-(-3t^3 - 10.5t^2 - 1602.75t)$

$144.5t^2 - 183.25t + 18{,}398$

$-(144.5t^2 + 505.75t + 77{,}199.125)$

$-689.00t - 58{,}801.125$

- How can you divide polynomials using long division?
- What is a divisor? What is a dividend? What is a remainder?

Wrap Up

Close

- Review all key terms and their definitions. Include the terms *area model, distributive property, divisor, dividend,* and *remainder.*
- You may also want to review any other vocabulary terms that were discussed during the lesson which may include *percent, percent equation, proportion, decimal form, product, quotient, geometric model, domain,* and *long division.*
- Remind the students to write the key terms and their definitions in the notes section of their notebooks. You may also want the students to include examples.
- Ask the students to compare and contrast multiplying and dividing polynomials
- Have the students use the functions from this section to predict the number of students taking a foreign language in high schools this year. Have them do the same for the number of students taking Spanish as their foreign language.
- Have students develop polynomials as a class and list them on the front board. Practice multiplying and dividing some of the pairs of polynomials until it is clear that the students understand the processes.

Ties to the Cognitive Tutor Software

In the Cognitive Tutor software, students multiply and divide rational expressions using the same basic interface as they have previously used to work with simpler expressions. This approach helps students understand that the same basic mathematical operations apply.

Follow Up

Assignment

Use the Assignment for Lesson 10.3 in the Student Assignments book. See the Teacher's Resources and Assessments book for answers.

Assessment

See the Assessments provided in the Teacher's Resources and Assessments book for Chapter 10.

Open-Ended Writing Task

Have the students write a short paragraph explaining how to multiply and how to divide polynomials. Remind them to be precise in their vocabulary. For instance, they should correctly use the words *product, quotient, divisor, dividend,* and *remainder* in their explanation. They should also include a detailed example of each.

Reflections

Insert your reflections on the lesson as it played out in class today.

What went well?

What did not go as well as you would have liked?

How would you like to change the lesson in order to improve the things that did not go well and capitalize on the things that did go well?

10

Notes

10.4

Making Stained Glass

Multiplying Binomials

Learning By Doing Lesson Map

Get Ready

Objectives

In this lesson, you will:

- Use the FOIL pattern to multiply binomials.
- Use formulas to find special products.

Key Terms

- FOIL pattern
- square of a binomial sum
- square of a binomial difference

NCTM Content Standards

Grades 9–12 Expectations

Number and Operations Standards

- Use number-theory arguments to justify relationships involving whole numbers.
- Judge the effects of such operations as multiplication, division, and computing powers and roots on the magnitude of quantities.
- Develop fluency in operations with real numbers, vectors, and matrices, using mental computation or paper-and-pencil calculations for simple cases and technology for more complicated cases.

Algebra Standards

- Generalize patterns using explicitly defined and recursively defined functions.
- Understand and perform transformations, such as arithmetically combining, composing, and inverting commonly used functions, using technology to perform such operations on more complicated symbolic expressions.
- Use symbolic algebra to represent and explain mathematical relationships.

Lesson Overview

Within the context of this lesson, students will be asked to:

- Multiply binomials.
- Develop patterns and formulas for multiplying binomials.
- Use the FOIL Pattern to multiply binomials efficiently.
- Use formulas to find the products of the most common types of special binomials.

Essential Questions

The following key questions are addressed in this section:

1. What is a binomial?
2. What is the product of two binomials?
3. How can you use the FOIL Pattern to multiply two binomials?
4. What special binomial products are common?
5. What patterns for multiplying binomials exist?

Show The Way

Warm Up

Place the following questions or an applicable subset of these questions on the board before students enter class. Students should begin working as soon as they are seated.

Simplify each expression. Where applicable, $x \neq 0$.

1. 3^0 **1**

2. $(-2)^0$ **1**

3. $-(2)^0$ **−1**

4. -2^0 **−1**

5. $(5x)^2$ **$25x^2$**

6. $5(x)^2$ **$5x^2$**

7. $\frac{4x^3}{2x^3}$ **2**

8. -3^3 **−27**

9. $\frac{12x^3}{3x}$ **$4x^2$**

10. $(-2)^5$ **−32**

11. $x^2(x^5)$ **x^7**

12. $(x^2)^5$ **x^{10}**

13. $2(x^2)^3$ **$2x^6$**

14. $-10(2^2)^3$ **$-10(2^6) = -640$**

15. $-10(-2^2)^3$ **$-10(-4)^3 = 640$**

16. $-3(2^2)^3$ **$-3(4)^3 = -192$**

17. $(4x)^2$ **$16x^2$**

18. $(-3x^5)^2$ **$9x^{10}$**

Motivator

10

Begin the lesson with the motivator to get students thinking about the topic of the upcoming problem. This lesson is about an artist making designs in stained glass. The motivating questions are about making stained glass.

Ask the students the following questions to get them interested in the lesson.

- What is stained glass?
- Do you know how stained glass is made?
- Where have you seen stained glass?
- How difficult do you think it is to make stained glass?

Explore Together

Problem 1

Students will represent the area of a rectangular region as the product of two binomials. The students will then multiply the binomials to find the area in a simplified manner.

Grouping

Ask for a student volunteer to read the Scenario and Problem 1 aloud. Have a student restate the problem. Pose the Guiding Questions below to verify student understanding. Begin by asking the students to complete parts (A) through (C) of Problem 1 individually. Walk around the room to get a sense of how well the students understand the information in the problem. After the students have had a few minutes to think about the situation, call the class back together to discuss and present their work for parts (A) through (C) of Problem 1. Some students may not have been able to complete these correctly on their own.

Guiding Questions

- What information is given in this problem?
- What does "the ratio of the length to the width is 7 to 5" mean?
- If the length of the stained glass is 7 inches, what is the width of the stained glass?
- If the length of the stained glass is 21 inches, what is the width?
- If the length of the stained glass is 21 inches and the width is 15 inches, what is the total length and width of the frame?
- How can you find the area of a rectangle? How can you find the area of the rectangular frame?
- How wide is the wooden frame on each side of the stained glass?

SCENARIO An artist creates designs in stained glass. One of the artist's more popular designs is a rectangle in which the ratio of the length to the width of the stained glass is 7:5. The artist also surrounds the stained glass with a two-inch wide carved wooden frame.

Problem 1 Stained Glass Design

A. Complete the diagram below that models the stained-glass design in its frame.

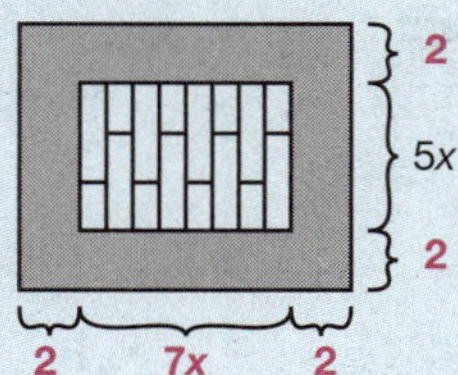

B. Write an expression for the total length of the stained-glass design with its frame.

$7x + 4$

C. Write an expression for the total width of the stained-glass design with its frame.

$5x + 4$

D. Write a function for the area of the stained-glass design with the frame.

$f(x) = (7x + 4)(5x + 4)$

E. Use a distributive property to simplify the function in part (D). Show your work.

$$\begin{aligned} f(x) &= (7x + 4)(5x + 4) \\ &= (7x + 4)(5x) + (7x + 4)(4) \\ &= 7x(5x) + 4(5x) + 7x(4) + 4(4) \\ &= 35x^2 + 20x + 28x + 16 \\ &= 35x^2 + 48x + 16 \end{aligned}$$

F. What type of function is in part (E)? Use a complete sentence in your answer.

The function is a quadratic function.

Grouping

Have students work together in small groups to complete parts (D) through (F). Then call the class back together to have the students discuss and present their work for parts (D) through (F) of Problem 1.

10

Explore Together

Investigate Problem 1

Students will investigate multiplication of two binomials using the **FOIL pattern**.

Grouping

Ask for a student volunteer to read Question 1 aloud. Have a student restate the problem. Pose the Guiding Questions below to verify student understanding. Complete Question 1 together as a whole class.

Guiding Questions

- What is standard form?
- What kind of function has a degree of 1?
- If a linear function has a positive coefficient, is it an increasing or decreasing function? What will the graph look like?
- If a linear function has a negative coefficient, is it an increasing or decreasing function? What will the graph look like?
- What kind of function has a degree of 2?
- If a quadratic function has a positive coefficient, what will the graph of the function look like?
- If a quadratic function has a negative coefficient, what will the graph of the function look like?
- What kind of function has a degree of 3?

Grouping

Ask for a student volunteer to read Question 2 aloud. Have a student restate the problem. Pose the Guiding Questions at the right to verify student understanding, then solve Question 2 together as a whole class. Have students work together in small groups to complete Question 3, then call the class back together to have the students discuss and present their work for Question 3.

Investigate Problem 1

1. What is the value of writing the function in standard form in part (E)? Show your work and use a complete sentence in your answer.

   Sample Answer: You can tell whether the graph, a parabola, opens upward or downward. You can also easily determine the *y*-intercept and the vertex.

2. In part (E), you found the product of two binomials by using the distributive property. Consider the product after two uses of the distributive property:

   $(2x + 3)(x + 8) = (2x + 3)(x) + (2x + 3)(8)$ First use of distributive property

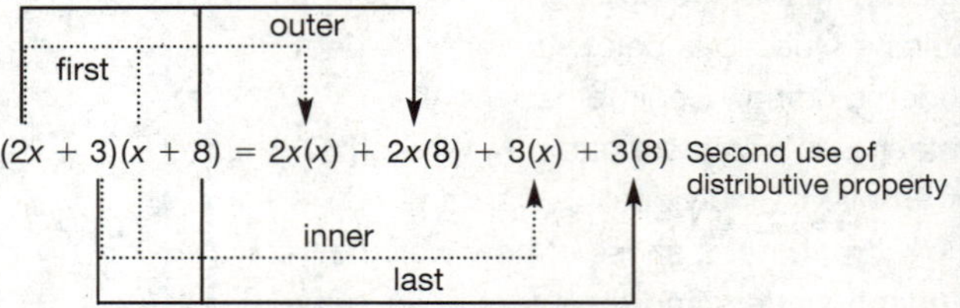

   You can quickly find the product of two binomials by remembering the word FOIL. FOIL stands for multiplying the **f**irst terms, the **o**uter terms, the **i**nner terms, and the **l**ast terms. Then the products are added together as shown above. Finish finding the product of $2x + 3$ and $x + 8$.

   $2x(x) + 2x(8) + 3(x) + 3(8) = 2x^2 + 16x + 3x + 24$

   $= 2x^2 + 19x + 24$

3. Use the FOIL pattern to find each product. Show your work.

   $(x + 2)(x + 9)$

   $(x + 2)(x + 9)$
   $= x^2 + 9x + 2x + 18$
   $= x^2 + 11x + 18$

   $(x - 1)(x + 7)$

   $(x - 1)(x + 7)$
   $= x^2 + 7x - x - 7$
   $= x^2 + 6x - 7$

   $(x - 3)(x - 6)$

   $(x - 3)(x - 6)$
   $= x^2 - 6x - 3x + 18$
   $= x^2 - 9x + 18$

   $(4x + 3)(x + 5)$

   $(4x + 3)(x + 5)$
   $= 4x^2 + 20x + 3x + 15$
   $= 4x^2 + 23x + 15$

   $(3x - 4)(x + 2)$

   $(3x - 4)(x + 2)$
   $= 3x^2 + 6x - 4x - 8$
   $= 3x^2 + 2x - 8$

   $(5x + 1)(3x - 2)$

   $(5x + 1)(3x - 2)$
   $= 15x^2 - 10x + 3x - 2$
   $= 15x^2 - 7x - 2$

Guiding Questions

- What is the distributive property?
- What is distributed in the first step of the example in Question 2? What is distributed in the second step?

10

Explore Together

Problem 2

Students will work to develop a pattern for squaring a binomial.

Grouping

Ask for a student volunteer to read Problem 2 aloud. Have a student restate the problem. Pose the Guiding Questions below to verify student understanding. Have students work together in small groups to complete parts (A) through (D) of Problem 2.

Guiding Questions

- What information is given in this problem?
- How is this problem different from Problem 1? How are the problems the same?
- How can you calculate the area of a rectangle? How can you calculate the area of a square?
- What is the length of a side of the stained glass in Problem 2? What is the width of the wooden frame?
- What is the length of each side of the stained glass including the frame?

Call the class back together to have the students discuss and present their work for parts (A) through (D) of Problem 2.

Problem 2 New Stained Glass Design

The stained glass artist is working on a new design that is square with a carved wood frame that is three inches wide.

A. Draw a diagram that models the stained-glass design with its frame. Let x be the width of the stained glass without the frame.

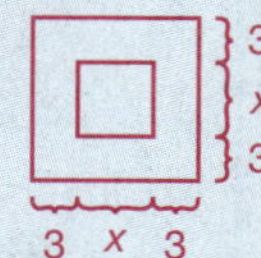

B. Write a function for the area of the stained-glass design with its frame.

$f(x) = (x + 6)(x + 6)$

C. Use the FOIL pattern to simplify your function in part (B). Show all your work.

$f(x) = (x + 6)(x + 6) = x^2 + 6x + 6x + 36 = x^2 + 12x + 36$

D. Suppose that the artist changes the width of the frame to five inches wide. Write and simplify a new function for the area of the stained glass with the frame. Show all your work.

$f(x) = (x + 10)(x + 10) = x^2 + 10x + 10x + 100 = x^2 + 20x + 100$

Investigate Problem 2

1. Use complete sentences to describe the similarities and differences between the functions in parts (C) and (D).

   Sample Answer: The functions are both squares of binomials. The constants are different.

2. **Just the Math: Square of a Binomial Sum** In Problem 2, you wrote and simplified the square of a binomial. In each simplified product, the middle term related to the product of the linear term and the constant term in the binomial. For instance, the square of $(x + 3)$ is $x^2 + 6x + 9$. How is the middle term $6x$ related to x and 3? Use a complete sentence to explain.

   The middle term $6x$ is two times the product of the linear term x and the constant term 3 in the factors: $2(x)(3) = 6x$.

Investigate Problem 2

Just the Math

Students will identify the pattern for squaring a binomial. This pattern is a time saving tool that will reduce tedium and frustration for students. Squaring values is an important skill used in many applications for problem solving.

Grouping

Ask for a student volunteer to read Question 1 aloud. Have a student restate the problem. Complete Question 1 together as a whole class. Ask for a student volunteer to read Question 2 aloud. Have a student restate the problem. Have students work together in small groups to complete Question 2.

10

Explore Together

Investigate Problem 2

Students will continue to investigate the pattern for squaring a binomial.

Grouping

Students will be working together in small groups to complete Question 2.

Common Student Errors

Students may have difficulty recognizing the patterns in squaring a binomial. If students get frustrated with this you may want to call the class back together to pose the Guiding Questions below to help the students better understand the question. You may also want to work together as a whole class to complete Question 2.

Guiding Questions

- What would correctly complete $(x + 5)^2 = (\square)(\square)$?
- How is $(x + 5)(x + 5)$ special?
- What are the outside terms? What is their product?
- What are the inside terms? What is their product?
- What do you notice about the inside and outside terms and their products?
- What are the first terms? What are the last terms?
- How are the first and the last terms special? How is each of their products special?

Call the class back together to have the students discuss and explain their work for Question 2.

Grouping

Ask for a student volunteer to read Question 3 aloud. Have a student restate the problem. Have students work together in small groups to complete Questions 3 and 4.

10

Investigate Problem 2

Does this relationship hold true for the square of the binomials in parts (C) and (D)? Use complete sentences to explain.

Yes, this is true for $(x + 6)(x + 6)$, because the middle term $12x$ is two times the product of the linear term x and the constant term 6. This is also true for $(x + 10)(x + 10)$, because the middle term $20x$ is two times the product of the linear term x and the constant term 10.

In $(x + 3)(x + 3)$, which is $x^2 + 6x + 9$, how is the last term 9 related to the constant term 3? Use a complete sentence to explain.

The last term 9 is the square of the constant term 3.

Does this relationship hold true for the square of the binomials in parts (C) and (D)? Use complete sentences to explain.

Yes, this is true for $(x + 6)(x + 6)$, because the last term 36 is the square of the constant term 6. This is also true for $(x + 10)(x + 10)$, because the last term 100 is the square of the constant term 10.

Complete the formula below for simplifying the square of a binomial sum of the form $(a + b)^2$. Then use the FOIL pattern to check your answer.

$(a + b)^2 = \boxed{a}^2 + 2\boxed{ab} + \boxed{b}^2$

FOIL:

$$\begin{aligned}(a + b)^2 &= (a + b)(a + b)\\ &= a^2 + ab + ab + b^2\\ &= a^2 + 2ab + b^2\end{aligned}$$

3. Use the FOIL pattern to find the products $(x - 4)^2$ and $(4x - 3)^2$. Show your work.

$$\begin{aligned}(x - 4)^4 &= (x - 4)(x - 4)\\ &= x^2 - 4x - 4x + 16\\ &= x^2 - 8x + 16\end{aligned}$$

$$\begin{aligned}(4x - 3)^2 &= (4x - 3)(4x - 3)\\ &= 16x^2 - 12x - 12x + 9\\ &= 16x^2 - 24x + 9\end{aligned}$$

Guiding Questions

- What is the FOIL pattern and how is it used?
- Complete the statement: $(x - 4) = (x + \square)$.

Explore Together

Investigate Problem 2

Students will investigate various applications of the FOIL pattern.

Grouping

Students will be working together in small groups to complete Questions 3 and 4.

Common Student Errors

Students will often confuse $(5x)^2$ with the expression $5x^2$. Remind students of the difference between these expressions. If necessary, refer back to the Warm Up for this lesson to refresh the students' understanding of this relationship.

Call the class back together to have the students discuss and present their work for Questions 3 and 4.

Grouping

Ask for a student volunteer to read Question 5 aloud. Have a student restate the problem. Then have the students complete Questions 5 and 6 individually.

Call the class back together to have the students discuss and present their work for Questions 5 and 6.

Have students work together in small groups to complete Question 7, then call the class back together to have the students discuss and present their work for Question 7.

Key Formative Assessments

- In your own words, describe the term FOIL pattern.
- How can you use the FOIL pattern?
- How can you find the square of a binomial sum?
- How can you find the square of a binomial difference?
- What other FOIL pattern(s) did you discover?
- How is the FOIL pattern related to the distributive property?

Investigate Problem 2

4. Complete the formula for simplifying the **square of a binomial difference** of the form $(a - b)^2$. Then use the FOIL pattern to show that your formula works.

$(a - b)^2 = \boxed{a}^2 - 2\boxed{ab} + \boxed{b}^2$

FOIL:

$(a - b)^2 = (a - b)(a - b)$

$= a^2 - ab - ab + b^2$

$= a^2 - 2ab + b^2$

5. Use the FOIL pattern to find the products $(x - 3)(x + 3)$ and $(2x - 5)(2x + 5)$. Show your work.

$(x - 3)(x + 3) = x^2 + 3x - 3x - 9$

$= x^2 - 9$

$(2x - 5)(2x + 5) = 4x^2 + 10x - 10x - 25$

$= 4x^2 - 25$

6. Write a formula for simplifying the product of a sum and a difference of the form $(a - b)(a + b)$. Then use the FOIL pattern to show that your formula works.

$(a - b)(a + b) = a^2 - b^2$

FOIL:

$(a - b)(a + b) = a^2 + ab - ab - b^2$

$= a^2 - b^2$

7. Find each product. Show your work.

$(x + 5)(x + 3)$

$(x + 5)(x + 3) = x^2 + 3x + 5x + 15$

$= x^2 + 8x + 15$

$(x - 6)^2$

$(x - 6)^2 = x^2 - 2(6)(x) + 36$

$= x^2 - 12x + 36$

$(3x + 4)(x - 2)$

$(3x + 4)(x - 2) = 3x^2 - 6x + 4x - 8$

$= 3x^2 - 2x - 8$

$(2x + 1)^2$

$(2x + 1)^2 = (2x)^2 + 2(2x)(1) + 1$

$= 4x^2 + 4x + 1$

$(x + 9)(x - 9)$

$(x + 9)(x - 9) = x^2 - 81$

$(3x + 2)(3x - 2)$

$(3x + 2)(3x - 2) = 9x^2 - 4$

10

Wrap Up

Close

- Review all key terms and their definitions. Include the terms *FOIL pattern, square of a binomial sum,* and *square of a binomial difference.*
- You may also want to review any other vocabulary terms that were discussed during the lesson which may include *sum, difference, polynomial, monomial, binomial, trinomial, distributive property, standard form, coefficient, square,* and *cube.*
- Remind the students to write the key terms and their definitions in the notes section of their notebooks. You may also want the students to include examples.
- Pose the following questions for the students.
  - Summarize the FOIL pattern.
  - What do the letters in FOIL represent?
  - How can you find the square of a binomial sum?
  - How can you find the square of a binomial difference?
  - What binomials give a product that is the difference of two squares, such as $x^2 - 4$?
  - Why is it easier to use the FOIL pattern than to use the distributive property to multiply two binomials?
  - Can you use the FOIL pattern to multiply two trinomials? Why or why not?

Ties to the Cognitive Tutor Software

In the Cognitive Tutor software, students multiply and divide rational expressions using the same basic interface as they have previously used to work with simpler expressions. This approach helps students understand that the same basic mathematical operations apply.

Follow Up

Assignment

Use the Assignment for Lesson 10.4 in the Student Assignments book. See the Teacher's Resources and Assessments book for answers.

Assessment

See the Assessments provided in the Teacher's Resources and Assessments book for Chapter 10.

Open-Ended Writing Task

Have the students explain in a detailed, but short paragraph, the steps to use to multiply by using FOIL.

Reflections

Insert your reflections on the lesson as it played out in class today.

What went well?

____

____

What did not go as well as you would have liked?

____

____

How would you like to change the lesson in order to improve the things that did not go well and capitalize on the things that did go well?

____

____

____

____

____

10

Notes

10.5 Suspension Bridges

Factoring Polynomials

Learning By Doing Lesson Map

Get Ready

Objectives

In this lesson, you will:

- Factor a polynomial by factoring out a common factor.
- Factor a polynomial of the form $x^2 + bx + c$.
- Factor a polynomial of the form $ax^2 + bx + c$.

Key Terms

- factor
- linear factor
- trinomial
- FOIL pattern

NCTM Content Standards

Grades 9–12 Expectations

Number and Operations Standard

- Develop fluency in operations with real numbers, vectors, and matrices, using mental computation or paper-and-pencil calculations for simple cases and technology for more complicated cases.

Algebra Standards

- Generalize patterns using explicitly defined and recursively defined functions.
- Understand and perform transformations, such as arithmetically combining, composing, and inverting commonly used functions, using technology to perform such operations on more complicated symbolic expressions.
- Understand the meaning of equivalent forms of expressions, equations, inequalities, and relations.
- Use symbolic algebra to represent and explain mathematical relationships.
- Identify essential quantitative relationships in a situation and determine the class or classes of functions that might model the relationships.

Geometry Standard

- Draw and construct representations of two- and three-dimensional geometric objects using a variety of tools.

Lesson Overview

Within the context of this lesson, students will be asked to:

- Solve a quadratic equation using the quadratic formula.
- Factor out a common monomial factor to solve a quadratic function.
- Factor trinomials of the form $x^2 + bx + c$.
- Factor trinomials of the form $ax^2 + bx + c$.

Essential Questions

The following key questions are addressed in this section:

1. What is a suspension bridge?
2. What is a factor?
3. What is a linear factor?
4. How can you factor a trinomial of the form $x^2 + bx + c$?
5. How can you factor a trinomial of the form $ax^2 + bx + c$?

Show The Way

Warm Up

Place the following questions or an applicable subset of these questions on the board before students enter class. Students should begin working as soon as they are seated.

Find the greatest common factor of each set of numbers.

1. 63 and 72 **9**

2. 18 and 24 **6**

3. 75 and 250 **25**

4. 15, 45, and 80 **5**

Classify each polynomial as a monomial, a binomial, or a trinomial.

5. $2x + 3$ **binomial**

6. 8 **monomial**

7. $x^2 - 2x - 4$ **trinomial**

8. $5x$ **monomial**

9. $x - 9$ **binomial**

Multiply each binomial by using the FOIL pattern.

10. $(x + 2)(x + 5)$ $x^2 + 7x + 10$

11. $(x - 3)(x + 1)$ $x^2 - 2x - 3$

12. $(x + 4)(x - 4)$ $x^2 - 16$

13. $(x - 8)(x - 8)$ $x^2 - 16x + 64$

10

Motivator

Begin the lesson with the motivator to get students thinking about the topic of the upcoming problem. This lesson is about a preparing a presentation about suspension bridges. The motivating questions are about bridges.

Ask the students the following questions to get them interested in the lesson.

- What is a suspension bridge?
- What other kinds of bridges have been constructed?
- Where are the nearest bridges to our school?
- What materials do you think are used to create most bridges?

Explore Together

Problem 1

Students will use the quadratic formula to solve a quadratic equation.

Grouping

Ask for a student volunteer to read the Scenario and Problem 1 aloud. Have a student restate the problem. Pose the Guiding Questions below to verify student understanding. Have students work together in small groups to complete parts (A) through (D) of Problem 1.

Guiding Questions

- What information is given in this problem?
- What is a suspension bridge?
- What does the phrase "high water level" mean?
- What does y represent in this situation? What is the unit for y?
- What does x represent in this situation? What is the unit for x?
- How far above the high water level is the suspension cable attached to the left tower on the bridge?
- How can you find the height above the high water level at a point 10 meters to the right of the left tower?
- How many times between the left and right tower do you expect the cable to be 90 meters above the high water level? Why?
- What is the highest point that the cable will be above the high water level?

Common Student Errors

Students may have difficulty recognizing the need to use the quadratic formula to complete part (C). You may also have to remind students that the quadratic formula is $x = \frac{-b \pm \sqrt{b^2 - 4ac}}{2a}$ and that the formula is used when the quadratic equation is written in the form $ax^2 + bx + c = 0$.

SCENARIO In your science class, you have to make a presentation on a "feat of engineering." You have decided to make your presentation on suspension bridges, such as the one shown below. During the research for the presentation, you have discovered that the cables that are used to carry the weight on a bridge form the shape of a parabola. Because you have recently studied parabolas in your math class, you decide to use what you know about parabolas in your presentation.

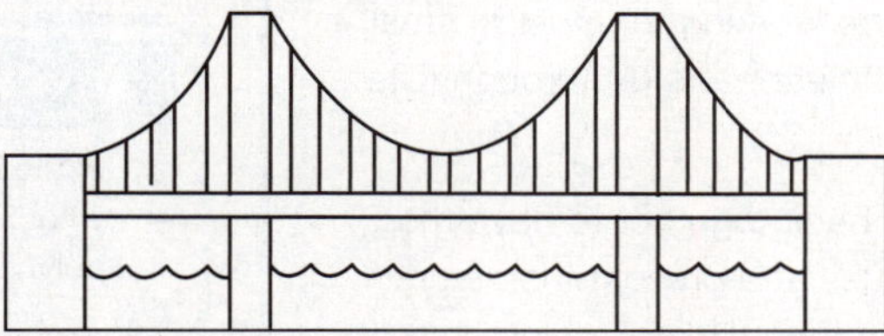

Problem 1 Modeling the Cable Shape

You have modeled the main cable's shape of a particular suspension bridge by using the function $y = 0.002x^2 - 0.4x + 96$, where x is the distance in meters from the left tower and y is the height in meters of the cable above the high water level.

A. Write and simplify an equation that you can use to find the cable's horizontal distance from the left tower at 96 meters above the high water level. Show your work.

$$96 = 0.002x^2 - 0.4x + 96$$
$$0 = 0.002x^2 - 0.4x$$

B. How can you solve the equation in part (A)? Use a complete sentence in your answer.

Sample Answer: You can solve the equation by using the Quadratic Formula.

C. Solve the equation in part (A). Show all your work.

Sample Answer: $a = 0.002$, $b = -0.4$, $c = 0$

$$x = \frac{-(-0.4) \pm \sqrt{(-0.4)^2 - 4(0.002)(0)}}{2(0.002)}$$

$$x = \frac{0.4 \pm 0.4}{0.004}$$

$$x = \frac{0.4 + 0.4}{0.004} = \frac{0.8}{0.004} = 200 \text{ or } x = \frac{0.4 - 0.4}{0.004} = 0$$

10

Explore Together

Problem 1

Students will investigate factors of a polynomial.

Grouping

Students will be working together in small groups to complete parts (A) through (D) of Problem 1.

Call the class back together to have the students discuss and present their work for parts (A) through (D) of Problem 1.

Key Formative Assessments

- What does an x-value of zero represent in this situation?
- What does an x-value of 200 represent in this situation?
- How far apart are the left and right towers?

Investigate Problem 1

Grouping

Ask for a student volunteer to read Question 1 aloud. Have a student restate the problem. Pose the Guiding Questions below to verify student understanding. Have students work together in small groups to complete Question 1.

Guiding Questions

- What is a quadratic function?
- What other types of functions have you learned?
- What is the degree of a quadratic function? What is the degree of a linear function? What is the degree of a cubic function?
- How can you solve the equation $3x = 12$ for x?
- Divide each side of the equation $3x - 12 = 0$ by 3. What is the result? If $3x - 12 = 0$, is it also true that $x - 4 = 0$?

10

Problem 1 Modeling the Cable Shape

D. Interpret your solutions from part (C) in the problem situation. Use a complete sentence in your answer.

When the cable is at a height of 96 feet, the cable is 0 feet and then 200 feet from the left tower of the bridge.

Investigate Problem 1

1. In the last lesson, you multiplied polynomials. The product of what kinds of polynomials gives you a quadratic (second-degree) polynomial? Use a complete sentence in your answer.

Sample Answer: The product of two linear polynomials is a quadratic polynomial.

Consider your equation again from part (A):

$0.002x^2 - 0.4x = 0$

Let's simplify this equation. Because 0.002 evenly divides 0.4, divide each side by 0.002. What is the new equation?

$x^2 - 200x = 0$

If we could write the polynomial $x^2 - 200x$ as the product of two linear polynomials, how would this help us be able to solve the equation $x^2 - 200x = 0$? Use complete sentences in your answer.

Sample Answer: If the product of two linear polynomials is zero, then one or both of the polynomials must be zero. So, we could set each linear polynomial equal to zero and find the values for x that are solutions of the equations.

What expression do the terms of $x^2 - 200x$ have in common?

x

Rewrite your equation so that each term is written as a product with this common expression as a factor.

$x(x) - x(200) = 0$

The product of what two linear expressions gives you $x^2 - 200x$?

x and $x - 200$

So now you have the following.

$x^2 - 200x = 0$

$x(x) - x(200) = 0$

$x(x - 200) = 0$

What are the solutions of the equation?

$x = 0$ and $x = 200$

- Why might we prefer to work with the equation $x - 4 = 0$ than with the equation $3x - 12 = 0$?

Common Student Errors

Question 1 may be confusing for some students. If many students struggle, you may want to complete Question 1 together as a whole class.

Explore Together

Investigate Problem 1

Students will be introduced formally to factoring polynomials.

Call the class back together to have the students discuss and present their work for Question 1.

Just the Math

Students will factor out common monomial factors from polynomials.

Take Note

Recall the Symmetric Property from Lesson 4.5: For any real numbers a and b, if $a = b$, then $b = a$.

Grouping

Ask for a student volunteer to read Question 2 aloud. Have a student restate the problem. Pose the Guiding Questions below to verify student understanding. Have students work together in small groups to complete Questions 2 and 3.

Take Note

An expression is *factored completely* when none of the factors of the expression can be factored further.

Guiding Questions

- What is a common factor?
- What are the common factors of 15 and 45? What is the greatest common factor of 15 and 45?
- What is the common factor in the function $x^2 - 200x$?
- Complete each of the following multiplicative distributive properties.

$a(b + c) = ab + \square$

$a(b - c) = ab - \square$

$(b + c)a = ba + \square$

$(b - c)a = \square - \square$

Investigate Problem 1

You have just solved the equation by *factoring* the polynomial. How do these solutions compare to the solutions from part (C)? Use a complete sentence in your answer.

They are the same.

Which method of finding the solutions was easier? Why? Use a complete sentence in your answer.

Sample Answer: The factoring method was easier because it took fewer steps to find the answer.

2. **Just the Math: Factoring Out a Common Factor** In Question 1, you solved the equation by **factoring out a common factor** of x. Which mathematical rule was used to factor the polynomial? Use a complete sentence to explain your reasoning.

Sample Answer: The distributive property $a(b - c) = ab - ac$ was used, just in the reverse direction.

So, because the equality is *symmetric*, you can write your multiplicative distributive properties as follows:

$ab + ac = a(b + c)$ $\quad$ $ab - ac = a(b - c)$

$ba + ca = (b + c)a$ $\quad$ $ba - ca = (b - c)a$

When the distributive properties are used in this manner, you are *factoring* expressions.

Factor each expression completely. Show your work.

$x^2 + 12x$

$x^2 + 12x = x(x) + x(12)$
$= x(x + 12)$

$x^3 - 5x$

$x^3 - 5x = x(x^2) - x(5)$
$= x(x^2 - 5)$

$3x^2 - 9x$

$3x^2 - 9x = 3x(x) - 3x(3)$
$= 3x(x - 3)$

$10x^2 - 6$

$10x^2 - 6 = 2(5x^2) - 2(3)$
$= 2(5x^2 - 3)$

3. What is the domain of the function in Problem 1? Use complete sentences to explain your reasoning.

Sample Answer: The domain is all real numbers from 0 to 200. Because the cable's shape is a parabola and the towers are the same height, the cable height should be the same at both towers. The cable starts 0 meters from the left tower and is at a height of 96 meters and the cable again reaches a height of 96 meters at 200 meters from the left tower.

Notes You may want to ask classes of advanced students to calculate the vertex for the bridge situation. When they find the x-value of 100 for the vertex, they can substitute into the function to get the range for y-values to be between 76 and 96 meters inclusive.

Explore Together

Problem 2

Students will continue to investigate factoring polynomials.

Grouping

Ask for a student volunteer to read Problem 2 aloud. Have a student restate the problem. Pose the Guiding Questions below to verify student understanding. Have students work together in small groups to complete parts (A) through (C) of Problem 2.

Guiding Questions

- What is the original function relating the horizontal distance from the left tower and the height of the cable above the high water level?
- Which variable is equal to 81 meters in Problem 2?

Call the class back together to have the students discuss and present their work for parts (A) through (C) of Problem 2.

Investigate Problem 2

Ask for a student volunteer to read Question 1 aloud. Have a student restate the problem. Pose the Guiding Questions below to verify student understanding. Have students work together in small groups to complete Question 1.

Guiding Questions

- What is the product of $(x + a)(x + b)$ using the FOIL pattern?
- What is the polynomial that you wrote in part (A) of Problem 2?
- What is a product? What is a sum?
- Is the product of 2 positive numbers positive or negative? Is the product of two negative numbers positive or negative? Is the product of one positive and one negative number positive or negative?

10

Problem 2 Using the Cable Shape Model

A. Write and simplify an equation that you can use to find the cable's horizontal distance from the left tower at 81 meters above the high water level. Show your work.

$81 = 0.002x^2 - 0.4x + 96$

$0 = 0.002x^2 - 0.4x + 15$

Now divide each side by 0.002.

$0 = x^2 - 200x + 7500$

B. How is this equation the same as or different from the equation in Problem 1? Use a complete sentence in your answer.

Sample Answer: The polynomial in this equation has a constant term. Both polynomials have the same x^2-term and x-term.

C. Can you factor out a common factor?

No.

Investigate Problem 2

1. If we want to factor the polynomial from part (A) as the product of two linear expressions $(x + a)$ and $(x + b)$, what must be true about the product ab? Explain your reasoning. Use a complete sentence in your answer.

Sample Answer: The product ab must be 7500 because $(x + a)(x + b) = x^2 + bx + ax + ab$ and we want $x^2 + bx + ax + ab = x^2 - 200x + 7500$.

What must be true about the sum $a + b$? Explain your reasoning. Use a complete sentence in your answer.

Sample Answer: The sum $a + b$ must be −200 because $(x + a)(x + b) = x^2 + bx + ax + ab$ and we want $x^2 + bx + ax + ab = x^2 - 200x + 7500$.

Make a list of pairs of negative integers whose product is 7500. Then find the sum of each pair of numbers. Stop when you find the pair of numbers that has a sum of −200.

−1, −7500;	Sum: −7501	−10, −750;	Sum: −760
−2, −3750;	Sum: −3752	−12, −625;	Sum: −637
−3, −2500;	Sum: −2503	−15, −500;	Sum: −515
−4, −1875;	Sum: −1879	−20, −375;	Sum: −395
−5, −1500;	Sum: −1505	−25, −300;	Sum: −325
−6, −1250;	Sum: −1256	−30, −250;	Sum: −280
		−50, −150;	Sum: −200

Explore Together

Investigate Problem 2

Students will investigate factoring polynomials of the form $x^2 + bx + c$.

Grouping

Students will be working together in small groups to complete Question 1.

Call the class back together to have the students discuss and present their work for Question 1.

Key Formative Assessments

- What is the product of any number and zero?
- If the product of two numbers is zero, what do we know about at least one of the factors?
- Why did we set the factors of $(x + (-50))$ and $(x + (-150))$ equal to zero in Question 1?

Take Note

Your special product rules from Lesson 10.4 can help you factor special products:

$a^2 + 2ab + b^2 = (a + b)^2$

$a^2 - 2ab + b^2 = (a - b)^2$

$a^2 - b^2 = (a + b)(a - b)$

Grouping

Ask for a student volunteer to read Question 2 aloud. Have a student restate the problem. Pose the Guiding Questions below to verify student understanding. Have students work together in small groups to complete Questions 2 through 4.

Guiding Questions

- How can you factor a trinomial of the form $ax^2 + bx + c$?
- How can you factor a trinomial of the form $ax^2 - bx + c$?
- How can you factor a trinomial of the form $ax^2 - bx - c$?
- How can you verify your factorization using the FOIL pattern?

Investigate Problem 2

Which pair of numbers has a product of 7500 and a sum of –200?

–50 and –150

Complete the factorization of $x^2 - 200x + 7500$.

$x^2 - 200x + 7500 = (x + \boxed{(-50)})(x + \boxed{(-150)})$

What are the solutions of the equation $x^2 - 200x + 7500 = 0$? Show your work.

$x - 50 = 0$ or $x - 150 = 0$

$x = 50$ $\quad$ $x = 150$

What is the cable's horizontal distance from the left tower at 81 meters above high water level? Use a complete sentence in your answer.

The cable's horizontal distances from the left tower at 81 meters above high water level are 50 meters and 150 meters.

2. Factor each trinomial as a product of linear factors. Then use the FOIL pattern to verify your answer. Show your work.

$x^2 + 4x + 3$

$x^2 + 4x + 3 = (x + 3)(x + 1)$

$(x + 3)(x + 1) = x^2 + 4x + 3$

$x^2 - 4x - 5$

$x^2 - 4x - 5 = (x - 5)(x + 1)$

$(x - 5)(x + 1) = x^2 - 4x - 5$

$x^2 - x - 6$

$x^2 - x - 6 = (x - 3)(x + 2)$

$(x - 3)(x + 2) = x^2 - x - 6$

$x^2 - 9x + 20$

$x^2 - 9x + 20 = (x - 4)(x - 5)$

$(x - 4)(x - 5) = x^2 - 9x + 20$

$x^2 + 18x + 81$

$x^2 + 18x + 81 = (x + 9)(x + 9)$

$(x + 9)(x + 9) = x^2 + 18x + 81$

$x^2 - 36$

$x^2 - 36 = (x - 6)(x + 6)$

$(x - 6)(x + 6) = x^2 - 36$

3. How does the sign of the constant in the trinomial determine the signs of the constants in the linear factors? Use complete sentences in your answer.

Sample Answer: If the trinomial constant is positive, then the constants in the factors are either both positive or both negative. If the trinomial constant is negative, then in the factors, one constant is positive and one is negative.

10

Explore Together

Investigate Problem 2

Students will investigate factoring of polynomials of the form $ax^2 + bx + c$.

Grouping

Students will be working together in small groups to complete Questions 2 through 4.

Call the class back together to have the students discuss and present their work for Questions 2 through 4.

Grouping

Ask for a student volunteer to read Question 5 aloud. Have a student restate the problem. Complete Question 5 together as a whole class. Ask for a student volunteer to read Question 6 aloud. Have a student restate the problem. Have students work together in small groups to complete Question 6.

Call the class back together to have the students discuss and present their work for Question 6.

Key Formative Assessments

10

- What is a common factor?
- Can a common factor be a constant?
- Can a common factor be a monomial expression?
- How can you factor out a common monomial factor from a polynomial?
- How can you factor a trinomial of the form $x^2 + bx + c$?
- How can you factor a trinomial of the form $ax^2 + bx + c$?
- How can you factor a trinomial of the form $ax^2 - bx + c$?

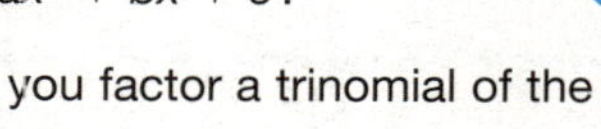

- How can you factor a trinomial of the form $ax^2 - bx - c$?
- Are parentheses necessary for writing a factorization?

Investigate Problem 2

4. In Questions 1 and 2, what do all the trinomials have in common? How are they different? Use complete sentences in your answer.

Sample Answer: All the coefficients of the x^2-terms are 1 and the coefficients of the other terms are different.

5. Consider the trinomial $2x^2 + 11x + 9$. If we want to factor this expression as a product of linear factors, what must be true about the coefficients of the x-terms of the linear factors? Use complete sentences to explain your reasoning.

Sample Answer: One of the x-terms must have a coefficient of one and the other must have a coefficient of two because the coefficient of the product of these x-terms must be 2.

What must be true about the constants of the linear factors? Use complete sentences to explain your reasoning.

Sample Answer: Because the product of the constants must be nine, the constants could be 1 and 9 or 3 and 3.

Now that you have identified all the possible coefficients of the x-terms and the constants for the linear factors, list the possible pairs of linear factors below that could result from the different positions of the numbers in the factors. The first one is listed.

$(2x + 3)(1x + 3) = \boxed{2x^2 + 9x + 9}$

$(2x + 1)(1x + 9) = 2x^2 + 19x + 9$

$(2x + 9)(1x + 1) = 2x^2 + 11x + 9$

Now find the product of each pair of linear factors and write your answer above. Which pair of linear factors gives the factorization of $2x^2 + 11x + 9$?

$2x^2 + 11x + 9 = (2x + 9)(1x + 1)$

6. Factor each trinomial as a product of linear factors. Then use the FOIL pattern to verify your answer. Show your work.

$3x^2 + 7x + 2 = \boxed{(3x + 1)(x + 2)}$

$(3x + 1)(x + 2) = 3x^2 + 6x + x + 2$
$= 3x^2 + 7x + 2$

$5x^2 + 17x + 6 = \boxed{(5x + 2)(x + 3)}$

$(5x + 2)(x + 3) = 5x^2 + 15x + 2x + 6$
$= 5x^2 + 17x + 6$

$6x^2 + 19x + 10 = \boxed{(3x + 2)(2x + 5)}$

$(3x + 2)(2x + 5) = 6x^2 + 15x + 4x + 10$
$= 6x^2 + 19x + 10$

Wrap Up

Close

- Review all key terms and their definitions. Include the terms *factor, linear factor, trinomial,* and *FOIL pattern.*
- You may also want to review any other vocabulary terms that were discussed during the lesson which may include *feat of engineering, quadratic formula, standard form, degree of a polynomial, linear function, quadratic function, distributive property,* and *symmetric property.*
- Remind the students to write the key terms and their definitions in the notes section of their notebooks. You may also want the students to include examples.
- Pose the following questions for the students.
  - How can you factor out a common monomial factor from a polynomial?
  - How can you factor a polynomial of the form $x^2 + bx + c$? How can you factor a polynomial of the form $ax^2 + bx + c$?
  - How are factoring a polynomial of the form $x^2 + bx + c$ and factoring a polynomial of the form $ax^2 + bx + c$ different? How are they the same?
  - How can you use the factored form of a trinomial to solve an equation?
  - It is easier to solve a quadratic equation using the quadratic formula or the factored form of the equation? Explain your reasoning.
  - When would you use the quadratic formula to solve a quadratic equation rather than factoring?

Ties to the Cognitive Tutor Software

In the Cognitive Tutor software, students multiply and divide rational expressions using the same basic interface as they have previously used to work with simpler expressions. This approach helps students understand that the same basic mathematical operations apply.

10

Follow Up

Assignment

Use the Assignment for Lesson 10.5 in the Student Assignments book. See the Teacher's Resources and Assessments book for answers.

Assessment

See the Assessments provided in the Teacher's Resources and Assessments book for Chapter 10.

Open-Ended Writing Task

Ask the students to respond to the following writing prompt. In Question 2 of Investigate Problem 1, you factored out common monomial factors like the $3x$ from each term of the function $(3x^2 - 9x)$. In Question 2 of Investigate Problem 2, you factored out binomial factors like $(x + 3)$ from the trinomial $(x^2 + 4x + 3)$. How can you combine these skills to factor the polynomial $(2x^3 + 16x^2 + 30x)$?

Reflections

Insert your reflections on the lesson as it played out in class today.

What went well?

What did not go as well as you would have liked?

How would you like to change the lesson in order to improve the things that did not go well and capitalize on the things that did go well?

10

Notes

10.6

Swimming Pools

Rational Expressions

Learning By Doing Lesson Map

Get Ready

Objectives

In this lesson, you will:

- Find the domains of rational expressions.
- Simplify rational expressions.
- Add, subtract, multiply, and divide rational expressions.

Key Terms

- rational expression
- domain
- excluded value
- restricting the domain

NCTM Content Standards

Grades 9–12 Expectations

Number and Operations Standards

- Compare and contrast the properties of numbers and number systems, including the rational and real numbers, and understand complex numbers as solutions to quadratic equations that do not have real solutions.
- Use number-theory arguments to justify relationships involving whole numbers.
- Judge the effects of such operations as multiplication, division, and computing powers and roots on the magnitude of quantities.
- Develop fluency in operations with real numbers, vectors, and matrices, using mental computation or paper-and-pencil calculations for simple cases and technology for more complicated cases.

Algebra Standards

- Understand and perform transformations, such as arithmetically combining, composing, and inverting commonly used functions, using technology to perform such operations on more complicated symbolic expressions.
- Write equivalent forms of expressions, inequalities, and systems of equations and solve them with fluency—mentally or with paper and pencil in simple cases and using technology in all cases.
- Use symbolic algebra to represent and explain mathematical relationships.

Lesson Overview

Within the context of this lesson, students will be asked to:

- Represent a ratio as a rational function.
- Determine the domains for rational expressions.
- Simplify rational expressions.
- Add, subtract, multiply, and divide rational expressions.

Essential Questions

The following key questions are addressed in this section:

1. What is a rational expression?
2. How are rational expressions similar to fractions? How are they different?
3. How can you add rational expressions? How can you subtract rational expressions? How can you multiply rational expressions? How can you divide rational expressions?
4. How can you restrict the domain for rational expressions?
5. Why must you restrict the domain for rational expressions?

Show The Way

Warm Up

Place the following questions or an applicable subset of these questions on the board before students enter class. Students should begin working as soon as they are seated.

Factor each expression.

1. $4x + 12$ $4(x + 3)$
2. $5x^3 - 5x^2 + 15x$ $5x(x^2 - x + 3)$
3. $x^2 + 13x + 42$ $(x + 6)(x + 7)$
4. $x^2 - 6x + 9$ $(x - 3)^2$
5. $x^2 - 4$ $(x + 2)(x - 2)$
6. $-3x^3 + 3x^2 + 6x$ $-3x(x + 1)(x - 2)$

Motivator

Begin the lesson with the motivator to get students thinking about the topic of the upcoming problem. This lesson is about designing a swimming pool. The motivating questions are about swimming pools.

Ask the students the following questions to get them interested in the lesson.

- What is a custom in-ground pool?
- What other kinds of pool can exist?
- What shapes of in-ground pools have you seen?
- Do you have a swimming pool?
- Do you like to swim?

Explore Together

Problem 1

Students will determine the domain for a rational expression.

Grouping

Ask for a student volunteer to read the Scenario and Problem 1 aloud. Have a student restate the problem. Pose the Guiding Questions below to verify student understanding. Have students work together in small groups to complete parts (A) through (D) of Problem 1. Then call the class back together to have the students discuss and present their work for parts (A) through (D).

Take Note

Recall that the *domain* of an expression is all possible numbers that x can be so that either the expression is a real number or the value of x makes sense in a real-life problem situation.

Also remember that division by zero does not result in a real number. To express this, we say that division by zero is "undefined."

Guiding Questions

- What information is given in this problem?
- What is surface area for a pool?
- How can you calculate the area of a rectangular surface?
- What is a ratio?
- When is a fraction undefined?
- Why can't we divide a number by zero? Can we divide zero by another number? Can we divide zero by zero? Why not?

Note You may want to ask the students for the domain for the actual situation in part (C) as well as for the expression.

SCENARIO A swimming pool company designs and installs custom in-ground swimming pools. A swimming pool design is based on the location, the amount of area available, and client preferences. In-ground swimming pools can have any shape and can be covered in tiles, fabric, or a concrete coating.

Problem 1 Pool Length and Area

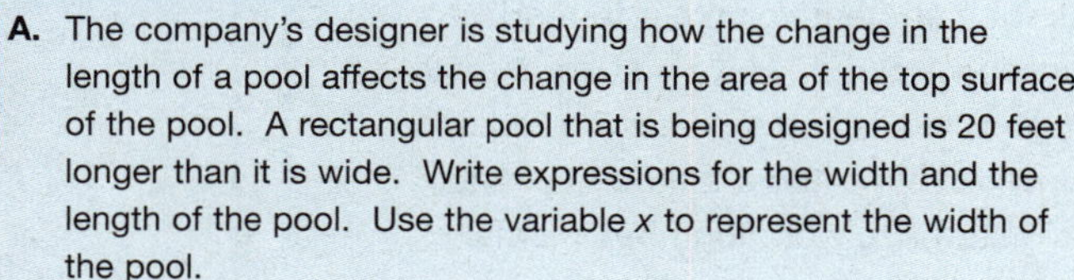

A. The company's designer is studying how the change in the length of a pool affects the change in the area of the top surface of the pool. A rectangular pool that is being designed is 20 feet longer than it is wide. Write expressions for the width and the length of the pool. Use the variable x to represent the width of the pool.

Width: x; Length: $x + 20$

B. Write a proportion that compares the length of the pool to the area of the surface of the pool.

$$\frac{x + 20}{x(x + 20)}$$

C. What is the domain of the expression in part (B)? Do not consider the problem situation when finding the domain. Use complete sentences to explain your reasoning.

The domain is all real numbers except 0 and –20. When x is 0 or –20, the denominator is 0, which is not defined.

D. How are the numerator and denominator of your expression the same? How are they different? Use complete sentences in your answer.

Sample Answer: The numerator and denominator are both polynomials, but the polynomials are different. The numerator and denominator both contain the expression $x + 20$.

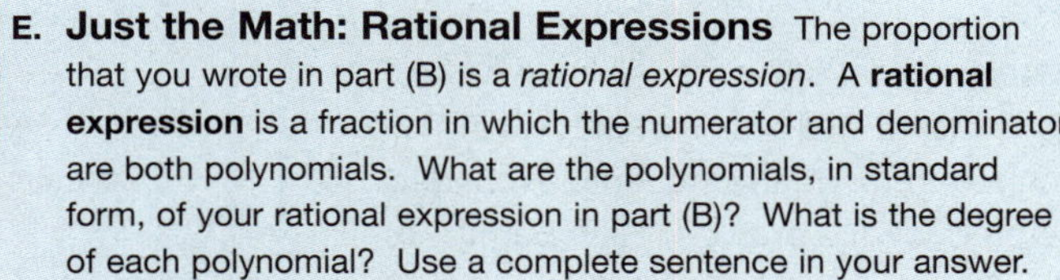

E. Just the Math: Rational Expressions The proportion that you wrote in part (B) is a *rational expression*. A **rational expression** is a fraction in which the numerator and denominator are both polynomials. What are the polynomials, in standard form, of your rational expression in part (B)? What is the degree of each polynomial? Use a complete sentence in your answer.

The numerator is a first-degree polynomial $x + 20$ and the denominator is a second-degree polynomial $x^2 + 20x$.

Just the Math

The students will be introduced to the concept of rational expressions in part (E) of Problem 1.

Grouping

Ask for a student volunteer to read part (E) of Problem 1 aloud. Have a student restate the problem. After students understand the situation, complete part (E) together as a whole class.

10

Explore Together

Investigate Problem 1

Students will determine the domain for several different rational expressions.

Grouping

Ask for a student volunteer to read Question 1 aloud. Have a student restate the problem. Pose the Guiding Questions below to verify student understanding. Have students work together in small groups to complete Questions 1 through 3.

Guiding Questions

- When is a rational expression undefined?
- How can you find the values where a rational expression is undefined?
- Once you know the set of values where a rational expression is undefined, how can you determine the domain for the rational expression?
- When a product is equal to zero, what do you know about at least one factor of the product?

Call the class back together to have the students discuss and present their work for Questions 1 through 3.

Notes Most students will have written their rational expression in part (B) in factored form. If any students expanded the product of x and $x + 20$ in part (B) to $x^2 + 20x$, they will have a different answer to Question 2.

Grouping

Have students work together in small groups to complete Questions 4 and 5. Then call the class back together to have the students discuss and present their work for Questions 4 and 5.

Common Student Errors

Students may not find the correct signs of the excluded values for some domains. Remind the students to set each factor of the denominator equal to zero, and solve the resulting equations.

Investigate Problem 1

1. Use your work from Problem 1 to describe how to find the domain of any rational expression. Use a complete sentence in your answer.

Sample Answer: Look at the denominator of the expression and determine which values of x make the denominator zero.

2. What is the form of the polynomials in your rational expression in part (B)? Use a complete sentence in your answer.

The polynomials are in factored form.

3. Do you think that it is easier to identify the domain of a rational expression when the numerator and denominator are in standard form or in factored form? Use a complete sentence to explain your reasoning.

Sample Answer: It is easier to identify the domain when a rational expression is in factored form because it is easier to see which values of x make the denominator zero.

4. Factor the numerator and denominator of each rational expression, if possible. Then identify the domain of the rational expression.

$\frac{4}{x + 1}$

cannot factor; domain: all real numbers except −1

$\frac{x + 3}{x^2 - 5x}$

$\frac{x + 3}{x(x - 5)}$; domain: all real numbers except 0 and 5

$\frac{x - 3}{x^2 + 5x + 4}$

$\frac{x - 3}{(x + 1)(x + 4)}$; domain: all real numbers except −1 and −4

$\frac{x - 1}{x^2 + x - 6}$

$\frac{x - 1}{(x + 3)(x - 2)}$; domain: all real numbers except −3 and 2

5. Can you simplify your rational expression in part (B)? If so, what is your simplified expression? Explain how you found your answer. Use complete sentences in your answer.

Sample Answer: Yes. The simplified expression is $\frac{1}{x}$ which is the result of dividing out the factor $(x + 20)$ from both the numerator and denominator.

Explore Together

Investigate Problem 1

Students will investigate restrictions for the domains of rational expressions.

Grouping

Ask for a student volunteer to read Question 6 aloud. Have a student restate the problem. Pose the Guiding Questions below to verify student understanding. Have students work together in small groups to complete Questions 6 and 7.

Guiding Questions

- What is being asked in Question 6?
- Evaluate the rational expression $\frac{x + 20}{x(x + 20)}$ when $x = 1$.
- Evaluate the rational expression $\frac{1}{x}$ when $x = 1$.
- How did we develop the rational expression $\frac{x + 20}{x(x + 20)}$?
- How did we develop the rational expression $\frac{1}{x}$?
- What is a domain?

Call the class back together to have the students discuss and present their work for Questions 6 and 7.

Key Formative Assessments

- What is a restricted domain?
- How can you restrict a domain for a rational expression?
- How are the domains for the expressions $\frac{x + 20}{x(x + 20)}$ and $\frac{1}{x}$ similar? How are the domains different?
- What does it mean to be excluded from an activity?
- What is exclusion?

Investigate Problem 1

6. Complete the table of values that compares your expression from part (B) to your expression from Question 5.

Expression x	$\frac{x + 20}{x(x + 20)}$	$\frac{1}{x}$
-25	$-\frac{1}{25}$	$-\frac{1}{25}$
-20	undefined	$-\frac{1}{20}$
-10	$-\frac{1}{10}$	$-\frac{1}{10}$
0	undefined	undefined
10	$\frac{1}{10}$	$\frac{1}{10}$
20	$\frac{1}{20}$	$\frac{1}{20}$

What do you notice about the values in the table? Use a complete sentence in your answer.

Sample Answer: The values of both expressions are the same at every x-value except $x = -20$.

7. What is the domain of your simplified expression in Question 5? Is this domain different from the domain in part (C)? If so, how is it different? Use complete sentences in your answer.

The domain is all real numbers except 0. The domain in part (C) excluded the x-values of 0 and -20 and this domain only excludes the x-value of 0.

In order for the original expression and the simplified expression to be equivalent, we must **restrict the domain** of the simplified expression so that it is the same as the original expression.

What value must be excluded from the domain of $\frac{1}{x}$ so that its domain is the same as the domain of $\frac{x + 20}{x(x + 20)}$?

$x = -20$

We indicate that the expressions are equal by writing $\frac{x + 20}{x(x + 20)} = \frac{1}{x}, x \neq -20$.

10

Explore Together

Investigate Problem I

Students will practice the skill of restricting the domain of rational expressions.

Grouping

Ask for a student volunteer to read Question 8 aloud. Have a student restate the problem. Have students work together in small groups to complete Question 8. Then call the class together to discuss and explain their work.

Take Note

Unless you are specifically instructed, you should leave your rational expressions in factored form. The factored form makes it easier for you to see the domain of the expression. Later on, when you study rational functions, you will see that it is easier to graph the functions when the rational expressions are left in factored form.

Common Student Errors

Students often are careless about the use and placement of parentheses when factoring, multiplying, and dividing rational expressions. They frequently do not understand when it is appropriate or necessary to use parentheses. Remind the students that expressions, such as $x + 6(x)$ and $(x + 6)(x)$ are not equal.

Just the Math

Students will multiply rational expressions in Question 9 and divide rational expressions in Question 10.

Take Note

In the *quotient* $a \div b$, a is the *dividend* and b is the *divisor*.

Grouping

Ask for a student volunteer to read Questions 9 and 10 aloud. Have students restate each problem. Complete Questions 9 and 10 together as a whole class.

Investigate Problem I

8. Simplify each rational expression. Then restrict the domain so that the simplified expression is equivalent to the original expression. Show your work.

$\dfrac{x}{x^2 - 3x}$

$\dfrac{x}{x(x-3)} = \dfrac{1}{x-3}, x \neq 0$

$\dfrac{x+2}{x^2 + 3x + 2}$

$\dfrac{x+2}{(x+1)(x+2)} = \dfrac{1}{x+1}, x \neq -2$

$\dfrac{x-6}{2x^2 - 11x - 6}$

$\dfrac{x-6}{(2x+1)(x-6)} = \dfrac{1}{2x+1}, x \neq 6$

$\dfrac{x^2 - 1}{x + 1}$

$\dfrac{(x+1)(x-1)}{x+1} = (x-1), x \neq -1$

$\dfrac{x^2 + x}{x^2 - 6x - 7}$

$\dfrac{x(x+1)}{(x-7)(x+1)} = \dfrac{x}{x-7}, x \neq -1$

$\dfrac{x^2 - 5x}{x^2 - 25}$

$\dfrac{x(x-5)}{(x-5)(x+5)} = \dfrac{x}{x+5}, x \neq 5$

9. Just the Math: Multiplying Rational Expressions You multiply rational expressions in the same way that you multiply fractions of numbers. For instance, to find the product of $\dfrac{x+1}{x-3}$ and $\dfrac{x}{x+1}$, you multiply numerators and you multiply denominators, and then simplify. Complete the steps below to find the product.

$\dfrac{x+1}{x-3} \cdot \dfrac{x}{x+1} = \dfrac{(x+1)(x)}{(x-3)(x+1)}$ Multiply numerators and multiply denominators.

$= \dfrac{x}{x-3}, x \neq -1$ Simplify.

10. Just the Math: Dividing Rational Expressions You divide rational expressions in the same way that you divide fractions of numbers. For instance, to find the quotient of $\dfrac{x}{x-5}$ and $\dfrac{x^2}{x+2}$, you multiply the dividend by the reciprocal of the divisor. Complete the steps below to find the quotient.

$\dfrac{x}{x-5} \div \dfrac{x^2}{x+2} = \dfrac{x}{x-5} \cdot \dfrac{x+2}{x^2}$ Multiply by reciprocal of divisor.

$= \dfrac{(x)(x+2)}{(x-5)(x^2)}$ Multiply numerators and multiply denominators.

$= \dfrac{x+2}{x(x-5)}, x \neq -2$ Simplify.

Notes In this text, we will limit the excluded values to be those values that are not explicit from the simplified rational expression. For instance, in the expression $\dfrac{x+2}{x(x-5)}$, it is implied that $x \neq 0$ and $x \neq 5$ because the denominator cannot be zero.

10

Explore Together

Investigate Problem 1

Students will practice the skill of multiplying and dividing rational expressions and restricting their domains.

Grouping

Ask for a student volunteer to read Question 11 aloud. Have a student restate the problem. Pose the Guiding Questions below to verify student understanding. Have students work together in small groups to complete Questions 11 and 12.

Guiding Questions

- What are you being asked to do in Question 11?
- How can you multiply rational expressions?
- How can you divide rational expressions?
- How can you write your answers in factored form?
- How can you restrict the domain for the product or quotient of rational expressions?
- Why is it important to consider the domain for every part of the solution as well as that of the original problem when restricting a domain?

Common Student Errors

Students will frequently forget to establish the domain based on the original and intermediate steps of the problem as well as the final product or quotient. Remind the students that if a value would make the denominator of a rational expression equal to zero, then even if that factor is not part of the final result, the original expression would not be defined, and therefore the domain must be restricted.

Call the class back together to have the students discuss and present their work for Questions 11 and 12.

Investigate Problem 1

11. Find the product or quotient. Show your work and leave your answer in simplified form. Remember to restrict the domain, if necessary.

$\frac{x+6}{x} \cdot \frac{x}{x-5}$

$$\frac{x+6}{x} \cdot \frac{x}{x-5} = \frac{(x+6)(x)}{x(x-5)}$$
$$= \frac{x+6}{x-5}, x \neq 0$$

$\frac{x+3}{x-2} \div \frac{x}{x-4}$

$$\frac{x+3}{x-2} \div \frac{x}{x-4} = \frac{x+3}{x-2} \cdot \frac{x-4}{x}$$
$$= \frac{(x+3)(x-4)}{x(x-2)}$$

$\frac{x^2+3x}{x-4} \cdot \frac{x+5}{x+3}$

$$\frac{x^2+3x}{x-4} \cdot \frac{x+5}{x+3} = \frac{(x^2+3x)(x+5)}{(x-4)(x+3)}$$
$$= \frac{x(x+3)(x+5)}{(x-4)(x+3)}$$
$$= \frac{x(x+5)}{x-4}, x \neq -3$$

$\frac{x-1}{x^2-4x} \div \frac{x+1}{x}$

$$\frac{x-1}{x^2-4x} \div \frac{x+1}{x} = \frac{x-1}{x^2-4x} \cdot \frac{x}{x+1}$$
$$= \frac{(x-1)(x)}{(x^2-4x)(x+1)}$$
$$= \frac{(x-1)(x)}{x(x-4)(x+1)}$$
$$= \frac{x-1}{(x-4)(x+1)}, x \neq 0$$

$\frac{x^2-7x}{x+3} \cdot \frac{x^2+4x+3}{x-7}$

$$\frac{x^2-7x}{x+3} \cdot \frac{x^2+4x+3}{x-7} = \frac{(x^2-7x)(x^2+4x+3)}{(x+3)(x-7)}$$
$$= \frac{x(x-7)(x+3)(x+1)}{(x+3)(x-7)}$$
$$= x(x+1), x \neq -3, 7$$

12. Do you think that it matters whether you factor your expressions before you multiply or divide? Use complete sentences to explain your reasoning.

Answers will vary.

10

Explore Together

Problem 2

Students will create fractional expressions to represent the amount of a job completed in one hour by each of two workers. They will then add to find the amount of the job completed per hour with both people working together.

Grouping

Ask for a student volunteer to read Problem 2 aloud. Have a student restate the problem. After students understand the situation, complete part (A) together as a whole class. Have students work together in small groups to complete parts (B) through (E) of Problem 2.

Call the class back together to have the students discuss and present their work for parts (B) through (E) of Problem 2.

Common Student Errors

Students may not recognize the fact that the entire job can be represented by the value of 1. You may want to pose the following Guiding Questions to verify student understanding of this concept.

Guiding Questions

- What percent is the same as the whole amount of something?
- What number is equal to 100%?
- What value would you use to represent 100% of the job?
- If someone can do $\frac{1}{3}$ of a job in one hour, how many hours would it take the person to complete the entire job?
- If someone can do a job in 5 hours, how much of the job could they do in one hour? Is that a rate? What is the unit for the rate that you just found?
- What is a least common denominator and how can you find one?

Problem 2 Installing a Pool

Two workers are responsible for cutting and placing steel reinforcement bars (rebar) on the bottom and sides of a standard size pool.

A. It takes one of the workers nine hours to cut and place the rebar for a standard-sized pool by himself. Write a fraction that represents the amount of the job that is completed by this worker in one hour if the worker is doing the job by himself.

$\frac{1}{9}$

B. It takes the other worker an unknown number of hours to cut and place the rebar for a standard-sized pool by herself. Write a fraction that represents the amount of the job that is completed by this worker in one hour if the worker is doing the job by herself. Use t to be the number of hours it takes this worker to complete the job by herself.

$\frac{1}{t}$

C. Use your answers from parts (A) and (B) to write an expression that shows the amount of the job that is completed in one hour by the workers if they are doing the job together.

$\frac{1}{9} + \frac{1}{t}$

D. Suppose that it takes the second worker eight hours to complete the job by herself. If the workers work together, what fraction of the job is completed in one hour? Show your work and use a complete sentence in your answer.

$\frac{1}{9} + \frac{1}{8} = \frac{8}{72} + \frac{9}{72} = \frac{17}{72}$

In one hour, $\frac{17}{72}$ of the job is completed.

E. Use complete sentences to explain the steps that you took to find your answer to part (D).

Sample Answer: The answer was found by adding the fractions $\frac{1}{9}$ and $\frac{1}{8}$. First, the least common denominator (LCD) was found and then each fraction was written by using the LCD. Then the numerators were added and the sum was written over the common denominator.

Explore Together

Investigate Problem 2

Students will add and subtract rational expressions by using the same approach as was used for adding and subtracting fractions.

Grouping

Ask for a student volunteer to read Question 1 aloud. Have a student restate the problem. Pose the Guiding Questions below to verify student understanding. Have students work together in small groups to complete Questions 1 and 2. Then call the class back together to have the students discuss and present their work for Questions 1 and 2.

Take Note

The *least common denominator* (LCD) of a pair of fractions is the least of the common multiples of the denominators of the fractions. For instance, the common multiples of the denominators of $\frac{3}{4}$ and $\frac{5}{6}$ are 12, 24, 36, … . The least of these multiples is 12, so the LCD of $\frac{3}{4}$ and $\frac{5}{6}$ is 12.

Guiding Questions

- Do you have to restrict the domain when you are adding or subtracting fractions?
- Do you think that you will have to restrict the domain when you are adding or subtracting rational expressions?
- Why do we need to restrict the domain when adding or subtracting rational expressions but not for fractions?

Key Formative Assessments

- How can you determine the domain for a rational expression?
- How can you multiply, divide, add, or subtract rational expressions?

Investigate Problem 2

1. You can add and subtract rational expressions in the same way that you add and subtract fractions. Consider your sum in part (C). Before you can add numerators, what must you do first? Use a complete sentence in your answer.

   Sample Answer: You must write each fraction with the LCD.

   What is the least common denominator (LCD) of the rational expressions in part (C)? Use a complete sentence to explain how you found your answer.

   9*t*; The first expression's denominator is 9 and the second expression's denominator is *t*, so the product 9*t* is the least common denominator because 9 and *t* contain no common factors.

   By what expression do you have to multiply $\frac{1}{t}$ so that the denominator is the LCD?

   $\frac{9}{9}$

   By what expression do you have to multiply $\frac{1}{9}$ so that the denominator is the LCD?

   $\frac{t}{t}$

   Simplify the sum from part (C) by completing each step below.

   $\frac{1}{9}+\frac{1}{t}=\frac{1}{9}\cdot\frac{\boxed{t}}{\boxed{t}}+\frac{1}{t}\cdot\frac{\boxed{9}}{\boxed{9}}$ Multiply each expression by an appropriate form of 1.

   $=\frac{\boxed{t}}{\boxed{9t}}+\frac{\boxed{9}}{\boxed{9t}}$ Multiply.

   $=\frac{\boxed{t+9}}{\boxed{9t}}$ Add.

2. Find each sum or difference. Show your work and write your answer in simplified form.

   $\frac{4}{x}+\frac{2}{3x}$

   $\frac{4}{x}+\frac{2}{3x}=\frac{12}{3x}+\frac{2}{3x}$

   $=\frac{12+2}{3x}$

   $=\frac{14}{3x}$

   $\frac{1}{3}-\frac{8}{x-2}$

   $\frac{1}{3}-\frac{8}{x-2}=\frac{x-2}{3(x-2)}-\frac{24}{3(x-2)}$

   $=\frac{x-2-24}{3(x-2)}$

   $=\frac{x-26}{3(x-2)}$

10

Wrap Up

Close

- Review all key terms and their definitions. Include the terms *rational expression, domain, excluded value,* and *restricting the domain.*
- You may also want to review any other vocabulary terms that were discussed during the lesson which may include *undefined, range, grouping symbols, parentheses, sum, difference, product, quotient, dividend, divisor, least common denominator, LCD, multiple,* and *factor.*
- Remind the students to write the key terms and their definitions in the notes section of their notebooks. You may also want the students to include examples.
- Ask the students to explain the meaning of the term rational expression.
- Have the students explain how to add fractions. Then, have the students explain how to add rational expressions. Repeat for subtraction, multiplication, and division.
- Ask the students to explain how to restrict domains.
- Ask the students to explain why we need to restrict domains.

Follow Up

Assignment

Use the Assignment for Lesson 10.6 in the Student Assignments book. See the Teacher's Resources and Assessments book for answers.

Assessment

See the Assessments provided in the Teacher's Resources and Assessments book for Chapter 10.

Open-Ended Writing Task

Have the students write a short paragraph to compare and contrast operations with fractions and operations with rational expressions.

Reflections

Insert your reflections on the lesson as it played out in class today.

What went well?

____

____

What did not go as well as you would have liked?

____

____

How would you like to change the lesson in order to improve the things that did not go well and capitalize on the things that did go well?

____

____

____

____

____

10

Notes

Looking Ahead to Chapter 11

Focus

In Chapter 11, you will be introduced to probability, including theoretical and experimental probabilities, making predictions by using probabilities, independent and dependent events, and geometric probabilities. You will also learn about counting techniques.

Chapter Warm-up

Answer these questions to help you review skills that you will need in Chapter 11.

Simplify each expression. Write your answers in simplest form.

1. $\frac{2}{3} + \frac{4}{5}$ $\frac{22}{15} = 1\frac{7}{15}$

2. $\frac{7}{18} - \frac{1}{6}$ $\frac{2}{9}$

3. $\frac{2}{9} + \frac{12}{15} - \frac{3}{5}$ $\frac{19}{45}$

Solve each proportion.

4. $\frac{5}{x} = \frac{14}{350}$ $x = 125$

5. $\frac{x}{12} = \frac{4}{16}$ $x = 3$

6. $\frac{9}{22} = \frac{63}{x}$ $x = 154$

Read the problem scenario below.

You are conducting a survey at your school to see what everyone's favorite subject is. You take a random sample of 50 students in your school. Eighteen students say that math is their favorite subject; thirteen students say that psychology is their favorite subject; eleven students say that English is their favorite subject, and eight students say that art is their favorite subject. Assume that there are 2700 students in your school.

7. Based on your sample, how many students would you expect to say that math is their favorite subject? **972 students**

8. Based on your sample, how many students would you expect to say that English is their favorite subject? **594 students**

11

Key Terms

outcomes ■ p. 493
sample space ■ p. 493
event ■ p. 493
probability ■ p. 493
probability of the event ■ p. 494
favorable outcome ■ p. 494
odds in favor ■ p. 497
complementary events ■ p. 497
odds against ■ p. 498
theoretical probability ■ p. 500
experimental probability ■ p. 500
experiment ■ p. 500
trial ■ p. 500
success ■ p. 500
line plot ■ p. 507
tree diagram ■ p. 513
Fundamental Counting Principle ■ p. 517
permutation ■ p. 517
factorial ■ p. 518
permutation of *n* distinct objects taken *r* at a time ■ p. 522
combinations ■ p. 526
combinations of *n* distinct objects taken *r* at a time ■ p. 522
compound events ■ p. 528
independent events ■ p. 528
dependent events ■ p. 528
triangle ■ p. 533
parallelogram ■ p. 533
trapezoid ■ p. 533
congruent ■ p. 533
area ■ p. 533
geometric probability ■ p. 536
fair game ■ p. 542

CHAPTER

11

Probability

The word *sock* is derived from the Latin word *soccus*. A soccus was a shoe made of matted animal hair that was worn by Greek and Roman comedians. In Lesson 11.7, you will find the probability of picking two socks of the same color from a drawer.

11

Your Best Guess

Introduction to Probability

Learning By Doing Lesson Map

Get Ready

Objective

In this lesson, you will:

- Find the probability of an event.
- Find the odds in favor of an event.
- Find the odds against an event.

Key Terms

- outcomes
- sample space
- event
- probability
- probability of the event
- favorable outcome
- odds in favor
- odds against
- complementary events

Materials

- Coins for each student to toss in Problem 1
- Number cubes for each student to roll in Problem 1

NCTM Content Standards

Grades 9–12 Expectations

Data Analysis and Probability Standards

- Use simulations to explore the variability of sample statistics from a known population and to construct sampling distributions.
- Understand the concepts of sample space and probability distribution and construct sample spaces and distributions in simple cases.

Lesson Overview

Within the context of this lesson, students will be asked to:

- Simulate guessing for a true or false question with the toss of a coin and guessing for a 6 choice multiple-choice question with the rolling of a number cube.
- Calculate probabilities for events.
- Calculate the probability of a complementary event.
- Calculate the odds in favor of an event occurring.
- Calculate the odds against an event occurring

Essential Questions

The following key questions are addressed in this section:

1. What is an event?
2. What is a sample space?
3. What is a probability?
4. What are odds?

Show The Way

Warm Up

Place the following questions or an applicable subset of these questions on the board before students enter class. Students should begin working as soon as they are seated.

Simplify each expression.

1. $\frac{12}{100}$ $\frac{3}{25}$ or 0.12

2. $\frac{3}{5} \cdot \frac{15}{6}$ $\frac{3}{2}$ or 1.5

3. $\frac{5}{8} + \frac{2}{8}$ $\frac{7}{8}$

4. $\frac{12}{14} - \frac{3}{7}$ $\frac{3}{7}$

5. $\frac{\left(\frac{1}{2}\right)}{\left(\frac{3}{2}\right)}$ $\frac{1}{3}$

6. $\frac{\left(\frac{2}{7}\right)}{\left(\frac{10}{14}\right)}$ $\frac{2}{5}$

7. $\frac{1}{5} + \frac{4}{5}$ 1

8. $\frac{2}{3} \cdot \frac{3}{2}$ 1

9. $1 - \frac{4}{9}$ $\frac{5}{9}$

10. $\frac{2}{3} + \frac{3}{2}$ $\frac{13}{6}$ or $2\frac{1}{6}$

Motivator

Begin the lesson with the motivator to get students thinking about the topic of the upcoming problem. This lesson is about taking a test. The motivating questions are about test questions.

Ask the students the following questions to get them interested in the lesson.

- What does it mean to guess on a true-or-false question?
- How many choices do most multiple-choice questions have?
- What does it mean to guess on a multiple-choice question?
- Which do you think that you would be more likely to guess correctly: a true-or-false question or a multiple-choice question? Why?
- Why do you think that most standardized tests award 1 point for a correct multiple-choice answer, but only deduct a fraction of a point for each wrong answer?

11

Explore Together

Problem 1

Students will simulate guessing for a true-or-false question by tossing a coin.

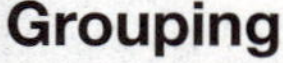

Grouping

Ask for a student volunteer to read the Scenario and Problem 1 aloud. Have a student restate the problem. Pose the Guiding Questions below to verify student understanding. Have the students work together in small groups to complete parts (A) and (B) of Problem 1. Then call the class back together to have the students discuss and present their work for parts (A) and (B).

Guiding Questions

- What do we mean by the phrase "the probability that an event will occur?"
- What is an outcome? What is a sample space?
- How many possible outcomes exist for a coin toss? What are they?
- How many possible outcomes exist for a true-or-false question? What are the 2 possible outcomes?
- What does it mean to simulate a situation or activity?
- How can you simulate a guess on a true-or-false question with a coin toss?
- How can choosing heads or tails before tossing a coin simulate a guess on a true-or-false question? Why do we assign heads as correct or incorrect rather than as true or false and tails as the other possible outcome?
- Why would we toss the coin more than once to simulate this situation?

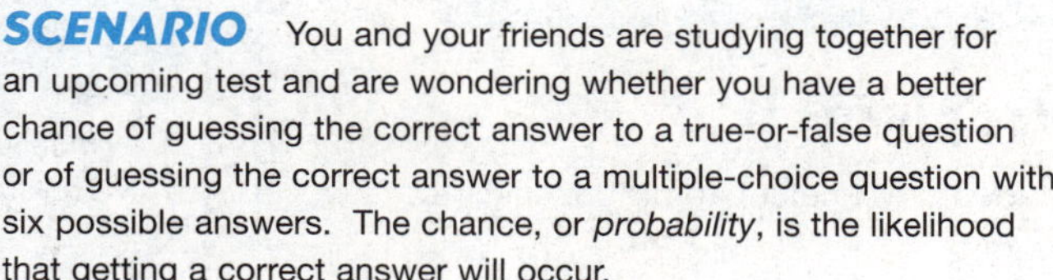

SCENARIO You and your friends are studying together for an upcoming test and are wondering whether you have a better chance of guessing the correct answer to a true-or-false question or of guessing the correct answer to a multiple-choice question with six possible answers. The chance, or *probability*, is the likelihood that getting a correct answer will occur.

Problem 1 Toss and Roll

A. Choosing true or choosing false are the **outcomes** of answering a true-or-false question. The collection of possible outcomes is called a **sample space.** An **event** consists of one or more outcomes. For instance, choosing true and being correct is an event. The chance that the event will happen is the **probability** of the event.

Because the sample space consists of two outcomes, you can model guessing the answer to a true-or-false question by tossing a coin. The coin landing heads up can represent choosing true and the coin landing tails up can represent choosing false. Suppose that a multiple-choice question has six possible answers: A, B, C, D, E, and F. What is the sample space for the multiple choice question? Use a complete sentence in your answer.

The possible outcomes are choosing A, choosing B, choosing C, choosing D, choosing E, and choosing F.

Because the sample space consists of six outcomes, you can model guessing the answer to this multiple-choice question by rolling a six-sided number cube. Rolling the number 1 can represent choosing A, rolling the number 2 can represent choosing B, and so on.

B. Choose one of the outcomes from a coin toss. Toss a coin 12 times and keep track of the number of times that a toss results in heads and the number of times that the toss results in tails in the table below. How many times did your choice of outcome appear during the 12 tosses? Use a complete sentence in your answer.

Possible outcomes	
Heads	**Tails**
7	5

Sample Answer: I chose heads and heads appeared seven times.

11

Notes Rather than using coins in part (B), you can use two colored counters if you have them. These are inexpensive round disks and have one color on each side. For instance, some are yellow on one side and red on the other side. Students would then have the possible outcomes of yellow and red. Also, the answers given in parts (B) and (C) are sample results only. Student answers will vary.

Explore Together

Problem 1

Students will calculate probabilities for events.

Note The 3 circled in part (C) is a sample student response only.

Grouping

Ask for a student volunteer to read part (C) aloud. Have a student restate the problem. Pose the Guiding Questions below to verify student understanding. Have the students work together in small groups to complete part (C) of Problem 1. Then call the class back together to have the students discuss and present their work for part (C).

Guiding Questions

- How many possible outcomes exist for rolling a number cube?
- Are each of the 6 possible outcomes equally likely?
- What does it mean to say that the 6 different outcomes are equally likely?
- What is the outcome from rolling a number cube used to simulate?
- Why do we repeat the roll of the number cube many times?
- What are you asked to do in part (C)?

Investigate Problem 1

Just the Math

Students will be formally introduced to the calculation of the probability of an event and to favorable outcomes.

Grouping

Ask for a student volunteer to read Question 1 aloud. Have a student restate the problem. Pose the Guiding Questions on the right to verify student understanding. Have students work together in small groups to complete Questions 1 and 2.

Problem 1 Toss and Roll

C. Choose a possible outcome from rolling a number cube. Circle your choice in the table below. Roll a number cube 12 times and record the results in the table below. How many times did the number that you chose appear out of the 12 rolls? Use a complete sentence in your answer.

Possible outcomes					
1	2	(3)	4	5	6
2	0	3	3	2	2

Sample Answer: My choice, 3, appeared 3 times.

Based on your results from parts (B) and (C), would you have a better chance of correctly guessing the outcome of a coin toss or the roll on a number cube? Use a complete sentence to explain your reasoning.

Sample Answer: Because there are only two possible outcomes in a coin toss rather than the six outcomes in the roll of a number cube, you will have a better chance of guessing correctly on the outcome of a coin toss.

Investigate Problem 1

1. **Just the Math: Probability** The chance that an event will occur is the **probability of the event.** You can find the probability of an event by finding the ratio of the number of **favorable outcomes** (the outcomes that you want) to the number of possible outcomes.

$$\text{Probability} = \frac{\text{Number of favorable outcomes}}{\text{Number of possible outcomes}}$$

You guess that a coin toss will result in heads. What is the probability that your guess is correct? Use complete sentences to explain.

There is one favorable outcome, tossing heads, and two possible outcomes, tossing heads or tails. The probability of guessing correctly is the ratio of one favorable outcome to two possible outcomes, or $\frac{1}{2}$.

You guess that a roll of a six-sided number cube will result in 1. What is the probability that your guess is correct? Use complete sentences to explain.

There is one favorable outcome, rolling a 1. There are six possible outcomes, rolling 1, 2, 3, 4, 5, or 6. The probability of guessing correctly is the ratio of one favorable outcome to six possible outcomes, or $\frac{1}{6}$.

Guiding Questions

- What are favorable outcomes?
- In how many ways can you roll a 2 on a number cube? What is the probability of rolling a 2?
- How can you calculate the probability of any specific outcome for a number cube?

11

Explore Together

Investigate Problem 1

Students will investigate the characteristics of probabilities.

Grouping

Students will be working in small groups to complete these questions.

Common Student Errors

Students may incorrectly generalize the number of favorable outcomes to always equal 1 based on the first two parts of Question 1. The second two parts of the question should address that error. Be ready for students to incorrectly think the number of favorable outcomes for getting a 3 or 6 is 1. Students may think that the probability is $\frac{1}{2}$. If that happens, redirect the student by asking how many possible numbers exist for rolling a number cube. Then ask how many of those 6 numbers are either 3 or 6. Ask them which of those numbers represents the number of favorable outcomes and which represents the total number of possible outcomes.

Call the class back together to have the students discuss and present their work for Questions 1 and 2.

Key Formative Assessments

- What is probability?
- How can you calculate a probability?
- What is the smallest possible probability?
- What does a probability of zero mean?
- What is the largest possible probability?
- What does a probability of 100% mean?
- What does a probability of 1 mean?
- What is an outcome for an experiment?
- What is a sample space for an experiment?
- What is a favorable outcome?
- How can you calculate the number of possible outcomes?

Investigate Problem 1

You guess that a roll of a six-sided number cube will result in an odd number. What is the probability that your guess is correct? Use complete sentences to explain.

There are three favorable outcomes, rolling a 1, 3, or 5. There are six possible outcomes, rolling 1, 2, 3, 4, 5, or 6. The probability of guessing correctly is the ratio of three favorable outcomes to six possible outcomes, or $\frac{3}{6} = \frac{1}{2}$.

You guess that a roll of a six-sided number cube will result in a number that is evenly divisible by three. What is the probability that your guess is correct? Use complete sentences to explain.

There are two favorable outcomes, rolling a 3 or 6. There are six possible outcomes, rolling 1, 2, 3, 4, 5, or 6. The probability of guessing correctly is the ratio of two favorable outcomes to six possible outcomes, or $\frac{2}{6} = \frac{1}{3}$.

2. In general, which is greater, the number of favorable outcomes or the number of possible outcomes? Use complete sentences to explain your reasoning.

Sample Answer: The number of possible outcomes is greater because the favorable outcomes are selected from the collection of possible outcomes.

Can the number of favorable outcomes and the number of possible outcomes be the same? If so, what would the probability be? Use complete sentences to explain your reasoning.

Sample Answer: Yes. The probability would be 1 because the numerator and denominator in the ratio of the number of favorable outcomes to the number of possible outcomes would be the same.

What is the largest probability? What is the smallest probability? Give an example of each situation. Use complete sentences to explain.

Sample Answer: The largest probability is 1, when the number of favorable outcomes is equal to the number of possible outcomes. An example would be the probability of correctly guessing a roll that results in a number between 1 and 6 when rolling a six-sided number cube. The smallest probability is 0, when there are no favorable outcomes. An example would be the probability of correctly guessing a roll that results in a 7 when rolling a number cube.

Explore Together

Problem 2

Students will compare the probability in favor of an event, the probability against an event, and the ratio of the probabilities.

Grouping

Ask for a student volunteer to read Problem 2 aloud. Have a student restate the problem. Pose the Guiding Questions below to verify student understanding. Have students work together in small groups to complete parts (A) through (C).

Guiding Questions

- What are you asked to find in part (A) of Problem 2?
- What are you asked to find in part (B) of Problem 2? How is that different than what you are asked to find in part (A)?
- What is a ratio?
- How can you simplify the ratio of two fractions?

Common Student Errors

Some students have difficulty in simplifying a ratio of two fractions. The Warm Up activity has this type of ratio. If you did not use the Warm Up activity, you may want to remind the students that dividing by a fraction is equivalent to multiplying by the reciprocal of the fraction in the denominator.

Call the class back together to have the students discuss and present their work for parts (A) through (C) of Problem 2.

Key Formative Assessments

- How does the ratio in part (C) compare to the probabilities in parts (A) and (B)?
- What does the ratio in part (C) represent?

Problem 2 What are the Odds?

You and your friends are still thinking about the likelihood of answering a multiple-choice question correctly. Again, suppose that the multiple-choice question has six possible answers: A, B, C, D, E, and F.

A. What is the probability that you would guess the answer to this multiple-choice question correctly? Use a complete sentence in your answer.

The probability that the guess is correct is $\frac{1}{6}$.

B. What is the probability that you would guess the answer to this multiple-choice question incorrectly? Explain how you found the probability. Use complete sentences in your answer.

Sample Answer: Because there is only one correct answer, there are five other answers that are incorrect. So the probability that the guess is incorrect is $\frac{5}{6}$.

C. Compare the probabilities in parts (A) and (B) by writing the ratio of the probability that you would correctly answer the question to the probability that you would incorrectly answer the question. Simplify your ratio and use a complete sentence in your answer.

$\frac{\frac{1}{6}}{\frac{5}{6}} = \frac{1}{6} \div \frac{5}{6} = \frac{1}{6} \cdot \frac{6}{5} = \frac{1}{5}$; **The ratio of the probabilities is $\frac{1}{5}$.**

This ratio tells you that for every one chance that you would *correctly* answer the question, there are five chances that you would *incorrectly* answer the question. What does this ratio tell you about the likelihood of you answering the question correctly? Use a complete sentence in your answer.

Sample Answer: It is not likely that I will answer the question correctly by guessing.

Explore Together

Investigate Problem 2

Students will investigate the odds in favor of an event occurring.

Just the Math

Students will be introduced formally to the concept of odds in Question 1.

Grouping

Ask for a student volunteer to read Question 1 aloud. Have a student restate the problem. Pose the Guiding Questions below to verify student understanding. Work together as a class to complete Question 1. Then have students work together in small groups to complete Questions 2 and 3. Call the class back together to have the students discuss and present their work for Questions 2 and 3.

Guiding Questions

- How can you calculate the odds in favor of an event?
- What are complementary events?
- What is the complement to odd numbers?
- What is the complement to a correct guess?
- Why is the complement to the numbers less than 3 not just the numbers that are more than 3? What number is not represented in either event?

Common Student Errors

Some students confuse complementary events with opposites. They have a strong relationship, but are not identical. For instance, if we look at 10 markers and notice that there are 5 black, 2 blue, 1 green, and 2 orange markers, the complement of a blue marker is a green, black, or orange marker. However, we could say the complement of a blue marker is a marker that is not blue. That would include the other 3 colors that are not blue.

Investigate Problem 2

1. **Just the Math: Odds in Favor** The ratio that you wrote in part (C) represents your *odds in favor* of correctly answering the question. The **odds in favor** of an event occurring can be defined as the ratio of the number of favorable outcomes to the number of unfavorable outcomes:

$$\text{Odds in favor} = \frac{\text{Number of favorable outcomes}}{\text{Number of unfavorable outcomes}}.$$

What can you conclude about the sum of the number of favorable outcomes and the number of unfavorable outcomes? Use a complete sentence in your answer.

Sample Answer: The sum must be the number of all possible outcomes.

The event that consists of the favorable outcomes and the event that consists of the unfavorable outcomes are *complementary events*. Two events are **complementary events** when one event or the other event must occur, but not both. What is the sum of the probabilities of two complementary events? Use complete sentences to explain your reasoning.

Sample Answer: The sum is one; because the sum of the number of outcomes of complementary events is the number of possible events.

2. You guess that the roll of a six-sided number cube will result in an even number. What are the odds in favor of your guess being correct? Show your work and use a complete sentence in your answer.

Favorable outcomes: 2, 4, 6

Unfavorable outcomes: 1, 3, 5

The odds in favor of the guess being correct is $\frac{3}{3} = \frac{1}{1}$, or 1 to 1.

3. You guess that the roll of a six-sided number cube will result in a number that is divisible by three. What are the odds in favor of your guess being correct? Show your work and use a complete sentence in your answer.

Favorable outcomes: 3, 6;

Unfavorable outcomes: 1, 2, 4, 5

The odds in favor of the guess being correct is $\frac{2}{4} = \frac{1}{2}$, or 1 to 2.

Explore Together

Investigate Problem 2

Students will investigate the odds against events occurring.

Grouping

Ask for a student volunteer to read Question 4 aloud. Pose the Guiding Questions below to verify student understanding. Have students work together in small groups to complete Questions 4 through 6. Then call the class back together to discuss and present their work.

Take Note

Composite numbers are numbers that are greater than 1 that have more than two whole number factors.

Guiding Questions

- What do the odds in favor of an event represent? What do the odds against an event represent?
- How can you calculate the odds against an event?
- What do you think is the relationship between the odds for an event and the odds against an event?
- Why do you think that the product of the odds for an event and against an event is 1?
- What is the sum of the probability for an event and the probability against an event? Why do you think that the sum is 1?

Take Note

Whenever you see the share with the class icon, your group should prepare a short presentation to share with the class that describes how you solved the problem. Be prepared to ask questions during other groups' presentations and to answer questions during your presentation.

Investigate Problem 2

4. Another ratio that is used to compare probabilities is the **odds against** an event occurring. For instance, the odds against your correctly guessing the answer to the multiple-choice question is the ratio of the probability that you will guess the answer incorrectly to the probability that you will guess the answer correctly. Write a rule for the odds against an event occurring in terms of the number of favorable and unfavorable outcomes. Use complete sentences to explain your reasoning.

   Odds against $= \dfrac{\text{Number of unfavorable outcomes}}{\text{Number of favorable outcomes}}$

   The probability of an event not occurring is the ratio of the number of unfavorable outcomes to the number of possible outcomes. The probability of an event occurring is the ratio of the number of favorable outcomes to the number of possible outcomes. The ratio of these probabilities simplifies as the ratio of the number of unfavorable outcomes to the number of favorable outcomes.

5. You guess that the roll of a six-sided number cube will result in a composite number. What are the odds against you being correct? Show your work and use a complete sentence in your answer.

   Favorable outcomes: 4, 6;

   Unfavorable outcomes: 1, 2, 3, 5

   The odds against the guess being correct is $\frac{4}{2} = \frac{2}{1}$, or 2 to 1.

6. You guess that the roll of a six-sided number cube will result in an even number. What are the odds against your guess being correct? Show your work and use a complete sentence in your answer.

   Favorable outcomes: 2, 4, 6;

   Unfavorable outcomes: 1, 3, 5

   The odds against the guess being correct is $\frac{3}{3} = \frac{1}{1}$, or 1 to 1.

   Does this answer surprise you? Use complete sentences to explain your reasoning.

   Sample Answer: No, because the odds in favor were 1 to 1 and the odds against is the reciprocal of the odds in favor.

Key Formative Assessments

- What are events for an experiment?
- What is a sample space?
- How can you calculate probability?
- How can you calculate the odds for an event? How can you calculate the odds against an event?
- What are complementary events?

Wrap Up

Close

- Review all key terms and their definitions. Include the terms *outcomes, sample space, event, probability, favorable outcome, odds in favor, odds against,* and *complementary events.*
- You may also want to review any other vocabulary terms that were discussed during the lesson, which may include *chance, simulate, simulation, coin toss, number cube, experiment, ratio, factor, reciprocal, even, odd, prime numbers,* and *composite numbers.*
- Remind the students to write the key terms and their definitions in the notes section of their notebooks. You may also want the students to include examples.
- Discuss the relationship between probabilities and odds.
- Have the students work in groups to define each of the key terms and create an example of each. Then, call the class back together to have the students discuss and present their definitions and examples for each key term.

Ties to the Cognitive Tutor Software

The Cognitive Tutor software provides students with numeric and diagramming tools to help determine the number of possible outcomes in simple and compound probability situations. Students should come to understand that, though there may be different methods to determine the number of outcomes, the probability always amounts to the number of desired outcomes divided by the number of possible outcomes.

Follow Up

Assignment

Use the Assignment for Lesson 11.1 in the Student Assignments book. See the Teacher's Resources and Assessments book for answers.

Assessment

See the Assessments provided in the Teacher's Resources and Assessments book for Chapter 11.

Open-Ended Writing Task

Ask the students to write a short explanation of the difference between probabilities and odds.

Reflections

Insert your reflections on the lesson as it played out in class today.

What went well?

___

___

What did not go as well as you would have liked?

___

___

How would you like to change the lesson in order to improve the things that did not go well and capitalize on the things that did go well?

___

___

___

___

___

Notes

11

11.2 What's in the Bag

Theoretical and Experimental Probabilities

Learning By Doing Lesson Map

Get Ready

Objective

In this lesson, you will:

- Find theoretical probabilities.
- Find experimental probabilities.
- Compare theoretical and experimental probabilities.

Key Terms

- theoretical probability
- experimental probability
- experiment
- trial
- success

Materials

- 30 small slips of paper for each student
- 1 bag or container for each student
- A marker, pen, or pencil for each student

NCTM Content Standards

Grades 9–12 Expectations

Number and Operations Standard

- Judge the reasonableness of numerical computations and their results.

Algebra Standard

- Draw reasonable conclusions about a situation being modeled.

Data Analysis and Probability Standards

- Use simulations to explore the variability of sample statistics from a known population and to construct sampling distributions.
- Understand how sample statistics reflect the values of population parameters and use sampling distributions as the basis for informal inference.
- Understand the concepts of sample space and probability distribution and construct sample spaces and distributions in simple cases.

Lesson Overview

Within the context of this lesson, students will be asked to:

- Calculate the theoretical probability for an experiment.
- Prepare and conduct an experiment.
- Calculate experimental probabilities.
- Compare theoretical and experimental probabilities.
- Consider changes to the experiment that would likely result in experimental probabilities that would more closely approximate the theoretical probabilities.

Essential Questions

The following key questions are addressed in this section:

1. What is an experiment?
2. What is a trial for an experiment?
3. What is a success?
4. What is a theoretical probability?
5. What is an experimental probability?

Show The Way

Warm Up

Place the following questions or an applicable subset of these questions on the board before students enter class. Students should begin working as soon as they are seated.

While you are waiting for your friend to pick you up after school, you notice that there are exactly 20 cars remaining in the parking lot. You count them by color and find that 7 are white, 4 are blue, 3 are tan, 2 are red, 2 are green, 1 is black, and 1 is yellow. You remember learning probability in your algebra class that day and decide to find the probability for the color of the car that will be driven away next. Find each probability for the next car driven out of the parking lot. Write each probability as a fraction and then as a percent.

1. Probability the car is white $\frac{7}{20} = 35\%$
2. Probability the car is blue $\frac{4}{20} = \frac{1}{5} = 20\%$
3. Probability the car is tan $\frac{3}{20} = 15\%$
4. Probability the car is red $\frac{2}{20} = \frac{1}{10} = 10\%$
5. Probability the car is green $\frac{2}{20} = \frac{1}{10} = 10\%$
6. Probability the car is black $\frac{1}{20} = 5\%$

Find the complementary event for each event.

7. Earning enough money to buy a stereo **Not earning enough money to buy the stereo.**
8. Rolling a 4 on a number cube with 6 sides **Rolling a 1, 2, 3, 5, or 6.**

Find the probability of the complementary event for each event.

9. The probability that it will rain is 37%.

   The probability that it will not rain is 63%.

10. The probability of choosing the digit 3 is 10%.

    The probability of choosing a digit different from 3 is 90%.

11

Motivator

Begin the lesson with the motivator to get students thinking about the topic of the upcoming problem. This lesson is about conducting an experiment. The motivating questions are about probability experiments.

Ask the students the following questions to get them interested in the lesson.

- Why would a person or company conduct probability experiments?
- Have you ever observed or taken part in a probability experiment?
- What did the experiment study?
- What were the results of the experiment?

Explore Together

Problem 1

Students will calculate theoretical probabilities for an experiment.

Grouping

Ask for a student volunteer to read the Scenario and Problem 1 aloud. Have a student restate the problem. Pose the Guiding Questions below to verify student understanding. Have students work together in small groups to complete parts (A) through (E) of Problem 1. Then call the class back together to have the students discuss and present their work for parts (A) through (E).

Guiding Questions

- What is an experiment?
- What is an outcome of an experiment?
- What is the sample space for this experiment?
- What is a probability?
- Why is it important to make sure that you can not see inside the bag or container?
- Why do you have to shake the bag or container? What is the goal in doing so?
- Would you have to shake the bag or container each time you draw a slip of paper and replace it in the bag or container? Why?
- Why do you think this experiment is designed with 10 slips of paper of each of the 3 types rather than just 3 slips of paper with one of each?

Notes Rather than using slips of paper with marks on them, you can use colored chips or marbles if you have access to them. Change the activity to use these manipulatives instead of the paper slips.

SCENARIO You decide to learn about and apply probabilities in your own hands-on experiment. Take 30 slips of paper that are all the same size and mark an "X" on 10 of the slips, mark an "O" on 10 of the slips, and leave the last 10 slips blank. Then put these slips into a paper bag (or some other bag or container that you cannot see inside of) and shake up the bag.

Problem 1 What Will Come Out of the Bag?

A. If you choose a slip of paper from the bag, how many different outcomes are possible? What are the possible outcomes? Use a complete sentence in your answer.

There are three possible outcomes: the slip is blank, the slip has an "X" on it, or the slip has an "O" on it.

B. Determine the probability of each outcome in part (A). Write each probability in simplest form. Use a complete sentence in your answer.

The probability of each outcome is $\frac{10}{30} = \frac{1}{3}$.

C. What do you notice about the probabilities? Why is this so? Use complete sentences in your answer.

Sample Answer: The probabilities are the same because there are the same number of each kind of slip in the bag.

D. Which kind of slip are you most likely to choose? Use complete sentences to explain your reasoning.

Because the probabilities are the same, you are equally likely to choose any one of the three kinds of slips.

E. Suppose that the bag contains 16 "X" slips, 6 "O" slips, and 8 blank slips. Would your answer to part (D) change? If your answer would change, what would your new answer be? Use a complete sentence to explain your reasoning.

Sample Answer: Yes. There are still 30 slips of paper in the bag, but the probabilities would change because there is a different number of each kind of slip in the bag. In this case, you would be most likely to choose an "X" slip because there are more of this slip than either or both of the other kinds of slips.

11

Explore Together

Investigate Problem 1

Students will perform an experiment with 10 trials to gather data.

Grouping

Ask for a student volunteer to read Question 1 aloud. Have a student restate the problem. Pose the Guiding Questions below to verify student understanding. Have the students complete Question 1 individually.

Guiding Questions

- How will you perform this experiment?
- What is a trial? How many trials of this experiment will you conduct?
- What do you have to do between each trial of the experiment?
- Do you expect that your results will be exactly the same as you predicted in parts (A) through (E)? Why or why not?
- Do you think every student will get exactly the same results as the other students in the class?
- Why should we have more than one student conduct this experiment?

Common Student Errors

Students may forget to mix the slips of paper between trials. Also, some students will not recognize that the actual results will not necessarily be the same as the predicted results.

Notes Throughout this lesson, sample experimental results are given for the purpose of showing realistic student answers. The student answers will vary from these answers and from each other's answers.

Just the Math

Theoretical and experimental probabilities are explained in Question 2.

Grouping

Ask for a student volunteer to read Question 2 aloud. Have a student restate the problem. Pose the Guiding Questions below to verify student understanding. Have the students complete Questions 2 and 3 individually.

Guiding Questions

- What is the difference between theoretical probability and experimental probability?
- How can you calculate a theoretical probability? How can you calculate an experimental probability?

Investigate Problem 1

1. Perform an experiment by choosing a slip from the bag that you used in part (A). Record the kind of slip that you choose in the table below and put the slip back into the bag. Repeat this process 9 more times.

Slip	X	O	blank
Number of times slip was chosen	4	3	3

2. **Just the Math: Theoretical and Experimental Probabilities** The probabilities that you found in Lesson 11.1 and in Problem 1 are *theoretical probabilities*. A **theoretical probability** is a probability that is based on knowing all of the possible outcomes that are equally likely to occur.

$$\text{Theoretical probability} = \frac{\text{Number of favorable outcomes}}{\text{Number of possible outcomes}}$$

An **experimental probability** is based on running an **experiment** in the same way many times; each run is called a **trial.** Each time that the desired event occurs is called a **success.** You can find the experimental probability by finding the ratio of the number of successes to the number of trials:

$$\text{Experimental probability} = \frac{\text{Number of successes}}{\text{Number of trials}}.$$

Use your data from Question 1 to find the experimental probability of each possible outcome. Use complete sentences in your answer.

The experimental probability of choosing an "X" slip is $\frac{4}{10} = \frac{2}{5}$.

The experimental probability of choosing an "O" slip is $\frac{3}{10}$.

The experimental probability of choosing a blank slip is $\frac{3}{10}$.

How do the experimental probabilities compare to the theoretical probabilities in part (B)? Use complete sentences in your answer.

Sample Answer: The experimental and theoretical probabilities for each possible outcome were different. However, in each case, the two probabilities were close.

3. Would your experimental probabilities have been different if you had not replaced the slips of paper? Use complete sentences to explain your reasoning.

Sample Answer: Yes, the number of possible outcomes would be different, and the number of successes might be different for each trial.

Explore Together

Investigate Problem 1

Students will compare their theoretical and experimental probabilities.

Grouping

Ask for a student volunteer to read Question 4 aloud. Have a student restate the problem. Pose the Guiding Questions below to verify student understanding. Complete Question 4 together as a class. Then, have the students complete Questions 5 and 6 individually. Call the class back together to have the students discuss and present their work for Questions 5 and 6.

Guiding Questions

- What are you asked to do in Question 4?
- Why is it important to gather the results from the entire class?
- What do you think that you will find for the class results compared to your own results? Why?
- How can we gather the class results efficiently?

Key Formative Assessments

- What is a theoretical probability?
- What is an experimental probability?
- How are theoretical probability and experimental probability similar? How are they different?
- Why are the theoretical probability and experimental probability usually not exactly the same?
- Why would they be closer to each other if we conduct more trials for the experiment?

- What is an experiment?

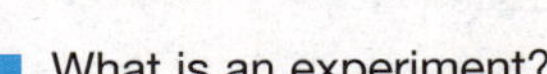

- What is a trial for an experiment?
- What were the possible outcomes for the experiment?
- Were the outcomes equally likely?

Investigate Problem 1

4. Gather the class results from the experiment in Question 1 and record the results in the table below.

	Your results	Class results
X	4	34
O	3	31
Blank	3	35
Total	10	100

5. Find the experimental probabilities of each outcome by using the results of the class. Use complete sentences in your answer.

Sample Answer:

The experimental probability of choosing an "X" slip is $\frac{38}{100} = \frac{19}{50}$.

The experimental probability of choosing an "O" slip is $\frac{27}{100}$.

The experimental probability of choosing a blank slip is $\frac{35}{100} = \frac{7}{20}$.

How do these experimental probabilities compare to the theoretical probabilities in part (B)? Use a complete sentence in your answer.

Sample Answer: These experimental probabilities are not the same as the theoretical probabilities, but they are close.

6. Are the experimental probabilities in Question 5 closer to the theoretical probabilities than the experimental probabilities in Question 2? What can you conclude about experimental and theoretical probabilities? Explain your reasoning. Use complete sentences in your answer.

Sample Answer: The experimental probabilities in Question 5 are closer to the theoretical probabilities than the experimental probabilities in Question 2. Since more trials made the experimental probabilities closer to the theoretical probabilities, you can conclude that even more trials will make the experimental probabilities even closer to the theoretical probabilities.

11

Wrap Up

Close

- Review all key terms and their definitions. Include the terms *theoretical probability, experiment, trial, success,* and *experimental probability.*
- You may also want to review any other vocabulary terms that were discussed during the lesson, which may include *sample space, outcome, probability, equally likely,* and *container.*
- Remind the students to write the key terms and their definitions in the notes section of their notebooks. You may also want the students to include examples.
- Discuss the relationship between theoretical and experimental probabilities.
- Ask the students to review the methods used in the experiment in this lesson. Have them explain the importance of each step in the process.
- Have the students discuss how increasing the number of trials is likely to make the experimental probabilities more similar to the theoretical probabilities.
- Have the students explain how you can determine the number of favorable outcomes for an experiment and how you can determine the number of successes for an experiment.
- If more work with this concept is needed, have the students determine the theoretical probabilities for additional situations as practice.

Ties to the Cognitive Tutor Software

In the abstract, students will struggle with the idea that the experimental probability differs from the theoretical probability. Through experience with repeated events, students can come to understand that the experimental probability will approach the theoretical probability, when there is a large number of trials. The Cognitive Tutor software provides students with the ability to "experience" a large number of trials in different contexts relatively quickly.

Follow Up

Assignment

Use the Assignment for Lesson 11.2 in the Student Assignments book. See the Teacher's Resources and Assessments book for answers.

Assessment

See the Assessments provided in the Teacher's Resources and Assessments book for Chapter 11.

Open-Ended Writing Task

Ask the students to design another experiment that could be conducted to gather data that would allow the class to compare the theoretical probabilities to the experimental probabilities.

Reflections

Insert your reflections on the lesson as it played out in class today.

What went well?

____

____

What did not go as well as you would have liked?

____

____

How would you like to change the lesson in order to improve the things that did not go well and capitalize on the things that did go well?

____

____

____

____

____

Notes

11

11.3 A Brand New Bag

Using Probabilities to Make Predictions

Learning By Doing Lesson Map

Get Ready

Objective

In this lesson, you will:

- Use experimental probabilities to make predictions.

Key Terms

- trial
- experimental probability

Materials

- 30 small slips of paper for each student
- 1 bag or container for each student
- A marker, pen, or pencil for each student

NCTM Content Standards

Grades 9–12 Expectations

Number and Operations Standard

- Judge the reasonableness of numerical computations and their results.

Algebra Standard

- Draw reasonable conclusions about a situation being modeled.

Data Analysis and Probability Standards

- Use simulations to explore the variability of sample statistics from a known population and to construct sampling distributions.
- Understand how sample statistics reflect the values of population parameters and use sampling distributions as the basis for informal inference.
- Understand the concepts of sample space and probability distribution and construct sample spaces and distributions in simple cases.

Lesson Overview

Within the context of this lesson, students will be asked to:

- Prepare and conduct an experiment.
- Calculate experimental probabilities.
- Use experimental probabilities to make predictions about the contents of the bag or container.
- Consider changes to the experiment that would likely result in experimental probabilities that would more closely approximate the actual probabilities.

Essential Questions

The following key questions are addressed in this section:

1. What is an experiment?
2. What is a trial for an experiment?
3. What is an experimental probability?

Show The Way

Warm Up

Place the following questions or an applicable subset of these questions on the board before students enter class. Students should begin working as soon as they are seated.

A bag contains 12 quarters, 20 dimes, 8 nickels, and 10 pennies. Calculate the theoretical probability for each event if one coin is randomly selected from the bag. Write your answer as a fraction and as a percent.

1. Probability it is a quarter $\frac{12}{50} = \frac{6}{25} = 24\%$
2. Probability it is a dime $\frac{20}{50} = \frac{2}{5} = 40\%$
3. Probability it is a nickel $\frac{8}{50} = \frac{4}{25} = 16\%$
4. Probability it is a penny $\frac{10}{50} = \frac{1}{5} = 20\%$
5. Probability it is worth more than 5 cents

   $\frac{32}{50} = \frac{16}{25} = 64\%$
6. Probability it is circular

   $\frac{50}{50} = 1 = 100\%$

An experiment is conducted on a bag of coins that may or may not have the same number of coins as the bag above. The results from 200 trials in which a coin is drawn, identified, replaced, and then mixed in with the other coins are as follows: a quarter was drawn 56 times, a dime 74 times, a nickel 26 times, and a penny 44 times. Find the experimental probability for each event.

7. Probability for a quarter $\frac{56}{200} = \frac{7}{25} = 28\%$
8. Probability for a dime $\frac{74}{200} = \frac{37}{100} = 37\%$
9. Probability for a nickel $\frac{26}{200} = \frac{13}{100} = 13\%$
10. Probability for a penny $\frac{44}{200} = \frac{11}{50} = 22\%$
11. Probability it is worth more than 5 cents

    $\frac{130}{200} = \frac{13}{20} = 65\%$
12. Probability it is circular.

    $\frac{200}{200} = 1 = 100\%$
13. Do you think that the bags contained the same number of coins, or do you think that the second bag is likely to have a different number of coins than the first bag? Explain your answer.

    They are likely to be the same bags or have the same contents because the theoretical and experimental probabilities are very close.

11

Motivator

Begin the lesson with the motivator to get students thinking about the topic of the upcoming problem. This lesson is about making predictions from experimental probabilities. The motivating questions are about probability experiments.

Ask the students the following questions to get them interested in the lesson.

- Have you ever seen a television game show that involved probability experiments?
- What television game show was it?
- What games involved probability?

Explore Together

Problem 1

Students will calculate experimental probabilities for an experiment.

Grouping

Ask for a student volunteer to read the Scenario and Problem 1 aloud. Have a student restate the problem. Pose the Guiding Questions below to verify student understanding. Have students work together in small groups to complete parts (A) through (D).

Guiding Questions

- What are the possible outcomes of this experiment?
- What is the sample space for this experiment?
- How is this experiment different than the experiment in the previous lesson? How are they similar?
- Is the intent of this experiment for you to know the number of each mark your friend will make on the papers? Why not?
- Do you expect each student to put the same number of each type of mark in the bags?
- What do you think that you will be able to estimate with your data in this experiment?
- When would being able to estimate the number of each type of something be helpful in real life?
- What type of probabilities will you be able to calculate from your data, theoretical or experimental?

Notes Rather than using slips of paper with marks on them, you can use colored chips or marbles if you have access to them. Change the activity to use these manipulatives instead of the paper slips.

Throughout this lesson, sample experimental results are given for the purpose of showing realistic student answers.

SCENARIO Now you are ready to use probabilities. Have a friend take 30 slips of paper that are all the same size. Your friend can divide the slips up any way he or she likes by marking an "X" on some of the slips, an "O" on some of the slips, and leaving the rest of the slips blank. Then have your friend put the slips into a paper bag (or some other bag or container that you cannot see inside of) and shake up the bag.

Problem 1 What's in Your Bag?

A. Perform an experiment by choosing a slip from the bag. Record the kind of slip you choose in the table below and put the slip back into the bag. Repeat this process 9 more times.

Slip	X	O	blank
Number of times slip was chosen	2	2	6

B. What is the experimental probability of choosing each kind of slip? Use a complete sentence in your answer.

Sample Answer:

The experimental probability of choosing an "X" slip is $\frac{2}{10} = \frac{1}{5}$.

The experimental probability of choosing an "O" slip is $\frac{2}{10} = \frac{1}{5}$.

The experimental probability of choosing a blank slip is $\frac{6}{10} = \frac{3}{5}$.

C. What fraction of the slips in the bag do you think are "X" slips? Use complete sentences to explain your reasoning.

Sample Answer: Because the experimental probability is $\frac{1}{5}$, there is one success (choosing an "X" slip) for every 5 trials. So, $\frac{1}{5}$ of the slips are "X" slips.

What fraction of the slips in the bag do you think are "O" slips? Use complete sentences to explain your reasoning.

Sample Answer: Because the experimental probability is $\frac{1}{5}$, there is one success (choosing an "O" slip) for every 5 trials. So, $\frac{1}{5}$ of the slips are "O" slips.

11

The student answers will vary from these answers and from each other's answers.

Explore Together

Problem 1

Students will analyze the data that they collected in their experiment to predict the percent of each type of mark in the bag or container.

Students will be completing parts (C) and (D) in small groups.

Notes Emphasize that the students must not tell each other how many of each mark they put in the bag during this experiment. Additional questions will be asked of the students to predict the number of each type of mark in the bag in Problem 1.

Call the class back together to have the students discuss and present their work for parts (A) through (D).

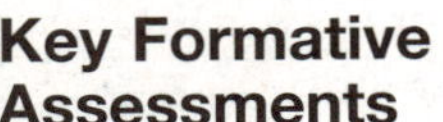

Key Formative Assessments

- What did you find in parts (A) through (D)?
- Did you expect each student to get the same sample data as the other students? Why or why not?
- Do the results surprise you or are they what you expected?
- How did you find the number of each slip you predicted to be in the bag?
- Would the number of trials have to be a factor of 30 to be able to use this method to predict the types of slips of paper in the bag or container? Why or why not?

11

Investigate Problem 1

Grouping

Ask for a student volunteer to read Question 1 aloud. Have a student restate the problem. Have the students solve Questions 1 through 6 individually.

Problem 1 What's in Your Bag?

What fraction of the slips in the bag do you think are blank slips? Use complete sentences to explain your reasoning.

Sample Answer: Because the experimental probability is $\frac{3}{5}$, there are three success (choosing a blank slip) for every 5 trials. So, $\frac{3}{5}$ of the slips are blank slips.

D. How many of each kind of slip do you think is in the bag? Show your work and use a complete sentence in your answer.

Sample Answer: "X" slips: $30\left(\frac{1}{5}\right) = 6$; "O" slips: $30\left(\frac{1}{5}\right) = 6$;

Blank slips: $30\left(\frac{3}{5}\right) = 18$

There are 6 "X" slips, 6 "O" slips, and 18 blank slips.

Investigate Problem 1

1. How accurate do you think your prediction from part (D) is? Use a complete sentence to explain your reasoning.

   Sample Answer: Because there are only 30 slips in the bag, the prediction should be fairly accurate.

2. Could you increase the accuracy of your prediction without looking in the bag? If so, explain how you would increase the accuracy. Use a complete sentence in your answer.

   Sample Answer: Yes, I would increase the accuracy by running more trials.

3. Run 40 more trials of the experiment and combine these results with the results from part (A) in the table below.

Slip	X	O	Blank
Number of times slip was chosen	22	13	15

4. Can you use the results from Question 3 to predict the number of each kind of slip that is in the bag? Why or why not? Use complete sentences in your answer.

   Sample Answer: Yes, even though 50 is not divisible by 30, we can still multiply the probability by 30, the number of slips in the bag.

Common Student Errors

Students may not quickly recognize that the product of the probability and the number of slips of paper in the bag gives a prediction of the number of each type of mark in the bag. Be sure to emphasize this relationship in part (D) and Question 4.

Explore Together

Investigate Problem 1

Students will analyze their experimental probabilities and predictions.

Common Student Errors

Students may not recognize that the results in Question 5 should be rounded to the nearest whole number. Discuss appropriate answers with the students. Suggest that their answers could include the words "about," "approximately," "nearly," "a little less than," or "a little more than" to make it clear in their answer that they used rounding.

Call the class back together to have the students discuss and present their work for Questions 1 through 6.

Grouping

Ask for a student volunteer to read Question 7 aloud. Have a student restate the problem. Pose the Guiding Questions below to verify student understanding. Have the students complete Question 7 individually. Then have the students work together in small groups to complete Questions 8 and 9. Call the class back together to have the students discuss and present their work for Questions 7 through 9

Guiding Questions

- What are you asked to do in Question 7?
- Why is it important to actually count the number of each mark in the bag if you already estimated them?

Key Formative Assessments

- What is an experimental probability?
- What is an experiment?
- What is a trial for an experiment?
- How accurate is an experimental probability to approximate an actual probability?
- What can be done to make this type of approximation quite accurate?

Investigate Problem 1

5. Use your results from Question 3 to predict the number of each kind of slip that is in the bag. Show your work and use complete sentences in your answer.

Sample Answer: "X" slips: $30\left(\frac{22}{50}\right) = 13.2$;

"O" slips: $30\left(\frac{13}{50}\right) = 7.8$; Blank slips: $30\left(\frac{15}{50}\right) = 9$

There are approximately 13 "X" slips, approximately 8 "O" slips, and approximately 9 blank slips.

6. Do you think that the prediction in Question 5 is better than the prediction in part (D)? Why or why not? Use complete sentences in your answer.

Sample Answer: The prediction should be better because the number of trials has been increased.

7. Open the bag and count each kind of slip. Compare the actual numbers to the predicted numbers from part (D) and Question 5. How close were the predictions? Which prediction was better? Use complete sentences in your answer.

Sample Answer: In the bag, there are 16 "X" slips, 7 "O" slips, and 7 blank slips. The first prediction was not very close, but the second prediction was very close.

8. What could account for any differences between the predicted numbers and the actual numbers? Use complete sentences to explain your reasoning.

Sample Answer: The differences could occur because of the number of trials. As the number of trials were increased, the accuracy increased.

9. If you are using experimental probabilities to make predictions, how do you think you should determine the number of trials that are necessary to get a fairly accurate outcome? Use complete sentences to explain your reasoning.

Sample Answer: Consider the number of possible outcomes. As the number of possible outcomes increases, the number of possible trials should increase.

11

Wrap Up

Close

- Review all key terms and their definitions. Include the terms *experiment, trial,* and *experimental probability.*
- You may also want to review any other vocabulary terms that were discussed during the lesson, which may include *sample space, outcome, probability, theoretical probability, randomly selected,* and *container.*
- Remind the students to write the key terms and their definitions in the notes section of their notebooks. You may also want the students to include examples.
- Discuss the importance of being able to estimate actual probabilities with experimental probabilities.
- Ask the students to review the methods used in the experiment in this lesson. Have them explain the importance of each step in the process.
- Have the students discuss how increasing the number of trials is likely to make the experimental probabilities more similar to the actual probabilities.

Ties to the Cognitive Tutor Software

Dependent probabilities can be difficult for students, because they require students to consider the number of outcomes both before and after an event. The diagramming strategies used in the Cognitive Tutor software and the text can help make this abstract reasoning concrete.

Follow Up

Assignment

Use the Assignment for Lesson 11.3 in the Student Assignments book. See the Teacher's Resources and Assessments book for answers.

11

Assessment

See the Assessments provided in the Teacher's Resources and Assessments book for Chapter 11.

Open-Ended Writing Task

Ask the students to design another experiment that could be conducted to estimate the percent of defective disks in a large bin of 2000 disks. This is a common quality control application of probability. A sample is taken and the entire shipment is accepted or rejected based on the results of the sample.

Reflections

Insert your reflections on the lesson as it played out in class today.

What went well?

____

____

What did not go as well as you would have liked?

____

____

How would you like to change the lesson in order to improve the things that did not go well and capitalize on the things that did go well?

____

____

____

____

____

Notes

11

11.4 Fun with Number Cubes

Graphing Frequencies of Outcomes

Learning By Doing Lesson Map

Get Ready

Objective

In this lesson, you will:

- Use a line plot to graph frequencies of outcomes.
- Find and compare probabilities.

Key Terms

- line plot

Materials

- A pair of 6-sided number cubes for each student

NCTM Content Standards

Grades 9–12 Expectations

Number and Operations Standard

- Judge the reasonableness of numerical computations and their results.

Algebra Standard

- Draw reasonable conclusions about a situation being modeled.

Data Analysis and Probability Standards

- Understand histograms, parallel box plots, and scatter plots and use them to display data.
- Use simulations to explore the variability of sample statistics from a known population and to construct sampling distributions.
- Understand how sample statistics reflect the values of population parameters and use sampling distributions as the basis for informal inference.
- Understand the concepts of sample space and probability distribution and construct sample spaces and distributions in simple cases.

Lesson Overview

Within the context of this lesson, students will be asked to:

- Create a line plot for data based on theoretical probabilities.
- Create a line plot for the data from an actual experiment.
- Calculate and compare theoretical and experimental probabilities.

Essential Questions

The following key questions are addressed in this section:

1. What is a line plot?
2. What type of information can be determined from a line plot?
3. How can the shape of a line plot for theoretical data differ from the shape of a line plot for experimental data?

Show The Way

Warm Up

Place the following questions or an applicable subset of these questions on the board before students enter class. Students should begin working as soon as they are seated.

Calculate the number of successes that you would expect for each situation. The probability of a success for one trial and the number of trials are given.

	Probability of Success	Number of Trials	Expected Number of Successes
1.	$\frac{5}{9}$	27	15
2.	25%	40	10
3.	0.2	150	30
4.	$\frac{2}{3}$	60	40
5.	$\frac{5}{8}$	100	About 63
6.	0.01	12	About 0
7.	0.07	500	35
8.	17.2%	50	About 9
9.	95%	15,286	About 14,522
10.	100%	28	28
11.	0%	81	0

Classify each situation as a theoretical probability or an experimental probability.

12. We found that 14 out of 32 students in the class were wearing tennis shoes. **Experimental**

13. The probability of choosing the correct answer for a 5 choice multiple-choice question without reading the question is 20%. **Theoretical**

11

Motivator

Begin the lesson with the motivator to get students thinking about the topic of the upcoming problem. This lesson is about performing experiments with number cubes. The motivating questions are about experiments.

Ask the students the following questions to get them interested in the lesson.

- What is an experiment?
- What types of situations might be modeled with an experiment?
- What is a number cube?
- How many sides are on most number cubes?

Explore Together

Problem 1

Students will create a line plot to display the predicted outcomes of an experiment in which they roll a number cube 30 times. The students will then conduct the experiment and create a line plot for the actual outcomes.

Grouping

Ask for a student volunteer to read the Scenario and Problem 1 aloud. Have a student restate the problem. Pose the Guiding Questions below to verify student understanding. Have students work together in small groups to complete parts (A) and (B), and Question 1 of Problem 1.

Guiding Questions

- What is a simulation?
- What is a number cube? How many sides are on most number cubes?
- What is the probability that any one side of a number cube will be at the top when it is rolled?
- What is a line plot?
- How can you create a line plot?
- Why is it important to space each X above each previous X consistently when creating a line plot?
- In what way is a line plot a graph?
- Why will you conduct 30 trials for this experiment?
- Would this experiment be as accurate if you were to roll 30 number cubes at the same time?
- Why do we roll only one number cube 30 times rather than 30 number cubes simultaneously?

Notes Throughout this lesson, sample experimental results are given for the purpose of showing realistic student answers. The student answers will vary from these answers and from each other's answers.

SCENARIO In this chapter, you have found that number cubes can be used to model problem situations. So, let's run an experiment that uses number cubes.

Problem 1 One Number Cube Experiment

In this experiment, you will roll a six-sided number cube and record the result. However, this time we will use a data display called a *line plot* to record the results.

A **line plot** displays the frequency of numerical data on a number line. To make a line plot of your results, first draw a number line that includes all the possible outcomes of rolling a six-sided number cube. Then draw an X above the number line for the outcome of each roll. If there is more than one occurrence of an outcome, draw another X above the first occurrence.

A. Before you perform the experiment, guess what your line plot will look like after recording the results of 30 rolls of a six-sided number cube. Create a line plot of your guess. Use a complete sentence to explain why the results will turn out this way.

Sample Answer:

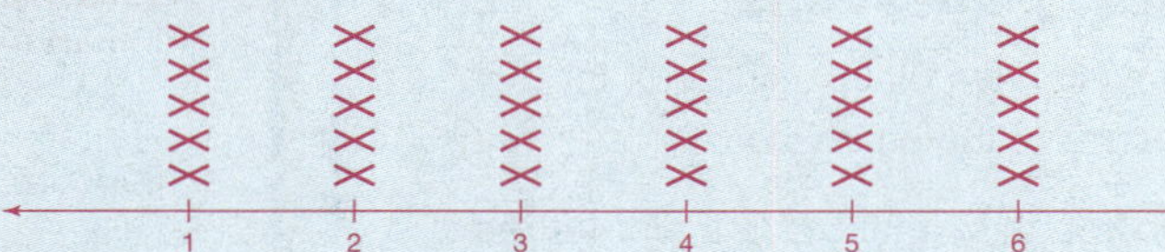

The probability for each outcome is the same, so each outcome should occur five times.

B. Perform the experiment and record the results of 30 rolls on the line plot below.

Sample Answer:

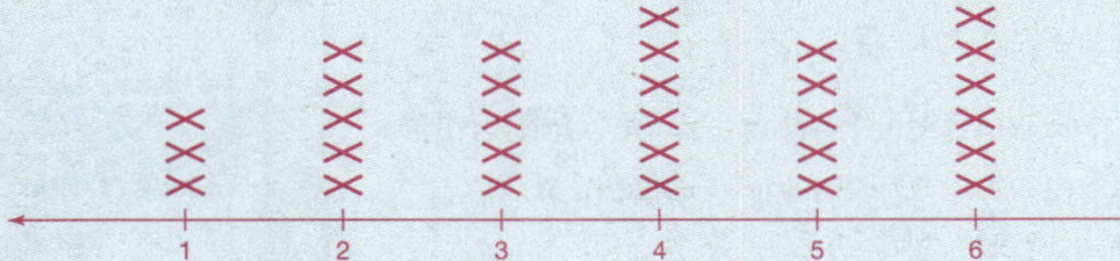

Explore Together

Investigate Problem 1

Students will compare and analyze their line plots for the theoretical data and the experimental data.

Call the class back together to have the students discuss and present their work for parts (A), and (B), and Question 1.

Grouping

Ask for a student volunteer to read Question 2 aloud. Have a student restate the problem. Pose the Guiding Questions below to verify student understanding. Complete Question 2 together as a class.

Guiding Questions

- How can we make a line plot for the results of the entire class?
- Do you think that the line plot for the class will appear closer to the one that you created in part (A) or the one that you created in part (B)? Why?
- How many students conducted the experiment? How many rolls did each student record?
- How many total trials for the experiment were recorded?
- Approximately how many of each of the six numbers would you expect to have occurred? Why?

Notes To create a class line plot for Problem 2, you will need to use the overhead projector or the front board. You may want each student to take a turn adding his or her data while the class is working on Question 1 to save some class time.

Problem 2

Grouping

Ask for a student volunteer to read Problem 2 aloud. Have a student restate the problem. Pose the Guiding Questions on the right to verify student understanding. Complete part (A) together as a class.

Investigate Problem 1

1. Does the shape of your line plot in part (B) agree with your prediction in part (A)? Use a complete sentence in your answer.

   Sample Answer: They are similar, but they are not exactly the same.

   Compare your line plot in part (B) with the line plots of other students in your class. What do you notice? Use a complete sentence in your answer.

   Sample Answer: All of the line plots are similar.

2. Now consider a line plot of the results of the class. How does the class line plot compare to your line plot in part (B)? Use complete sentences in your answer.

   Sample Answer: The class line plot and my line plot have a similar shape. The class line plot is closer to my prediction in part (A).

3. Suppose that you perform this experiment again. Do you think that your results will be the same as or different from the results in Problem 1? Use complete sentences to explain your reasoning.

   Sample Answer: Both sets of results should be similar because the probability for each possible outcome is the same.

Problem 2 Number Cube Pair Experiment

In this next experiment, you will roll a pair of six-sided number cubes and record the sum of the results of the roll on a line plot.

A. What are the possible sums? How many possible sums are there? Use complete sentences in your answer.

   The possible sums are 2, 3, 4, 5, 6, 7, 8, 9, 10, 11, and 12. There are 11 possible sums.

Guiding Questions

- What is a sum?
- How will you calculate the sum for the pair of number cubes?
- What numbers are possible to roll for each of the individual number cubes?

Explore Together

Problem 2

Students will conduct 50 trials of the experiment to find the sum for the roll of a pair of number cubes. They will create a line plot to represent their data.

Grouping

Ask for a student volunteer to read part (B) of Problem 2 aloud. Have a student restate the problem. Pose the Guiding Questions below to verify student understanding. Have the students complete parts (B), and (C), and Question 1 individually. Then, call the class back together to have the students discuss and present their work.

Guiding Questions

- Do you think that each of the possible sums is equally likely to occur?
- What 2 sums are the least likely to occur? Why?
- What sums do you think are the most likely to occur? Why?
- What does it mean to draw the rough shape of the line plot?
- Why are we not asked to draw the line plot rather than just the rough shape in part (B)?

Investigate Problem 2

Grouping

Ask for a student volunteer to read Question 2 aloud. Have a student restate the problem. Pose the Guiding Questions below to verify student understanding. Complete Question 2 together as a class.

Guiding Questions

- How can we make a line plot for the results of the entire class?
- Do you think that the line plot for the class will appear closer to what you predicted in part (A) or to what you created in part (B)? Why?

Problem 2 Number Cube Pair Experiment

B. Before you perform the experiment, guess what your line plot will look like after recording the results of 50 rolls of a pair of number cubes. Draw the rough shape of the line plot. Use complete sentences to explain why the results will turn out this way.

Sample Answer:

The sums between 4 and 10 can occur with different pairs of numbers, so the probability is greater. The sums less than 4 and greater than 10 can occur with fewer pairs of numbers, so the probability is less.

C. Perform the experiment and record the results of 50 rolls on the line plot below.

Sample Answer:

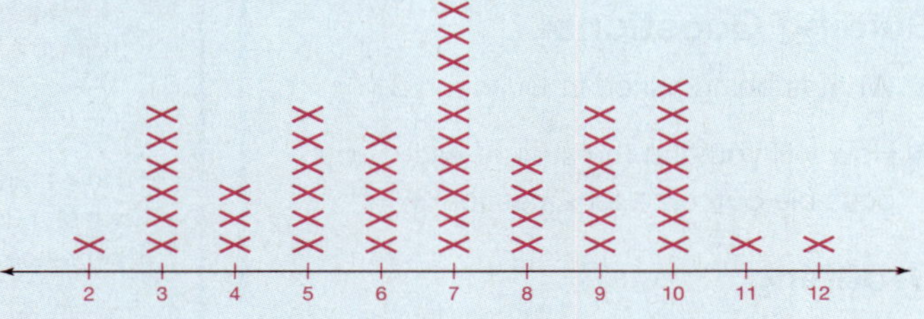

Investigate Problem 2

1. Describe the shape of your line plot. Use a complete sentence in your answer.

   Sample Answer: The line plot generally rises and then falls as you look at it from left to right.

   How does the shape of your line plot compare with your prediction in part (B)? Use complete sentences in your answer.

   Sample Answer: The shape is not exactly as I expected. The shape of the line plot rises and falls more than I thought it would as you look at it from left to right.

2. Now consider a line plot of the results of the entire class. How does the class line plot compare to your line plot in part (C)? Use complete sentences in your answer.

   Sample Answer: The class line plot and my line plot have a similar shape. The class line plot rises and then falls more smoothly as you look at it from left to right.

- How many students conducted the experiment?
- How many sums did each student record?
- How many total trials for the experiment were recorded?

11

Explore Together

Investigate Problem 2

Students will consider the sample space of all possible sums for the roll of a pair of number cubes; then, they will investigate the probability of each.

Grouping

Ask for a student volunteer to read Question 3 aloud. Have a student restate the problem. Pose the Guiding Questions below to verify student understanding. Have the students complete Questions 3 through 5 individually. Then, call the class back together to have the students discuss and present their work.

Guiding Questions

- What is being asked in Question 3?
- How will you find the sum of each possible outcome for Question 3?

Grouping

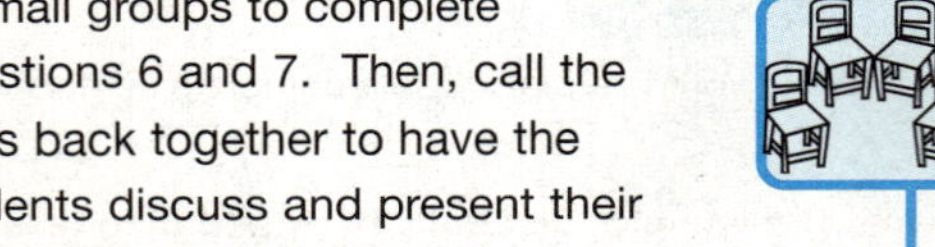

Ask for a student volunteer to read Question 6 aloud. Have a student restate the problem. Have students work together in small groups to complete Questions 6 and 7. Then, call the class back together to have the students discuss and present their work for Questions 6 and 7.

Key Formative Assessments

- What is the sample space for the possible sums?
- Why are there sums that occur more ways than the other sums?
- Are the different sums equally likely?
- How can you calculate the probability of each possible sum?
- What is a pattern?
- How did you find the pattern to answer Question 7?

Investigate Problem 2

3. Complete the table below that shows all of the possible rolls of pairs of six-sided number cubes and the sums of the results.

		First number cube					
		1	2	3	4	5	6
Second number cube	1	2	3	4	5	6	7
	2	3	4	5	6	7	8
	3	4	5	6	7	8	9
	4	5	6	7	8	9	10
	5	6	7	8	9	10	11
	6	7	8	9	10	11	12

4. Describe a possible outcome of one run of the experiment. Use a complete sentence in your answer.

Sample Answer: The roll of the first number cube is 3 and the roll of the second number cube is 4. The sum of the roll is 7.

5. How many possible outcomes are there? Note: It is important which number cube is first and which is second. Use a complete sentence in your answer.

There are 36 possible outcomes.

6. For each sum, find the number of different ways in which the sum can occur.

Sum is 2: 1 way Sum is 3: 2 ways Sum is 4: 3 ways

Sum is 5: 4 ways Sum is 6: 5 ways Sum is 7: 6 ways

Sum is 8: 5 ways Sum is 9: 4 ways Sum is 10: 3 ways

Sum is 11: 2 ways Sum is 12: 1 way

7. Are there any patterns in the number of ways that each sum can occur? Use complete sentences in your answer.

Sample Answer: Yes. The number of ways increases by one way as the sum of the numbers increases by one until you reach the sum of seven. Then, the number of ways decreases by one way as the sum of the numbers increases by one.

Explore Together

Investigate Problem 2

Students will calculate probabilities for the possible outcomes of rolling a pair of number cubes and recording the sum.

Grouping

Ask for a student volunteer to read Question 8 aloud. Have a student restate the problem. Pose the Guiding Questions below to verify student understanding. Have the students complete Questions 8 through 12 together in small groups. Then, call the class back together to have the students discuss and present their work for Questions 8 through 12.

Guiding Questions

- What is being asked in Question 8?
- How many total outcomes exist for the roll of a pair of number cubes?
- How will you find the probability for each possible sum for Question 8?
- What does it mean for an outcome to be most likely to occur?
- What does it mean for an outcome to be least likely to occur?
- What is an odd number? What is an even number?

Key Formative Assessments

- What is a line plot?
- How can you create a line plot?
- What type of information can be determined by analyzing a line plot?
- What was the relationship between the line plot that you predicted based on theoretical probabilities and the line plot that you created based on your experimental probabilities?

Investigate Problem 2

8. For each sum, find the probability that the sum will occur.

Probability that sum is 2: $\frac{1}{36}$ Probability that sum is 3: $\frac{2}{36}$

Probability that sum is 4: $\frac{3}{36}$ Probability that sum is 5: $\frac{4}{36}$

Probability that sum is 6: $\frac{5}{36}$ Probability that sum is 7: $\frac{6}{36}$

Probability that sum is 8: $\frac{5}{36}$ Probability that sum is 9: $\frac{4}{36}$

Probability that sum is 10: $\frac{3}{36}$ Probability that sum is 11: $\frac{2}{36}$

Probability that sum is 12: $\frac{1}{36}$

9. Which sums(s) are you most likely to roll? Which sum(s) are you least likely to roll? Use complete sentences in your answer.

Sample Answer: Seven is the most likely sum because the probability of the sum is the highest. Two and 12 are the least likely sums because the probability of these sums is the lowest.

10. Is a sum of 5 or a sum of 8 more likely? Why? Use a complete sentence in your answer.

It is more likely that a sum of 8 is rolled because the probability of rolling a sum of 8 is greater than the probability of rolling a sum of 5.

11. Is an even sum or an odd sum more likely? Why? Show your work and use complete sentences to explain your reasoning.

Probability of an even sum: $\frac{1+3+5+5+3+1}{36} = \frac{18}{36}$

Probability of an odd sum: $\frac{2+4+6+4+2}{36} = \frac{18}{36}$

It is equally likely to roll an even sum as it is to roll an odd sum because the probabilities are the same.

12. Does the information that you gathered in Questions 3 through 11 help you explain the shape of your line plot in Problem 2? If so, how? Use complete sentences in your answer.

Sample Answer: Yes, you can see that the probabilities increase and then decrease as the values of the sums increase.

11

Wrap Up

Close

- Review all key terms and their definitions. Include the term *line plot.*
- You may also want to review any other vocabulary terms that were discussed during the lesson, which may include *frequency, sample space, outcome, probability, equally likely, most likely, least likely, pattern, sum, experiment, odd, even, trial,* and *bell-shaped or normal distribution of data.*
- Remind the students to write the key terms and their definitions in the notes section of their notebooks. You may also want the students to include examples.
- Discuss the relationship between the line plots of the theoretical and experimental data.
- Ask the students to review the methods used to create a line plot.
- Have the students discuss the importance of spacing between the numbers both horizontally and vertically in a line plot.

Ties to the Cognitive Tutor Software

Dependent probabilities can be difficult for students, because they require students to consider the number of outcomes both before and after an event. The diagramming strategies used in the Cognitive Tutor software and the text can help make this abstract reasoning concrete.

Follow Up

Assignment

Use the Assignment for Lesson 11.4 in the Student Assignments book. See the Teacher's Resources and Assessments book for answers.

11

Assessment

See the Assessments provided in the Teacher's Resources and Assessments book for Chapter 11.

Open-Ended Writing Task

Ask the students to gather data of interest to them. They should then create a line plot to display their data and present it to the class at the beginning of the next class session.

Reflections

Insert your reflections on the lesson as it played out in class today.

What went well?

What did not go as well as you would have liked?

How would you like to change the lesson in order to improve the things that did not go well and capitalize on the things that did go well?

Notes

11

11

11.5 Going to the Movies

Counting and Permutations

Learning By Doing Lesson Map

Get Ready

Objective

In this lesson, you will:

- Find the number of permutations of n objects.
- Simplify expressions that involve factorials.

Key Terms

- tree diagram
- Fundamental Counting Principle
- permutation
- factorial

Materials

- The students may need a calculator for the warm up exercises.

NCTM Content Standards

Grades 9–12 Expectations

Algebra Standards

- Generalize patterns using explicitly defined and recursively defined functions.
- Use symbolic algebra to represent and explain mathematical relationships.
- Identify essential quantitative relationships in a situation and determine the class or classes of functions that might model the relationships.
- Use symbolic expressions, including iterative and recursive forms, to represent relationships arising from various contexts.
- Draw reasonable conclusions about a situation being modeled.

Data Analysis and Probability Standard

- Understand the concepts of sample space and probability distribution and construct sample spaces and distributions in simple cases.

Lesson Overview

Within the context of this lesson, students will be asked to:

- Calculate the number of ways that 5 friends can sit beside each other in a row.
- Calculate the number of permutations of a given number n of objects.
- Simplify expressions that involve factorials.

Essential Questions

The following key questions are addressed in this section:

1. What is a tree diagram?
2. How can a tree diagram help you to find the number of possible outcomes for a situation?
3. What is a permutation?
4. How can you calculate the number of permutations of n objects?
5. What is a factorial?

Show The Way

Warm Up

Place the following questions or an applicable subset of these questions on the board before students enter class. Students should begin working as soon as they are seated.

A local animal shelter has the following animals available for adoption: 8 small dogs, 12 medium dogs, 17 large dogs, 21 cats, and 7 other animals. Of the dogs, 13 have a mix of several colors, 9 are black, 7 are brown, 5 are white, and the other 3 have spots. Of the cats, 6 have a mix of several colors, 8 are black, 4 are brown, and the other 3 are white. Of the other animals, 2 are black, 2 are white, and the other 3 are a mix of several colors. One of the 65 animals is randomly chosen to be featured in a commercial for the shelter. Answer each question. Write your answers as fractions and as percents rounded to the nearest whole percent.

1. Probability it is a dog $\frac{37}{65} \approx 57\%$
2. Probability it is a cat $\frac{21}{65} \approx 32\%$
3. Probability it is a small dog $\frac{8}{65} \approx 12\%$
4. Probability it is a medium dog $\frac{12}{65} \approx 18\%$
5. Probability it is a large dog $\frac{17}{65} \approx 26\%$
6. Probability it is black $\frac{19}{65} \approx 29\%$
7. Probability it is brown $\frac{11}{65} \approx 17\%$
8. Probability it is white $\frac{10}{65} = \frac{2}{13} \approx 15\%$
9. Probability it is spotted $\frac{3}{65} \approx 5\%$
10. Probability it has a mix of several colors $\frac{22}{65} \approx 34\%$
11. Probability it is some other animal than a cat or dog $\frac{7}{65} \approx 11\%$
12. Probability it is cat with a mix of several colors $\frac{6}{65} \approx 9\%$

Motivator

11

Begin the lesson with the motivator to get students thinking about the topic of the upcoming problem. This lesson is about sitting in a row at a movie with your friends. The motivating questions are about going to a movie with your friends.

Ask the students the following questions to get them interested in the lesson.

- When you go to a movie with your friends, how many people usually go together?
- Do you usually sit in the same row?
- How do you decide who will sit beside whom?
- Do you prefer to sit near the front, middle, or back of the theater?

Explore Together

Problem 1

Students will create a list and a tree diagram of possible seating arrangements to represent the situation.

Grouping

Ask for a student volunteer to read the Scenario and Problem 1 aloud. Have a student restate the problem. Pose the Guiding Questions below to verify student understanding. Have students work individually for about one minute on part (A). Then stop the students and discuss how difficult it is to make a list this way. Suggest that it would be helpful to have a more efficient and organized way to find the possible seating orders. Then have the students work individually on part (B) for 1 minute and stop them. Again, discuss the reality that the task is tedious and lengthy and it would be unreasonable to have the students create the entire tree diagram.

Guiding Questions

- What is a seating order?
- How are you asked to represent the names of each person in your list? Why would you want to abbreviate each person's name with only the first letter of his or her name in this list?
- How many letters will be listed in each possible outcome for this experiment?
- Does the order in which you list the letters matter? For instance, is YACDK different than YACKD?
- Do you think that there will be more or less than 5 ways to arrange the friends in a row? Do you think that there will be more or less than 10 ways? Do you think that there will be more or less than 25 ways?

SCENARIO You and your four friends, Alex, Chris, Daryl, and Kelly, are going to see a movie. You and Daryl are in the same science class. You wonder how likely it is that you will sit next to Daryl. In order to determine this, you must first find out how many different seating orders there are if you all sit in a row together.

Problem 1 Making a List

A. Begin an organized list that shows the different seating orders that can occur. In your list, use the first letter of each person's name and use "Y" for yourself to list the orders. Hint: Start your list by considering the orders that occur if you are sitting in the first seat. Your teacher will tell you when you can stop working on the list.

Sample Answer:

YACDK	YKACD	YCAKD	YDACK
YACKD	YKADC	YCADK	YDAKC
YAKDC	YKDCA	YCKDA	YDKCA
YAKCD	YKDAC	YCKAD	YDKAC
YADCK	YKCAD	YCDKA	YDCKA
YADKC	YKCDA	YCDAK	YDCAK

B. Another way you can list the seating orders is by completing a *tree diagram*. A **tree diagram** is a visual display of an ordering. A tree diagram is started for you below. Begin to complete the diagram. Your teacher will tell you when you can stop working on the diagram.

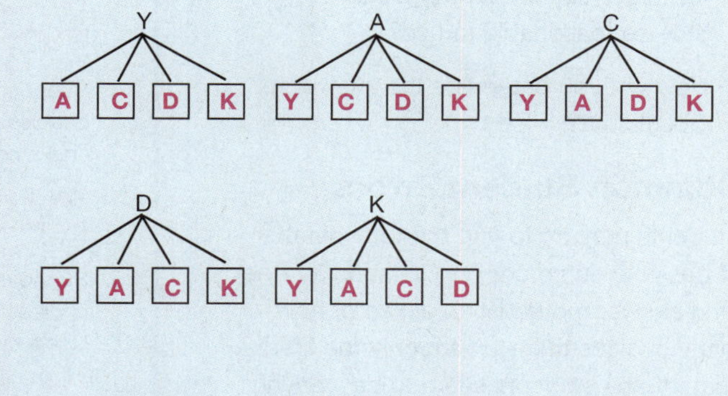

Notes The sample answers given here are typical for what you might expect a student to answer in the short amount of time provided for parts (A) and (B). The complete set of 120 possibilities is not provided for either part. However, you may want to make a full tree diagram on a transparency that you can show to the students when you feel it is appropriate during the lesson.

Explore Together

Investigate Problem 1

Students will investigate ordered arrangements of people.

Grouping

Ask for a student volunteer to read Questions 1 and 2 aloud. Have a student restate the problem. Pose the Guiding Questions below to verify student understanding. Have the students complete Questions 1 and 2 individually. Then, call the class back together to have the students discuss and explain their work for Questions 1 and 2.

Guiding Questions

- Were you able to find more arrangements in part (A) or in part (B)? Why?
- How many possible seating arrangements did you find in parts (A) and (B)?
- Do you think that the methods in parts (A) and (B) were convenient for this situation? When would they be reasonable to use?
- How will you determine the answer to Question 2?

Common Student Errors

Students may try to find the total number of possible outcomes for Question 2 rather than answer the question asked of how many choices there are for only the first seat. If any students seem to be working extensively on this question, be ready to refocus them on the question that was asked.

The students may also ignore the request to write in complete sentences. Be sure that they remember the importance of answering the questions in complete sentences when instructed to do so.

Investigate Problem 1

1. Were you able to find all of the possible seating orders by using a list or a tree diagram? If so, how many seating orders are possible? If not, why were you not able to complete the list or tree diagram and what made the task difficult? Use complete sentences in your answer.

   Sample Answer: Yes, there are 120 possible seating orders.

   or

   Sample Answer: No, because there are too many possible seating orders.

If necessary, use the seating chart below to help you answer the following questions.

First seat	Second seat	Third seat	Fourth seat	Fifth seat

2. How many possible choices are there for the person who sits in the first seat? Use a complete sentence to explain your reasoning.

   There are five possible choices because there are five people sitting together.

3. Suppose that Alex sits in the first seat. How many possible choices are there for the person who sits in the second seat? Use a complete sentence to explain your reasoning.

   There are four possible choices because there are only four people left: myself, Chris, Daryl, and Kelly.

4. Suppose that you sit in the first seat. How many possible choices are there for the person who sits in the second seat? Use a complete sentence to explain your reasoning.

   There are four possible choices because there are only four people left: Alex, Chris, Daryl, and Kelly.

5. Suppose that Chris sits in the first seat. How many possible choices are there for the person who sits in the second seat? Use a complete sentence to explain your reasoning.

   There are four possible choices because there are only four people left: myself, Alex, Daryl, and Kelly.

6. Suppose that Daryl sits in the first seat. How many possible choices are there for the person who sits in the second seat? Use a complete sentence to explain your reasoning.

   There are four possible choices because there are only four people left: myself, Alex, Chris, and Kelly.

Grouping

Ask for a student volunteer to read Question 3 aloud. Have a student restate the problem. Have the students work in small groups to complete Questions 3 through 8.

Explore Together

Investigate Problem 1

Students will continue to investigate ordered arrangements of people.

Grouping

Students will be working together in small groups to complete Questions 3 through 8.

Call the class back together to have the students discuss and present their work for Questions 3 through 8.

Grouping

Ask for a student volunteer to read Question 9 aloud. Have a student restate the problem. Pose the Guiding Questions below to verify student understanding. Have students work together in small groups to complete Question 9 through 11.

Guiding Questions

- What are you asked to do in Question 9?
- What information is given in this Question?
- How does this question differ from the previous questions?

Call the class back together to have the students discuss and present their work for Questions 9 through 11.

Key Formative Assessments

- How did you determine the number of people who could sit in the first seat?
- After one person sits in the first seat, how many people are able to sit in the second seat?
- After two people sit in the first and second seats, how many people are able to sit in the third seat?
- Why did you multiply the factors 5, 4, and 3 for Question 11?
- What would a tree diagram look like for the possible people to sit in the first 3 seats?

Investigate Problem 1

7. Suppose that Kelly sits in the first seat. How many possible choices are there for the person who sits in the second seat? Use a complete sentence to explain your reasoning.

There are four possible choices because there are only four people left: myself, Alex, Chris, and Daryl.

8. How many different ways can the first two seats be filled? Use complete sentences to explain your reasoning.

Sample Answer: In the first seat, there are five different choices of people. For each of the five people that sit in the first seat, there are four choices for the second seat. So there are 4 + 4 + 4 + 4 + 4 = 20 different ways the first two seats can be filled.

Write a multiplication problem that represents the number of different ways that the first two seats can be filled. Indicate what each factor represents in the problem situation.

5(4) = 20; The factor 5 indicates the number of choices for the first seat and the factor 4 indicates the number of choices for the second seat once the first seat is filled.

9. Suppose that Alex sits in the first seat and Daryl sits in the second seat. How many possible choices are there for the person who sits in the third seat? Use a complete sentence to explain your reasoning.

There are three possible choices because there are only three people left: myself, Chris, and Kelly.

10. For each different way that the first two seats can be filled, how many possible choices are there for the person who sits in the third seat? Use a complete sentence to explain your reasoning.

Sample Answer: No matter how the first two seats are filled, there are always three choices for the third seat because there are three people left to be seated.

11. Write an expression that represents the number of different ways that the first three seats can be filled. Then simplify your expression. Indicate what the parts of your expression represent in the problem situation.

5(4)(3) = 60; The factor 5 indicates the number of choices for the first seat, the factor 4 indicates the number of choices for the second seat once the first seat is filled, and the factor 3 indicates the number of choices for the third seat once the first and second seats are filled.

Explore Together

Investigate Problem 1

Students will continue to investigate ordered arrangements of people and will write an expression for the number of ways that 5 people can sit in a row.

Grouping

Ask for a student volunteer to read Questions 12 and 13 aloud. Have a student restate the problem. Pose the Guiding Questions below to verify students' understanding. Have students work together in small groups to complete Questions 12 through 16. Then, call the class back together to have the students discuss and present their work for Questions 12 through 16.

Guiding Questions

- What are you asked to find in Question 12?
- What are you asked to find in Question 13?
- What is an expression?
- How can you complete these questions?

Common Student Errors

Some students may incorrectly think that because there is only 1 person available to sit in the last seat that there are zero options and will multiply by zero rather than 1. Other students may incorrectly continue the pattern to say that there are zero people available to be seated after all 5 friends have been seated and multiply the factors of 5, 4, 3, 2, 1, and 0 to get 0.

Key Formative Assessments

- How do you think that you could find the number of ways that a total of 8 people could be arranged in order in a row?
- How do you think that you could find the number of ways that a total of 15 people could be arranged in order in a row?

Investigate Problem 1

12. For each of the 60 different ways that the first three seats can be filled, how many possible choices are there for the person who sits in the fourth seat? Use a complete sentence to explain your reasoning.

Sample Answer: No matter how the first three seats are filled, there are always two choices for the fourth seat because there are two people left to be seated.

13. Write an expression that represents the number of different ways that the first four seats can be filled. Then simplify your expression. Indicate what the parts of your expression represent in the problem situation.

5(4)(3)(2) = 120; The factor 5 indicates the number of choices for the first seat, the factor 4 indicates the number of choices for the second seat once the first seat is filled, the factor 3 indicates the number of choices for the third seat once the first and second seats are filled, and the factor 2 indicates the number of choices for the fourth seat once the first, second, and third seats are filled.

14. For each different way that the first four seats can be filled, how many possible choices are there for the person who sits in the fifth seat? Use a complete sentence to explain your reasoning.

Sample Answer: No matter how the first four seats are filled, there is always one choice for the fifth seat because there is one person left to be seated.

15. Does filling the fifth seat with the person in the group that remains change the number of possible seating orders? Why or why not? Use a complete sentence in your answer.

Sample Answer: No, because there are no choices left once you reach the last person.

16. Write an expression that represents the number of different ways that the five seats can be filled. Then simplify your expression. Indicate what the parts of your expression represent in the problem situation.

5(4)(3)(2)(1) = 120; The factor 5 indicates the number of choices for the first seat, the factor 4 indicates the number of choices for the second seat once the first seat is filled, the factor 3 indicates the number of choices for the third seat once the first and second seats are filled, the factor 2 indicates the number of choices for the fourth seat once the first, second, and third seats are filled, and the factor 1 indicates the number of choices for the fifth and last seat.

Explore Together

Investigate Problem 1

Students will calculate the number of permutations of people and of objects.

Just the Math

Students will be formally introduced to the concept of the Fundamental Counting Principle.

Grouping

Ask for a student volunteer to read Question 17 aloud. Have a student restate the problem. Pose the Guiding Questions below to verify student understanding. Have the students complete Question 17 individually. Then, have the students discuss and explain their work for Question 17.

Guiding Questions

- What is the Fundamental Counting Principle?
- How can the Fundamental Counting Principle be applied to more than two events?
- How can the Fundamental Counting Principle be applied to the work that we did to find the number of arrangements of 5 people in a row at the theater?

Just the Math

Students will be formally introduced to the concept of permutations.

Grouping

Ask for a student volunteer to read Question 18 aloud. Have a student restate the problem. Pose the Guiding Questions at the right to verify student understanding. Have the students work together in small groups to complete Question 18. Then, call the class back together to have the students discuss and present their work for Question 18.

Investigate Problem 1

17. Just the Math: Fundamental Counting Principle
You used the *Fundamental Counting Principle* to find the answer to Problem 1. The **Fundamental Counting Principle** states that if you have m choices for one event and n choices for another event, then the number of choices for both events is $m \cdot n$. This rule extends to more than two events, as in the case of Problem 1.

You decide to get a beverage and a snack at the movie. The movie theater offers six different beverages and nine different snacks. How many different combinations of beverage and snack could you have? Show your work and use a complete sentence in your answer.

6(9) = 54; There are 54 different beverage and snack combinations.

18. Just the Math: Permutations In Problem 1, you were finding the number of *permutations* of five objects. A **permutation** is an ordering of a set of objects. In Problem 1, you found the number of possible permutations of five people sitting together in a row. Use a tree diagram or organized list to find all the possible permutations of the numbers 1, 2, and 3. How many permutations are there? Use a complete sentence in your answer.

Sample Answer: 123, 132, 213, 231, 312, 321; There are six permutations.

Use a tree diagram or organized list to find all the possible permutations of the letters D, F, and G. How many permutations are there? Use a complete sentence in your answer.

Sample Answer: DFG, DGF, FDG, FGD, GDF, GFD; There are six permutations.

How does the number of permutations of the numbers 1, 2, and 3 compare to the number of permutations of the letters D, F, and G? Use a complete sentence in your answer.

The numbers of permutations are the same.

Does it make any difference whether you are finding the number of permutations of three numbers, three letters, or any other three objects? Use complete sentences to explain your reasoning.

Sample Answer: No, because it is not important what kinds of objects are being ordered. It is important how many objects are being ordered.

Guiding Questions

- What is a permutation?
- How can you use a tree diagram to find the number of permutations for n people or objects?

11

Explore Together

Investigate Problem 1

Students will calculate permutations of n objects by finding the value of n factorial.

Grouping

Ask for a student volunteer to read Question 19 aloud. Have a student restate the problem. Pose the Guiding Questions below to verify student understanding. Have students work together in small groups to complete Question 19. Then, call the class back together to have the students discuss and present their work for Question 19.

Guiding Questions

- What are you asked to do in Question 19?
- What is a factorial?
- How can you evaluate a factorial?
- How can you evaluate an expression involving more than one factorial value?
- How does a factorial value relate to the number of permutations of n objects that can be formed?

Note You may want to explain to more advanced students that 1! and 0! are both defined to be 1.

Key Formative Assessments

- What is a permutation?
- What is a factorial?
- How can you evaluate a factorial expression?
- How can you evaluate a product of 2 factorial expressions?
- How can you evaluate a ratio of 2 factorial expressions?
- Why do we restrict factorials to be for whole numbers only and not negative values?

Investigate Problem 1

19. You can find the number of permutations of n objects by using the Fundamental Counting Principle. The number of permutations of n objects is $n! = n \cdot (n - 1) \cdot (n - 2) \cdot \ldots \cdot 3 \cdot 2 \cdot 1$. The expression $n!$ stands for ***n* factorial** and represents the product of the integers between 1 and n. For example, $5! = 5 \cdot 4 \cdot 3 \cdot 2 \cdot 1 = 120$. Write the number of permutations of the numbers 1, 2, and 3 by using factorial notation. Then simplify the expression.

$3! = 3 \cdot 2 \cdot 1 = 6$

Simplify each expression by first writing each factorial as a product.

$6!$

$6 \cdot 5 \cdot 4 \cdot 3 \cdot 2 \cdot 1 = 720$

$8!$

$8 \cdot 7 \cdot 6 \cdot 5 \cdot 4 \cdot 3 \cdot 2 \cdot 1 = 40{,}320$

$4!\,2!$

$(4 \cdot 3 \cdot 2 \cdot 1)(2 \cdot 1) = 48$

$\frac{5!}{3!}$

$$\frac{5 \cdot 4 \cdot 3 \cdot 2 \cdot 1}{3 \cdot 2 \cdot 1} = 5 \cdot 4$$
$$= 20$$

$\frac{8!}{2!4!}$

$$\frac{8!}{2!4!} = \frac{8 \cdot 7 \cdot 6 \cdot 5 \cdot 4 \cdot 3 \cdot 2 \cdot 1}{(2 \cdot 1)(4 \cdot 3 \cdot 2 \cdot 1)}$$
$$= 7 \cdot 6 \cdot 5 \cdot 4$$
$$= 840$$

$\frac{2!3!}{4!}$

$$\frac{2!3!}{4!} = \frac{(2 \cdot 1)(3 \cdot 2 \cdot 1)}{4 \cdot 3 \cdot 2 \cdot 1}$$
$$= \frac{2}{4} = \frac{1}{2}$$

11

Explore Together

Problem 2

Students will calculate the number of possible ways in which 2 specific people can sit beside each other and have 3 others in the same row of 5 people.

Grouping

Ask for a student volunteer to read part (A) of Problem 2 aloud. Have a student restate the problem. Pose the Guiding Questions below to verify student understanding. Have students work together in small groups to complete parts (A) through (E). Then, call the class back together to have the students discuss and present their work for parts (A) through (E).

Guiding Questions

- What is being asked in part (A)?
- How many seats will you and your friends use?
- What is the sample space for this experiment?
- If you sit in the first seat on the left, and the second seat is to your right, why can't your friend sit on your left instead? Then who would be sitting in the first seat?
- If you sit in the first seat of the 5 seats, why is there only one seat where your friend from science class can sit?

Common Student Errors

Some students will be confused in parts (A) and (E) as to why the friend can only sit in one seat with the requirements given. It may help if you draw a diagram on the front board representing the 5 possible seats. This could be drawn simply as 5 horizontal segments in a row. Then when you discuss part (A) you can write the word "self" above the first seat and "friend from science" above the second seat.

Problem 2 What's the Probability?

A. If you sit in the first seat, where can your friend from science class sit so that you are sitting next to each other? How many seating orders are possible for the other three people? Show your work and use a complete sentence in your answer.

The friend can sit in the second seat.

$3! = 3 \cdot 2 \cdot 1 = 6$; There are six possible seating orders for the other three people.

B. If you sit in the second seat, where can your friend from science class sit so that you are sitting next to each other? How many seating orders are possible for the other three people? Show your work and use a complete sentence in your answer.

The friend can sit in the first or third seats.

$2(3!) = 2(3 \cdot 2 \cdot 1) = 12$; There are twelve possible seating orders for the other three people.

C. If you sit in the third seat, where can your friend from science class sit so that you are sitting next to each other? How many seating orders are possible for the other three people? Show your work and use a complete sentence in your answer.

The friend can sit in the second or fourth seats.

$2(3!) = 2(3 \cdot 2 \cdot 1) = 12$; There are twelve possible seating orders for the other three people.

D. If you sit in the fourth seat, where can your friend from science class sit so that you are sitting next to each other? How many seating orders are possible for the other three people? Show your work and use a complete sentence in your answer.

The friend can sit in the third or fifth seats.

$2(3!) = 2(3 \cdot 2 \cdot 1) = 12$; There are twelve possible seating orders for the other three people.

E. If you sit in the last seat, where can your friend from science class sit so that you are sitting next to each other? How many seating orders are possible for the other three people? Show your work and use a complete sentence in your answer.

The friend can sit in the fourth seat.

$3! = 3 \cdot 2 \cdot 1 = 6$; There are six possible seating orders for the other three people.

Explore Together

Investigate Problem 2

Students will calculate the theoretical probability that 2 specific people in a group of 5 friends will sit beside each other.

Grouping

Ask for a student volunteer to read Question 1 aloud. Have a student restate the problem. Pose the Guiding Questions below to verify student understanding. Complete Question 1 together as a class. Then, have the students complete Questions 2 through 4 together in a small groups. Then, call the class back together to have the students discuss and present their work.

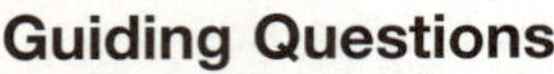

Guiding Questions

- How can you find the total number of ways that you and your friend would be able to sit beside each other if you and 4 of your friends were to sit in a row of 5 people?
- What is a probability?
- How can you calculate a theoretical probability?
- What is a favorable outcome for this situation? What is an unfavorable outcome for this situation?
- In what part of Problem 1 did you write an expression that you were able use to calculate the total number of possible outcomes for the situation of having 5 people sit in a row?

Key Formative Assessments

- What is a tree diagram?
- What is the Fundamental Counting Principle?
- When can you calculate the number of possible outcomes for a situation using the Fundamental Counting Principle?
- What is a permutation?
- What is a factorial?
- How can you evaluate a factorial expression?

Investigate Problem 2

1. In how many ways can you and your friend sit together? Use a complete sentence in your answer.

   Sample Answer: There are 6 + 12 + 12 + 12 + 6 = 48 ways my friend and I can sit next to each other.

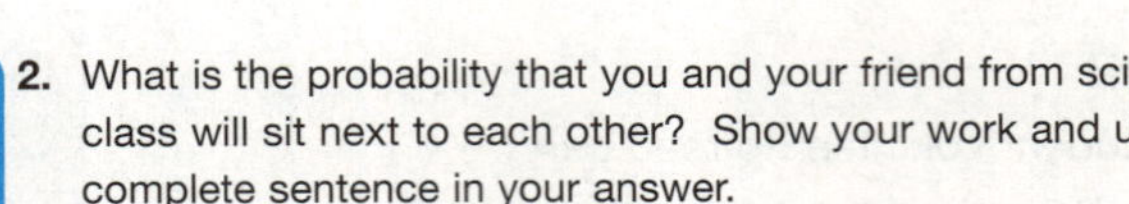

2. What is the probability that you and your friend from science class will sit next to each other? Show your work and use a complete sentence in your answer.

   $$\frac{\text{Number of favorable outcomes}}{\text{Number of possible outcomes}} = \frac{48}{120} = \frac{2}{5};$$

   The probability is $\frac{2}{5}$.

   What is the probability that you and your friend from science class will not sit next to each other? Show your work and use a complete sentence in your answer.

   $1 - \frac{2}{5} = \frac{3}{5}$; **The probability is $\frac{3}{5}$.**

3. What is the probability that you will sit in the last seat and your friend from science class will sit in the fourth seat? Show your work and use a complete sentence in your answer.

   $$\frac{\text{Number of favorable outcomes}}{\text{Number of possible outcomes}} = \frac{6}{120} = \frac{1}{20};$$

   The probability is $\frac{1}{20}$.

4. What is the probability that you will sit in the second seat and your friend from science class will sit in the third seat? Show your work and use a complete sentence in your answer.

   $$\frac{\text{Number of favorable outcomes}}{\text{Number of possible outcomes}} = \frac{6}{120} = \frac{1}{20};$$

   The probability is $\frac{1}{20}$.

11

Wrap Up

Close

- Review all key terms and their definitions. Include the terms *tree diagram, Fundamental Counting Principle, permutation,* and *factorial.*
- You may also want to review any other vocabulary terms that were discussed during the lesson, which may include *factor* and *theoretical probability.*
- Remind the students to write the key terms and their definitions in the notes section of their notebooks. You may also want the students to include examples.
- As a class, draw a complete tree diagram for a situation of interest to the students. For instance, represent the situation for choosing one book (either math, science, history, or German) then choosing a notebook (either red, blue, or orange), then choosing a pencil or a pen.
- Discuss the relationship between a factorial and the number of permutations for n objects.
- Ask the students to summarize how to find the number of permutations of n objects.
- Ask the students to summarize how to evaluate an expression involving a factorial.

Ties to the Cognitive Tutor Software

Dependent probabilities can be difficult for students, because they require students to consider the number of outcomes both before and after an event. The diagramming strategies used in the Cognitive Tutor software and the text can help make this abstract reasoning concrete.

Follow Up

Assignment

Use the Assignment for Lesson 11.5 in the Student Assignments book. See the Teacher's Resources and Assessments book for answers.

Assessment

See the Assessments provided in the Teacher's Resources and Assessments book for Chapter 11.

Open-Ended Writing Task

Ask the students to create a full tree diagram for the possible outcomes for the answers to 3 true or false questions.

Reflections

Insert your reflections on the lesson as it played out in class today.

What went well?

What did not go as well as you would have liked?

How would you like to change the lesson in order to improve the things that did not go well and capitalize on the things that did go well?

Notes

11

11.6

Going Out for Pizza

Permutations and Combinations

Learning By Doing Lesson Map

Get Ready

Objective

In this lesson, you will:

- Find the number of permutations of n objects taken r at a time.
- Find the number of combinations of n objects taken r at a time.

Key Terms

- permutations
- permutation of n distinct objects taken r at a time
- combinations
- combination of n distinct objects taken r at a time

Materials

- The students may need a calculator for the warm up exercises.

NCTM Content Standards

Grades 9–12 Expectations

Number and Operations Standards

- Use number-theory arguments to justify relationships involving whole numbers.
- Judge the reasonableness of numerical computations and their results.

Algebra Standards

- Generalize patterns using explicitly defined and recursively defined functions.
- Use symbolic algebra to represent and explain mathematical relationships.
- Use a variety of symbolic representations, including recursive and parametric equations, for functions and relations.
- Draw reasonable conclusions about a situation being modeled.

Lesson Overview

Within the context of this lesson, students will be asked to:

- Calculate the number of permutations of n objects taken r objects at a time.
- Calculate the number of permutations of n objects taken without concern for order r at a time.
- Determine the relationship between permutations and combinations.

Essential Questions

The following key questions are addressed in this section:

1. What is a permutation?
2. What is a combination?
3. Is a permutation or a combination an ordered arrangement of objects?
4. Is a permutation or a combination a group of objects without concern for order?
5. How can you calculate the number of permutations? How can you calculate the number of combinations?

Show The Way

Warm Up

Place the following questions or an applicable subset of these questions on the board before students enter class. Students should begin working as soon as they are seated.

You and your friends decide to listen to different music CDs in a car on a long trip. How many different orders of the CDs can be played if you don't repeat any of the CDs for each number of CDs?

1. 5 CDs $_5P_5 = 5! = 120$
2. 7 CDs $_7P_7 = 7! = 5040$
3. 10 CDs $_{10}P_{10} = 10! = 3{,}628{,}800$
4. 9 CDs $_9P_9 = 9! = 362{,}880$
5. 6 CDs $_6P_6 = 6! = 720$
6. 4 CDs $_4P_4 = 4! = 24$
7. 8 CDs $_8P_8 = 8! = 40{,}320$
8. 12 CDs $_{12}P_{12} = 12! = 479{,}001{,}600$

If you have one particular order that you would most enjoy to listen to the music, what is the probability that the order you most prefer will be randomly chosen?

9. 5 CDs $\frac{1}{120}$
10. 7 CDs $\frac{1}{5040}$
11. 10 CDs $\frac{1}{3{,}628{,}800}$
12. 9 CDs $\frac{1}{362{,}880}$
13. 6 CDs $\frac{1}{720}$
14. 4 CDs $\frac{1}{24}$
15. 8 CDs $\frac{1}{40{,}320}$
16. 12 CDs $\frac{1}{479{,}001{,}600}$

Motivator

Begin the lesson with the motivator to get students thinking about the topic of the upcoming problem. This lesson is about choosing pizza toppings. The motivating questions are about pizza toppings.

Ask the students the following questions to get them interested in the lesson.

- What toppings do you like on pizza?
- In your opinion, what restaurant has the best pizza?
- What toppings have you seen on pizzas?
- What toppings do you think are most popular for pizza?

Explore Together

Problem 1

Students will calculate the number of ways that four out of six friends can sit together in a row at a movie theater.

Grouping

Ask for a student volunteer to read the Scenario and Problem 1 part (A) aloud. Have a student restate the problem. Pose the Guiding Questions below to verify student understanding. Have students work together in small groups to complete parts (A) through (F) of Problem 1. Then, call the class back together to have the students discuss and present their work.

Guiding Questions

- What information is given in this problem?
- Will all 6 people be able to sit together in the theater?
- How many people will sit together?
- How is this situation different than the situation in the previous lesson?
- Can the Fundamental Counting Principle help you to calculate the number of ways that 4 of the 6 people can sit in a row? How?

Common Student Errors

Students may incorrectly think that they can calculate the number of ways to arrange the 4 out of 6 people by finding ${}_6P_6$ or 6!, neither of which is correct. Other students will incorrectly think that they can simply find 4! or ${}_4P_4$ to find the number of possible outcomes. Be ready to refocus students who make such mistakes. You may want to draw 4 horizontal segments in a row on the front board to represent the 4 possible seats. Doing this will help visual learners to better understand the situation.

SCENARIO The next time you go to the movies, you are part of a group of six people. Because your group arrives late to the movie, you cannot all sit together. At most, you can find four seats together.

Problem 1 Guessing Again

A. In the four seats, how many possible choices are there for the person who sits in the first seat? Use a complete sentence to explain your reasoning.

There are six possible choices because there are six people in your group.

B. Once the first person is seated, how many possible choices are there for the person who sits in the second seat? Use a complete sentence to explain your reasoning.

There are five possible choices because there are five people yet to be seated.

C. Once the first and second persons are seated, how many possible choices are there for the person who sits in the third seat? Use a complete sentence to explain your reasoning.

There are four possible choices because there are four people yet to be seated.

D. Once the first three people are seated, how many possible choices are there for the person who sits in the fourth seat? Use a complete sentence to explain your reasoning.

There are three possible choices because there are three people yet to be seated.

E. How can you determine the number of seating orders that are possible for the four seats? Use a complete sentence to explain your reasoning.

Sample Answer: You can use the Fundamental Counting Principle.

F. Find the number of different seating orders that are possible for the four seats when six people are present. Show your work and use a complete sentence in your answer.

$6 \cdot 5 \cdot 4 \cdot 3 = 360$; There are 360 different seating orders for the four seats.

Explore Together

Investigate Problem 1

Students will investigate the number of ways to choose n objects taken r at a time in order.

Grouping

Ask for a student volunteer to read Question 1 aloud. Have a student restate the problem. Pose the Guiding Questions below to verify student understanding. Have the students complete Questions 1 through 5 in small groups. Then call the class back together to have the students discuss and present.

Guiding Questions

- What is a permutation?
- Is the order of the arrangement important in a permutation?
- Is the order that the friends sit important in this situation for the movie theater?
- How can you calculate the number of permutations for n objects if all n are selected in order?
- How can you simplify a factorial expression like 3!?

Just the Math

The number of possible permutations of n objects taken r at a time in order will be introduced formally in Question 6.

Notes Some students may be confused by the term distinct in Question 6. You may need to explain that the word distinct means different.

Grouping

Ask for a student volunteer to read Question 6 aloud. Have a student restate the problem. Pose the Guiding Questions at the right to verify student understanding. Have the students complete Question 6 individually after the entire class discusses the beginning of Question 6 and the Guiding Questions.

Investigate Problem 1

1. Do you think that the problem situation uses permutations? Use a complete sentence to explain your reasoning.

   Sample Answer: Yes, the problem situation involves an ordering of objects.

2. Is the number of possible seating orders for the four seats equal to 6!? Use a complete sentence to explain your reasoning.

   Sample Answer: No, because only four of the six people are being seated together.

3. What does 6! represent in terms of this problem situation? Use a complete sentence in your answer.

   Sample Answer: It represents the number of possible seating orders if the six people sit together.

4. How many people in the group did not get seated? Use a complete sentence in your answer.

   Two people were not seated.

   How many different seating arrangements do these people represent? Use a complete sentence in your answer.

   These people represent 2!, or 2 seating arrangements.

5. Complete the expression below that gives the number of possible seating orders in four seats when six people are present. Show that this expression will simplify to your answer in part (F).

$$\frac{\boxed{6}!}{\boxed{2}!} = \frac{6 \cdot 5 \cdot 4 \cdot 3 \cdot 2 \cdot 1}{2 \cdot 1} = \frac{\boxed{6 \cdot 5 \cdot 4 \cdot 3}}{\boxed{1}} = \boxed{360}$$

6. **Just the Math: Permutations of n Objects Taken r at a Time** In Problem 1, you found the number of permutations of six people taken four at a time. Because choices were not made for two of the people in the group, the last two numbers in 6!, 2 and 1, were not used. A **permutation of n distinct objects taken r at a time** is a permutation that contains only r objects. The number of permutations of n distinct objects taken r at a time is given by

$$_nP_r = \frac{n!}{(n - r)!}.$$

Guiding Questions

- What is the meaning of $n!$ in this question? What is the meaning of the quantity of $(n - r)!$ in this question?
- Why do we find the ratio of the quantities of $n!$ and of $(n - r)!$ to find the number of permutations?
- In Question 6, when you are asked to calculate $_6P_4$, what is the value of n? What is the value of r? How do these values relate to the original scenario for the movie theater in this lesson?

Explore Together

Investigate Problem 1

Students will evaluate the number of permutations of n objects taken r at a time.

Call the class back together to have the students discuss and present their work for Question 6.

Grouping

Ask for a student volunteer to read Question 7 aloud. Have a student restate the problem. Pose the Guiding Questions below to verify student understanding. Have students work together in small groups to complete Question 7. Then call the class back together to have the students discuss and present their work for Question 7.

Guiding Questions

- What are you asked to do in Question 7?
- What formula are you asked about in the third part of Question 7?
- If you begin with the formula and simplify it to a single factorial expression, will that meet the criteria of what is asked in Question 7?

Note If class time is limited, you can assign Questions 8 and 9 as part of a homework assignment.

Grouping

Have the students work together in small groups to complete Questions 8 and 9. Then call the class back together to have the students discuss and present their work for Questions 8 and 9.

Key Formative Assessments

- How can you calculate the number of permutations if the number of objects chosen in order is equal to the number of objects to be chosen from?
- How can you calculate the number of permutations if the number of objects chosen in order is less than the number of objects to be chosen from?

Investigate Problem 1

Use the formula to find ${}_6P_4$, which represents the number of different seating orders that can occur if four people from a group of six can sit in a row together. Show your work.

$${}_6P_4 = \frac{6!}{(6-4)!} = \frac{6!}{2!} = \frac{6 \cdot 5 \cdot 4 \cdot 3 \cdot 2 \cdot 1}{2 \cdot 1} = 360$$

7. What does the expression ${}_6P_6$ represent? Use a complete sentence in your answer.

   The expression represents the number of permutations of six objects taken six at a time.

   What might this expression represent in the problem situation? Use a complete sentence in your answer.

   Sample Answer: The expression represents the number of different seating orders of six people altogether in a row.

   Can you write ${}_6P_6$ as a single factorial without using the formula? If so, write the expression.

   Yes, ${}_6P_6 = \frac{6!}{0!} = 6!$.

   Consider ${}_6P_6 = \frac{6!}{(6-6)!} = \frac{6!}{0!}$. What must be the value of 0!? Use a complete sentence to explain your reasoning.

   $0! = 1$ because ${}_6P_6 = 6!$.

8. Consider ${}_nP_r$ when r is n and consider ${}_nP_r$ when r is less than n. How are these situations the same? How are they different? Use complete sentences to explain your reasoning.

   Sample Answer: Both situations represent the number of permutations that involve n objects. In the case when $r = n$, the permutations of the entire group of objects are considered. In the case when $r < n$, the permutations of r objects from the group of objects are considered.

9. Find the value of each expression. Show your work.

   ${}_4P_3$

   $${}_4P_3 = \frac{4!}{(4-3)!} = \frac{4!}{1!} = 4! = 24$$

   ${}_6P_2$

   $${}_6P_2 = \frac{6!}{(6-2)!} = \frac{6!}{4!} = \frac{6 \cdot 5 \cdot 4 \cdot 3 \cdot 2 \cdot 1}{4 \cdot 3 \cdot 2 \cdot 1} = 6 \cdot 5 = 30$$

- Is it possible to calculate the number of permutations if the number of objects chosen in order is more than the number of objects to be chosen from? Why or why not?

11

Explore Together

Problem 2

Students will calculate the number of ways to top pizzas when order is not important.

Grouping

Ask for a student volunteer to read the Scenario and part (A) of Problem 2 aloud. Have a student restate the problem. Pose the Guiding Questions below to verify student understanding. Have students work together in small groups to complete parts (A) through (C). Then, call the class back together to have the students discuss and present their work.

Guiding Questions

- What information is given in Problem 2?
- How many toppings are available?
- How many of the toppings are meat toppings?
- How many of the toppings are vegetable toppings?
- Does the order in which the toppings are placed on the pizza matter?
- Is the number of sets of 2 vegetable toppings for a pizza a permutation? Why or why not?

Problem 2 Let's Eat!

After the movie, your group decides to go to a new pizza shop that is becoming known for its tasty pizza crust. The following toppings are available on the menu: peppers, onions, olives, mushrooms, pepperoni, sausage, and ham.

A. Suppose that your group wants to get a pizza with two vegetable toppings. List the possible pizzas that your group could order. Use letters to represent the toppings and identify what each letter stands for. How many pizzas are possible? Use a complete sentence in your answer.

Sample Answer: Use P for peppers, N for onions, L for olives, and M for mushrooms.

PN, PL, PM, NL, NM, LM; There are six possible pizzas.

B. Suppose that your group wants to get a pizza with three vegetable toppings. List the possible pizzas that your group could order. Use letters to represent the toppings and identify what each letter stands for. How many pizzas are possible? Use a complete sentence in your answer.

Sample Answer: Use P for peppers, N for onions, L for olives, and M for mushrooms.

PNL, PNM, PLM, NLM; There are four possible pizzas.

C. Can you count the number of possible pizzas in parts (A) and (B) by using the formula for the permutations of n objects taken r at a time? Why or why not? Use a complete sentence in your answer.

Sample Answer: No, because order is not important.

Investigate Problem 2

Grouping

Ask for a student volunteer to read Question 1 aloud. Have a student restate the problem. Have students work together in small groups to complete Questions 1 and 2.

Investigate Problem 2

1. Suppose that order is important; that is, a pepper and onion pizza is different from an onion and pepper pizza. List the possible two-topping vegetable pizzas that your group could order. Use letters to represent the toppings and identify what each letter stands for.

Sample Answer: Use P for peppers, N for onions, L for olives, and M for mushrooms.

PN, NP, PL, LP, PM, MP, NL, LN, NM, MN, LM, ML

How does this list compare to the list in part (A)? Use a complete sentence in your answer.

Sample Answer: This list includes the list in part (A) and is twice as long.

Common Student Errors

Students may be confused about why they are being asked to find the number of ways that pizzas can be created in which the order matters for toppings. You may need to reassure students that you agree that the order in which toppings are placed on a pizza above the sauce and cheese is not really important. The reason that they are asked to find the number of ordered arrangements of the toppings is so that they can divide the number of permutations for toppings by the number of ways to put the chosen toppings on the pizza in order. They will find this ratio to be called a combination later in the lesson.

Explore Together

Investigate Problem 2

Students will calculate the number of ways to top a pizza if order is important.

Call the class back together to have the students discuss and present their work for Questions 1 and 2.

Key Formative Assessments

- What did you find in Questions 1 and 2?
- Why was the answer for Question 1 twice the answer in part (A)?
- Why was the answer for Question 2 six times the answer in part (B)?

Grouping

Ask for a student volunteer to read Question 3 aloud. Have a student restate the problem. Pose the Guiding Questions below to verify student understanding. Have students work together in small groups to complete Questions 3 and 4. Then call the class back together to have the students discuss and present their work.

Guiding Questions

- What are you asked to do in Question 3?
- How many possible pizzas did you find in part (A)? How many possible pizzas did you find in Question 1?
- How many possible pizzas did you find in part (B)? How many possible pizzas did you find in Question 2?
- What is the relationship between the 2 topping pizza in which order matters and the 2 topping pizza in which order does not matter? What is the relationship between the 3 topping pizza in which order matters and the 3 topping pizza in which order does not matter?

Key Formative Assessments

- What does the ratio of the number of permutations of n objects taken r at a time divided by the number of permutations of n objects taken n at a time represent in this situation?
- Is it a permutation?
- Does the order in which toppings are put on a pizza really matter?

Investigate Problem 2

2. List the possible three-topping vegetable pizzas that your group could order if the order of the toppings is important. Use letters to represent the toppings and identify what each letter stands for.

Sample Answer: Use P for peppers, N for onions, L for olives, and M for mushrooms.

PNL, PLN, NPL, NLP, LPN, LNP, PNM, PMN, MPN, MNP, NPM, NMP, PLM, PML, LMP, LPM, MPL, MLP, NLM, NML, LNM, LMN, MLN, MNL

How does this list compare to the list in part (B)? Use a complete sentence in your answer.

Sample Answer: This list includes the list in part (B) and is six times as long.

3. How could you find the number of possible pizzas in part (A) by using the number of pizzas in Question 1? Use a complete sentence in your answer.

Sample Answer: Divide the number of pizzas in Question 1 by 2 because there are twice as many pizzas in Question 1 as there are in part (A).

How could you find the number of possible pizzas in part (B) by using the number of pizzas in Question 2? Use a complete sentence in your answer.

Sample Answer: Divide the number of pizzas in Question 2 by 6 because there are six times as many pizzas in Question 2 as there are in part (B).

4. Complete the expression below that gives the number of two-topping vegetable pizzas that can be made from a choice of four vegetable toppings. Then show that this expression will simplify to your answer in part (A).

$$\frac{{}_4P_2}{\boxed{2}!} = \frac{4!}{2!2!} = \frac{4 \cdot 3 \cdot 2 \cdot 1}{(2 \cdot 1)(2 \cdot 1)} = 3 \cdot 2 = 6$$

Complete the expression below that gives the number of three-topping vegetable pizzas that can be made from a choice of four vegetable toppings. Then show that this expression will simplify to your answer in part (B).

$$\frac{{}_4P_3}{\boxed{3}!} = \frac{4!}{3!1!} = \frac{4 \cdot 3 \cdot 2 \cdot 1}{(3 \cdot 2 \cdot 1)(1)} = 4$$

11

Explore Together

Investigate Problem 1

Students will develop a formula to calculate the number of combinations of n objects taken r at a time.

Just the Math

Students will be formally introduced to combinations in Question 5.

Grouping

Ask for a student volunteer to read Question 5 aloud. Have a student restate the problem. Pose the Guiding Questions below to verify student understanding. Have students work together in small groups to complete Questions 5 and 6. Then, call the class back together to have the students discuss and present their work for Questions 5 and 6.

Guiding Questions

- What is a combination? How does a combination differ from a permutation?
- What is the formula to calculate the number of possible combinations?
- What does n represent in the formula for combinations?
- What does r represent in the formula for combinations?
- What is the formula for the number of permutations of n objects taken r at a time?
- How is the formula for combinations different from the formula for permutations?
- Why is the formula for the number of combinations divided by $r!$?

Notes It may help the students if you ask them to associate permutations with order and combinations with groups or committees. It is important for the students to understand that both permutations and combinations calculate the number of ways to choose objects when none of the objects can be chosen more than once. In other words, the objects are chosen without replacement.

5. **Just the Math: Combinations of n Objects Taken r at a Time** In Problem 2, you found the number of possible groupings of objects when the order of the objects was *not* important. These groupings that do not need to be ordered are called **combinations.** The number of **combinations of n distinct objects taken r at a time** is the number of groupings of n objects taken r at a time and is given by

$${}_nC_r = \frac{n!}{(n-r)!r!}.$$

Use complete sentences to explain how ${}_nC_r$ is related to ${}_nP_r$.

Sample Answer: The number of combinations of n objects taken r at a time is the quotient of the number of permutations of n objects taken r at a time and $r!$.

6. Your group wants to get a three-topping pizza and will choose from all of the toppings. Find the number of three-topping pizzas that are possible. Show your work and use a complete sentence in your answer.

$${}_7C_3 = \frac{7!}{(7-3)!3!}$$
$$= \frac{7!}{4!3!}$$
$$= \frac{7 \cdot 6 \cdot 5 \cdot 4 \cdot 3 \cdot 2 \cdot 1}{(4 \cdot 3 \cdot 2 \cdot 1)(3 \cdot 2 \cdot 1)}$$
$$= 7 \cdot 5$$
$$= 35$$

There are 35 three-topping pizzas possible.

When the waiter comes to your table, he tells your group that the shop has a special. The special is that the cost of a five-topping pizza is the same as a three-topping pizza. Find the number of five-topping pizzas that are possible. Show your work and use a complete sentence in your answer.

$${}_7C_5 = \frac{7!}{(7-5)!5!}$$
$$= \frac{7!}{2!5!}$$
$$= \frac{7 \cdot 6 \cdot 5 \cdot 4 \cdot 3 \cdot 2 \cdot 1}{(2 \cdot 1)(5 \cdot 4 \cdot 3 \cdot 2 \cdot 1)}$$
$$= 7 \cdot 3$$
$$= 21$$

There are 21 five-topping pizzas possible.

Key Formative Assessments

- What is a permutation?
- What is a combination?
- How are permutations and combinations similar? How are they different?

Wrap Up

Close

- Review all key terms and their definitions. Include the terms *permutations* and *combinations.*
- You may also want to review any other vocabulary terms that were discussed during the lesson, which may include the *Fundamental Counting Principle, factorial, arrangements, orders, groups, sets, distinct, replacement,* and *ratio.*
- Remind the students to write the key terms and their definitions in the notes section of their notebooks. You may also want the students to include examples.
- Discuss the relationship between permutations and combinations.
- Ask the students to explain the formula to find the number of permutations of n objects taken r at a time.
- Ask the students to explain the formula to find the number of combinations of n objects taken r at a time.
- Ask the students to explain which they think would be a larger number, the number of permutations of 10 objects taken out of 15 objects or the number of combinations of 10 objects taken out of 15 objects. Then have the students test their hypothesis by calculating each of the values.
- Have the students respond to the following claim: The formula for the number of permutations of r objects chosen from n objects is ${}_nP_r = \frac{n!}{(n-r)!}$ because the numerator gives the product of n times each counting number less than n through 1 and the denominator stops the product after r factors in the numerator by dividing the remaining factors from the numerator. You decide to test this hypothesis with ${}_{12}P_5$. So, you write the expanded expression for 12! in the numerator and the expanded expression for 7! in the denominator. This gives you $\frac{12 \cdot 11 \cdot 10 \cdot 9 \cdot 8 \cdot 7 \cdot 6 \cdot 5 \cdot 4 \cdot 3 \cdot 2 \cdot 1}{7 \cdot 6 \cdot 5 \cdot 4 \cdot 3 \cdot 2 \cdot 1}$. Then you divide the common factors finding $12 \cdot 11 \cdot 10 \cdot 9 \cdot 8$. How does the product of $12 \cdot 11 \cdot 10 \cdot 9 \cdot 8$ represent the number of ways that you can choose 5 objects in order out of 12 total objects? How many objects can you choose for the first position? How many objects can you choose for the second position? How many objects can you choose for the third position? How many objects can you choose for the fourth position? How many objects can you choose for the fifth position?

Ties to the Cognitive Tutor Software

Dependent probabilities can be difficult for students, because they require students to consider the number of outcomes both before and after an event. The diagramming strategies used in the Cognitive Tutor software and the text can help make this abstract reasoning concrete.

11

Follow Up

Assignment

Use the Assignment for Lesson 11.6 in the Student Assignments book. See the Teacher's Resources and Assessments book for answers.

Assessment

See the Assessments provided in the Teacher's Resources and Assessments book for Chapter 11.

Open-Ended Writing Task

Ask the students to design and conduct an experiment to find the probability that a pizza ordered will be plain with cheese only. Then find the probability that a pizza ordered will have the topping that the students feel is the most common topping ordered. Have the students analyze their sample to determine whether their prediction of the most common topping was accurate.

Reflections

Insert your reflections on the lesson as it played out in class today.

What went well?

____

____

What did not go as well as you would have liked?

____

____

How would you like to change the lesson in order to improve the things that did not go well and capitalize on the things that did go well?

____

____

____

____

____

Notes

11

11.7 Picking Out Socks

Independent and Dependent Events

Learning By Doing Lesson Map

Get Ready

Objective

In this lesson, you will:

- Find the probability of independent events.
- Find the probability of dependent events.
- Use a tree diagram to find probabilities.

Key Terms

- compound events
- independent event
- dependent event
- tree diagram

Materials

- You may want to bring in a spinner from a game to help students understand Investigate Problem 1 Question 1 more easily.

NCTM Content Standards

Grades 9–12 Expectations

Algebra Standards

- Generalize patterns using explicitly defined and recursively defined functions.
- Use symbolic algebra to represent and explain mathematical relationships.
- Draw reasonable conclusions about a situation being modeled.

Data Analysis and Probability Standards

- Use simulations to explore the variability of sample statistics from a known population and to construct sampling distributions.
- Understand how sample statistics reflect the values of population parameters and use sampling distributions as the basis for informal inference.
- Understand the concepts of sample space and probability distribution and construct sample spaces and distributions in simple cases.

Lesson Overview

Within the context of this lesson, students will be asked to:

- Create organized lists and tree diagrams to represent the random selection of socks from a drawer with replacement and without replacement.
- Calculate probabilities of compound events for independent events.
- Calculate probabilities of compound events for dependent events.
- Use a tree diagram and an organized list to more easily calculate probabilities.

Essential Questions

The following key questions are addressed in this section:

1. What is a tree diagram?
2. What is a compound event?
3. What are independent events?
4. What are dependent events?

Show The Way

Warm Up

Place the following questions or an applicable subset of these questions on the board before students enter class. Students should begin working as soon as they are seated.

You use the letters from a word to create a password for your school computer account. You use every letter in the word, and you can put the letters in any order. How many different passwords can be created from each word?

1. MONEY $_5P_5 = 5! = 120$
2. OUTLINE $_7P_7 = 7! = 5040$
3. THE $_3P_3 = 3! = 6$
4. COVER $_5P_5 = 5! = 120$
5. GRAPHS $_6P_6 = 6! = 720$
6. OPEN $_4P_4 = 4! = 24$
7. COUNTERS $_8P_8 = 8! = 40{,}320$
8. SPECIAL $_7P_7 = 7! = 5040$
9. DISK $_4P_4 = 4! = 24$
10. CLOSED $_6P_6 = 6! = 720$

For another password, you decide to use only 5 different letters from the word. How many different 5 letter passwords can be created from each word?

11. MONEY $_5P_5 = 5! = 120$
12. OUTLINE $_7P_5 = \frac{7!}{2!} = \frac{5040}{2} = 2520$
13. GRAPHS $_6P_5 = \frac{6!}{1!} = \frac{720}{1} = 720$
14. COVER $_5P_5 = 5! = 120$
15. COUNTERS $_8P_5 = \frac{8!}{3!} = \frac{40{,}320}{6} = 6720$
16. SPECIAL $_7P_5 = \frac{7!}{2!} = \frac{5040}{2} = 2520$
17. DISK **Zero, it is impossible**
18. CLOSED $_6P_5 = \frac{6!}{1!} = \frac{720}{1} = 720$
19. Why is $_6P_5$ equal to $_6P_6$? Use complete sentences in your answer.

    Because if you choose only 5 of the items, there will be one that remains not chosen and there is only 1 way for the 1 out of 1 item to not be chosen or to be chosen if you choose all 6 items.

Motivator

11

Begin the lesson with the motivator to get students thinking about the topic of the upcoming problem. This lesson is about getting dressed for school. The motivating questions are about socks in a drawer.

Ask the students the following questions to get them interested in the lesson.

- Have you ever had unpaired socks in your sock drawer?
- Does it seem like there are always socks that are the wrong color, but never the color you are looking for?
- Do you lose socks and have a lot of single unpaired socks?
- Where do all of the lost socks go?

Explore Together

Problem 1

Students will investigate sampling with and without replacement.

Grouping

Ask for a student volunteer to read the Scenario and Problem 1 aloud. Have a student restate the problem. Pose the Guiding Questions below to verify student understanding. Have students work together in small groups to complete parts (A) through (C) of Problem 1. Then, call the class back together to have the students discuss and present their work.

Guiding Questions

- What information is given in this problem?
- How many socks are in the drawer?
- How many of the socks are black? How many of the socks are gray?
- What do you have to assume about looking in the drawer and about the location of the socks in the drawer to answer these questions?
- What is an outcome of an experiment?
- What is an event?
- What is the sample space for this experiment?
- What is a probability?
- How can you calculate a probability?

Notes Several assumptions must be made in order to answer these questions. First, you are not looking into the drawer when you chose the socks. Also, if you replace a sock in the drawer, then the socks are mixed well enough again that you have the same chance of selecting any of the socks in the drawer. Finally, you don't have to worry about the socks matching each other. The only concern is of the color of the socks.

SCENARIO It is Monday morning and you are late for school. You did not do any laundry over the weekend, so there are only six socks in your sock drawer. You also do not have the socks combined into pairs. In the drawer there are two black socks and four gray socks.

Problem 1 A Tale of Two Socks

A. You reach into the drawer without looking and pull out a sock. What is the probability that the sock is black?

$\frac{2}{6} = \frac{1}{3}$

What is the probability that the sock is gray?

$\frac{4}{6} = \frac{2}{3}$

B. Suppose that you pull out a gray sock. Then you pull out another sock without putting the gray sock back. What is the probability that the second sock is gray? Is this probability the same as or different from the probability of pulling out a gray sock in part (A)? Use complete sentences to explain your reasoning.

$\frac{3}{5}$; This probability is different from the probability in part (A) because there is one less gray sock and one less sock in the drawer.

C. Suppose that instead, you pull out a black sock. Then you pull out another sock, but this time you put the black sock back into the drawer first. What is the probability that the second sock is black? Is the probability the same as or different from the probability of pulling out a black sock from part (A)? Use complete sentences to explain your reasoning.

$\frac{2}{6} = \frac{1}{3}$; This probability is the same as the probability in part (A) because there are the same number of socks in the drawer.

11

Explore Together

Problem 1

Students will classify compound events independent or dependent events.

Grouping

Ask for a student volunteer to read part (D) aloud. Have a student restate the problem. Complete part (D) together as a class.

Key Formative Assessments

- What is sampling?
- What does it mean to sample with replacement?
- What does it mean to sample without replacement?
- How does calculating probabilities for sampling with replacement compare to sampling without replacement?

Investigate Problem 1

Just the Math

Students will be formally introduced to the terminology for compound, dependent, and independent events.

Notes Question 1 refers to a "spinner." You may want to sketch a spinner on the front board, describe a spinner from a game, or bring in a spinner from an actual game.

To help students understand the term compound event, you may want to compare the concept to compound words in English. A compound word is a combination of two or more words put together to create one word. For instance, yearbook, sidewalk, eyeglass, playground, overpass, etc.

Grouping

Ask for a student volunteer to read Question 1 aloud. Have a student restate the problem. Pose the Guiding Questions on the right to verify student understanding. Have the students work together in small groups to complete Questions 1 through 3.

Problem 1 A Tale of Two Socks

D. In part (B), you "sampled without replacement." In part (C), you "sampled with replacement." How is the way in which you found the probability in part (B) different from the way in which you found the probability in part (C)? Use complete sentences in your answer.

Sample Answer: Unlike the probability in part (C), the number of socks and the number of gray socks in the drawer had to be reduced by one before the probability in part (B) could be calculated.

Investigate Problem 1

1. **Just the Math: Independent and Dependent Events** Drawing one sock and getting a particular sock and then drawing another sock and getting a particular sock out of a drawer are **compound events.** There are two kinds of compound events. **Independent events** are events in which the occurrence of one event does not affect the probability of the other event. **Dependent events** are events in which the occurrence of one event does affect the probability of the other event.

   Determine whether the compound events are independent or dependent. Use a complete sentence in your answer.

   You spin a spinner and land on red and you roll a six-sided number cube and you get a 5.

   The events are independent.

   You spin two different spinners and both spinners land on blue.

   The events are independent.

   You choose one card from a stack of cards that are contained in a board game, then you choose a second card, without replacing the first card and the two cards are the same.

   The events are dependent.

2. When you sample a compound event with replacement, are the events independent or dependent? Use a complete sentence in your answer.

   The events are independent.

   When you sample a compound event without replacement, are the events independent or dependent? Use a complete sentence in your answer.

   The events are dependent.

Guiding Questions

- What are you asked to do in Question 1?
- What is a compound event?
- What are independent events?
- What are dependent events?

Explore Together

Investigate Problem 1

Students will calculate probabilities for compound events with replacement.

Grouping

Students will be working together in small groups to complete Questions 1 through 3.

Call the class back together to have the students discuss and present their work for Questions 1 through 3.

Key Formative Assessments

- When are compound events classified as independent events?
- When are compound events classified as dependent events?

Grouping

Ask for a student volunteer to read Question 4 aloud. Have a student restate the problem. Pose the Guiding Questions below to verify student understanding. Have students work together in small groups to complete Questions 4 through 6. Then, call the class back together to have the students discuss and present their work.

Guiding Questions

- What are you asked to do in Question 4?
- What is an organized list?
- What is a tree diagram?
- How can you create an organized list or a tree diagram?
- How can you use your list or tree diagram to calculate probabilities?
- Are the events in this situation compound events?
- Are they dependent or independent events? Why?

Key Formative Assessments

- Summarize how you can find the probability for independent events.
- Why do some people call this a multiplication rule for probability of independent events?

Investigate Problem 1

3. In part (B) are the events independent or dependent? Use a complete sentence in your answer.

The events are dependent.

In part (C) are the events independent or dependent?

The events are independent.

4. Create an organized list or tree diagram to show the sample space of pulling two socks out of the drawer when you replace the first sock before pulling out the second sock. If you use letters or symbols for each sock, be sure to indicate what each letter or symbol represents.

Sample Answer: Use B_1 and B_2 for the black socks and use G_1, G_2, G_3, and G_4 for the gray socks.

$B_1 B_1$, $B_1 B_2$, $B_1 G_1$, $B_1 G_2$, $B_1 G_3$, $B_1 G_4$,
$B_2 B_1$, $B_2 B_2$, $B_2 G_1$, $B_2 G_2$, $B_2 G_3$, $B_2 G_4$,
$G_1 B_1$, $G_1 B_2$, $G_1 G_1$, $G_1 G_2$, $G_1 G_3$, $G_1 G_4$,
$G_2 B_1$, $G_2 B_2$, $G_2 G_1$, $G_2 G_2$, $G_2 G_3$, $G_2 G_4$,
$G_3 B_1$, $G_3 B_2$, $G_3 G_1$, $G_3 G_2$, $G_3 G_3$, $G_3 G_4$,
$G_4 B_1$, $G_4 B_2$, $G_4 G_1$, $G_4 G_2$, $G_4 G_3$, $G_4 G_4$

5. What is the probability of pulling out two gray socks if you replace the first sock?

$\frac{16}{36} = \frac{4}{9}$

What is the probability of pulling out a gray sock on the first pull?

$\frac{4}{6} = \frac{2}{3}$

What is the probability of pulling out a gray sock on the second pull when you replace the first sock?

$\frac{4}{6} = \frac{2}{3}$

How does the first probability relate to the second probability and the third probability that you wrote above? Use complete sentences in your answer.

Sample Answer: The probability of pulling out two gray socks is equal to the product of the probability of pulling out a gray sock on the first pull and the probability of pulling out a gray sock on the second pull.

6. Use the result of Question 5 to write a rule for finding the probability of two independent events. Use a complete sentence in your answer.

Sample Answer: To find the probability that two independent events will occur, multiply the probability that the first event will occur by the probability that the second event will occur.

11

Explore Together

Investigate Problem 1

Students will calculate probabilities for compound events without replacement.

Grouping

Ask for a student volunteer to read Question 7 aloud. Have a student restate the problem. Pose the Guiding Questions below to verify student understanding. Have the students work together in small groups to complete Question 7 through 9. Then, call the class back together to have the students discuss and present their work.

Guiding Questions

- Will your organized list or tree diagram for this situation be the same as or different from the one that you created in Question 4? Why or why not?
- How will they differ? Which one is larger or has more possible outcomes? Why?
- How can you use this list or tree diagram to calculate probabilities?
- Are the events in this situation compound events? Are they dependent or independent events? Why?

Common Student Errors

Students may incorrectly think that because the only change is not replacing the first sock in this new situation that the list or tree diagram will be the same. They may think that only the probabilities will change. You may need to help them to recognize that there are 6 more possible outcomes for the situation with replacement than for the situation without replacement.

Notes The probability for compound events that are dependent is often called conditional probability. This is because the probability of the second event is based on the condition of the experiment after the first event has already occurred.

Key Formative Assessments

- How were the first and second experiments different? How were they similar?
- How were the probability calculations different? How were they similar?
- What is a general rule for how to find the probability for compound events that are independent?
- What is a general rule for how to find the probability for compound events that are dependent?

11

Investigate Problem 1

7. Create an organized list or tree diagram to show the sample space of pulling two socks out of the drawer when you do not replace the first sock. If you use letters or symbols for each sock, be sure to indicate what each letter or symbol represents.

Sample Answer: Use B_1 and B_2 for the black socks and use G_1, G_2, G_3, and G_4 for the gray socks.

B_1 B_2,	B_1 G_1,	B_1 G_2,	B_1 G_3,	B_1 G_4,
B_2 B_1,	B_2 B_1,	B_2 G_2,	B_2 G_3,	B_2 G_4,
G_1 B_1,	G_1 B_2,	G_1 G_2,	G_1 G_3,	G_1 G_4,
G_2 B_1,	G_2 B_2,	G_2 G_1,	G_2 G_3,	G_2 G_4,
G_3 B_1,	G_3 B_2,	G_3 G_1,	G_3 G_2,	G_3 G_4,
G_4 B_1,	G_4 B_2,	G_4 G_1,	G_4 G_2,	G_4 G_3

8. What is the probability of pulling out two gray socks if you *do not* replace the first sock? Show your work and use a complete answer in your answer.

$\frac{12}{30} = \frac{2}{5}$; **The probability is** $\frac{2}{5}$.

How does this probability of pulling out two gray socks relate to the probability of pulling out a gray sock on the first pull and the probability of pulling out a gray sock on the second pull when you do not replace the first sock? Use complete sentences in your answer.

Sample Answer: The probability of pulling out two gray socks is equal to the product of the probability of pulling out a gray sock on the first pull and the reduced probability of pulling out a gray sock on the second pull.

9. Use the result of Question 8 to write a rule for finding the probability of two dependent events. Use a complete sentence in your answer.

Sample Answer: To find the probability that two dependent events will occur, multiply the probability that the first event will occur by the probability that the second event will occur given that the first event occurred.

Explore Together

Investigate Problem 1

Students will create a tree diagram to help calculate probabilities for compound events without replacement.

Grouping

Ask for a student volunteer to read Question 10 aloud. Have a student restate the problem. Pose the Guiding Questions below to verify student understanding. Have the students work together in small groups to complete Questions 10 through 12. Then, call the class back together to have the students discuss and present their work.

Guiding Questions

- What are the benefits of creating a tree diagram for this situation?
- How many black socks are in the drawer? How many blue socks are in the drawer? How many gray socks are in the drawer?
- How many total socks are in the drawer?
- After you remove the first sock, how many socks will remain in the drawer?
- How can you calculate the probability of each individual sock color?
- Will the probabilities for each color remain the same for both the first and second sock chosen? Why or why not?
- Are the possible events for randomly choosing 2 socks equally likely?
- Do you think it is likely that you will randomly choose 2 black socks? Why or why not?
- What do you think is the most likely event? Why? How can you test your hypothesis?

Key Formative Assessments

- What are compound events?
- What are independent events? What are dependent events?

Investigate Problem 1

10. Suppose that you have six blue socks, two black socks, and four gray socks in your sock drawer. You can model the probabilities of drawing two socks by using a tree diagram. The probabilities are given on the "branches" of the tree. Complete the tree diagram below if you pull the first sock and then pull the second sock without replacing the first sock. The probability of drawing a black sock first is already filled in for you.

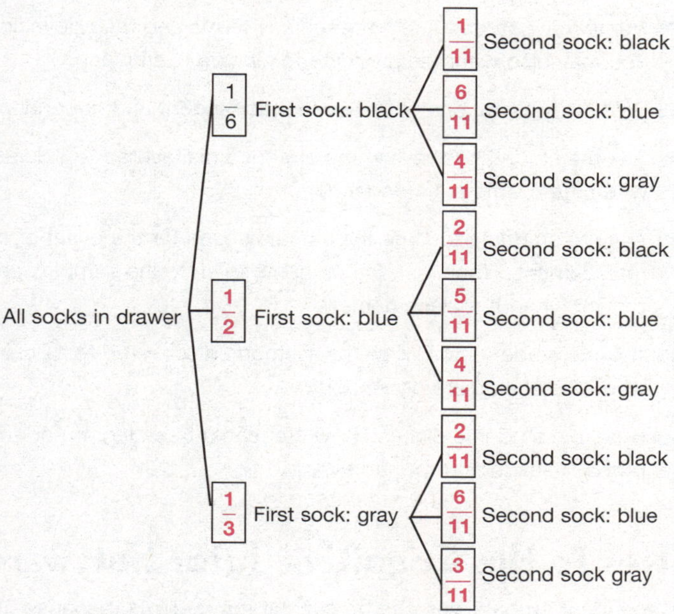

11. What is the probability that you draw two black socks? Use a complete sentence in your answer.

The probability of drawing two black socks is $\frac{1}{66}$.

What is the probability that you draw two blue socks? Use a complete sentence in your answer.

The probability of drawing two blue socks is $\frac{5}{22}$.

12. What is the probability that one sock is gray and one sock is blue? Show your work and use a complete sentence in your answer.

$\frac{1}{2}\left(\frac{4}{11}\right) + \frac{1}{3}\left(\frac{6}{11}\right) = \frac{2}{11} + \frac{2}{11} = \frac{4}{11}$;

The probability that one sock is gray and one sock is blue is $\frac{4}{11}$.

- How can you calculate the probability of compound events that are dependent?
- How can you calculate the probability of compound events that are independent?
- How can a tree diagram help you to calculate probabilities for compound events?

11

Wrap Up

Close

- Review all key terms and their definitions. Include the terms *compound events, independent events, dependent events,* and *tree diagram.*
- You may also want to review any other vocabulary terms that were discussed during the lesson, which may include *sampling, event, outcome, random selection, sample space, product, likely, equally likely, theory,* and *hypothesis.*
- Remind the students to write the key terms and their definitions in the notes section of their notebooks. You may also want the students to include examples.
- Discuss the relationship between independent and dependent events.
- Ask the students to review the methods discovered to calculate the probabilities of compound events when the events are independent.
- Ask the students to state the sample space for the event of choosing 2 socks from a drawer with replacement. Then, ask the students to state the sample space for the event of choosing 2 socks from a drawer without replacement.
- Ask the students to review the methods discovered to calculate the probabilities of compound events when the events are dependent.
- Have the students discuss how creating a tree diagram or an organized list can help them to calculate probabilities for compound events.

Ties to the Cognitive Tutor Software

Dependent probabilities can be difficult for students, because they require students to consider the number of outcomes both before and after an event. The diagramming strategies used in the Cognitive Tutor software and the text can help make this abstract reasoning concrete.

Follow Up

Assignment

Use the Assignment for Lesson 11.7 in the Student Assignments book. See the Teacher's Resources and Assessments book for answers.

11

Assessment

See the Assessments provided in the Teacher's Resources and Assessments book for Chapter 11.

Open-Ended Writing Task

Ask the students to create an experiment for a compound event of at least 3 steps that is of interest to them. It should be able to be modeled using a tree diagram. Then have the students create the tree diagram to represent the event with replacement and another tree diagram to represent the event without replacement. The students should then write a brief paragraph explaining and comparing the probabilities of some of the possible events with and without replacement.

Reflections

Insert your reflections on the lesson as it played out in class today.

What went well?

What did not go as well as you would have liked?

How would you like to change the lesson in order to improve the things that did not go well and capitalize on the things that did go well?

Notes

11

11.8

Probability on the Shuffleboard Court

Geometric Probabilities

Learning By Doing Lesson Map

Get Ready

Objective

In this lesson, you will:

- Find and use geometric probabilities.

Key Terms

- triangle
- parallelogram
- trapezoid
- congruent
- area
- geometric probability

NCTM Content Standards

Grades 9–12 Expectations

Number and Operations Standard

- Judge the reasonableness of numerical computations and their results.

Algebra Standard

- Draw reasonable conclusions about a situation being modeled.

Data Analysis and Probability Standards

- Use simulations to explore the variability of sample statistics from a known population and to construct sampling distributions.
- Understand how sample statistics reflect the values of population parameters and use sampling distributions as the basis for informal inference.
- Understand the concepts of sample space and probability distribution and construct sample spaces and distributions in simple cases.

Lesson Overview

Within the context of this lesson, students will be asked to:

- Compare areas and find ratios of areas.
- Calculate the geometric probabilities for a shuffleboard court.
- Use the concept of congruent shapes and areas to find geometric probabilities.
- Identify the geometric shapes of a triangle, parallelogram, and a trapezoid.

Essential Questions

The following key questions are addressed in this section:

1. What is a geometric probability?
2. What is a triangle? What is a parallelogram? What is a trapezoid?
3. What are congruent figures?
4. What is area?

Show The Way

Warm Up

Place the following questions or an applicable subset of these questions on the board before students enter class. Students should begin working as soon as they are seated.

You are too tired to choose what to eat, so you decide to randomly select food using the following guidelines. You will eat a piece of fruit, a canned vegetable, and a frozen entrée. There are 2 bananas, 3 oranges, and 4 apples in your fruit bowl. There are 7 cans of corn, 4 cans of carrots, 2 cans of beans, and 1 can of spinach in the pantry. In your freezer, there are 4 chicken dinners and 2 pasta dinners.

1. Create a tree diagram or an organized list for the possible meals for choices.

banana, corn, chicken	orange, corn, chicken	apple, corn, chicken
banana, corn, pasta	orange, corn, pasta	apple, corn, pasta
banana, carrots, chicken	orange, carrots, chicken	apple, carrots, chicken
banana, carrots, pasta	orange, carrots, pasta	apple, carrots, pasta
banana, beans, chicken	orange, beans, chicken	apple, beans, chicken
banana, beans, pasta	orange, beans, pasta	apple, beans, pasta
banana, spinach, chicken	orange, spinach, chicken	apple, spinach, chicken
banana, spinach, pasta	orange, spinach, pasta	apple, spinach, pasta

You randomly select a piece of fruit, a canned vegetable, and a frozen entrée. Calculate each probability.

2. Probability your fruit is a banana $\frac{2}{9} \approx 22.2\%$

3. Probability your fruit is an orange $\frac{3}{9} \approx 33.3\%$

4. Probability your fruit is an apple $\frac{4}{9} \approx 44.4\%$

5. Probability your entrée is pasta $\frac{2}{6} \approx 33.3\%$

6. Probability of choosing an apple, a can of corn, and a pasta entrée $\frac{4}{9} \cdot \frac{7}{14} \cdot \frac{2}{6} = \frac{2}{27} \approx 7.4\%$

7. Probability of choosing an orange, a can of spinach, and a chicken entrée $\frac{3}{9} \cdot \frac{1}{14} \cdot \frac{4}{6} = \frac{1}{63} \approx 1.6\%$

8. Probability of choosing an apple, a can of beans, and a chicken entrée $\frac{4}{9} \cdot \frac{2}{14} \cdot \frac{4}{6} = \frac{8}{189} \approx 4.2\%$

Motivator

Begin the lesson with the motivator to get students thinking about the topic of the upcoming problem. This lesson is about determining probability on a shuffleboard court. The motivating questions are about shuffleboard.

Ask the students the following questions to get them interested in the lesson.

- What is shuffleboard?
- Have you ever played or watched someone else play shuffleboard? If so, where?
- How is shuffleboard played? How is shuffleboard scored?

Explore Together

Problem 1

Students will predict geometric probabilities based on the area shaded in a shuffleboard design.

Grouping

Ask for a student volunteer to read the Scenario and Problem 1 aloud. Have a student restate the problem. Pose the Guiding Questions below to verify student understanding. Have students work together in small groups to complete parts (A) through (D). Then, call the class back together to have the students discuss and present their work for parts (A) through (D).

Take Note

A trapezoid is a four-sided geometric figure with exactly two parallel sides. A parallelogram is a four-sided geometric figure with its opposite sides parallel.

Guiding Questions

- What is shuffleboard?
- What is the shape of the entire shuffleboard court design?
- What is the shape of the gray space?
- What is the shape of the black space?
- What is the shape of each of the white spaces?
- How many white triangles are in the design?
- What is an area for a shape?
- Which color appears to cover the least amount of area on the design?
- Which color appears to cover the greatest amount of area on the design?

SCENARIO A local community center often holds shuffleboard tournaments. Shuffleboard is a game in which players use a long stick to push a small disk onto the playing area. Scoring is determined by the part of the playing area on which the disk lands. A diagram of a modified shuffleboard design is shown below.

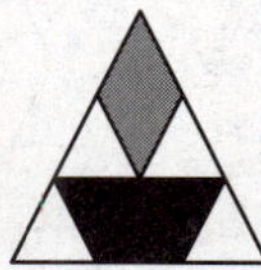

Problem 1 Where Will the Disk Land?

A. If the disk lands on the playing area, describe the areas on which the disk can land. Use a complete sentence in your answer.

Sample Answer: The disk could land on the white area, the gray area, or the black area.

B. Do you think that it is more likely for the disk to land on the gray parallelogram or the black trapezoid? Use a complete sentence to explain your reasoning.

Sample Answer: It is more likely for the disk to land on the trapezoid because the area of the trapezoid is greater than the area of the parallelogram.

C. Do you think that it is more likely for the disk to land on one of the four white triangles or the black trapezoid? Use a complete sentence to explain your reasoning.

Sample Answer: It is more likely for the disk to land on one of the white triangles because the combined area is larger than the area of the trapezoid.

D. Do you think that it is more likely for the disk to land on the triangle composed of the parallelogram and the two top white triangles or the trapezoid composed of the black trapezoid and the lower two white triangles? Use a complete sentence to explain your reasoning.

Sample Answer: It is more likely for the disk to land on the large trapezoid because the trapezoid's area is greater.

11

Notes Some students may have difficulty determining by sight which area is the greatest. The exact areas are not calculated in this lesson because the concept of calculating the area of these shapes has not yet been addressed in this course. If your students are familiar and comfortable with such calculations, you may want to include an initial activity to calculate the area of each shape and of each color for the design. The area of the shuffleboard court is given in Question 2 of this lesson.

Explore Together

Investigate Problem 1

Students will determine the amount of the area shaded with each of the 3 colors in the shuffleboard court.

Grouping

Ask for a student volunteer to read Question 1 aloud. Have a student restate the problem. Pose the Guiding Questions below to verify student understanding. Have the students work together in small groups to complete Question 1.

Guiding Questions

- What are you asked to do in Question 1?
- Do you have to know the area of each individual triangle to be able to answer this question?
- If 2 shapes are congruent, what does that mean?
- How will the areas of 2 congruent shapes compare to each other? Why?
- How many congruent triangles are used to form the entire shuffleboard court?
- How many triangles combine to form the gray parallelogram?
- How many of the triangles combine to form the black trapezoid?
- How many triangles are white?

Common Student Errors

Students may have difficulty understanding the concept of congruent figures. You may need to stop and discuss this concept as a class briefly before the students complete Question 1.

Note Question 1 continues onto the next page.

11

Investigate Problem 1

1. The shuffleboard court in Problem 1 can be divided into nine congruent triangles, as shown below.

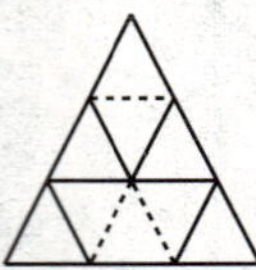

What portion of the shuffleboard is taken up by one white triangle? Use a complete sentence to explain your reasoning.

Sample Answer: Because 1 of the 9 triangles that make up the shuffleboard make up one white triangle, $\frac{1}{9}$ of the shuffleboard is taken up by one white triangle.

What portion of the shuffleboard is taken up by the white triangles? Use a complete sentence to explain your reasoning.

Sample Answer: Since 4 of the 9 triangles that make up the shuffleboard make up the white triangles, $\frac{4}{9}$ of the shuffleboard is taken up by the white triangles.

What portion of the shuffleboard is taken up by the parallelogram? Use a complete sentence to explain your reasoning.

Sample Answer: Since 2 of the 9 triangles that make up the shuffleboard make up the parallelogram, $\frac{2}{9}$ of the shuffleboard is taken up by the parallelogram.

What portion of the shuffleboard is taken up by the trapezoid? Use a complete sentence to explain your reasoning.

Sample Answer: Since 3 of the 9 triangles that make up the shuffleboard make up the trapezoid, $\frac{3}{9}$, or $\frac{1}{3}$, of the shuffleboard is taken up by the trapezoid.

Explore Together

Investigate Problem 1

Students will calculate geometric probabilities based on area.

Grouping

The students will be working together in small groups to complete Question 1.

Call the class back together to have the students discuss and present their work for Question 1.

Grouping

Ask for a student volunteer to read Question 2 aloud. Have a student restate the problem. Pose the Guiding Questions below to verify student understanding. Have the students work together in small groups to complete Question 2. Then, call the class back together to have the students discuss and present their work for Question 2.

Guiding Questions

- What are you asked to find in Question 2?
- What information are you given in Question 2 to help you complete the question?
- How can you find the area of each shape asked?
- Can you think of any other way to find these areas? If so, how? Is either way significantly better than the other way?

Notes The students may complete Question 2 as shown in the sample answers, or they may use the ratios that they found in Question 1 to answer these questions. Both ways are correct. While the students are working in their groups, look to see if different groups use different methods. If so, when the students present their work, you can take the opportunity to have students from different groups present both methods and discuss how they are mathematically equivalent.

Investigate Problem 1

What portion of the shuffleboard is taken up by the parallelogram and the top two triangles? Use a complete sentence to explain your reasoning.

Sample Answer: Because 4 of the 9 triangles that make up the shuffleboard make up the larger triangle, $\frac{4}{9}$ of the shuffleboard is taken up by this triangle.

What portion of the shuffleboard is taken up by the trapezoid and the bottom two triangles? Use a complete sentence to explain your reasoning.

Sample Answer: Because 5 of the 9 triangles that make up the shuffleboard make up the larger trapezoid, $\frac{5}{9}$ of the shuffleboard is taken up by this trapezoid.

Does your answer to Question 1 change your answers to parts (B) through (D)? If so, explain why. Use complete sentences in your answer.

Answers will vary.

2. The area of the shuffleboard court is 6.93 square feet. Find the area of one of the nine congruent triangles. Show your work and use a complete sentence in your answer.

Sample Answer: $\frac{1}{9}(6.93) = 0.77$

The area of one of the triangles is 0.77 square feet.

Find the area of the gray parallelogram. Show your work and use a complete sentence in your answer.

Sample Answer: 2(0.77) = 1.54;

The area of the parallelogram is 1.54 square feet.

Find the area of the black trapezoid. Show your work and use a complete sentence in your answer.

Sample Answer: 3(0.77) = 2.31;

The area of the trapezoid is 2.31 square feet.

Key Formative Assessments

- What were the areas of the shapes of each color?
- Were your original thoughts in parts (A) through (D) about which shapes were largest correct or do you have to revise them?

Explore Together

Investigate Problem 1

Students will calculate geometric probabilities.

Just the Math

Students will be formally introduced to the concept of geometric probability in Question 3.

Grouping

Ask for a student volunteer to read Question 3 aloud. Have a student restate the problem. Pose the Guiding Questions below to verify student understanding. Have the students work together in small groups to complete Question 3.

Guiding Questions

- What is a geometric probability?
- What is a ratio?
- Will you need to find the ratio of lengths, areas, or volumes for these questions?
- What areas do you expect to need to compare to calculate the probabilities?
- How can you calculate the ratio of these areas?
- What is the total area of the court?
- What does it mean for events to be more likely? What does it mean for events to be less likely?
- Is this type of probability a theoretical or an experimental probability? Why?

Common Student Errors

Students may have difficulty determining which area values they need to compare to calculate the probabilities. You may need to refocus students who are confused with this concept.

Key Formative Assessments

- What is a geometric probability?
- How can you calculate a geometric probability?
- What benefits exist for being able to calculate geometric probabilities?

Investigate Problem 1

Find the total area of the four white triangles. Show your work and use a complete sentence in your answer.

Sample Answer: 4(0.77) = 3.08;

The total area of the four triangles is 3.08 square feet.

3. **Just the Math: Geometric Probability** You can find the probability of landing on one of the shapes on the shuffleboard court by using a *geometric probability*. A geometric probability is a ratio of lengths, areas, or volumes. To find the **geometric probability** of landing on one of the shapes on the shuffleboard court, find the ratio of the area of the shape to the total area of the court.

   Find the geometric probability of landing on the gray area. Write your answer as a fraction in simplest form. Show your work and use a complete sentence in your answer.

   $\frac{1.54 \text{ square feet}}{6.93 \text{ square feet}} = \frac{2}{9}$;

   The geometric probability of landing on the gray area is $\frac{2}{9}$.

   Find the geometric probability of landing on any white area. Write your answer as a fraction in simplest form. Show your work and use a complete sentence in your answer.

   $\frac{3.08 \text{ square feet}}{6.93 \text{ square feet}} = \frac{4}{9}$;

   The geometric probability of landing on the white area is $\frac{4}{9}$.

   Use the probabilities above to find the probability of landing on the black area. Write your answer as a fraction in simplest form. Show your work and use a complete sentence in your answer.

   $1 - \frac{2}{9} - \frac{4}{9} = \frac{9}{9} - \frac{6}{9} = \frac{3}{9} = \frac{1}{3}$;

   The geometric probability of landing on the black area is $\frac{1}{3}$.

   On which area are you most likely to land? On which area are you least likely to land? Use complete sentences to explain your reasoning.

   Sample Answer: You are most likely to land on the white area because its area is greater than the black or gray areas. You are least likely to land on the gray area because its area is less than the white or black areas.

Wrap Up

Close

- Review all key terms and their definitions. Include the terms *triangle, parallelogram, trapezoid, congruent, area,* and *geometric probability.*
- You may also want to review any other vocabulary terms that were discussed during the lesson, which may include *shuffleboard court, design, modified, disk, congruent figures, shapes, theoretical probability,* and *experimental probability.*
- Remind the students to write the key terms and their definitions in the notes section of their notebooks. You may also want the students to include examples.
- Discuss the relationship between theoretical and geometric probabilities.
- Ask the students to summarize the method used to calculate geometric probabilities.
- Have the students explain the meaning of congruent figures.
- Have the students describe and draw the characteristics of a triangle, a parallelogram, and a trapezoid. Then have the students compare the similarities in the shapes and contrast the differences.

Follow Up

Assignment

Use the Assignment for Lesson 11.8 in the Student Assignments book. See the Teacher's Resources and Assessments book for answers.

Assessment

See the Assessments provided in the Teacher's Resources and Assessments book for Chapter 11.

Open-Ended Writing Task

Ask the students to write an exit pass summarizing geometric probability and how to calculate geometric probabilities. Have the students include any questions that they have about the concept that they would like to have answered at the beginning of the next class period. Collect the exit slips as the students leave and answer any questions at the beginning of the next class session.

Reflections

Insert your reflections on the lesson as it played out in class today.

What went well?

What did not go as well as you would have liked?

How would you like to change the lesson in order to improve the things that did not go well and capitalize on the things that did go well?

Notes

11

11.9 Game Design

Geometric Probabilities and Fair Games

Learning By Doing Lesson Map

Get Ready

Objective

In this lesson, you will:

- Use geometric probabilities to find values in a game.
- Determine whether a game is fair.
- Alter the rules of a game so that it is fair.

Key Terms

- geometric probability
- fair game

Materials

- You may want to create a spinner to simulate the spinner in Problem 1.
- You will need a pair of number cubes for each group of 2 students in the class.

NCTM Content Standards

Grades 9–12 Expectations

Number and Operations Standard

- Judge the reasonableness of numerical computations and their results.

Algebra Standards

- Draw reasonable conclusions about a situation being modeled.
- Use simulations to explore the variability of sample statistics from a known population and to construct sampling distributions.
- Understand how sample statistics reflect the values of population parameters and use sampling distributions as the basis for informal inference.

Data Analysis and Probability Standard

- Understand the concepts of a sample space and probability distribution and construct sample spaces and distributions in simple cases.

Lesson Overview

Within the context of this lesson, students will be asked to:

- Calculate geometric probabilities.
- Calculate the expected values for the outcomes of a game of chance.
- Compare probabilities to determine whether the possible outcomes are equally likely and whether the game is fair.
- Alter the rules of a game that are unfair to develop a related game that is fair.

Essential Questions

The following key questions are addressed in this section:

1. What is a geometric probability?
2. What is a fair game?
3. When is a game determined to be fair?

Show The Way

Warm Up

Place the following questions or an applicable subset of these questions on the board before students enter class. Students should begin working as soon as they are seated.

Simplify each fractional expression.

1. $\frac{2}{3} + \frac{8}{9}$ $\frac{14}{9}$

2. $\frac{1}{4} + \frac{2}{16} + \frac{4}{8}$ $\frac{7}{8}$

3. $\frac{1}{2} + \frac{1}{3} + \frac{1}{6}$ 1

4. $\frac{19}{30} - \frac{12}{27}$ $\frac{17}{90}$

5. $\frac{15}{40} - \frac{3}{5}$ $\frac{-9}{40}$

6. $\frac{3}{12} - \frac{1}{8}$ $\frac{1}{8}$

7. $\frac{3}{7} \cdot \frac{14}{12}$ $\frac{1}{2}$

8. $\frac{5}{6} \cdot \frac{22}{15}$ $\frac{11}{9}$

9. $\frac{7}{11} \cdot \frac{33}{47}$ $\frac{21}{47}$

10. $\frac{7}{2} \div \frac{1}{12}$ 42

11. $\frac{5}{8} \div \frac{3}{24}$ 5

12. $\frac{13}{27} \div \frac{6}{9}$ $\frac{13}{18}$

Motivator

Begin the lesson with the motivator to get students thinking about the topic of the upcoming problem. This lesson is about game designs. The motivating questions are about games.

Ask the students the following questions to get them interested in the lesson.

- Do you enjoy playing games of chance? Do you enjoy playing games of strategy?
- Have you ever developed a game yourself?
- What are your favorite games to play?
- Do you enjoy playing video or arcade games? Do any of those games seem to be unfair?

11

Explore Together

Problem 1

Students will investigate a game of chance that uses a spinner divided into different size sections with different point values.

Grouping

Ask for a student volunteer to read the Scenario and Problem 1 aloud. Have a student restate the problem. Pose the Guiding Questions below to verify student understanding. Have students work together in small groups to complete parts (A) and (B).

Guiding Questions

- What information is given in this problem?
- What is a spinner?
- Is this game based on luck (called a game of chance) or is this game based on strategy (called a game of strategy)?
- What is the sum of the 3 areas in the spinner?
- Why must the sum of the areas be equal to one?
- What are the possible point values for one spin in this game?
- Why are the point values different for the 3 different areas?
- What type of area is assigned the smallest point value? Why?
- What type of area is assigned the largest point value? Why?
- Does this game seem to be a fair game? Why or why not?

Notes You may want to create a spinner to model this situation. Spinners that are transparent are available at little cost to use on an overhead projector, interactive spinners are available for interactive white-boards, and spinners are included in many children's games of chance.

SCENARIO Every Saturday night, you and your friends get together to make up games to play, and then you play the games. The first game this Saturday night involves spinning a spinner.

Problem 1 Spinning a Spinner

You and your friends create the spinner shown below. You divide the spinner below so that $\frac{1}{2}$ of the spinner is worth 15 points, $\frac{1}{3}$ of the spinner is worth 25 points, and $\frac{1}{6}$ of the spinner is worth 40 points.

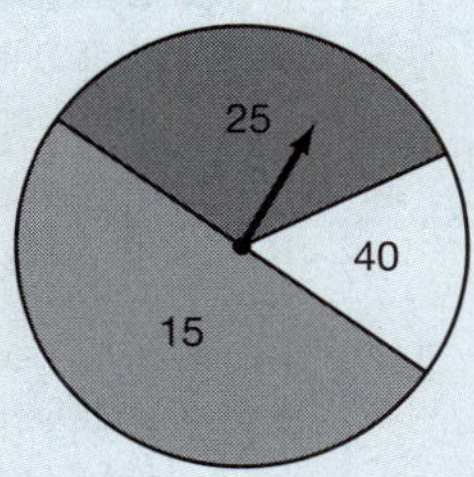

A. If you spin the spinner eight times, what is the greatest number of points that you could earn? Explain your reasoning. Use complete sentences in your answer.

Sample Answer: Because the greatest number of points on the spinner is 40 points, you could earn at most 8 • 40 = 320 points.

Do you think that it is likely that you could earn this number of points from eight spins? Why or why not? Use complete sentences in your answer.

Sample Answer: It is not likely because the part of the spinner that is worth 40 points has the smallest of the three areas.

B. If you spin the spinner eight times, what is the least number of points you could earn? Explain your reasoning. Use complete sentences in your answer.

Sample Answer: Because the least number of points on the spinner is 15 points, you could earn at least 8 • 15 = 120 points.

Explore Together

Problem 1

Students will calculate the probability of each possible spin outcome using geometric probabilities.

Grouping

Students will be working together in small groups to complete parts (A) and (B).

Call the class back together to have the students discuss and present their work for parts (A) and (B).

Investigate Problem 1

Grouping

Ask for a student volunteer to read Question 1 aloud. Have a student restate the problem. Pose the Guiding Questions below to verify student understanding. Have the students complete Question 1 individually. Then, call the class together to have the students discuss and present their work for Question 1.

Guiding Questions

- What are you asked to do in Question 1?
- How can you calculate the probabilities?
- How do the probabilities relate to the areas for the regions on the spinner?
- What do you expect the sum of the probabilities of all the possible outcomes to be?
- How can you check to see if your probabilities are logical using the sum of the probabilities? Why?

Common Student Errors

Students may have difficulty working with these probabilities because they involve fractional values. The Warm Up activity for this section requires the students to add and subtract fractional values. If you expect your students to have difficulty with such operations, be sure to use the Warm Up activity for this lesson.

11

Problem 1 Spinning a Spinner

Do you think that it is likely that you could earn this number of points from eight spins? Why or why not? Use complete sentences in your answer.

Sample Answer: It is likely because the part of the spinner that is worth 15 points has the largest of the three areas.

Investigate Problem 1

1. What is the probability of landing on the part worth 15 points? Use a complete sentence to explain your reasoning.

   The probability of landing on the part worth 15 points is $\frac{1}{2}$ because this part takes up $\frac{1}{2}$ of the spinner's area.

   What is the probability of landing on the part worth 25 points? Use a complete sentence to explain your reasoning.

   The probability of landing on the part worth 25 points is $\frac{1}{3}$ because this part takes up $\frac{1}{3}$ of the spinner's area.

   What is the probability of landing on the part worth 40 points? Use a complete sentence to explain your reasoning.

   The probability of landing on the part worth 40 points is $\frac{1}{6}$ because this part takes up $\frac{1}{6}$ of the spinner's area.

2. Suppose that you spin the spinner 18 times. How many times can you expect to land on the part of the spinner that is worth 15 points? Show your work and use a complete sentence in your answer.

   $18\left(\frac{1}{2}\right) = 9;$

   The spinner should land on the part of the spinner that is worth 15 points nine times.

   How many points would you earn in the number of spins described above? Show your work and use a complete sentence in your answer.

   $9(15) = 135;$

   You would earn one hundred thirty-five points.

Explore Together

Investigate Problem 1

Students will calculate average or expected values for given situations.

Grouping

Ask for a student volunteer to read Question 2 aloud. Have a student restate the problem. Pose the Guiding Questions below to verify student understanding. Have students work together in small groups to complete Questions 2 and 3.

Guiding Questions

- What amount of the spinner counts for 15 points?
- What is the probability that the spinner will land on the 15-point section for one spin?
- What portion of the total number of spins would you predict that the spinner will land in the 15-point section? Why?
- How can you calculate the number of spins out of 18 spins that you should expect to earn 15 points?
- What portion of the spins do you predict that the spinner will land on the 25-point section?
- How can you calculate the number of spins out of 18 spins that you should expect to earn 25 points?

Notes It is important in this situation to assume that the spinner is moving freely, that it is not leaning toward any number area, etc. You may want to talk to the students about the assumptions required for this problem.

Key Formative Assessments

- How did you calculate the probabilities for the spinner outcomes?
- Were the outcomes equally likely? How do you know?
- How did you decide whether you could make each outcome equally likely?
- What is a geometric probability?

Investigate Problem 1

In 18 spins, how many times can you expect to land on the part of the spinner that is worth 25 points? Show your work and use a complete sentence in your answer.

$18\left(\frac{1}{3}\right) = 6$;

The spinner should land on the part of the spinner that is worth 25 points six times.

How many points would you earn in the number of spins described above? Show your work and use a complete sentence in your answer.

$6(25) = 150$;

You would earn one hundred fifty points.

In 18 spins, how many times can you expect to land on the part of the spinner that is worth 40 points? Show your work and use a complete sentence in your answer.

$18\left(\frac{1}{6}\right) = 3$;

The spinner should land on the part of the spinner that is worth 40 points three times.

How many points would you earn in the number of spins described above? Show your work and use a complete sentence in your answer.

$3(40) = 120$;

You would earn one hundred twenty points.

Use complete sentences to explain how you found the total number of points that were likely to be earned from landing on each part of the spinner in 18 spins.

Sample Answer: Multiply the probability of landing on the specific part of the spinner by the number of spins. Then, multiply the result by the point value of the part.

How many total points are you likely to earn from 18 spins? Show your work and use a complete sentence in your answer.

$135 + 150 + 120 = 405$; You are likely to earn 405 points.

3. How would you make it equally likely to land on each part of the spinner? Use a complete sentence in your answer.

Divide the spinner into thirds.

- If the outcomes are equally likely, should you change the point values on the spinner? Why or why not?
- If you make all the point values on the spinner the same, how will that affect the game?

11

Explore Together

Problem 2

Grouping

Ask for a student volunteer to read Problem 2 and part (A) aloud. Have a student restate the problem. Pose the Guiding Questions below to verify student understanding. Have the students complete part (A) individually. Then, call the class together to discuss and present their work.

Guiding Questions

- What are the possible sums when two number cubes are rolled?
- What sum do you think is most likely to occur when a pair of number cubes are rolled and the sum is found? How can you check your hypothesis?
- The "roller" earns 3 times as many points when a 7 is rolled than as the "recorder" earns for any other sum. So, what has to be true about the probability of the "roller" rolling a sum of 7 compared to any other sum for the game to end in a tie?

Notes You will need a pair of number cubes for each pair of students. One student will roll the number cubes and the other will record the points for the rolls as well as keep track of the score for each player.

Grouping

Ask for a student volunteer to read part (B) aloud. Have a student restate the problem. Pose the Guiding Questions to verify student understanding. Have students work together in pairs of 2 students to complete parts (B) and (C). Then, call the class back together to have the students discuss and present their work. Complete part (D) and Question 1 together as a class.

Guiding Questions

- What materials will you need?
- How will you play the game? What are the rules of the game?

Problem 2 Rolling the Number Cubes

The second game this particular Saturday night involves two six-sided number cubes and is a two-player game. In the first round, one player is the "roller" and the other player is the "recorder." In successive rounds, the players switch roles. At the beginning of the game, each player starts with 12 points. If the sum of the numbers on a roller's roll is 7, then the recorder must give the roller 3 of his or her points. If the sum is not 7, then the roller must give the recorder 1 of his or her points. The winner of the round is the player who has the most points after 12 rolls. If one of the players runs out of points before 12 rolls, then the other player wins.

A. Who do you think is more likely to win a round of the game, the roller or the recorder? Use complete sentences to explain your reasoning.

Answers will vary.

B. Play two rounds of the game with a person in your class. Record whether or not the sum of the roll is 7 and record the number of points each player has after each roll in the table below.

			Roll 1	Roll 2	Roll 3	Roll 4	Roll 5	Roll 6	Roll 7	Roll 8	Roll 9	Roll 10	Roll 11	Roll 12
Round 1	Sum of 7?		N	Y	Y	N	N	N	N	N	N	N	N	Y
	Roller	12	11	14	17	16	15	14	13	12	11	10	9	12
	Recorder	12	13	10	7	8	9	10	11	12	13	14	15	12
Round 2	Sum of 7?		N	N	N	N	N	N	N	N	Y	N	N	N
	Roller	12	11	10	9	8	7	6	5	4	7	6	5	4
	Recorder	12	13	14	15	16	17	18	19	20	17	18	19	20

C. Identify the winner of each game, the roller or the recorder. Use complete sentences in your answer.

Sample Answer: The first game resulted in a tie. The recorder won the second game.

D. How many total games were played in your class? How many rounds were won by the roller? How many rounds were won by the recorder?

Answers will vary.

- How will you determine who is the "roller" for the first round and who is the "roller" for the second round?

Notes Throughout this lesson, sample experimental results are given for the purpose of showing realistic student answers. The student answers will vary from these answers and from each other's answers.

11

Explore Together

Investigate Problem 2

Students will create a table of possible outcomes for the experiment of rolling 2 number cubes. They will then calculate expected values based on the theoretical geometric probabilities.

Notes The table in Question 2 is identical to the table completed in Lesson 11.4 in Investigate Problem 2 Question 3. You may prefer to have the students complete this table as part of a homework assignment or copy it directly from Lesson 11.4 and their previous work.

Grouping

Ask for a student volunteer to read Question 2 aloud. Have a student restate the problem. Pose the Guiding Questions below to verify student understanding. Have the students work in pairs to complete Question 2.

Guiding Questions

- What are you asked to do in Question 2?
- How can you complete this table?
- Have you ever completed a similar table? When?
- How many possible roll combinations exist?
- Will each roll combination result in a different sum?
- Which sum(s) do you expect to occur most frequently?
- Which sum(s) do you expect to occur least frequently?

Common Student Errors

Students may confuse the concept of each of the possible rolls being equally likely with the concept of the probability of each sum. Be ready to refocus any students who blend the fact that there are 36 equally likely roll combinations, but the possible sums are not equally likely.

Investigate Problem 2

1. Do you think that the roller and the recorder have an equal chance of winning the game in Problem 2? Use complete sentences to explain your reasoning.

   Sample Answer: The recorder has a better chance of winning because it is more likely that a sum other than 7 will be rolled.

2. Complete the table below that shows all of the possible rolls of pairs of six-sided number cubes and the sums of the results.

		First number cube					
		1	2	3	4	5	6
Second number cube	1	2	3	4	5	6	7
	2	3	4	5	6	7	8
	3	4	5	6	7	8	9
	4	5	6	7	8	9	10
	5	6	7	8	9	10	11
	6	7	8	9	10	11	12

What is the probability of rolling the number cubes and the sum is 7? What is the probability of rolling the number cubes and the sum is not 7? Show your work.

Probability that roll results in sum of 7: $\frac{6}{36} = \frac{1}{6}$;

Probability that roll does not result in sum of 7: $1 - \frac{1}{6} = \frac{5}{6}$

In 12 rolls, how many times can you expect to roll numbers that result in a sum of 7? Use a complete sentence in your answer.

You can expect to roll numbers that result in a sum of 7 two times.

11

Explore Together

Investigate Problem 2

Students will classify a game as a fair game or as an unfair game.

Grouping

The students will be working in pairs to complete Question 2.

Call the class back together to have the students discuss and present their work for Question 2.

Just the Math

The concept of a fair game is introduced formally in Question 3.

Grouping

Ask for a student volunteer to read Question 3 aloud. Have a student restate the problem. Pose the Guiding Questions below to verify student understanding. Have students work together in small groups to complete Question 3. Then, call the class back together to have the students discuss and present their work.

Note A sample answer and method is given in Question 3. Students could also correctly alter the rules of the game so that different sums result in different numbers of points being given to the players. Each solution that yields a game that is considered to be fair based on each player having an equally likely chance of winning the game is correct.

Guiding Questions

- What is a fair game?
- Can the probabilities for the outcomes be different for the "roller" and for the "recorder" and still have the game be considered fair? How?
- Does having the "roller" earn 3 points for a favorable outcome compensate for the "recorder" only earning 1 point for a favorable outcome, but having that outcome more likely than the "roller" getting a sum of 7?

11

Investigate Problem 2

In 12 rolls, how many times can you expect to roll numbers that *do not* result in a sum of 7? Use a complete sentence in your answer.

You can expect to roll numbers that do not result in a sum of seven 10 times.

Suppose that the game plays out as you expect during a round. How many points does each player gain or lose? Use complete sentences in your answer.

The recorder gives 6 points to the roller and the roller gives 10 points to the recorder.

3. **Just the Math: Fair Game** A game is considered to be a **fair game** if each player has an equally likely chance of winning. Is the game in Problem 2 fair? Use complete sentences to explain your reasoning.

   Sample Answer: No, because there is a much greater probability that the result of a roll will not be a sum of 7. This results in the recorder earning more points than the roller.

   How would you make the game fair? Explain your answer. Use complete sentences in your answer.

   Sample Answer: If a roll results in a sum of 7, then the roller gets 5 points. If a roll results in a sum other than 7, then the recorder gets 1 point. The probability that the result is a sum of 7 is $\frac{1}{6}$. Then in a round of 12 rolls, it is likely that a sum of 7 will occur twice, resulting in the roller getting 2(5 points) = 10 points. The probability that the result of the roll is not a sum of 7 is $\frac{5}{6}$. Then in 12 rolls, it is likely that a sum that is not 7 will occur 10 times, resulting in the recorder getting 10(1 point) = 10 points.

Key Formative Assessments

- What is a geometric probability?
- What is a fair game?
- How can you determine whether a game is a fair game?

Wrap Up

Close

- Review all key terms and their definitions. Include the terms *theoretical probability, experiment, trial, success,* and *experimental probability.*
- You may also want to review any other vocabulary terms that were discussed during the lesson, which may include *likely, equally likely, games of chance, games of strategy, mean, average values, expected values, fractions, sum,* and *frequency.*
- Remind the students to write the key terms and their definitions in the notes section of their notebooks. You may also want the students to include examples.
- Ask the students to review the process to calculate geometric probabilities.
- Have the students discuss how a game can be classified as a fair game or an unfair game.
- Have the students explain how you can alter a game in order to have the game be fair.

Follow Up

Assignment

Use the Assignment for Lesson 11.9 in the Student Assignments book. See the Teacher's Resources and Assessments book for answers.

Assessment

See the Assessments provided in the Teacher's Resources and Assessments book for Chapter 11.

Open-Ended Writing Task

Ask the students to design a game of chance that is not fair. Then have the students explain how they could alter their game to make it a fair game. You may want to have the students create actual fair games as a project and have the opportunity to spend a class period playing the games that they created.

Reflections

Insert your reflections on the lesson as it played out in class today.

What went well?

____

____

What did not go as well as you would have liked?

____

____

How would you like to change the lesson in order to improve the things that did not go well and capitalize on the things that did go well?

____

____

____

____

____

Notes

11

Looking Ahead to Chapter 12

Focus In Chapter 12, you will learn how to analyze data sets including calculating and comparing different measures of central tendency as well as the sample variance and standard deviation of a data set. You will also learn how to display data using a variety of graphs.

Chapter Warm-up

Answer these questions to help you review skills that you will need in Chapter 12.

Simplify each expression.

1. $\frac{12 + 18}{6}$

5

2. $\frac{4(25 - 3 \cdot 7)}{5}$

3.2

3. $\frac{13 - 25 + 8 \cdot 9}{7 - 4}$

20

Solve each inequality. Then graph the solution on the number line.

4. $\frac{x}{6} + 3 > -1$

5. $2x - 5 < 3$

6. $-\frac{2}{9}x \geq 4$

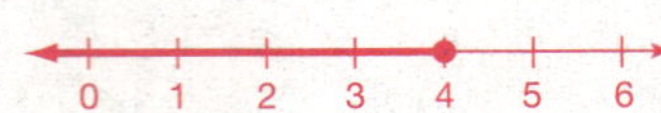

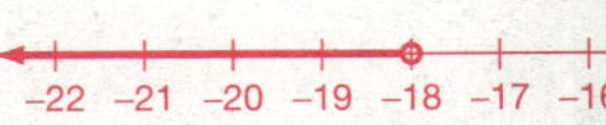

Read the problem scenario below.

You arrived 10 minutes late to a basketball game between your favorite teams, the Tigers and the Lions. When you arrived, you were shocked to see that the Tigers had a score of 20 and the Lions had a score of 3. Throughout the rest of the game, the Tigers averaged 3 points per minute and the Lions averaged 4 points per minute.

7. Write an equation for each team that models their score y as a function of the number of minutes you have been at the game x.

Tigers: $y = 3x + 20$ Lions: $y = 4x + 3$

8. Which team will be winning 15 minutes after you arrive?

Fifteen minutes after you arrive, the Tigers have a score of 65 and the Lions have a score of 63, so the Tigers are winning.

Key Terms

stem-and-leaf plot ■ p. 547
distribution ■ p. 548
mean ■ p. 549
measure of central tendency ■ p. 549
median ■ p. 549
mode ■ p. 550
bimodal ■ p. 550
trimodal ■ p. 550
survey ■ p. 553
sample size ■ p. 556
range ■ p. 557
upper extreme ■ p. 557
lower extreme ■ p. 557
quartile ■ p. 558
box-and-whisker plot ■ p. 559
outlier ■ p. 560
percentile ■ p. 561
interquartile range ■ p. 561
line plot ■ p. 563
deviation ■ p. 564
absolute deviation ■ p. 565
mean absolute deviation ■ p. 565
sample variance ■ p. 567
sample standard deviation ■ p. 567

CHAPTER

12

Statistical Analysis

The game of basketball was created by James Naismith in 1891 in Massachusetts. Basketball was originally designed as an indoor game for college students to play during the winter. Naismith created 13 rules and used a soccer ball and peach baskets for the first game. In Lesson 12.4, you will compare the heights of the players on the home and visiting basketball teams.

12.1 Taking the PSAT

Measures of Central Tendency

Learning By Doing Lesson Map

Get Ready

Objectives

In this lesson, you will:

- Create a stem-and-leaf plot.
- Determine the distribution of a data set.
- Find the mean, median, and mode of a data set.
- Compare the mean and median for different distributions.

Key Terms

- stem-and-leaf plot
- distribution
- measure of central tendency
- median
- mode
- mean

NCTM Content Standards

Grades 9–12 Expectations

Number and Operations Standard

- Judge the reasonableness of numerical computations and their results.

Algebra Standards

- Judge the meaning, utility, and reasonableness of the results of symbol manipulation including those carried out by technology.
- Identify essential quantitative relationships in a situation and determine the class or classes of functions that might model the relationships.
- Draw reasonable conclusions about a situation being modeled.

Data Analysis and Probability Standards

- Understand histograms, parallel box plots, and scatter plots. Use them to display data.
- For univariate measurement data, be able to display the distribution, describe its shape, and select and calculate summary statistics.

Lesson Overview

Within the context of this lesson, students will be asked to:

- Create a stem-and-leaf plot.
- Determine the shape of the distribution of data.
- Calculate the mean, median, and mode for a data set and recognize these as measures of central tendency.
- Analyze and compare measures of central tendency for various distributions.

Essential Questions

The following key questions are addressed in this lesson:

1. What is a measure of central tendency?
2. What is a distribution?
3. How can you calculate the mean, median, and mode for a set of data?
4. How can you identify the shape of the distribution for a set of data?
5. How can you create a stem-and-leaf plot for a set of data?

12

Show The Way

Warm Up

Place the following questions or an applicable subset of these questions on the board before students enter class. Students should begin working as soon as they are seated.

Complete each statement.

1. A scatter plot or function is said to have a **positive** correlation if as the values of *x* increase, the values of *y* also increase.

2. A scatter plot or function is said to have a **negative** correlation if as the values of *x* increase, the values of *y* decrease.

3. A linear regression equation or line of best fit is the line that **most closely fits the data that is graphed in the scatter plot**.

4. A linear regression equation or line of best fit has a correlation coefficient close to **1** if the line has a positive slope and fits the data very well.

5. A linear regression equation or line of best fit has a correlation coefficient close to **−1** if the line has a negative slope and fits the data very well.

6. The amount of ice cream in a dish, or volume of ice cream, as a function of the amount of time since the ice cream was scooped has a **negative** correlation.

7. The amount of money earned as a function of the amount of time worked has a **positive** correlation.

8. The score on a test as a function of the number of problems solved correctly has a **positive** correlation.

9. The boldness of the printing from a printer as a function of the number of pages printed, without replacing the ink cartridge, has a **negative** correlation.

Motivator

Begin the lesson with the motivator to get students thinking about the topic of the upcoming problem. This lesson is about results of the PSAT test. The motivating questions are about the PSAT test.

Ask the students the following questions to get them interested in the lesson.

- What is the PSAT?
- Have you ever taken the PSAT?
- Why do students take the PSAT?
- The PSAT is a practice for the SAT. Why do over 1.2 million students find that taking the PSAT is an important step to taking and doing well on the SAT?

Explore Together

Problem 1

Students will create a stem-and-leaf plot for the average PSAT score for each state in the United States.

Grouping

Ask for a student volunteer to read the Scenario and Problem 1 aloud. Have a student restate the problem. Pose the Guiding Questions below to verify student understanding. Begin to complete part (A) of Problem 1 together as a whole class. If the students understand as you are completing this together as a class, you can have the students finish the problem individually and then review it with the students afterward.

PSAT Scores			
39.3	NV	46.9	IN
41.1	GA	47.3	AR
41.2	FL	47.4	AZ
41.6	ME	47.4	WV
41.7	SC	47.5	AL
42.0	MD	47.8	VT
42.3	DE	47.9	AK
42.3	MS	48.5	UT
42.3	OK	48.9	OR
42.5	RI	49.0	WA
42.8	LA	49.1	TN
43.0	NM	49.7	MI
43.6	TX	49.9	ID
43.9	KY	49.9	KS
44.1	PA	50.2	CO
44.8	CA	50.8	IL
45.0	VA	50.8	MT
45.1	NC	50.8	WY
45.4	CT	51.5	MO
45.9	NY	51.5	NE
46.0	NJ	51.5	WI
46.4	OH	51.7	SD
46.7	MA	52.7	IA
46.8	HI	52.9	MN
46.8	NH	53.2	ND

Guiding Questions

- What information is given in this problem?
- What is a PSAT?
- What is a stem-and-leaf plot?
- What does the stem of 39 represent?

SCENARIO You and your friends plan to take the PSAT (Preliminary Scholastic Aptitude Test). You learn that this test requires 2 hours and 10 minutes, is usually taken by students during their sophomore or junior years, and includes math, critical reading, and writing questions. You also find out that during the 2004–2005 school year, about 1.2 million sophomores took the PSAT.

Problem 1 How Did Your State Score?

A. The table at the left shows the average PSAT score for each state. To analyze the scores, you can use a *stem-and-leaf plot*. A **stem-and-leaf plot** is a data display that helps you to see how the data are spread out. The *leaves* of the data are made from the digits with the least place value. The *stems* of the data are made from the digits in the remaining place values. Each data value is listed once in the plot. Complete the plot. The first data value, 39.3, is done for you.

Stems	Leaves
39	3
40	
41	1 2 6 7
42	0 3 3 3 5 8
43	0 6 9
44	1 8
45	0 1 4 9
46	0 4 7 8 8 9
47	3 4 4 5 8 9
48	5 9
49	0 1 7 9 9
50	2 8 8 8
51	5 5 5 7
52	7 9
53	2

47 | 5 = 47.5 points

Be sure to include a key that shows what the stems and leaves indicate. Complete the key in your stem-and-leaf plot.

- What value does the leaf of 5 in the row with a stem of 47 represent?
- If there was a row with a stem of 58 and leaves of 4, 8, 8, and 9, then what values would be represented in that row?

Common Student Errors

Students may have difficulty understanding this concept. It is new to many students. Be ready to answer questions and refocus students who may be confused.

Explore Together

Problem 1

Students will investigate the shape of the distribution of data for PSAT scores. They will also be informally introduced to the concept of an average, or mean, as an expected value.

Grouping

Ask for a student volunteer to read part (B) of Problem 1 aloud. Have a student restate the problem. Complete part (B) of Problem 1 together as a whole class. Have students work together in small groups to complete parts (C) and (D) of Problem 1.

Call the class back together to have the students discuss and present their work for parts (C) and (D) of Problem 1.

Key Formative Assessments

- How did you determine the average for part (D)?
- What does the word average mean?

Investigate Problem 1

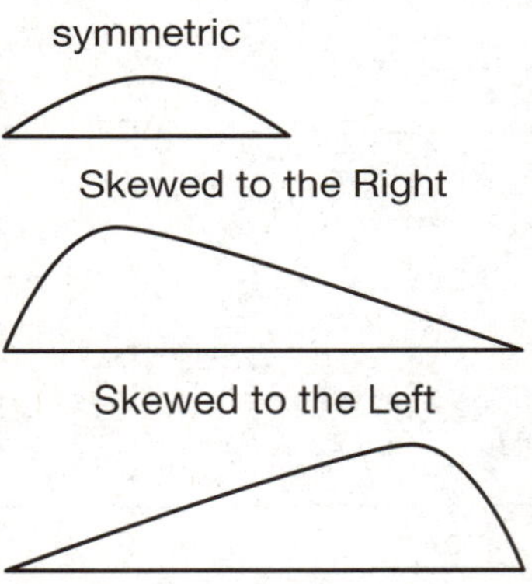

Just the Math

Students will be introduced formally to terminology for the shapes of distributions in Question 1.

Grouping

Ask for a student volunteer to read Question 1 aloud. Have a student restate the problem. Pose Guiding Questions to verify student understanding. Complete Questions 1 through 3 together as a whole class.

Problem 1 How Did Your State Score?

B. How does the stem-and-leaf plot display the data? Use a complete sentence in your answer.

Sample Answer: The stem-and-leaf plot shows where the data lie.

C. What is the highest average score for a state? What is the lowest average score for a state? Use complete sentences in your answers.

The highest average score for a state is 53.2. The lowest average score for a state is 39.3.

D. Based on your stem-and-leaf plot, what would you estimate is the average PSAT score for the United States?

Answers will vary, but should be approximately 47.

Investigate Problem 1

1. **Just the Math: Distributions** Rotate the page with your stem-and-leaf plot 90° in a counterclockwise direction so that the leaves go up instead of to the right. The way in which the data are *distributed*, such as being spread out or clustered together, is the **distribution** of the data. Describe the shape of the distribution that you see in your rotated stem-and-leaf plot. Use a complete sentence in your answer.

   Sample Answer: The distribution is low at both the left and right ends of the stem-and-leaf plot and is high in the middle.

   The shape of the distribution can reveal a lot of information about the data. There are many different distributions, but the most common are symmetric, skewed to the right, and skewed to the left, as shown at the left.

2. In your own words, explain how to draw each distribution at the left. Use complete sentences in your answers.

   Sample Answer: To draw a symmetric distribution, draw a curve that is a mirror image of itself such that it starts at zero, rises upward to a high point, and then descends downward toward zero. To draw a distribution that is skewed to the right, draw a curve that ascends quickly and has its high point towards the left. The curve then descends as you move toward the right. To draw a distribution that is skewed to the left, draw a curve that ascends moving left to right and then descends quickly. The high point is on the right.

3. What type of distribution does the PSAT score data have? Use a complete sentence in your answer.

   Sample Answer: The scores are slightly skewed to the right.

Guiding Questions

- What is a distribution?
- What are the important characteristics of the distributions shown in the left margin?
- What other shapes of distributions might exist?
- Will graphs of data look exactly like those distributions shown in the margin or close to them?

Explore Together

Investigate Problem 1

Students will be introduced formally to measures of central tendency by calculating the mean, median, and mode for a data set.

PSAT Scores			
39.3	NV	46.9	IN
41.1	GA	47.3	AR
41.2	FL	47.4	AZ
41.6	ME	47.4	WV
41.7	SC	47.5	AL
42.0	MD	47.8	VT
42.3	DE	47.9	AK
42.3	MS	48.5	UT
42.3	OK	48.9	OR
42.5	RI	49.0	WA
42.8	LA	49.1	TN
43.0	NM	49.7	MI
43.6	TX	49.9	ID
43.9	KY	49.9	KS
44.1	PA	50.2	CO
44.8	CA	50.8	IL
45.0	VA	50.8	MT
45.1	NC	50.8	WY
45.4	CT	51.5	MO
45.9	NY	51.5	NE
46.0	NJ	51.5	WI
46.4	OH	51.7	SD
46.7	MA	52.7	IA
46.8	HI	52.9	MN
46.8	NH	53.2	ND

Take Note

When you have an even number of values in a data set, you can find the median by finding the mean of the middle two numbers. For instance, in the data set 12, 13, 15, 16, 18, 19, the median is the mean of 15 and 16, which is $\frac{15 + 16}{2}$, or 15.5.

Grouping

Ask for a student volunteer to read Questions 4 and 5 and the take note box aloud. Have a student restate the problems. Pose the Guiding Questions below to verify student understanding. Complete Questions 4 and 5 together as a whole class.

Guiding Questions

- What is an average, or arithmetic mean? How can you calculate the mean?

Investigate Problem 1

4. Is it easier to see the distribution in the stem-and-leaf plot or the table? Use a complete sentence in your answer.

Sample Answer: The distribution is easier to see in the stem-and-leaf plot because the shape is shown by the leaves' heights.

5. Just the Math: Mean, Median, and Mode
The average PSAT score for each state is shown in order from least to greatest starting from the left. Locate your state's average PSAT score. Did the students that took the PSAT in your state do well? Use a complete sentence to explain your answer.

Answers will vary according to the student's state of residence, but should reference the average PSAT score from part (D).

One number that is often used to describe a set of data is the **mean** or **arithmetic mean.** The mean is also called the **average.** The mean is the sum of all the data values divided by the number of values in the data set. We write the mean as $\bar{x} = \frac{\Sigma x}{n}$, where Σx is the symbol for the sum of all the x-values (data values) and n is the number of values. What is the mean of the test scores in the table? Round your answer to the nearest hundredth. Use a complete sentence in your answer.

The mean of the test scores is 46.75.

Approximately what percent of the scores are above the mean? Approximately what percent of the scores are below the mean? Use a complete sentence in your answer.

There are 27 scores, or 54% of the scores, above the mean and 23 scores, or 46% of the scores, below the mean.

When we talk about a mean score, we are trying to determine a single value that best represents the performance of a group. This single value is a **measure of central tendency.** It is a value that represents a typical value in a data set.

6. Another measure of central tendency is the **median,** the middle score of the data, which is found by listing all the data values in order and finding the value that is exactly in the middle. Use your stem-and-leaf plot to find the median score. Interpret this value in terms of the problem situation. Where does your state fit? Use complete sentences in your answers.

Sample Answer: The median score is 46.85. Half of the states have average scores that are above this value and the other half of the states have average scores that are below this value. Answers will vary according to the state of residence of the student.

- What is the meaning of the capital letter sigma from the Greek alphabet, written as Σ?
- What does n represent?
- What is a percent?
- How can you round a number to the nearest hundredth?
- What is a median? How can you calculate the median? Why do we have to write the data in order from least to greatest before we find the median?

Explore Together

Investigate Problem 1

Students will continue to investigate the measures of central tendency and the shapes of distributions.

Grouping

Ask for a student volunteer to read Question 6 aloud. Have a student restate the problem. Pose the Guiding Questions below to verify student understanding. Have students work together in small groups to complete Question 6. Then call the class together to discuss and explain their work.

Guiding Questions

- What is a mode?
- How is a mode different from a mean? How is a mode different from a median?
- How is a mode similar to a mean? How is a mode similar to a median?
- What does bimodal data look like? What does trimodal data look like?

Note If a data set has more than 3 modes, we call the distribution multi-modal.

Just the Math

Students will consider skewed distributions as well as symmetric or bell-shaped distributions.

Grouping

Ask for a student volunteer to read Question 7 aloud. Have a student restate the problem. Pose the Guiding Questions below to verify student understanding. Have students work together in small groups to complete Question 7 then call the class together to discuss and explain their work.

Guiding Questions

- Which measure of central tendency is not affected by very small or very large data values?
- If there are very small numbers in the data set compared to most of the values, which measure of central tendency will be much smaller than most of the data values? Why?
- How is the mean affected if there are very small numbers in the data set compared to most of the data values?

Notes Symmetric data sets are often called bell-shaped or Normal Distributions. Numbers that are very different from most of the data values are also called outliers.

Investigate Problem 1

7. A third measure of central tendency is the **mode** of the data. The mode is the value in the data set that appears most often. If two values occur in the data set the same number of times, then each value is a mode and the data set is *bimodal*. If three values occur in the data set the same number of times, then each value is a mode and the data set is *trimodal*. Which test score appears the greatest number of times? Use a complete sentence in your answer.

   The test scores that appear most often are 42.3, 50.8, and 51.5.

   What is the mode of the test scores? Use a complete sentence in your answer.

   The data set is trimodal. The modes are 42.3, 50.8, and 51.5.

8. **Just the Math: Mean, Median, Mode, and Distributions** When a distribution is symmetric, the mean and median are equal. How do you think the mean compares to the median in a distribution that is skewed to the left? Use a complete sentence in your answer.

   In a distribution that is skewed to the left, the mean is less than the median.

   How do you think the mean compares to the median in a distribution that is skewed to the right? Use a complete sentence in your answer.

   In a distribution that is skewed to the right, the mean is greater than the median.

   Draw representations of two sets of data, one for a distribution that is skewed to the left and one for a distribution that is skewed to the right. Then mark the possible mean and median on each distribution.

   Skewed to the left:

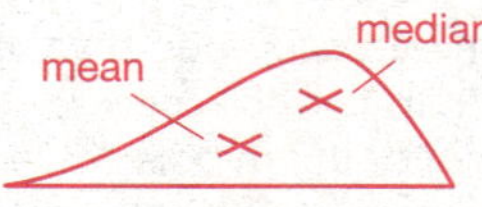

   Skewed to the right:

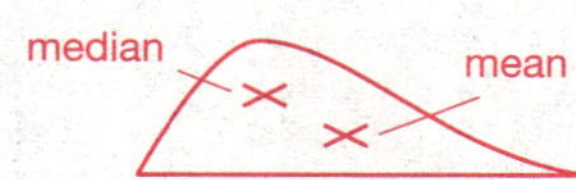

Explore Together

Investigate Problem 1

Students will analyze a data set in detail to apply the concepts of this lesson.

Grouping

Ask for a student volunteer to read Question 8 aloud. Have a student restate the problem. Have students work together in small groups to complete Questions 8 and 9.

Call the class back together to have the students discuss and present their work for Questions 8 and 9.

Notes All measures of central tendency are actually averages. The average we typically refer to is the mean. But especially in advertising, it is common to refer to any of the measures of central tendency as an average. Consumers are expected to know what an average is. It is helpful for the students to think of the measures of central tendency as expected values or typical values.

Key Formative Assessments

- How does a stem-and-leaf plot show more than just a list of data values?
- How can you interpret a stem-and-leaf plot as a graph?
- What benefits does a stem-and-leaf plot have compared to other graphs?
- What disadvantages does a stem-and-leaf plot have?
- What is a center of something? What does it mean for someone to tend to do something?
- Why are the mean, median, and mode called measures of central tendency?
- Which measure of central tendency would be most useful for stores to know the typical size of clothes or shoes that people buy, the typical color of suits that people buy, the typical style of shirts that people buy, etc?
- What distributions are common and what does each look like?

Investigate Problem 1

9. The set of data below is a set of PSAT scores from those students in a particular class at your school who took the test. Create a stem-and-leaf plot of the data and determine the distribution of the data. Then analyze the data by finding the mean, median, and mode. Show all your work. Finally, draw a representation of the data and mark the mean and median on the distribution.

Test scores: 36, 49, 16, 31, 21, 52, 29, 49, 48, 32, 42, 49, 44

Stem-and-leaf plot:

Stem	Leaf
1	6
2	1 9
3	1 2 6
4	2 4 8 9 9 9
5	2

3 | 5 = 35 points

The data are skewed to the left.

The mean is

$$\frac{16 + 21 + 29 + 31 + 32 + 36 + 42 + 44 + 48 + 49 + 49 + 49 + 52}{13}$$

$\approx$ 38.3.

The median value is 42 because it is the number in the middle of the data set.

The mode is 49 because 49 occurs most often in the data set.

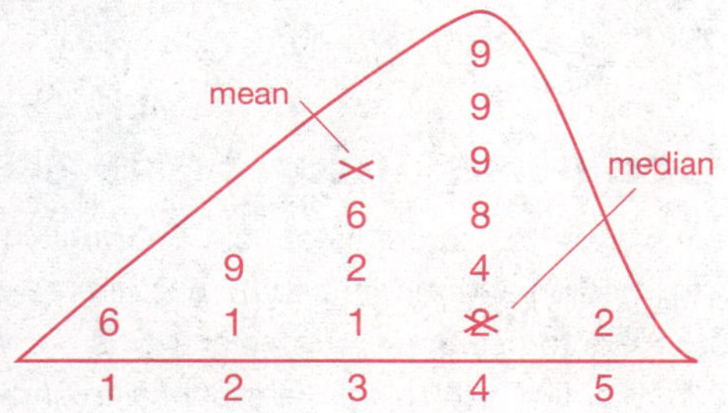

10. Based on your results in Question 9, decide which measure of central tendency, the median or the mean, is the better representation of the test scores of the class.

The median is the better representation because the data are skewed to the left and the median is greater than the mean.

Take Note

Whenever you see the share with the class icon, your group should prepare a short presentation to share with the class that describes how you solved the problem. Be prepared to ask questions during other groups' presentations and to answer questions during your presentation.

Wrap Up

Close

- Review all key terms and their definitions. Include the terms *stem-and-leaf plot, distribution, measure of central tendency, mean, median,* and *mode.*
- You may also want to review any other vocabulary terms that were discussed during the lesson, which may include *spread out, clustered, symmetric, bell-shaped, normal, skewed left, skewed right, multi-modal,* and *outliers.*
- Remind the students to write the key terms and their definitions in the notes section of their notebooks. You may also want the students to include examples.
- Ask the students to compare and contrast the measures of central tendency taught in this section.
- Ask the students to make a table to list the characteristics of the mean, the median, and the mode. Their table should include advantages and disadvantages of each. (For instance, the mean is well understood by many people and is most used by people as an average, but it is affected significantly by outliers in the data. The mean is a function of every data value, whereas the median is only the middle value or the mean of the 2 values in the center of the data. The median is not affected by outliers in the data. The mode is not significantly affected by outliers in the data.
- Ask the students to explain why they think that stem-and-leaf plots have the terms stem and leaf in the title.
- Have the students summarize the characteristics of the different distributions that were discussed in this lesson: skewed to the left, skewed to the right, and symmetric.
- Have the students identify the location of the mode for each distribution as well as the mean and median.

Ties to the Cognitive Tutor Software

The Cognitive Tutor software gives students lots of opportunity to consider how different measures of central tendency (mean, median and mode) can be used in real-world situations. An understanding of when it might be appropriate to use mean or median helps prepare students to understand concepts such as the form of a distribution, which will be essential in understanding probability and statistics.

Follow Up

Assignment

Use the Assignment for Lesson 12.1 in the Student Assignments book. See the Teacher's Resources and Assessments book for answers.

Assessment

See the Assessments provided in the Teacher's Resources and Assessments book for Chapter 12.

Open-Ended Writing Task

Have the students find a data set on the Internet for a topic of interest to them. Have the students create a stem-and-leaf plot for the data, find the measures of central tendency, and describe the distribution of the data. The students should then write a short essay about the data set they analyzed summarizing their findings.

Reflections

Insert your reflections on the lesson as it played out in class today.

What went well?

____

____

What did not go as well as you would have liked?

____

____

How would you like to change the lesson in order to improve the things that did not go well and capitalize on the things that did go well?

____

____

____

____

____

Notes

12.2

Compact Discs

Collecting and Analyzing Data

Learning By Doing Lesson Map

Get Ready

Objectives

In this lesson, you will:

- Collect and organize data.
- Find the mean and median of a data set.
- Determine how data values affect the mean and median of a data set.

Key Terms

- survey
- mean
- median
- sample size

Materials

- Small pieces of paper for the motivating activity
- Additional copies of the pages of this lesson on paper or transparency for each group if having the students conduct the survey and then each contributing data from some of their reviewers to create a combined set of data for the group

NCTM Content Standards

Grades 9–12 Expectations

Number and Operations Standards

- Develop fluency in operations with real numbers, vectors, and matrices, using mental computation or paper-and-pencil calculations for simple cases and technology for more complicated cases.
- Judge the reasonableness of numerical computations and their results.

Data Analysis and Probability Standards

- For univariate measurement data, be able to display the distribution, describe its shape, and select and calculate summary statistics.
- Understand how sample statistics reflect the values of population parameters and use sampling distributions as the basis for informal inference.
- Understand the concepts of sample space and probability distribution and construct sample spaces and distributions in simple cases.

Lesson Overview

Within the context of this lesson, students will be asked to:

- Conduct a survey of ratings for music CDs by students.
- Organize the data into a chart.
- Calculate the mean and median of the data.
- Determine how data values affect the mean and median of the data set.
- Discuss how sample size can affect the results of a survey.

You may want to do the motivating activity and assign part (B) and Questions 1 through 3 of Problem 1 as a homework assignment the day or two before you do this activity in class.

Essential Questions

The following key questions are addressed in this lesson:

1. What is a survey?
2. Why should we conduct surveys?
3. How can you conduct a survey?
4. What is a sample size and why is it important?
5. How can you calculate the mean and median for survey data?

12

Show The Way

Warm Up

Place the following questions or an applicable subset of these questions on the board before students enter class. Students should begin working as soon as they are seated.

Find the mean, median, and mode for each set of data. Round your answers to the nearest hundredth, if necessary.

1. 32, 40, 35, 29, 14, 32

mean = $30.\overline{3}$; median = 32; mode = 32

2. 11, –7, 16, –2, 11, 1, 4, 11, 8, 6, –7, –4

mean = 4; median = 5; mode = 11

3. 6, 1, 7, 6, 5, 5, 0, 1, 0, 8

mean = 3.9; median = 5;
The modes are 0, 1, 5, and 6.

4. 121, 143, 98, 144, 165, 118

mean = 131.5; median = 132; mode = none

5. $1.99, $2.99, $5.99, $12.99

mean = $5.99; median = $4.49; mode = none

6. 4, 4, 4, 4, 4, 4, 4, 4

mean = 4; median = 4; mode = 4

Motivator

Begin the lesson with the motivator to get students thinking about the topic of the upcoming problem. This lesson is about conducting a survey for favorite types of music. The motivating questions are about favorite music types.

Have the students participate in the following activity to get them interested in the lesson.

- As the students enter the classroom, hand each student a small piece of paper.
- Ask them to neatly print the name of 5 music CDs that they enjoy on the paper.
- Collect the papers.
- Have a student volunteer read the names from the papers while 1 or 2 other student volunteers write a list of the names on the front board and make tally marks to count the number of times each title is identified. So, if 4 different students identified "Abby Road" by the Beatles, then the volunteers would write Abby Road and make 4 tally marks beside the title.
- Find out which 5 CDs were the most popular.
- Use those 5 CDs in part (A) of Problem 1.

Explore Together

Problem 1

Students will conduct a survey to gather interesting data to analyze.

Grouping

Ask for a student volunteer to read the Scenario and Problem 1 aloud. Have a student restate the problem. Pose the Guiding Questions below to verify student understanding. Work together as a class to complete part (A) of Problem 1 if it was not already completed as a motivating activity.

Guiding Questions

- What is a survey?
- Why do we conduct surveys?
- What is a characteristic?
- What is a population?
- What characteristics are we trying to learn more about in this scenario? What population are we trying to learn more about?

Notes Sample data are shown in the table to demonstrate reasonable answers to the questions that follow. Students' answers will vary.

Common Student Errors

Students may have difficulty understanding this concept. It is new to many students. Be ready to answer questions and refocus students who may be confused.

Grouping

Ask for a student volunteer to read part (B) of Problem 1 aloud. Have a student restate the problem. Pose the Guiding Questions below to verify student understanding. Assign part (B) and Questions 1 through 3 of Problem 1 as part of a homework assignment.

Guiding Questions

- What type of data do you expect from the survey?
- How will you gather data for your survey?

SCENARIO You work at the school newspaper and want to write an article about the CDs that the students at your school think are the most popular. You decide to use a **survey,** which is an investigation of a characteristic of a population (in this case, the students at your school), to gather information for your article.

Problem 1 Rate the CDs

A. To set up the survey, you begin by making a list of five music CDs that are popular. Write your list below.

Answers will vary.

B. Ask six different people, including at least one adult if possible, to rate each CD based on the following numerical scale:

Excellent: 41–50

Good: 31–40

Fair: 21–30

Poor: 11–20

Awful: 0–10

Record the responses in the table below.

CD	Reviewer 1 rating	Reviewer 2 rating	Reviewer 3 rating	Reviewer 4 rating	Reviewer 5 rating	Reviewer 6 rating
CD 1	32	40	35	29	14	37
CD 2	5	31	23	37	19	38
CD 3	18	43	50	47	23	12
CD 4	19	40	9	35	14	4
CD 5	10	33	25	11	30	32

Explore Together

Investigate Problem 1

Students will calculate and compare the mean and median for each CD.

Take Note

Remember that when you have an even number of values in a data set, you can find the median by finding the mean of the middle two values. For instance, in the data set 12, 13, 15, 16, 18, 19, the median is the mean of 15 and 16, which is $\frac{15 + 16}{2}$, or 15.5.

Call the class back together to have the students discuss and present their work for part (B) and Questions 1 through 3. Remind the students that their answers will vary.

Notes Sample data are shown to demonstrate reasonable answers to the questions that follow. In the table for Question 4, the ratings for reviewer #6 have been replaced with a sample set of student ratings.

Grouping

Ask for a student volunteer to read Question 4 aloud. Have a student restate the problem.

Because the student answers for Questions 1 and 4 will vary, you may want to have the students complete Questions 4 through 7 individually. Another option is to have the students work together in small groups using the data from one student in the group. A third option is for each student in the group to give the ratings from 2 or 3 reviewers to combine to make a chart of data for the group. This option allows you to assign these questions again for part of a homework assignment for the students to complete individually with their own data in their own book.

Note It may help to have an additional copy of the pages of this lesson for the students on paper or transparency for them to work on as a group.

Investigate Problem 1

1. Find the mean rating for each CD. Then find the median rating for each CD. Record these measures of central tendency in the table below. Round your answers to the nearest hundredth if necessary.

CD	Mean rating	Median rating
CD 1	31.17	33.5
CD 2	25.5	27
CD 3	32.17	33
CD 4	20.17	16.5
CD 5	23.5	27.5

2. Are the mean rating and the median rating close together for any of the CDs? Use a complete sentence in your answer.

Sample Answer: The mean and median ratings are close for CD 3.

3. In each case, which measure of central tendency is more representative of the overall rating of the CD? Why? Use a complete sentence in your answer.

Answers will vary. Students may say that the mean is a better representation because it allows the very low or the very high ratings to influence the mean rating, while the median does not.

4. With which single reviewer do you agree the most? With which single reviewer do you agree the least? Having identified the reviewer whose ratings you like the least, replace all of his or her ratings with your own ratings in that reviewer's column. (If you agree with every reviewer, pretend that you do not and come up with another set of ratings for one of the reviewers.)

CD	Reviewer 1 rating	Reviewer 2 rating	Reviewer 3 rating	Reviewer 4 rating	Reviewer 5 rating	Reviewer 6 rating
CD 1	32	40	35	29	14	45
CD 2	5	31	23	37	19	40
CD 3	18	43	50	47	23	10
CD 4	19	40	9	35	14	30
CD 5	10	33	25	11	30	30

12

Explore Together

Investigate Problem 1

Students will calculate the mean and median for the data in the table with the changed values included.

Notes The answers to the questions that follow are sample answers only to demonstrate reasonable solutions. Student answers will vary. Questions 8 through 11 do not have sample answers because the student answers will vary considerably.

Call the class back together to have the students discuss and present their work for Questions 4 through 7. Have some students or groups put their charts on transparencies or copy them onto poster paper to share with the class. Choose students with different types of results and have them explain their answers to Questions 6 and 7 also. This may also give the class a chance to see several different types of distributions.

Grouping

Ask for a student volunteer to read Question 8 aloud. Have a student restate the problem. Pose the Guiding Questions below to verify student understanding. Have students work individually to complete Questions 8 through 11. These Questions can also be assigned as part of a homework assignment.

Guiding Questions

- What do you expect to happen if any ratings that you disagree with are replaced with your own ratings?
- Why would you be asked to make up new ratings for this question if you already agree with each rating given?
- If you replace data values with your own, is the sample data a valid set of data? Why or why not?
- Do you expect calculating the mean and the median of the new data to be easier, more difficult, or the same level of difficulty as finding the mean and median of the original data? Why?

Investigate Problem 1

5. Find the new mean rating and new median rating for each CD. Record your results in the table below. Round your answers to the nearest hundredth if necessary.

CD	Mean rating	Median rating
CD 1	32.5	33.5
CD 2	25.83	27
CD 3	31.83	33
CD 4	24.5	24.5
CD 5	23.17	27.5

6. Has the relationship between the mean and median changed for any of the CDs? If so, in what way has it changed? Use complete sentences in your answer.

Sample Answer: The relationship has changed between every mean and median. The mean increased for CDs 1, 2, and 4, and so the mean and median are closer. In fact, for CD 4, the mean and median are identical. The mean decreased for CDs 3 and 5, so the mean and median are farther apart.

7. How does including your rating instead of the original reviewer's change the distribution? Use a complete sentence in your answer.

Sample Answer: Including the new ratings instead of the ratings of Reviewer 6 skews the distribution slightly to the left, as there are more high ratings.

8. Identify each rating in the table in part (B) with which you disagree and replace it with your own. (If you agree with every rating, go through the exercise by making up new ratings for some of the original ratings and substituting these new ratings.)

CD	Reviewer 1 rating	Reviewer 2 rating	Reviewer 3 rating	Reviewer 4 rating	Reviewer 5 rating	Reviewer 6 rating
CD 1	32	40	35	29	39	37
CD 2	33	31	23	37	25	38
CD 3	18	22	22	22	23	22
CD 4	19	26	26	35	14	26
CD 5	10	15	25	11	15	15

12

Explore Together

Investigate Problem 1

Students will consider the importance of the sample size for a survey.

Call the class back together to have the students discuss and present their work for Questions 8 through 11.

Just the Math

The students will be introduced to the concept of a sample size in Question 12. Ask for a student volunteer to read Question 12 aloud. Have a student restate the problem. Pose the Guiding Questions below to verify student understanding. Complete Question 12 together as a whole class.

Guiding Questions

- What is a sample?
- What is a sample size?

Key Formative Assessments

- What sample size do you think would have been better to use for our survey rather than the sample size of 6 that we used? Why do you think it is a better sample size?
- What do you think is the largest reasonable sample size that would have been appropriate? Why?
- How might the mean and median change when you change any data for a very small sample?
- How much do you think the mean and median of a data set would change if you were to change any data for a large sample?
- Do you think that your survey was representative of the population of students at your school?
- Other than having a small sample size, what else might have made the sample not very representative of the population of students at your school?
- Other than increasing the sample size, what else might have made the sample more representative of the population of students at your school?

Investigate Problem 1

9. What is the new mean rating for each CD? What is the new median rating for each CD? Record the new measures of central tendency in the table below. Round your answers to the nearest hundredth if necessary.

CD	Mean rating	Median rating
CD 1	35.33	36
CD 2	31.17	32
CD 3	21.5	22
CD 4	24.33	26
CD 5	15.17	15

10. Has the relationship between the mean and median changed for any of the CDs? If so, in what way has it changed? Use complete sentences in your answer.

Answers will vary.

11. How does including your rating instead of the original reviewer's rating change the distribution? Use a complete sentence in your answer.

Answers will vary.

12. Just the Math: Sample Size In your survey, the **sample size,** or the number of people surveyed, was 6. Do you think that the sample size was large enough to accurately represent the opinions of the students at your school? Why or why not? Use complete sentences in your answer.

Sample Answer: The sample size should have been larger because just changing one set of responses altered the mean ratings and the median ratings significantly in many cases.

- Do you think that your sample is representative of the population of all students in our country? Do you think that your sample is representative of the population of all students in the world? Do you think that your sample is representative of the population of all people in our country? Why or why not?

12

Wrap Up

Close

- Review all key terms and their definitions. Include the terms *survey, mean, median,* and *sample size.*
- You may also want to review any other vocabulary terms that were discussed during the lesson, which may include *ratings, distribution, characteristic, sample,* and *population.*
- Remind the students to write the key terms and their definitions in the notes section of their notebooks. You may also want the students to include examples.
- Ask the students to explain how to conduct a survey.
- Ask the students how they think some survey results can be biased.
- Ask the students if they think that their survey may have been biased.
- Ask the students to name 3 ideas they learned in this lesson that they didn't know before doing the survey and answering the questions in the lesson.
- Ask the students how surveys help us in our lives.
- Ask the students to calculate the mode for each data set from this lesson. Then discuss how representative the mean is for the data as a measure of central tendency.

Ties to the Cognitive Tutor Software

The Cognitive Tutor software gives students lots of opportunity to consider how different measures of central tendency (mean, median and mode) can be used in real-world situations. An understanding of when it might be appropriate to use mean or median helps prepare students to understand concepts such as the form of a distribution, which will be essential in understanding probability and statistics.

Follow Up

Assignment

Use the Assignment for Lesson 12.2 in the Student Assignments book. See the Teacher's Resources and Assessments book for answers.

Assessment

See the Assessments provided in the Teacher's Resources and Assessments book for Chapter 12.

Open-Ended Writing Task

Ask the students to think about a topic that interests them. Have the students design and conduct a survey to investigate the opinions of others. The students should calculate the mean and median, if appropriate, and write a short paragraph to explain the answers to the questions below and on the next page.

- What is the population from which you are sampling?
- How will you choose the people to survey?

- What will you ask?
- What sample size will you use?
- What results do you think you will find?
- Will your sample be representative of the population?

Reflections

Insert your reflections on the lesson as it played out in class today.

What went well?

___

___

What did not go as well as you would have liked?

___

___

How would you like to change the lesson in order to improve the things that did not go well and capitalize on the things that did go well?

___

___

___

___

___

Notes

12.3

Breakfast Cereals

Quartiles and Box-and-Whisker Plots

Learning By Doing Lesson Map

Get Ready

Objectives

In this lesson, you will:

- Find the range and extremes of a data set.
- Find the first, second, and third quartiles of a data set.
- Represent a data set graphically by using a box-and-whisker plot.
- Identify outliers in a data set.
- Find percentiles of a data set.
- Find the IQR of a data set.

Key Terms

- range
- extreme
- median
- quartile
- *n*th percentile
- box-and-whisker plot
- outlier
- percentile
- interquartile range

NCTM Content Standards

Grades 9–12 Expectations

Number and Operations Standard

- Judge the reasonableness of numerical computations and their results.

Algebra Standard

- Draw reasonable conclusions about a situation being modeled.

Data Analysis and Probability Standards

- Understand histograms, parallel box plots, and scatter plots and use them to display data.
- For univariate measurement data, be able to display the distribution, describe its shape, and select and calculate summary statistics.

Lesson Overview

Within the context of this lesson, students will be asked to:

- Calculate the range and interquartile range for a given set of data.
- Calculate quartiles and percentiles for a given set of data.
- Construct a box-and-whisker plot for a given set of data.
- Identify the effect of an outlier in a data set on a box-and-whisker plot.

Essential Questions

The following key questions are addressed in this lesson:

1. What is a range for data?
2. What is an interquartile range?
3. How do you construct a box-and-whisker plot?
4. How do you find the quartiles for data?
5. How do you find a given percentile or find the percentile for a given number?
6. What are outliers?

Show The Way

Warm Up

Place the following questions or an applicable subset of these questions on the board before students enter class. Students should begin working as soon as they are seated.

Complete the following table by writing equivalent fractions, decimals, or percents.

	Fraction	Decimal	Percent
1.	$\frac{3}{4}$	**0.75**	**75%**
2.	$\frac{1}{2}$	0.50	**50%**
3.	$\frac{38}{100} = \frac{19}{50}$	0.38	**38%**
4.	$\frac{7}{100}$	**0.07**	**7%**
5.	$\frac{4}{20}$	**0.20**	**20%**
6.	$\frac{79}{10,000}$	0.0079	**0.79%**
7.	$\frac{12}{100} = \frac{3}{25}$	**0.12**	12%
8.	$\frac{1.5}{100} = \frac{3}{200}$	**0.015**	1.5%
9.	$\frac{4}{100} = \frac{1}{25}$	**0.04**	4%

Motivator

Begin the lesson with the motivator to get students thinking about the topic of the upcoming problem. This lesson is about the amounts of sugar in breakfast cereals. The motivating questions are about breakfast cereals.

Ask the students the following questions to get them interested in the lesson.

- Do you eat breakfast cereals?
- What is your favorite breakfast cereal?
- How much sugar do you think is in your favorite breakfast cereal?
- How can you find out how much sugar is in your favorite cereal?

Explore Together

Problem 1

Students will investigate measures of the spread of data and of the position for the data. Students will calculate the range for a set of data giving the number of grams of sugar in a serving of various breakfast cereals.

Grouping

Ask for a student volunteer to read the Scenario and Problem 1 aloud. Have a student restate the problem. Pose the Guiding Questions below to verify student understanding. Have students complete part (A) of Problem 1 individually.

Guiding Questions

- What information is given in this problem?
- What is a range?
- What information do you need to calculate the range?
- What is the largest amount of sugar in the listed cereals? What is the smallest amount of sugar?
- What do we call the largest amount? What do we call the smallest amount?
- Are any other data values used to find the range other than the upper and lower extreme values?

Call the class back together to have the students discuss and present their work for part (A) of Problem 1.

Key Formative Assessments

- What information can we learn from the range?
- If the students in an algebra class take a 100-point test and the range of scores on the test is 82, what does that mean?
- If the students in an algebra class take a 100-point test and the range of scores on the test is 7, what does that mean?
- Is the range considered to be a measure of central tendency? Why or why not?

SCENARIO Your aunt is a health teacher who is encouraging you to eat a healthier breakfast. She says that the amounts of sugar in different breakfast cereals vary widely. So, you decide to make a trip to the grocery store to see if what your aunt says is true.

Problem 1 How Much Sugar is Too Much?

A. At the store, you write down the amounts of sugar that are in one serving of different kinds of cereal. The table below shows your data. One way we can get a general idea of how the data varies is to find the *range* of the data. The **range** is the difference between the greatest value and the least value in the data set. The greatest value is called the **upper extreme** and the least value is called the **lower extreme.** What is the upper extreme of the amounts of sugar? What is the lower extreme? What is the range?

Cereal	Sugar in one serving (grams)
Cocoa Rounds	13
Flakes of Corn	4
Frosty Flakes	11
Grape Nuggets	7
Golden Nuggets	10
Honey Nut O's	10
Raisin Branola	9
Healthy Living Flakes	9
Wheatleys	8
Healthy Living Crunch	6
Multi-Grain O's	7
All Branola	5
Munch Crunch	12
Branola Flakes	5
Complete Flakes	4
Corn Chrisps	3
Rice Chrisps	4
Sugary Puffs	32
Shredded Wheatleys	1
Fruit Circles	11

The upper extreme is 32, the lower extreme is 1, and the range is 32 – 1, or 31.

12

Explore Together

Investigate Problem 1

Students will investigate measures of position for data by finding quartiles for data.

Grouping

Ask for a student volunteer to read part (B) aloud. Have a student restate the problem. Have students work together in small groups to complete part (B).

Ask for a student volunteer to read part (C) aloud. Have a student restate the problem. Pose the Guiding Questions below to verify student understanding. Have students work together in small groups to complete parts (C) and (D).

Guiding Questions

- If you divide a pizza into quarters, what does that mean?
- How many American quarters do you need to have the same amount of money as $1?
- If you divide a data set into quarters called quartiles, what does that mean?
- Which quartile separates the smallest 25% of the data from the largest 75% of the data?
- Which quartile separates the smallest 50% of the data from the largest 50% of the data?
- What is the third quartile?
- How can you find the median for a set of data?
- How can you find the quartiles for a set of data?

Key Formative Assessments

- How can knowing the values of the first, second, and third quartiles give you information about a data set? How can that information be useful or important?

12

Problem 1 How Much Sugar is Too Much?

B. In Lesson 12.1, you found that measures of central tendency help us to see what is typical in the data. Other aspects of data are important to identify, such as how the data are spread apart. To see this, we can divide the data into equal parts. Which measure of central tendency divides the data set into two equal parts? Use a complete sentence in your answer.

The median divides the data into two equal parts.

C. You can further divide a data set by using *quartiles*. The three **quartiles,** called Q_1, Q_2, and Q_3, divide an ordered data set into four equal parts. To find the quartiles of the cereal data, first arrange the values from least to greatest.

1 3 4 4 4 5 5 6 7 7 8 9 9 10 10 11 11 12 13 32

Next, find the median. The median is the middle or **second quartile,** Q_2. What is Q_2? Use a complete sentence in your answer. Then, draw a vertical line where Q_2 divides your ordered data.

The second quartile Q_2 is 7.5.

1 3 4 4 4 5 5 6 7 7 | 8 9 9 10 10 11 11 12 13 32

Now find the median of the values to the left of the line for Q_2. The median of the lower half of the data is the lower or **first quartile,** Q_1. What is Q_1? Use a complete sentence in your answer. Then draw a vertical line where Q_1 divides your ordered data.

The first quartile Q_1 is 4.5.

1 3 4 4 4 | 5 5 6 7 7 | 8 9 9 10 10 11 11 12 13 32

Finally, find the median of the values to the right of the line for Q_2. The median of the upper half of the data is the upper or **third quartile,** Q_3. What is Q_3? Use a complete sentence in your answer. Then draw a vertical line where Q_3 divides your ordered data.

The third quartile Q_3 is 10.5.

1 3 4 4 4 | 5 5 6 7 7 | 8 9 9 10 10 | 11 11 12 13 32

D. What fraction of the data falls on or below the first quartile Q_1? Use a complete sentence in your answer.

One fourth of the data falls on or below the first quartile Q_1.

What fraction of the data falls on or below the second quartile Q_2? Use a complete sentence in your answer.

One half of the data falls on or below the second quartile Q_2.

- If you are choosing between two breakfast cereals to buy and find that one of them has a sugar content that is smaller than the first quartile, and the other has a sugar content that is larger than the third quartile, but all the other nutrient values are the same, which would you choose? Why?

Explore Together

Investigate Problem 1

Students will be introduced formally to box-and-whisker plots.

Notes The interval from the first quartile to the third quartile is often called the interquartile. The difference between the third quartile and the first quartile is called the interquartile range. This is also called the IQR. These concepts will be investigated more fully in Question 7.

Just the Math

The box-and-whisker plot is given its name because it can look like the mouth of a cat with the whiskers off to each side. A box-and-whisker plot allows us to visually represent the data divided into quarters.

Grouping

Ask for a student volunteer to read Question 1 aloud. Have a student restate the problem. Pose the Guiding Questions below to verify student understanding. Have students work together in small groups to complete Question 1.

Guiding Questions

- What is a box-and-whisker plot?
- What data value is represented by the left endpoint for the box-and-whisker plot? What data value is represented by the right endpoint?
- How do you think that a box-and-whisker plot can represent the data in quartiles?

Key Formative Assessments

- How can you choose the left endpoint for the box-and-whisker plot? How can you choose the right endpoint?
- If one of the whiskers is very long compared to the other whisker, what would that mean about the data?
- How can each of the 4 parts of a box-and-whisker plot represent the same number of data values, yet they don't all have to be of equal length?

Problem 1 How Much Sugar is Too Much?

What fraction of the data falls on or below the third quartile Q_3? Use a complete sentence in your answer.

Three fourths of the data falls on or below the third quartile Q_3.

Between which two values does half of the data fall? Use a complete sentence to explain your reasoning.

Sample Answer: Half of the data falls between the third quartile and the first quartile. This is because three fourths of the data falls below the third quartile and one fourth of the data falls below the first quartile. The difference between these is one half.

Investigate Problem 1

1. **Just the Math: Box-and-Whisker Plots** A **box-and-whisker plot** (also known as a box plot) is used to represent a large data set (hundreds or thousands of values). Box-and-whisker plots allow you to easily make comparisons of data sets. The box-and-whisker plot below represents a set of data.

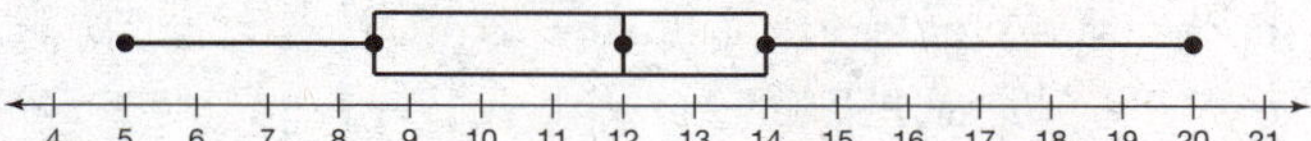

What do you think is the lower extreme of the data set? What do you think is the upper extreme? Use complete sentences in your answer.

The lower extreme is 5. The upper extreme is 20.

The vertical line inside the box represents the second quartile Q_2 of the data. What is the value of Q_2? Use a complete sentence in your answer.

The second quartile Q_2 is 12.

The point in the box directly to the left of the second quartile represents the first quartile Q_1. What is the value of Q_1? Use a complete sentence in your answer.

The first quartile Q_1 is 8.5.

Similarly, the point in the box directly to the right of the second quartile represents the third quartile Q_3. What is the value of Q_3? Use a complete sentence in your answer.

The third quartile Q_3 is 14.

The horizontal lines on both ends of the box are called *whiskers*. What do the dots at the end of the whiskers represent? Use a complete sentence in your answer.

The dots represent the lower and upper extremes.

- What do the vertical lines in the box-and-whisker plot represent?

12

Explore Together

Investigate Problem 1

Students will construct a box-and-whisker plot for the breakfast cereal data.

Grouping

Ask for a student volunteer to read Question 2 aloud. Have a student restate the problem. Pose the Guiding Questions below to verify student understanding. Have students work together in small groups to complete Questions 2 and 3 then call the class back together to have the students discuss and present their work for Questions 2 and 3.

Guiding Questions

- What information do you need to find or calculate in order to construct a box-and-whisker plot?
- Why would we use a number line to help construct the box-and-whisker plot?
- Where do you place each dot for a box-and-whisker plot?
- Where do you place the box for a box-and-whisker plot?

Common Student Errors

Students frequently find constructing box-and-whisker plots to be a difficult skill. While there are several steps to the construction, they are not very difficult and are all very logical. Be ready to reassure students and calmly guide them through the steps.

Just the Math

Students will consider outliers for data sets.

Grouping

Ask for a student volunteer to read Question 4 aloud. Have a student restate the problem. Pose the Guiding Questions on the right to verify student understanding. Have students work together in small groups to complete Questions 4 and 5 then call the class together to discuss and explain their work.

Investigate Problem 1

2. To create a box-and-whisker plot of the cereal data, use the extremes and quartiles that you found in Problem 1. Complete the information below from Problem 1.

Lower extreme	1
First quartile, Q_1	4.5
Second quartile, Q_2	7.5
Third quartile, Q_3	10.5
Upper extreme	32

Draw a number line below and label it to represent the full range of data values. Locate the second quartile, Q_2 on the number line. About an inch above the number line, draw a dot for Q_2 and label its value.

Repeat this process to draw dots for the first and third quartiles and for the upper and lower extremes.

Now, draw a box with sides at the first and third quartiles. Draw a vertical line through the median. Draw two whiskers from the sides of the box to the extremes.

3. From the box and whisker plot above, why is one whisker longer than the other? What does this mean? Use a complete sentence in your answer.

Sample Answer: The right whisker is longer because the value of 32 for Sugary Puffs is far to the right of all the other values.

4. Just the Math: Outliers An **outlier** is a data value that is much greater than or much less than the other values in the set. What is the outlier in the breakfast cereal data? Make a conjecture, or an educated guess, about which quartile would be the most affected by removing the outlier. Write your answer and your conjecture using complete sentences.

The outlier in the data is the value for Sugary Puffs, 32. Removing the outlier would affect the second quartile the most.

Guiding Questions

- What is an outlier for a data set?
- What value do you think might be an outlier for the sugar amounts?
- How might an outlier affect the box-and-whisker plot?

Explore Together

Investigate Problem 1

Just the Math

Students will extend their understanding of quartiles to see them as special percentiles.

Grouping

Ask for a student volunteer to read Question 6 aloud. Pose the Guiding Questions below to verify student understanding. Have students work together in small groups to complete Questions 5 and 6, then call the class back together to have the students discuss and present their work for Questions 5 and 6.

Guiding Questions

- What is a percentile?
- How many of the breakfast cereals had less than 10 grams of sugar?
- How many breakfast cereals were listed in the original table?
- What number do you get if you divide 13 by 20?
- What percentile ranking is 10 grams of sugar?
- How many cereals have less than 15 grams of sugar? What percentile ranking is 15?

Take Note

Percentiles are often used in health-related fields. For instance, if a 6-year-old child's height is in the 78th percentile, then 78 percent of all 6-year-old children's heights are less than the child's height.

Investigate Problem 1

5. Remove the outlier and find the values of the quartiles. Complete the table to see if you were correct.

	Original data	Data with outlier removed
First quartile, Q_1	4.5	4
Second quartile, Q_2	7.5	7
Third quartile, Q_3	10.5	10

6. **Just the Math: Percentiles** Quartiles divide the data set into four parts. **Percentiles** also divide the data set into parts. Based on the root of this word, into how many parts do you think percentiles divide the data set? Use a complete sentence in your answer.

   Percentiles divide the data set into 100 equal parts.

   The ***n*th percentile** for a data set is the value for which *n* percent of the numbers in the set are less than that value. For example, the median in a data set often represents the 50th percentile because 50% of the data are less than the median. In the original cereal data set, is the median the 50th percentile? Use a complete sentence in your answer.

   The median is the 50th percentile because 50% of the data are less than 7.5.

   What percentile ranking is the first quartile in this data set? Use a complete sentence to explain your reasoning.

   The first quartile is the 25th percentile because 5 of the 20 data values are below Q_1.

   The third quartile represents what percentile in this data set? Explain your answer using a complete sentence.

   The third quartile is the 75th percentile because 15 of the 20 data values are below Q_3.

7. **Just the Math: Interquartile Range** The **interquartile range,** or IQR, is the difference between the upper and lower quartiles ($Q_3 - Q_1$) and represents the range of approximately the middle 50% of the data. The IQR indicates the spread between the lower and upper quartiles. If it is a small number, then the middle 50% of the data are consistent. If it is a large number, then the middle 50% of the data are spread apart. Find the IQR for the original cereal data.

   The IQR is 10.5 – 4.5 = 6.

12

Wrap Up

Close

- Review all key terms and their definitions. Include the terms *range, upper extreme, lower extreme, quartile, first quartile, second quartile, third quartile, percentile, nth percentile, box-and-whisker plot, outlier, interquartile range,* and *IQR.*
- You may also want to review any other vocabulary terms that were discussed during the lesson, which may include *measures of central tendency, mean, median, mode, maximum, minimum, quarters, measures of dispersion, spread of data, measures of position, number line,* and *ranking.*
- Remind the students to write the key terms and their definitions in the notes section of their notebooks. You may also want the students to include examples.
- Ask the students the following:
  - Identify which measures of dispersion or spread of data you learned in this lesson.
  - What is the range? What information does the range tell us about a data set?
  - What is the IQR? What does the IQR tell us about a data set?
  - How is the IQR different from the range?
  - Identify which measures of position of data you learned in this lesson.
  - What measure of position tells us the least number in the data set?
  - What measure of position tells us the greatest number in the data set?
  - How can you find the percentile for a given number?
  - How can you find a number that is a given percentile?
  - How are these similar? How are they different?
  - What is a quartile?
  - How are quartiles and percentiles similar? How are quartiles and percentiles different?
  - What is a box-and-whisker plot?
  - How can you construct a box-and-whisker plot?
  - What information can we see quickly in a box-and-whisker plot?
  - How can you compare 2 or more data sets using box-and-whisker plots?

Follow Up

Assignment

Use the Assignment for Lesson 12.3 in the Student Assignments book. See the Teacher's Resources and Assessments book for answers.

Assessment

See the Assessments provided in the Teacher's Resources and Assessments book for Chapter 12.

Open-Ended Writing Task

Have the students make a list of 20 cereals and go to a store to find the amount of sugar in one serving of each cereal. They should then create a chart of the values and answer the questions from this lesson for their data. (You can choose a different ingredient, such as sodium, if you prefer.)

Alternatively, you can have them perform the following activity:

Have the students identify each of the measures of dispersion or spread of data that they learned in this lesson. They should also give a definition for and an example of each measure. (They should include both the range and the interquartile range in their answer. They should not include the percentiles because they are measures of position. The warm up activity for the next lesson will extend this understanding.)

Reflections

Insert your reflections on the lesson as it played out in class today.

What went well?

____

____

What did not go as well as you would have liked?

____

____

How would you like to change the lesson in order to improve the things that did not go well and capitalize on the things that did go well?

____

____

____

____

____

Notes

12

12.4

Home Team Advantage?

Sample Variance and Standard Deviation

Learning By Doing Lesson Map

Get Ready

Objectives

In this lesson, you will:

- Use a line plot to represent a data set.
- Find the deviation of a data set.
- Find the sample variance of a data set.
- Find the sample standard deviation of a data set.

Key Terms

- line plot
- deviation
- absolute deviation
- mean absolute deviation
- sample variance
- sample standard deviation

NCTM Content Standards

Grades 9–12 Expectations

Number and Operations Standard

- Judge the effects of such operations as multiplication, division, and computing powers and roots on the magnitudes of quantities.

Algebra Standard

- Generalize patterns using explicitly defined and recursively defined functions.

Data Analysis and Probability Standards

- For univariate measurement data, be able to display the distribution, describe its shape, and select and calculate summary statistics.
- Understand how sample statistics reflect the values of population parameters and use sampling distributions as the basis for informal inference.

Lesson Overview

Within the context of this lesson, students will be asked to:

- Create a line plot to represent a set of data.
- Calculate the deviations for each value in a data set.
- Calculate the square of each deviation, the sum of the squares and divide the sum by one less than the number of sample values to calculate the sample variance of a set of data.
- Calculate the sample standard deviation of a data set by finding the positive square root of the sample variance for the data set.

Essential Questions

The following key questions are addressed in this lesson:

1. What is a line plot?
2. What is a deviation?
3. How can you calculate the sample variance for a set of data?
4. How can you calculate the standard deviation for a set of data?
5. What do the deviations, the sample variance, and the sample standard deviation represent?

Show The Way

Warm Up

Place the following questions or an applicable subset of these questions on the board before students enter class. Students should begin working as soon as they are seated.

Use the data set shown below to complete each table.

12, 4, 8, 15, 7, 11, 9, 5, 4, 5, 8, 5, 4, 4, 3, 5, 4, 1, 3, 6, 8, 7, 4, 3, 2, 0, 1, 3, 7, 9, 9, 5, 14, 9

Measure of Central Tendency	Brief Definition	Value of Data
1. Mean	**Average, or sum of values divided by the number of values**	**6**
2. Median	**Middle value when the data is arranged from least to greatest**	**5**
3. Mode	**Most common data value; the value that occurs most often**	**4**

Measure of Position	Brief Definition	Value of Data
4. Lower extreme	**The least data value**	**0**
5. Upper extreme	**The greatest data value**	**15**
6. First Quartile	**The median of the lower half of the data**	**4**
7. Third Quartile	**The median of the upper half of the data**	**8**

Measure of Dispersion	Brief Definition	Value of Data
8. Range	**Difference of greatest and least data values**	**15**
9. IQR	**Difference between third and first quartile**	**4**

10. Which measure of central tendency, the mean, the median, or the mode, is also a measure of position? **The median**

Motivator

Begin the lesson with the motivator to get students thinking about the topic of the upcoming problem. This lesson is about analyzing the advantage of basketball teams. The motivating questions are about basketball.

Ask the students the following questions to get them interested in the lesson.

- Do you play basketball?
- Is it easier to play basketball if you are short, average height, or tall?
- Why would a coach want to know the heights of the players on an opposing team?

Explore Together

Problem 1

Students will create line plots for the heights of players on two different basketball teams, then use the line plots to compare the data sets.

Grouping

Ask for a student volunteer to read the Scenario and Problem 1 aloud. Have a student restate the problem. Pose the Guiding Questions below to verify student understanding. Begin part (A) of Problem 1 together as a whole class. If the students understand as you are completing this together as a class, you can have the students finish the problem individually and then review it with the students afterward.

Guiding Questions

- What information is given in this problem?
- What is a line plot?
- What does each X represent on a line plot?
- If 4 players on the same team have the same height of 67 inches, what would the line plot look like at 67?
- How can a line plot be considered to be a graph?
- How many players are on each team?
- How many X's should you have on each line plot?

Notes A line plot is sometimes called a dot plot. Instead of marking data values with an X, a dot is used.

Common Student Errors

Students sometimes mark an X on a line plot where there is no data to match. They don't recognize that if there are no players on the visiting team with a height of 70 inches, then the line plot should not have an X at that value. The Guiding Questions should help to prevent this error.

SCENARIO Your school's varsity basketball team has its first game of the season on their home court. They are playing a team about which they know very little. The coach does have a record of the heights of the players on the opposing team.

Problem 1 Height Advantage

A. To get a sense of which team may have an advantage, the coach asks you to analyze the heights of the players on both teams.

Home team heights (inches)	69	68	67	68	66	65	70	70	71	71
Visiting team heights (inches)	68	68	68	69	69	67	72	71	66	67

One way to represent this data is to draw a **line plot** for each team's set of heights. To make a line plot of the home team heights, first draw a number line that includes all the heights in both data sets. Then draw an X above the number line for each home team height. If there is more than one occurrence of a height, draw another X above the first occurrence.

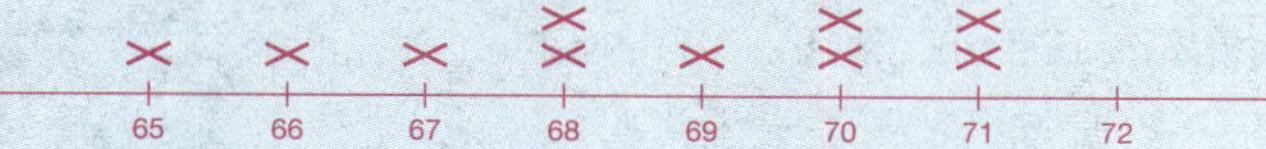

Use a similar number line to draw a line plot of the visiting team heights.

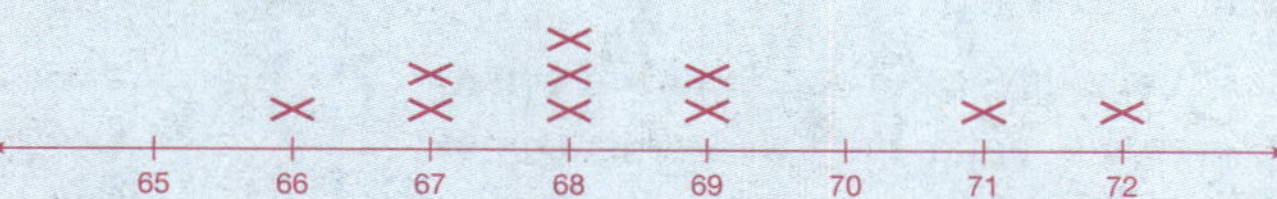

B. Compare the two line plots. Describe any difference that you see. Use complete sentences in your answer.

Sample Answer: The visiting team heights are clustered around 68 inches and the home team heights are spread out more evenly.

Explore Together

Problem 1

Students will investigate the variability of the heights of the players on the two teams.

Grouping

Ask for a student volunteer to read parts (B) and (C) of Problem 1 aloud. Have a student restate the problems. Pose the Guiding Questions below to verify student understanding. Have students work together in small groups to complete parts (B) and (C) of Problem 1.

Guiding Questions

- How can you calculate the mean?
- How can you calculate the range?

Call the class back together to have the students discuss and present their work for parts (B) and (C) of Problem 1.

Investigate Problem 1

Just the Math

In the previous lesson, students were introduced to the range and IQR as measures of dispersion. Students will be introduced formally to another measure of dispersion called deviation in Question 1. The sample mean called "x-bar" will also be formally presented in this question.

Grouping

Ask for a student volunteer to read Question 1 aloud. Have a student restate the problem. Pose the Guiding Questions below to verify student understanding. Complete Question 1 together as a whole class.

Guiding Questions

- What is a deviation?
- How can you calculate the difference between two values?
- What does the deviation represent in this problem situation?

Problem 1 Height Advantage

C. You decide that you will try to compare the heights by finding the mean and the range of each data set. Find these statistics. Show your work and use a complete sentence in your answer.

Home team mean:

$$\frac{69 + 68 + 67 + 68 + 66 + 65 + 70 + 70 + 71 + 71}{10} = 68.5$$

Visiting team mean:

$$\frac{68 + 68 + 68 + 69 + 69 + 67 + 72 + 71 + 66 + 67}{10} = 68.5$$

Home team range: $71 - 65 = 6$; Visiting team range: $72 - 66 = 6$

The mean height for both teams is 68.5 inches and the range of the heights for both teams is 6 inches.

Can you draw any conclusions about who has the advantage from the mean and the range? Why or why not? Use a complete sentence in your answer.

You cannot draw any conclusions from the mean and range because these values are the same for both teams.

Investigate Problem 1

1. **Just the Math: Deviation** Up until this point, you have observed differences in data sets by visually looking at their distributions, as you did in part (B). Other measures can help us to understand the distribution of a data set. One such measure is *deviation*. The **deviation** of a data value is the difference between the data value and the mean.

   We can use variables to represent the deviation by letting the variable x be the data value and by letting the variable $\bar{x}$, called "x-bar," be the mean. Write an expression for deviation.

   $(x - \bar{x})$

2. Complete the table by finding the deviation of each data value.

Home team heights (inches)	Deviation
69	0.5
68	−0.5
67	−1.5
68	−0.5
66	−2.5
65	−3.5
70	1.5
70	1.5
71	2.5
71	2.5

Visiting team heights (inches)	Deviation
68	−0.5
68	−0.5
68	−0.5
69	0.5
69	0.5
67	−1.5
72	3.5
71	2.5
66	−2.5
67	−1.5

Grouping

Ask for a student volunteer to read Question 2 aloud. Have a student restate the problem. Have students work together in small groups to complete Questions 2 and 3.

Explore Together

Investigate Problem 1

Students will find the sum of the deviations and calculate the mean deviation for the heights of the basketball players from each team.

Call the class back together to have the students discuss and present their work for Questions 2 and 3.

Key Formative Assessments

- Why is the mean deviation equal to zero for each team?
- Will the mean deviation for a data set always equal zero? Why?

Grouping

Ask for a student volunteer to read Question 4 aloud. Have a student restate the problem. Pose the Guiding Questions below to verify student understanding. Have students work together in small groups to complete Question 4. Then call the class back together to have the students discuss and present their work for Question 4.

Guiding Questions

- What is an absolute value?
- How can you calculate an absolute value?
- What is an absolute deviation?
- How can you calculate the absolute deviation for the data for each team?
- What is a mean?
- How can you calculate the mean absolute deviation for the data for each team?
- Do you expect the mean absolute deviation to be equal to zero as the mean deviation was equal to zero for each set of data? Why or why not?
- What is the importance of the mean absolute deviation for the data for each team?

Investigate Problem 1

3. Find the mean of the deviations for each team. Show your work and use a complete sentence in your answer.

The mean of the deviations of both teams is 0.

Is the mean of the deviations meaningful? Why or why not? Use a complete sentence in your answer.

For both teams, the deviation is not meaningful because the sum of the deviations is zero, which makes the mean of the deviation zero as well.

Were any of the deviations negative? If so, why? Use a complete sentence in your answer.

Some of the deviations were negative. When a data value is less than the mean, then the deviation is negative.

4. Although the mean of the deviations was not helpful in determining the spread of the data, you can find the absolute values of the deviations first and then find the mean of the results. These absolute values are called **absolute deviations.** Copy the deviations from the table on the previous page. Then complete the table by finding the absolute value of each deviation.

Home team heights (inches)	Deviation	Absolute value of deviation
69	0.5	0.5
68	−0.5	0.5
67	−1.5	1.5
68	−0.5	0.5
66	−2.5	2.5
65	−3.5	3.5
70	1.5	1.5
70	1.5	1.5
71	2.5	2.5
71	2.5	2.5

Visiting team heights (inches)	Deviation	Absolute value of deviation
68	−0.5	0.5
68	−0.5	0.5
68	−0.5	0.5
69	0.5	0.5
69	0.5	0.5
67	−1.5	1.5
72	3.5	3.5
71	2.5	2.5
66	−2.5	2.5
67	−1.5	1.5

Find the mean of the absolute values of the deviations, called the **mean absolute deviation,** for each team. Does the mean absolute deviation help you to determine the spread of the data? Why or why not? Use complete sentences in your answer.

The mean absolute deviation of the home team is 1.7 and the mean absolute deviation of the visiting team is 1.4. These values help to determine the spread because they indicate the average deviation of the data points from the mean for each data set.

Common Student Errors

The terminology in this lesson is new and overwhelming to some students. Be ready to ask students the Guiding Questions and have them rephrase the definitions of Key Terms as a class when needed to refocus confused students.

12

Explore Together

Investigate Problem 1

Students will calculate the numerator of the formula for the sample variance for each data set.

Grouping

Ask for a student volunteer to read Question 5 aloud. Have a student restate the problem. Pose the Guiding Questions below to verify student understanding. Complete the beginning of Question 5 together as a whole class. When the students seem to understand what is being asked and how to find the values, have them work individually to complete the table. Then have the students present and explain their table of values. Next, complete the remainder of Question 5 together as a whole class.

Guiding Questions

- What does it mean for something to vary?
- What is a sample?
- What do you think a sample variance might be a measure of?
- What is a square of a value?
- How can you find the square of a deviation that you already calculated?
- What symbol did we use earlier in this lesson to represent the sample mean?
- What do you think we mean when we say let x_1 be the first data value?

Common Student Errors

The notation used in the calculation of the sample variance is complicated and difficult for most students to understand. Be ready to review the notation and explain it in detail.

Notes The letter *i* in this notation represents the word *index*. It is the counter for the formula to tell us what term we are referring to. Ask the students what the value for *i* was when we wrote x_1, x_2, and x_3. Ask student why we would use the letter *i* to represent the subscript value in a formula rather than using the letter *x*. Explain that the formula given is read as "The sum of the squares of the deviations for each *x* value from the first *x*-value through the *n*th *x*-value." Ask the students what they think the formula tells us to do and how it relates to the values they found.

Investigate Problem 1

5. **Just the Math: Sample Variance** Even though the mean absolute deviation is meaningful, it is not commonly used to measure variability, or how the data values within a set differ. Let's consider a more common measure of variability by finding the *sample variance* of each data set. In the table below, copy the deviations that you found in Question 2. Then find the square of each deviation for both the home team and the visiting team.

Home team heights (inches)	Deviation	Square of deviation
69	0.5	0.25
68	–0.5	0.25
67	–1.5	2.25
68	–0.5	0.25
66	–2.5	6.25
65	–3.5	12.25
70	1.5	2.25
70	1.5	2.25
71	2.5	6.25
71	2.5	6.25

Visiting team heights (inches)	Deviation	Square of deviation
68	–0.5	0.25
68	–0.5	0.25
68	–0.5	0.25
69	0.5	0.25
69	0.5	0.25
67	–1.5	2.25
72	3.5	12.25
71	2.5	6.25
66	–2.5	6.25
67	–1.5	2.25

Let x_1 be the first data value and let $\bar{x}$ be the mean. Write an expression for the square of the deviation of the first data value.

$(x_1 - \bar{x})^2$

Let x_1 be the first data value, x_2 be the second data value, x_3 be the third data value, and so on. Write an expression for the sum of the squares of the deviations of a data set with 10 data values.

$$(x_1 - \bar{x})^2 + (x_2 - \bar{x})^2 + (x_3 - \bar{x})^2 + (x_4 - \bar{x})^2 + (x_5 - \bar{x})^2 + (x_6 - \bar{x})^2 + (x_7 - \bar{x})^2 + (x_8 - \bar{x})^2 + (x_9 - \bar{x})^2 + (x_{10} - \bar{x})^2$$

Another way to write this sum is to use the uppercase Greek letter sigma (Σ) which means "sum":

$$\sum_{i=1}^{n} (x_i - \bar{x})^2$$

The variable *n* represents the number of data values in the data set. In our problem situation, $n = 10$.

12

Explore Together

Investigate Problem 1

Students will calculate the sample variance for each data set of players' heights.

Grouping

Continue to complete Question 5 together as a whole class.

Notes Explain that the formula to calculate the sample variance has us divide the sum of the squares of the deviations by one less than the number of values in the sample. Students often will not recognize that this is very close to finding the mean of the square of the deviations. It may help to tell the students that dividing by n, rather than one less than n would give the value for the population variance which is the mean of the square of the deviations. We divide by a smaller number than n for a sample variance to make it a slightly larger number than the corresponding population variance would be.

Just the Math

The students will calculate the sample standard deviation from the sample variance in Question 6. The sample standard deviation is a very useful measure of the spread of the data. An advantage of the standard deviation over the variance is that the unit for the standard deviation is the same as the unit for the data. So the sample standard deviation for a set of times measured in seconds is also measured in seconds, while the sample variance for that data is measured in squared seconds. It is also important to tell students that most data values are within 2 standard deviations of the mean.

Grouping

Complete Questions 6 through 8 together as a whole class.

Key Formative Assessments

- What is a deviation?
- What does the sample standard deviation represent?

Investigate Problem 1

To find the **sample variance,** or s^2, divide the sum by one less than the number of data values in the data set. This is shown in symbols below.

Sample variance: $s^2 = \dfrac{\sum_{i=1}^{n} (x_i - \bar{x})^2}{n - 1}$

Find the sample variance for both the home team and the visiting team. Round your answers to the nearest hundredth. Show your work and use a complete sentence in your answer.

Home team: $s^2 = \dfrac{38.5}{9} \approx 4.28$

Visiting team: $s^2 = \dfrac{30.5}{9} \approx 3.39$

The sample variance of the home team is 4.28 and the sample variance of the visiting team is 3.39.

6. **Just the Math: Sample Standard Deviation** You can use the sample variance to find the **sample standard deviation,** or s, which is a common measure of the variability of a data set. To find the sample standard deviation, take the square root of the sample variance. This is shown in symbols below.

Sample standard deviation: $s = \sqrt{\dfrac{\sum_{i=1}^{n} (x_i - \bar{x})^2}{n - 1}}$

Find the sample standard deviation for the both the home team and the visiting team. Round your answers to the nearest hundredth. Use a complete sentence in your answer.

The sample standard deviation of the home team is 2.07 and the sample standard deviation of the visiting team is 1.84.

7. Use complete sentences to explain the meaning of the sample standard deviation of the home team and the sample standard deviation of the visiting team.

The sample standard deviation of the home team indicates that the heights of the players on the home team vary, on average, 2.07 inches from the mean of 68.5 inches and the heights of the players on the visiting team vary, on average, 1.84 inches from the mean of 68.5 inches.

8. How does finding the sample standard deviation help you to compare the heights of the players on the two teams? Use a complete sentence in your answer.

Using the sample standard deviation, you know that more of the players are closer to the mean height of 68.5 inches on the visiting team than on the home team.

- Which set of data values is more consistent, one with a larger standard deviation or one with a smaller standard deviation?

Wrap Up

Close

- Review all key terms and their definitions. Include the terms *line plot, deviation, absolute deviations, mean absolute deviation, sample variance,* and *variance.*
- You may also want to review any other vocabulary terms that were discussed during the lesson, which may include *dot plot, variability, sample, population, sample size n,* Σ, *index, mean, x-bar,* and *range.*
- Remind the students to write the key terms and their definitions in the notes section of their notebooks. You may also want the students to include examples.
- Ask the students to compare and contrast the sample variance and the sample standard deviation for data sets.
- Ask the students why they think we use the standard deviation more often than the variance to investigate how consistent or inconsistent data sets are.
- Ask the students to explain the difference between a population and a sample.
- You may want to tell the students that population standard deviation is represented by the lower case Greek symbol sigma that is written as σ. The population variance is represented by the square of the same symbol as σ^2.
- Have the students discuss the importance of the statement that typical data values are within 2 standard deviations of the mean.
- Ask the students to estimate the mean and standard deviation for a topic of interest to them. Ask the class to suggest data values that would be common and data values that would be uncommon or rare.
- Have the students summarize how to find the deviations for a data set, the sample variance, and the sample standard deviation.

Follow Up

Assignment

Use the Assignment for Lesson 12.4 in the Student Assignments book. See the Teacher's Resources and Assessments book for answers.

Assessment

See the Assessments provided in the Teacher's Resources and Assessments book for Chapter 12.

Open-Ended Writing Task

Have the students write a short essay to answer the following question. If we are told that most data values are within 2 standard deviations of the mean, and we know that the mean IQ score is 100 points and the standard deviation for IQ scores is 15 points, what interval of IQ values do you think is common? Explain your answer.

Reflections

Insert your reflections on the lesson as it played out in class today.

What went well?

What did not go as well as you would have liked?

How would you like to change the lesson in order to improve the things that did not go well and capitalize on the things that did go well?

Notes

12

Looking Ahead to Chapter 13

13

Focus

In Chapter 13, you will learn how to write quadratic equations in different forms. You will also learn how to work with and graph exponential functions, as well as how to perform transformation of graphs of functions. In addition, you will learn how to use the Pythagorean Theorem and the Midpoint and Distance Formulas.

Chapter Warm up

Answer these questions to help you review skills that you will need in Chapter 13.

Solve each quadratic equation.

1. $x^2 - x - 12 = 0$

$x = 4, x = -3$

2. $2x^2 - 128 = 0$

$x = 8, x = -8$

3. $2x^2 + 10x = 15$

$x \approx 1.208, x \approx -6.208$

Solve each equation for the indicated variable.

4. $3\pi + tc = h$; t

$t = \frac{h - 3\pi}{c}$

5. $rs + st = 4$; s

$s = \frac{4}{r + t}$

6. $\frac{3x + 4y}{z} = 8$; y

$y = \frac{8z - 3x}{4}$

Read the problem scenario below.

The equation that represents the height y of a model rocket in feet as a function of the time t in seconds after it has left the ground is given by $y = -16t^2 + 110t$.

7. When does the rocket reach its maximum height? What is the maximum height of the rocket?

The rocket reaches a maximum height of about 189 feet about 3.44 seconds after it takes off.

8. What are the x-intercepts of the graph of the function modeling the height of the rocket? What do they represent in the problem situation?

The x-intercepts are 0 and 6.875 seconds. They represent the times when the rocket is on the ground.

Key Terms

legs ■ p. 571
hypotenuse ■ p. 571
Pythagorean Theorem ■ p. 572
converse ■ p. 573
converse of the Pythagorean Theorem ■ p. 573
Distance Formula ■ p. 578
midpoint ■ p. 579
Midpoint Formula ■ p. 581
perfect square trinomial ■ p. 584
factor ■ p. 584
complete the square ■ p. 584
Quadratic Formula ■ p. 586
standard form ■ p. 587
factored form ■ p. 587
vertex form of a quadratic equation ■ p. 588
parent function ■ p. 598
translation ■ p. 598
reflected ■ p. 598
exponential function ■ p. 603
simple interest ■ p. 607
compound interest ■ p. 608
exponential growth model ■ p. 609
growth rate ■ p. 609
growth factor ■ p. 609
exponential decay model ■ p. 614
decay rate ■ p. 614
decay factor ■ p. 614
logical reasoning ■ p. 616
proof ■ p. 616
direct proof ■ p. 617
counterexample ■ p. 617
indirect proof ■ p. 618

CHAPTER

13

Quadratic and Exponential Functions and Logic

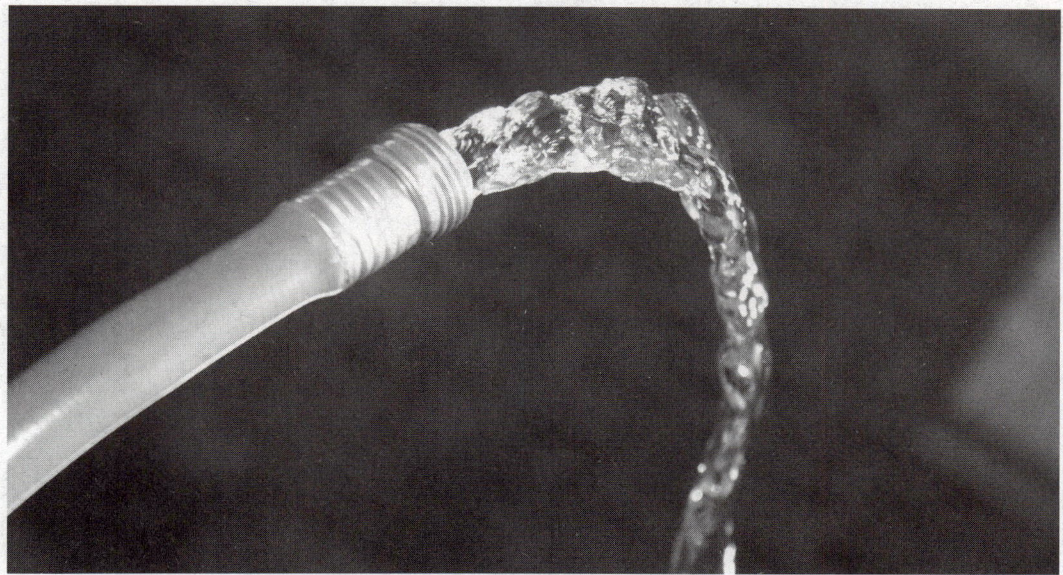

A typical garden hose can deliver between 9 and 17 gallons of water per minute. Most hoses are made from rubber, vinyl, or a rubber-vinyl combination. In Lesson 13.4, you will use quadratic functions to model the path of the water of a garden hose.

13

13.1 Solid Carpentry

The Pythagorean Theorem and Its Converse

13

Learning By Doing Lesson Map

Get Ready

Objectives

In this lesson, you will:

- Use the Pythagorean Theorem to find the side length of a right triangle.
- Use the converse of the Pythagorean Theorem to determine whether a triangle is a right triangle.

Key Terms

- legs
- hypotenuse
- converse of the Pythagorean Theorem
- Pythagorean Theorem
- converse

Materials

- graph paper
- rulers
- scissors

NCTM Content Standards

Grades 9–12 Expectations

Number and Operations Standards

- Judge the effects of such operations as multiplication, division, and computing powers and roots on the magnitude of quantities.
- Develop fluency in operations with real numbers, vectors, and matrices, using mental computation or paper-and-pencil calculations for simple cases and technology for more complicated cases.

Algebra Standards

- Generalize patterns using explicitly defined and recursively defined functions.
- Understand and perform transformations, such as arithmetically combining, composing, and inverting commonly used functions, using technology to perform such operations on more complicated symbolic expressions.
- Use symbolic algebra to represent and explain mathematical relationships.
- Identify essential quantitative relationships in a situation and determine the class or classes of functions that might model the relationships.
- Use symbolic expressions, including iterative and recursive forms, to represent relationships arising from various contexts.

Geometry Standard

- Use geometric models to gain insights into, and answer questions in, other areas of mathematics.

Measurement Standards

- Make decisions about units and scales that are appropriate for problem situations involving measurement.
- Understand and use formulas for the area, surface area, and volume of geometric figures, including cones, spheres, and cylinders.

Lesson Overview

Within the context of this lesson, students will be asked to:

- Verify the Pythagorean Theorem visually with constructions.
- Identify the legs and hypotenuse of a right triangle.
- Apply the Pythagorean Theorem.
- Apply the converse of the Pythagorean Theorem.

Essential Questions

The following key questions are addressed in this lesson:

1. What is a carpenter and what does a carpenter do?
2. What is the Pythagorean Theorem and how can you use it?
3. What is the converse of the Pythagorean Theorem and how can you use it?
4. What are the parts of a right triangle?

Show The Way

13

Warm Up

Place the following questions or an applicable subset of these questions on the board before students enter class. Students should begin working as soon as they are seated.

Evaluate each expression.

1. 2^2 **4**

2. 5^2 **25**

3. 1^4 **1**

4. $(-4)^2$ **16**

5. $(-2)^3$ **−8**

6. $(-2)^5$ **−32**

7. 3^2 **9**

8. 8^2 **64**

9. 4^2 **16**

10. 10^2 **100**

11. 10^5 **100,000**

12. 7^2 **49**

Find the area of each figure below.

13.

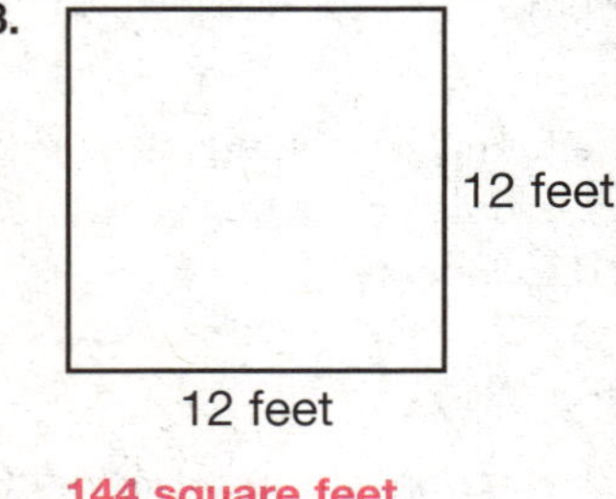

144 square feet

14.

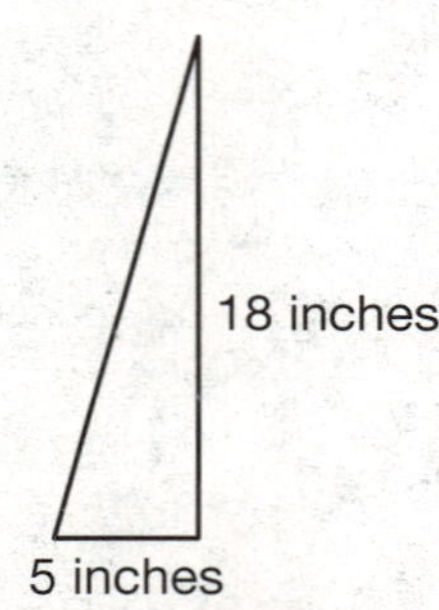

45 square inches

15.

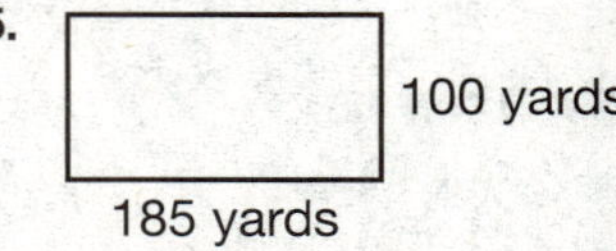

18,500 square yards

Motivator

Begin the lesson with the motivator to get students thinking about the topic of the upcoming problem. This lesson is about a carpenter working on a window. The motivating questions are about carpenters and their tools.

Ask the students the following questions to get them interested in the lesson.

- What is a carpenter?
- What does a carpenter do?
- What tools can you think of that a carpenter would use?
- What is a carpenter's square and what do they use it for?

Explore Together

Problem 1

Students will classify a triangle as a right triangle based on the lengths of its legs and hypotenuse.

Grouping

Ask for a student volunteer to read the Scenario, Problem 1, and part (A) aloud. Have a student restate the problem. Pose the Guiding Questions below to verify student understanding. Have students work together in small groups to complete parts (A) through (F).

Guiding Questions

- What information is given in this problem?
- What is a right angle?
- What is a right triangle?
- What is a leg of a right triangle?
- What is the hypotenuse of a right triangle?
- Which side do you think is the longest in a right triangle?
- How can you draw a right triangle with legs 3 units and 4 units on graph paper?
- How can you calculate the area of a square?
- What type of units will you find for the area of the squares?

Notes Many students will have already been introduced to the Pythagorean Theorem in a previous math class. These students may move through this lesson very quickly. However, they will still benefit from gaining a full and clear understanding of these concepts.

SCENARIO A carpenter is building a window frame and needs to make sure that the pieces of wood that form the frame meet at a right angle. However, he has dropped his metal square that he uses to check angles and the square has bent, so he needs another method for making sure that the angles are right angles.

13

Problem 1 Building a Window Frame

The carpenter measures four inches from a corner up the window frame and makes a mark. Then, he measures three inches from the same corner across the bottom of the window frame and makes a mark. Finally, he measures the straight-line distance from each of his marks and finds that this distance is five inches, as shown below. He concludes that this corner is a right angle. How does the carpenter know he is correct?

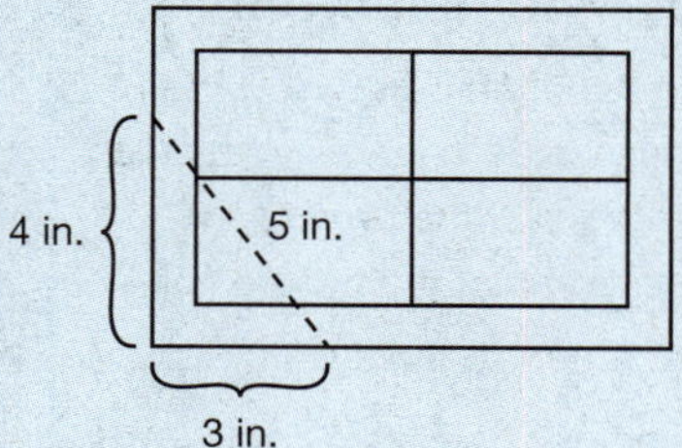

A. You will test the carpenter's method. First, on graph paper, draw a right triangle so that one leg is three units long and the other leg is four units long. Then draw a square on the hypotenuse of the triangle as shown.

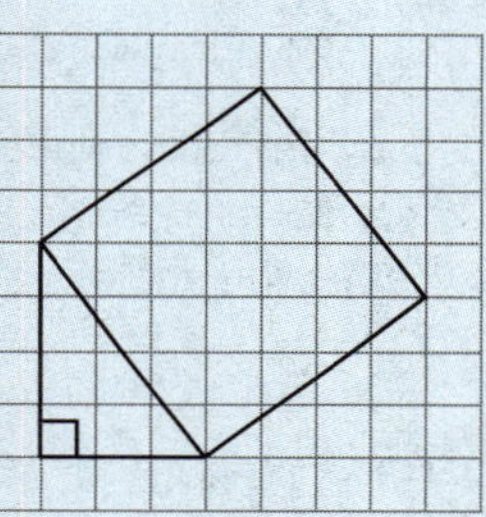

Now cut out a square from the graph paper that has a side length of three blocks. Then cut out another square from the graph paper that has a side length of four blocks. What is the area of each square? Use a complete sentence in your answer.

The areas of the squares are 9 square units and 16 square units.

Cut these squares into strips that are three blocks (or four blocks) long and one block wide. Arrange these strips on top of the square that is along the hypotenuse.

Take Note

In a right triangle, the sides that form the right angle are the **legs** and the side that is opposite the right angle is the **hypotenuse**.

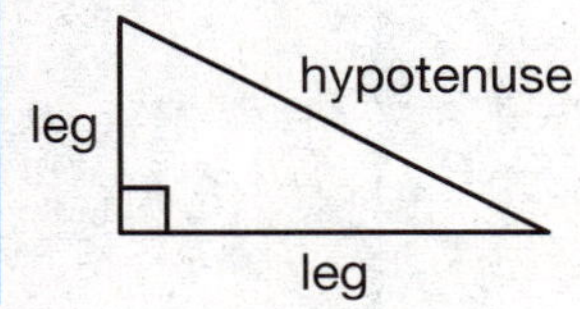

Explore Together

Problem 1

Students will investigate the Pythagorean Theorem.

Grouping

Students will be working together in small groups on these questions.

Call the class back together to have the students discuss and present their work for parts (A) through (F) of Problem 1.

Key Formative Assessments

- What was the relationship between the sum of the areas of the squares with side lengths of the legs and the area of the square with side length of the hypotenuse?
- Do you think this relationship will be true for all right triangles?
- Summarize this relationship.
- How would this relationship help the carpenter to check a frame to decide whether the angle is a right angle?

Investigate Problem 1

Students will be formally introduced to the Pythagorean Theorem.

Notes You may need to explain to the students that a theorem is a statement that has been proven to be true.

Grouping

Ask for a student volunteer to read Question 1 aloud. Have a student restate the problem. Pose Guiding Questions to verify student understanding. Complete Question 1 together as a class.

Guiding Questions

- What is a consequence?
- What is a theorem?
- What do a and b represent in the triangle? Does it matter which leg you call length a and which you call length b?
- What does c represent?

Problem 1 Building a Window Frame

B. What is the relationship between the areas of the squares that you cut into strips and the area of the square that is along the hypotenuse? Use a complete sentence in your answer.

The sum of the areas of the squares cut into strips equals the area of the square along the hypotenuse.

C. What is the area of the square that is along the hypotenuse? Show your work and use a complete sentence to explain your reasoning.

9 + 16 = 25; The area is 25 square units because the area is the sum of the areas of the smaller squares.

D. What is the length of the hypotenuse? Use a complete sentence to explain your reasoning.

The hypotenuse is five blocks long because it is the side length of the square that has an area of 25 square units.

E. Write the area of each of the three squares as a power. Then write an equation that relates these powers.

$3^2 + 4^2 = 5^2$

F. Was the carpenter's conclusion correct? Why or why not?

Yes, because the sum of the squares of the side lengths of the triangle is equal to the square of the hypotenuse.

Investigate Problem 1

1. In Problem 1, you demonstrated a consequence of the *Pythagorean Theorem.* The **Pythagorean Theorem** states that if a and b are the lengths of the legs of a right triangle and c is the length of the hypotenuse, then

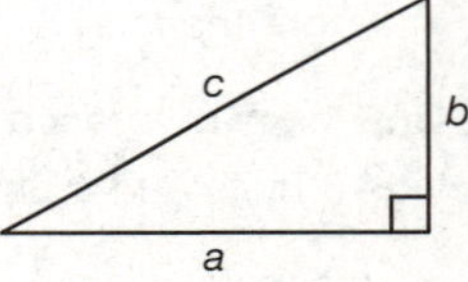

$a^2 + b^2 = c^2$.

Do you think that the Pythagorean Theorem works for triangles that are not right triangles? Use complete sentences to explain your reasoning.

No. Sample Answer: Consider a triangle that has sides that are three units long and four units long so that the angle that connects these sides is an angle that measures more than 90 degrees. The length of the third side is longer than five units, so the equality in the theorem does not hold.

Explore Together

Investigate Problem 1

Students will investigate the converse of the Pythagorean Theorem.

Grouping

Ask for a student volunteer to read Question 2 aloud. Have a student restate the problem. Pose the Guiding Questions below to verify student understanding. Have the students complete Question 2 individually. Then, call the class together to discuss Question 2.

Ask for a student volunteer to read Question 3 aloud. Have a student restate the problem.

Guiding Questions

- What information is given in this problem?
- For the statement, "If you work hard and do your algebra assignments carefully, then you will learn algebra and be able to get a better job," what is the original behavior? What is the consequence of the behavior?
- What is the converse of the statement, "If a rectangle has 4 equal sides, then it is a square?" Is that a true statement?
- Is the converse also true?
- State the Pythagorean Theorem as an "if-then" statement.
- What is the original condition for the Pythagorean Theorem? What is the consequence?

Grouping

Have students work together in small groups to complete Questions 3 through 7. Pose the Guiding Questions below to verify student understanding.

Guiding Questions

- How can you use the Pythagorean Theorem to determine the length of an unknown side of a right triangle?
- How can you use the converse of the Pythagorean Theorem to determine whether a triangle is a right triangle? How can you use this information to determine whether an angle is a right angle?

Investigate Problem 1

2. In Problem 1, you used the *converse* of the Pythagorean Theorem. The **converse** of a statement in "if-then" form is created by exchanging the "if" part and the "then" part of the statement. Complete the statement of the converse of the Pythagorean Theorem below by switching the "if" and "then" parts of the Pythagorean Theorem.

   Suppose that a, b, and c are the sides of a triangle.
   The **converse of the Pythagorean Theorem** states that if $a^2 + b^2 = c^2$, then the triangle is a right triangle.

3. The carpenter also has to repair some woodwork on a front porch and needs a ladder to do so. The work needs to be done on an area of the porch that is 12 feet above the ground. There is a bush in front of the porch where he needs to place the ladder, so he needs to place the ladder 5 feet from the porch. What length of ladder does the carpenter need to reach the area of the porch that needs repaired? Show your work and use a complete sentence in your answer.

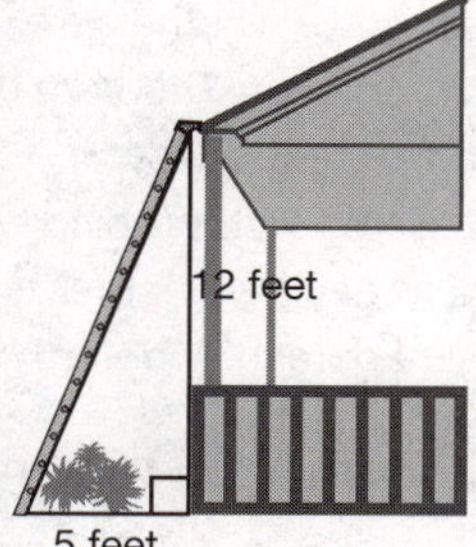

$5^2 + 12^2 = c^2$

$25 + 144 = c^2$

$169 = c^2$

$13 = c$

The ladder needs to be 13 feet long.

4. The carpenter is installing a door and is checking to see whether the opening in the wall forms a rectangle. The opening is 72 inches tall and 30 inches wide. The diagonal of the opening is 75 inches long. Does the opening form a rectangle? Show your work and use a complete sentence in your answer.

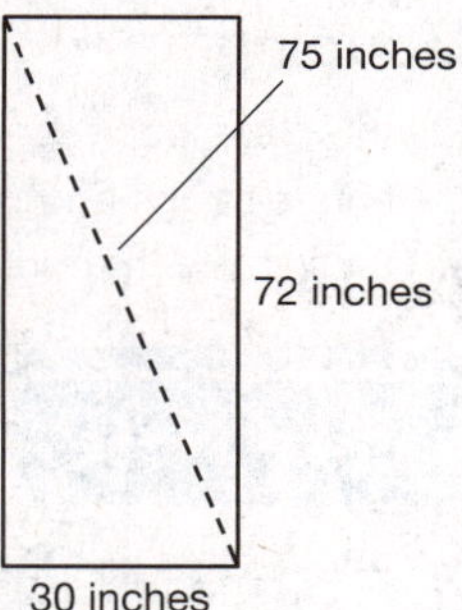

$72^2 + 30^2 \stackrel{?}{=} 75^2$

$5184 + 900 \stackrel{?}{=} 5625$

$6084 \neq 5625$

The opening does not form a rectangle.

Explore Together

Investigate Problem 1

13

Students will apply the Pythagorean Theorem and its converse.

Grouping

Students will be working in small groups to complete these questions.

Common Student Errors

Students will frequently not be able to recognize whether they are using the Pythagorean Theorem or its converse to complete a question. Be ready to guide students with probing questions to help them decide whether they are using the Pythagorean Theorem or its converse.

Common Student Errors

Students may ignore the diagram that is given and labeled for Question 5. Having them work together in small groups will usually help the students to alert others to this type of error when it happens. If students struggle to decide which side of the roof has what lengths, ask them what is shown in the diagram. It is vital for the students to learn to read diagrams.

Grouping

Call the class back together to have the students discuss and present their work for Questions 3 through 7.

Take Note

Whenever you see the share with the class icon, your group should prepare a short presentation to share with the class that describes how you solved the problem. Be prepared to ask questions during other groups' presentations and to answer questions during your presentation.

Key Formative Assessments

- What is the Pythagorean Theorem?
- What is the only type of triangle for which you can use the Pythagorean Theorem?
- What is the converse of the Pythagorean Theorem?
- What is a theorem?
- What are the important parts of a right triangle?
- How can a carpenter use the Pythagorean Theorem?

Investigate Problem 1

5. The carpenter is on another job building a shed. The roof of the shed will be triangular in shape, as shown. The roof will be ten feet wide and one side of the roof will be six feet long. How tall will the roof be? Show your work and use a complete sentence in your answer. Round your answer to the nearest tenth if necessary.

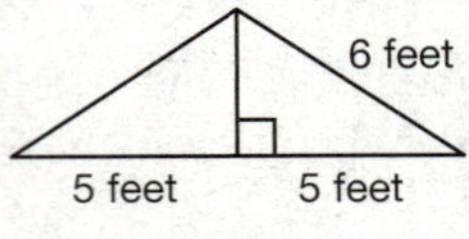

$6^2 = a^2 + 5^2$

$36 = a^2 + 5^2$

$11 = a^2$

$3.3 \approx a$

The roof will be about 3.3 feet tall.

6. In Questions 3 through 5, did you use the Pythagorean Theorem or the converse of the Pythagorean Theorem to answer the question? Use complete sentences to explain your reasoning.

Sample Answer: In Questions 3 and 5, I used the Pythagorean Theorem because I knew that the given triangle was a right triangle. In Question 4, I used the converse of the Pythagorean Theorem because I was trying to determine whether the triangle was a right triangle.

7. Use complete sentences to describe the situations in which you would use the Pythagorean Theorem to solve problems. Then use complete sentences to describe the situations in which you would use the converse of the Pythagorean Theorem to solve problems.

Sample Answer: When you know that a triangle is a right triangle, and you know the measures of two of the sides and you are trying to find the measure of the third side, you should use the Pythagorean Theorem. When you know the measures of the three sides of a triangle and you are trying to determine whether the triangle is a right triangle, you should use the converse of the Pythagorean Theorem.

Wrap Up

Close

- Review all key terms and their definitions. Include the terms *legs, hypotenuse, converse,* and *the Pythagorean Theorem.*
- You may also want to review any other vocabulary terms that were discussed during the lesson, which may include *theorem, right angle, right triangle, metal carpenters square,* and *Pythagoras.*
- Remind the students to write the key terms and their definitions in the notes section of their notebooks. You may also want the students to include examples.
- Ask the students to explain how they found the Pythagorean Theorem to be true.
- Ask the students to summarize the Pythagorean Theorem.
- Have the students state and explain the converse of the Pythagorean Theorem.
- Ask the students to explain all they know about the legs and hypotenuse of a right triangle.

Ties to the Cognitive Tutor Software

In the Cognitive Tutor software, students apply the Pythagorean Theorem to a number of different real-world situations. These applications help students to understand that the Pythagorean Theorem applies not only in cases where there is an object that is a right triangle but also in more abstract cases, where distances in orthogonal dimensions are known.

Follow Up

Assignment

Use the Assignment for Lesson 13.1 in the Student Assignments book. See the Teacher's Resources and Assessments book for answers.

Assessment

See the Assessments provided in the Teacher's Resources and Assessments book for Chapter 13.

Open-Ended Writing Task

One student makes the claim that the triangle checked by the carpenter worked because they measured 4 inches up the frame and 3 inches to the right. But if the carpenter had measured 3 inches up and 4 to the right, then it wouldn't have been a right angle. Ask the students to write a response to that statement explaining their opinion and why they feel that way.

13

Reflections

Insert your reflections on the lesson as it played out in class today.

What went well?

____

____

What did not go as well as you would have liked?

____

____

How would you like to change the lesson in order to improve the things that did not go well and capitalize on the things that did go well?

____

____

____

____

____

Notes

13.2

Location, Location, Location

The Distance and Midpoint Formulas

13

Learning By Doing Lesson Map

Get Ready

Objectives

In this lesson, you will:

- Find the distance between two points in the coordinate plane.
- Find the midpoint between two points in the coordinate plane.

Key Terms

- Distance Formula
- midpoint
- Midpoint Formula

NCTM Content Standards

Grades 9–12 Expectations

Number and Operations Standards

- Use number-theory arguments to justify relationships involving whole numbers.
- Develop fluency in operations with real numbers, vectors, and matrices, using mental computation or paper-and-pencil calculations for simple cases and technology for more complicated cases.

Algebra Standards

- Generalize patterns using explicitly defined and recursively defined functions.
- Use symbolic algebra to represent and explain mathematical relationships.
- Identify essential quantitative relationships in a situation and determine the class or classes of functions that might model the relationships.
- Draw reasonable conclusions about a situation being modeled.

Measurement Standard

- Use unit analysis to check measurement computations.

Lesson Overview

Within the context of this lesson, students will be asked to:

- Calculate the distance between 2 points by creating right triangles and using the Pythagorean Theorem to solve for the distance.
- Develop and apply the Distance Formula.
- Develop and apply the Midpoint Formula.

Essential Questions

The following key questions are addressed in this lesson:

1. What is a map?
2. What is a distance?
3. What is a midpoint?
4. What is a formula?

Show The Way

13

Warm Up

Place the following questions or an applicable subset of these questions on the board before students enter class. Students should begin working as soon as they are seated.

Use unit analysis or your knowledge of units to complete each statement.

1. 1 kilometer = **1000** meters

2. 1 cm = **0.01** meters

3. 3.2 kilometers = **3200** meters

Evaluate each absolute value.

4. $|-2|$ **2**

5. $|1282|$ **1282**

6. $|-42 + 3|$ **39**

Find the arithmetic mean, or average, of each pair of numbers.

7. −10, −15 **−12.5**

8. 13,248, 328,204 **170,726**

9. 5, −3 **1**

Use the Pythagorean Theorem to find the missing length of each right triangle.

10. **4**

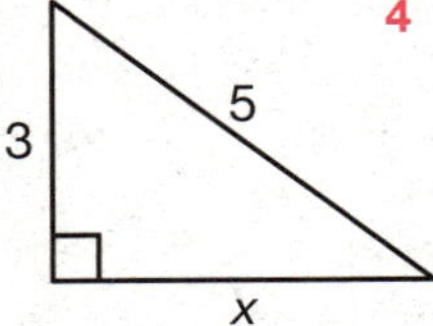

11. **13**

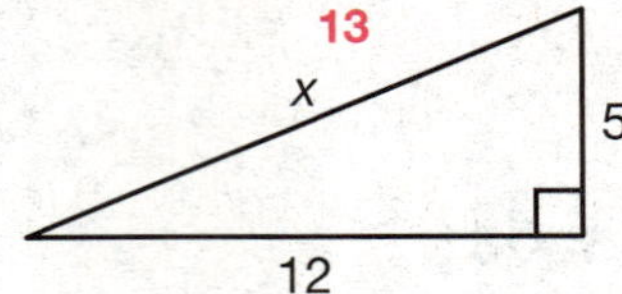

12. **15**

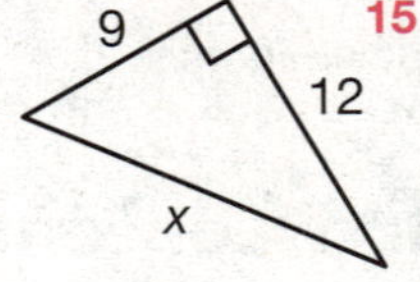

Motivator

Begin the lesson with the motivator to get students thinking about the topic of the upcoming problem. This lesson is about locations on maps. The motivating questions are about maps and how they are used.

Ask the students the following questions to get them interested in the lesson.

- What is represented on maps?
- What different types of maps have you seen?
- How could an excavation company use a map that represents the elevation of land at various points on a job site?
- What other uses for maps can you think of?
- Who makes maps?

Explore Together

Problem 1

Students will consider various maps and how to interpret locations on a map.

Grouping

Ask for a student volunteer to read the Scenario, Problem 1, and part (A) aloud. Have a student restate the problem. Pose the Guiding Questions below to verify student understanding. Have the students complete part (A) individually, then discuss the solution as a class.

Guiding Questions

- What information is given in this problem?
- What is a map? What is a globe?
- Why do we use maps?
- Why would people use the notation mentioned in this problem?
- What problem(s) might occur by writing the entire number out on the map for each coordinate.
- What point on the map do you think is considered the origin?
- What direction is north on a map? What direction is east? What direction is south? What direction is west?
- What type of line is a horizontal line? What type of line is a vertical line?
- Why do you think the horizontal distance from the origin in the bottom left corner to a point is called easting?
- What do you think is the term used to represent the vertical distance from the origin?

SCENARIO How many different kinds of maps of Earth do you think there are? The answer is many. You are probably most familiar with the map that is a globe. On this map, Earth is divided into sections by latitude and longitude lines that do not necessarily meet at right angles. Locations on a globe are measured in degrees, minutes, and seconds. Another kind of map is the Universal Transverse Mercator (UTM). This map of Earth uses a rectangular grid and the metric system of measurement.

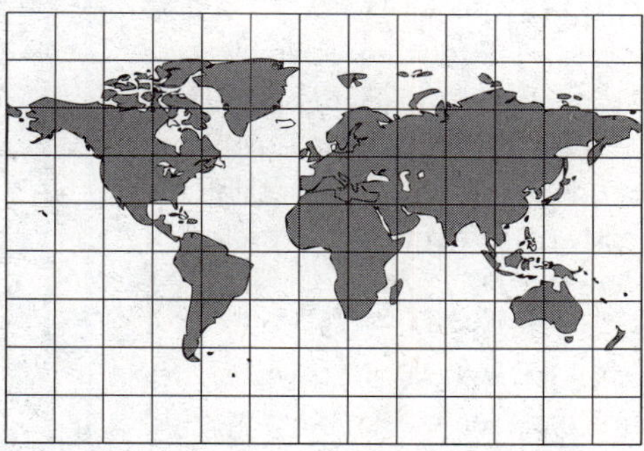

Problem 1 Where on Earth Are You?

Consider the map that shows elevations of an area in UTM coordinates at the right.

$^{3}88^{000\text{m}}$E $^{3}89$ $^{3}90$

$^{39}84^{000\text{m}}$N

$^{39}83$

$^{39}82$

B

A

A. On a UTM map, the *x*-coordinate is called an *easting* and is measured in meters. On the top portion of the map, you can see the label $^{3}88^{000\text{m}}$E. This means 388,000 meters east. The label to the right, $^{3}89$, is a shortened notation for $^{3}89^{000\text{m}}$E, or 389,000 meters east. What does the label $^{3}90$ mean?

$^{3}90^{000\text{m}}$E, or 390,000 meters east

Common Student Errors

Students sometimes are confused or other times careless when working with very large numbers. These numbers are difficult for students to envision and to check for reasonableness. Some incorrectly count the number of zeros in the notation, some will forget the numbers on the left, and some will ignore the unit. Remind the students to be careful and precise in working with large numbers.

13

Explore Together

Problem 1

Students will convert the notation for the map coordinates into coordinates measured in meters and in kilometers.

Grouping

Have students work together in small groups to complete parts (B) through (E) of Problem 1. Call the class back together to have the students discuss and present their work for parts (B) through (E) of Problem 1.

Common Student Errors

Students may forget how to convert coordinates from a distance measured in meters to a distance measured in kilometers. You may have to pose guiding questions to help the students convert measurements from meters to kilometers.

Guiding Questions

- How many meters are in 1 kilometer? How many meters are in 5 kilometers?
- How many kilometers are equal to 4000 meters? How many kilometers are equal to 400,000 meters?
- How can you convert a number from meters to kilometers?
- How can you convert a number from kilometers to meters?
- Why would we prefer to represent a number like 325,000 m as 325 km?

Investigate Problem 1

Grouping

Ask for a student volunteer to read Question 1 aloud. Have a student restate the problem. Pose the Guiding Questions at the right to verify student understanding. Have students work together in small groups to complete Questions 1 through 4.

Notes There are two correct answers to Question 2. In the answers to Question 3 through 6, the point (389, 3982) is used.

Problem 1 Where on Earth Are You?

B. The *y*-coordinate is called a *northing* and is also measured in meters. On the left portion of the map, you can see the label $^{39}84^{000\text{m}}$N. This means 3,984,000 meters north. The next label below, $^{39}83$, is a shortened notation for $^{39}83^{000\text{m}}$N, or 3,983,000 meters north. What does the label $^{39}82$ mean?

$^{39}82^{000\text{m}}$N, or 3,982,000 meters north

C. Write the coordinates of points *A* and *B* from the map.

easting of *A*: **388,000** meters
northing of *A*: **3,982,000** meters

easting of *B*: **389,000** meters
northing of *B*: **3,984,000** meters

D. Now write each set of coordinates in kilometers. Remember that 1 kilometer = 1000 meters.

easting of *A*: **388 kilometers**
northing of *A*: **3982 kilometers**

easting of *B*: **389 kilometers**
northing of *B*: **3984 kilometers**

E. Finally, write each point from part (D) in the coordinate notation (easting, northing).

Point *A*: (388, 3982); Point *B*: (389, 3984)

Investigate Problem 1

1. Label the points with their coordinates in kilometers on the map at the right. Then draw a line that connects the points.

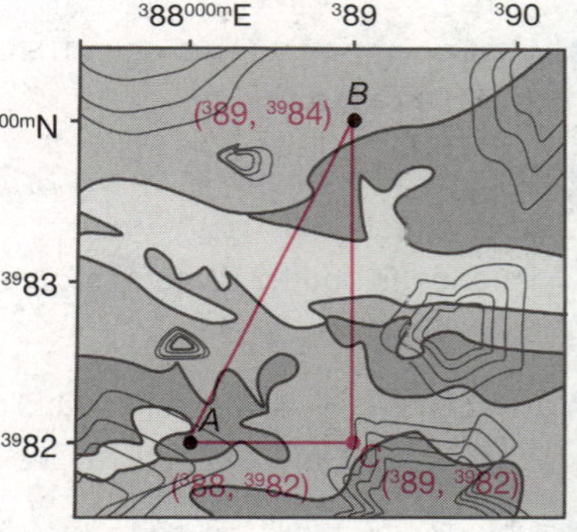

2. Let your line segment *AB* be the hypotenuse of a right triangle. Draw the legs of this right triangle and label the point where the sides meet as point *C*. What are the coordinates in kilometers of point *C*?

Point *C*: (389, 3982) or (388, 3984)

Guiding Questions

- How can you label the coordinates on the graph?
- What kind of line should you draw to connect the points, straight or curved?

Explore Together

Investigate Problem 1

Students will use the Pythagorean Theorem to find the distance between two locations on a map.

Grouping

Students will be working in small groups to complete these questions.

Call the class back together to have the students discuss and present their work for Questions 1 through 4.

Key Formative Assessments

- Why might we want to find the distance between 2 different locations?
- How can you find the distance between 2 points that are on the same horizontal line?
- How can you find the distance between 2 points that are on the same vertical line?
- Could we use the same process to find the distance between 2 points that are not on the same horizontal or vertical line?

Grouping

Ask for a student volunteer to read Question 5 aloud. Have a student restate the problem. Pose the Guiding Questions below to verify student understanding. Have students work together in small groups to complete Questions 5 and 6. Call the class back together to have the students discuss and present their work for Questions 5 and 6.

Guiding Questions

- What shape do the 3 lines connecting points *A*, *B*, and *C* create?
- How did you find the distance between points *A* and *C*? How did you find the distance between points *B* and *C*?
- Why can we not use the same process to find the distance between points *A* and *B*? How might you find the distance between points *A* and *B*?

Investigate Problem 1

How do the coordinates of point *C* relate to the coordinates of points *A* and *B*? Use complete sentences in your answer.

Sample Answer: The *x*-coordinate of point *C* is the same as the *x*-coordinate of point *B*. The *y*-coordinate of point *C* is the same as the *y*-coordinate of point *A*.

3. Write an expression that you can use to find the distance between point *A* and point *C* in kilometers. Then, find this distance. Use a complete sentence to explain how you found your answer.

   389 – 388 = 1;
   Subtract the *x*-coordinate of point *A* from the *x*-coordinate of point *C*.

4. Write an expression that you can use to find the distance between point *B* and point *C* in kilometers. Then find this distance. Use a complete sentence to explain how you found your answer.

   3984 – 3982 = 2;
   Subtract the *y*-coordinate of point *C* from the *y*-coordinate of point *B*.

5. Find the distance between point *A* and point *B*. Round your answer to the nearest tenth. Show your work and use a complete sentence in your answer.

   $AB^2 = AC^2 + BC^2$

   $AB^2 = 1^2 + 2^2$

   $AB^2 = 1 + 4$

   $AB^2 = 5$

   $AB = \sqrt{5}$

   $AB \approx 2.2$

   The distance between point *A* and point *B* is approximately 2.2 kilometers.

6. Complete the summary of the steps that you took to find the distance between points *A* and *B*.

$$AC^2 + BC^2 = AB^2$$

$$(\boxed{389} - \boxed{388})^2 + (\boxed{3984} - \boxed{3982})^2 = AB^2$$

$$\boxed{1}^2 + \boxed{2}^2 = AB^2$$

$$\boxed{1} + \boxed{4} = AB^2$$

$$\boxed{5} = AB^2$$

$$\boxed{\sqrt{5}} = AB$$

- What is the Pythagorean Theorem and how can it help to solve for this distance?
- Which side of the triangle *ABC* is the hypotenuse? Which 2 sides are the legs?

Explore Together

Investigate Problem 1

Students will calculate the distance between 2 points using the Distance Formula.

Just the Math

Students will be formally introduced to the Distance Formula. They will relate the formula to their work with the Pythagorean Theorem.

Grouping

Ask for a student volunteer to read Question 7 aloud. Have a student restate the problem. Pose the Guiding Questions below to verify student understanding. Complete Question 7 together as a class.

Guiding Questions

- What is a formula?
- Why might we want a formula for calculating the distance between 2 points rather than using the Pythagorean Theorem each time?
- Using the triangle given in the diagram for Question 7, what sum is equal to d^2?
- How can you use the fact that $d^2 = (x_2 - x_1)^2 + (y_2 - y_1)^2$ to calculate the value for d?
- Can d represent a negative number? Can d represent a positive number? Can d represent zero?
- What type of number is the square of a positive number? What type of number is the square of a negative number? What type of number is the square of zero?
- Can the square of a value ever be negative?
- Why don't we need the absolute value symbols that are shown in the diagram in the Distance Formula?

Investigate Problem 1

7. **Just the Math: The Distance Formula** We can use your method for finding the distance between points A and B to write the **Distance Formula:** if (x_1, y_1) and (x_2, y_2) are two points in the coordinate plane, then the distance d between (x_1, y_1) and (x_2, y_2) is given by

$$d = \sqrt{(x_2 - x_1)^2 + (y_2 - y_1)^2}.$$

The formula is illustrated in the figure below. We indicate that distance is positive by using the absolute value symbol.

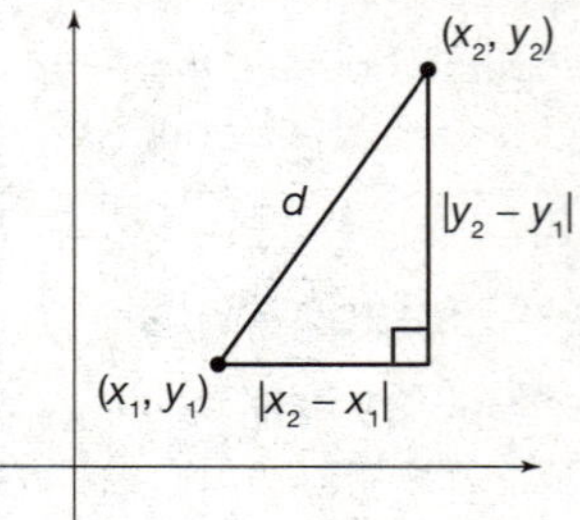

Do you think that it matters which point you identify as (x_1, y_1) and which point you identify as (x_2, y_2)? Use a complete sentence to explain your reasoning.

Sample Answer: No, because once you square each difference, the values are the same.

8. Find the distance between each pair of points. Round your answer to the nearest tenth if necessary. Show your work.

(0, 0) and (3, 4)

$d = \sqrt{(3 - 0)^2 + (4 - 0)^2}$
$= \sqrt{3^2 + 4^2}$
$= \sqrt{9 + 16}$
$= \sqrt{25}$
$= 5$

(1, 5) and (4, 8)

$d = \sqrt{(4 - 1)^2 + (8 - 5)^2}$
$= \sqrt{3^2 + 3^2}$
$= \sqrt{9 + 9}$
$= \sqrt{18}$
≈ 4.2

(0, –2), (3, 3)

$d = \sqrt{(3 - 0)^2 + [3 - (-2)^2]}$
$= \sqrt{3^2 + 5^2}$
$= \sqrt{9 + 25}$
$= \sqrt{34}$
≈ 5.8

(–1, 4), (3, 2)

$d = \sqrt{(-1 - 3)^2 + (4 - 2)^2}$
$= \sqrt{(-4)^2 + 2^2}$
$= \sqrt{16 + 4}$
$= \sqrt{20}$
≈ 4.5

Grouping

Have students work together in small groups to complete Question 8, then call the class back together to have the students discuss and present their work for this question.

Explore Together

Problem 2

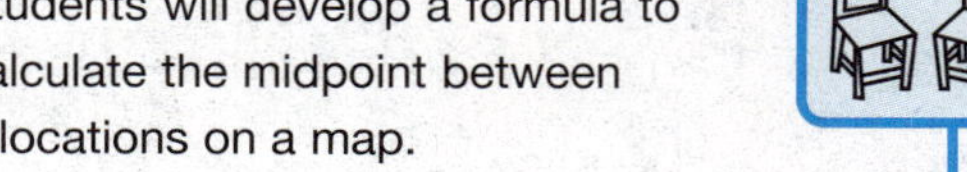

Students will develop a formula to calculate the midpoint between 2 locations on a map.

Grouping

Ask for a student volunteer to read Problem 2 aloud. Have a student restate the problem. Pose the Guiding Questions below to verify student understanding. Have students work together in small groups to complete parts (A) through (D) of Problem 2.

Guiding Questions

- What is meant by the term halfway? What is a midpoint?
- Why might we want to be able to identify the point that is halfway between 2 other points?
- How can you find the distance between 2 points on the same horizontal line? How can you find the distance between 2 points on the same vertical line?
- Is the midpoint located on the straight line that connects the 2 points? Will that always be true?

Common Student Errors

Some students may incorrectly add the coordinates to find the distance rather than subtracting the coordinates.

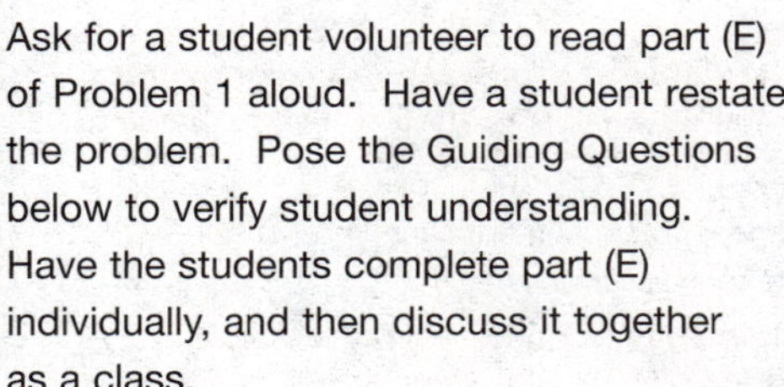

Ask for a student volunteer to read part (E) of Problem 1 aloud. Have a student restate the problem. Pose the Guiding Questions below to verify student understanding. Have the students complete part (E) individually, and then discuss it together as a class.

Guiding Questions

- If 2 points have the same y-coordinates, what kind of line connects those points, vertical or horizontal? What type of line connects 2 points with the same x-coordinates?
- If 2 points are 8 units apart, how far is the midpoint from each of those points?

Problem 2 Are We (Halfway) There Yet?

Consider the points A, B, and C labeled on the map at the right.

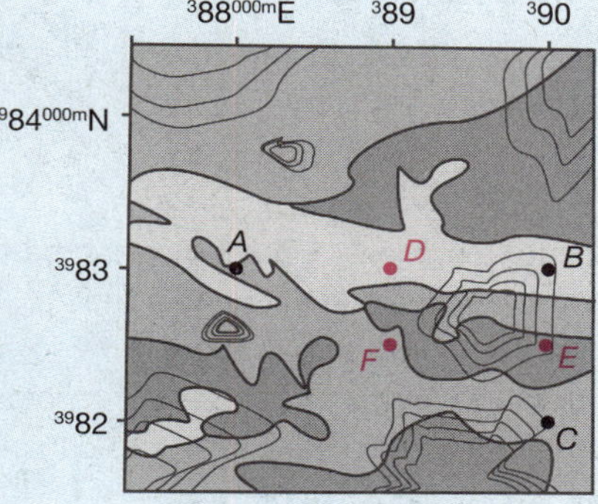

A. What is the distance between points A and B in kilometers? Show your work and use a complete sentence in your answer.

390 − 388 = 2; The distance between the points is two kilometers.

B. The point halfway between points A and B is called the **midpoint.** How many kilometers from point A is the midpoint? Show your work and use a complete sentence in your answer.

2 ÷ 2 = 1; The midpoint is one kilometer from point *A*.

C. What are the coordinates in kilometers of the point from part (B)? Plot this point and label it as point D on the map above. Show your work and use a complete sentence in your answer.

388 + 1 = 389; The coordinates of point *D* are (389, 3983).

D. Use complete sentences to describe how you would find the midpoint of a horizontal line segment.

Sample Answer: First find the horizontal distance between the points and divide the result by 2. Then, add this amount to the *x*-coordinate of the point on the left to get the *x*-coordinate of the midpoint. The *y*-coordinate is that of either endpoint because their *y*-coordinates are the same.

E. Consider the horizontal line segment with endpoints (x_1, y) and (x_2, y). Complete the steps below to find the x-coordinate of the midpoint.

$$\frac{x_2 - x_1}{2} + x_1 = \frac{x_2 - x_1}{2} + \frac{2x_1}{2}$$ Write expressions with common denominator.

$$= \frac{x_2 - x_1 + 2x_1}{2}$$ Add numerators.

$$= \frac{x_1 + x_2}{2}$$ Simplify.

The midpoint is located at $\left(\frac{x_1 + x_2}{2}, y\right)$.

- How can you find the value of a midpoint coordinate if you know the distance between the point to the left and the midpoint and you know the coordinate of the point?
- How can you find the value of a midpoint coordinate if you know the distance between the point to the right and the midpoint and you know the coordinate of the point?

Explore Together

Problem 2

13

Students will continue to investigate midpoints.

Grouping

Ask for a student volunteer to read part (F) aloud. Have a student restate the problem. Pose the Guiding Questions below to verify student understanding. Have students work together in small groups to complete parts (F) through (H) of Problem 2.

Guiding Questions

- What is being asked in part (F)?
- How can you find the coordinates of the midpoint between points B and C?

Common Student Errors

Students sometimes forget to add half the distance between the 2 points to the smaller value to find the coordinate of the midpoint. They may incorrectly use the distance between the 2 points as their midpoint coordinate. Other students will forget to find half the distance between the 2 points and claim that the midpoint is the second original point. If that happens, ask the students to explain the relationship between the points they started with and their midpoint value.

Call the class back together to have the students discuss and present their work for parts (F) through (H) of Problem 2.

Investigate Problem 2

Ask for a student volunteer to read Question 1 of Problem 2 aloud. Have a student restate the problem. Have the students complete Question 1 individually.

Note You may prefer to have the students work in small groups to complete Question 1.

Problem 2 Are We (Halfway) There Yet?

F. Now find the coordinates in kilometers of the midpoint between points B and C. Show all your work and use a complete sentence in your answer. Plot this point and label it as point E on the map on the previous page.

$\frac{3983 - 3982}{2} = \frac{1}{2} = 0.5$; $3982 + 0.5 = 3982.5$;

The coordinates of point E are (390, 3982.5).

G. Use complete sentences to describe how you would find the midpoint of a vertical line segment.

Sample Answer: First find the vertical distance between the points and divide the result by 2. Then add this amount to the y-coordinate of the lower point to get the y-coordinate of the midpoint. The x-coordinate is the x-coordinate of either endpoint because the x-coordinates of the endpoints are the same.

H. Consider the vertical line segment with endpoints (x, y_1) and (x, y_2). Complete the steps below to find the y-coordinate of the midpoint.

$$\frac{y_2 - y_1}{2} + y_1 = \frac{y_2 - y_1}{2} + \frac{2y_1}{2}$$ Write expressions with common denominator.

$$= \frac{y_2 - y_1 + 2y_1}{2}$$ Add numerators.

$$= \frac{y_1 + y_2}{2}$$ Simplify.

The midpoint is located at $\left(x, \frac{y_1 + y_2}{2} \right)$.

Investigate Problem 2

1. What do you think are the coordinates in kilometers of the midpoint between points A and C? Plot this point and label it as point F on the map on the previous page. Use complete sentences to explain your reasoning.

   Point F, the midpoint, should be at (389, 3982.5) because the x-coordinate should be halfway between the x-coordinates of points A and C and the y-coordinate should be halfway between the y-coordinates of points A and C.

   Use complete sentences to explain how can you use the Distance Formula to verify your answer.

   Use the Distance Formula to find the distance between points A and F and the distance between points C and F. The distances should be the same.

Explore Together

Investigate Problem 2

Students will calculate the midpoint between 2 points that are on neither a horizontal nor vertical line.

Call the class back together to have the students discuss and present their work for Question 1 of Problem 2.

Just the Math

Students will formally consider the Midpoint Formula in Question 2. Pose the Guiding Questions below to verify student understanding.

Grouping

Ask for a student volunteer to read Question 2 aloud. Have a student restate the problem. Have the students work in small groups to complete Question 2; then call the class back together to have the students discuss and present their work for Question 2.

Guiding Questions

- What is the mean of the x-coordinates of points A and B?
- How does the mean of the x-coordinates for points A and B compare to the x-coordinate of the midpoint of points A and B that you found in part (C)?
- Do you think the same relationship is true for the y-coordinates of points B and C and the y-coordinate of their midpoint?
- Compare the Midpoint Formula given in Question 2 to the mean of each pair of coordinates. What is the relationship?

Key Formative Assessments

- What is the Distance Formula?
- When can you use the Distance Formula?
- What is the Midpoint Formula?
- When can you use the Midpoint Formula?

13

Investigate Problem 2

Use your method described on the previous page to verify your answer to Question 1.

Distance between points A and F.

$$d = \sqrt{(389 - 388)^2 + (3982.5 - 3983)^2}$$
$$= \sqrt{1^2 + (-0.5)^2}$$
$$= \sqrt{1.25}$$

Distance between points C and F.

$$d = \sqrt{(390 - 389)^2 + (3982 - 3982.5)^2}$$
$$= \sqrt{1^2 + (-0.5)^2}$$
$$= \sqrt{1.25}$$

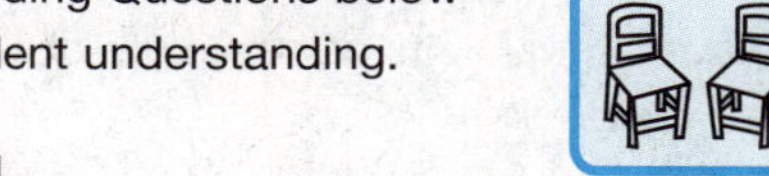

2. **Just the Math: The Midpoint Formula** If (x_1, y_1) and (x_2, y_2) are two points in the coordinate plane, then the **midpoint** of the line segment that joins these two points is given by $\left(\frac{x_1 + x_2}{2}, \frac{y_1 + y_2}{2}\right)$.

   Use the **Midpoint Formula** to find the midpoint of the line segment that connects points A and C on the map in Problem 2. Show all your work.

   $$\left(\frac{388 + 390}{2}, \frac{3983 + 3982}{2}\right) = \left(\frac{778}{2}, \frac{7965}{2}\right)$$
   $$= (389, 3982.5)$$

   How does this midpoint compare to the one you found in Question 1? Use a complete sentence in your answer.

   They are the same.

3. Find the midpoint of the line segment that has the given points as its endpoints. Show your work.

   (0, 4) and (2, 5)

   $$\left(\frac{0 + 2}{2}, \frac{4 + 5}{2}\right) = \left(\frac{2}{2}, \frac{9}{2}\right) = \left(1, \frac{9}{2}\right)$$

   (10, 5) and (8, 0)

   $$\left(\frac{10 + 8}{2}, \frac{5 + 0}{2}\right) = \left(\frac{18}{2}, \frac{5}{2}\right) = \left(9, \frac{5}{2}\right)$$

   (–1, 6) and (7, –2)

   $$\left(\frac{-1 + 7}{2}, \frac{6 - 2}{2}\right) = \left(\frac{6}{2}, \frac{4}{2}\right) = (3, 2)$$

   (–5, 6) and (–5, –8)

   $$\left(\frac{-5 + -5}{2}, \frac{6 - 8}{2}\right) = \left(\frac{-10}{2}, \frac{-2}{2}\right) = (-5, -1)$$

Wrap Up

13

Close

- Review all key terms and their definitions. Include the terms *distance formula* and *midpoint formula.*
- You may also want to review any other vocabulary terms that were discussed during the lesson, which may include *origin, latitude, longitude, easting, northing, right angle, right triangle, rectangular grid, metric, elevation, x-coordinate, y-coordinate, North, East, South, West, horizontal, vertical, Pythagorean Theorem, absolute value, halfway, mean,* and *average.*
- Remind the students to write the key terms and their definitions in the notes section of their notebooks. You may also want the students to include examples.
- Ask the students to summarize the development of the Distance Formula.
- Ask the students to summarize the development of the Midpoint Formula.
- Ask the students to make up two different ordered pairs. Then have the students calculate the distance between their points and the coordinates of the midpoint between their points.
- When they are finished, have the students trade ordered pairs and perform the calculations for the set they were given. Finally, have the students compare calculations to verify their answers. You can collect these to check them also.

Ties to the Cognitive Tutor Software

In the Cognitive Tutor software, students apply the Pythagorean Theorem to a number of different real-world situations. These applications help students to understand that the Pythagorean Theorem applies in cases where there is an object that is a right triangle but also in more abstract cases, where distances in orthogonal dimensions are known.

Follow Up

Assignment

Use the Assignment for Lesson 13.2 in the Student Assignments book. See the Teacher's Resources and Assessments book for answers.

Assessment

See the Assessments provided in the Teacher's Resources and Assessments book for Chapter 13.

Open-Ended Writing Task

Ask the students to write a short essay explaining the relationship between the Pythagorean Theorem and the Distance Formula.

Reflections

Insert your reflections on the lesson as it played out in class today.

What went well?

What did not go as well as you would have liked?

How would you like to change the lesson in order to improve the things that did not go well and capitalize on the things that did go well?

Notes

13

13.3 "Old Mathematics"

Completing the Square and Deriving the Quadratic Formula

13

Learning By Doing Lesson Map

Get Ready

Objectives

In this lesson, you will:

- Solve quadratic equations by completing the square.
- Derive the Quadratic Formula.

Key Terms

- perfect square trinomial
- factor
- complete the square
- Quadratic Formula

NCTM Content Standards

Grades 9–12 Expectations

Number and Operations Standards

- Compare and contrast the properties of numbers and number systems, including the rational and real numbers, and understand complex numbers as solutions to quadratic equations that do not have real solutions.
- Use number-theory arguments to justify relationships involving whole numbers.
- Judge the effects of such operations as multiplication, division, and computing powers and roots on the magnitude of quantities.
- Develop fluency in operations with real numbers, vectors, and matrices, using mental computation or paper-and-pencil calculations for simple cases and technology for more complicated cases.

Algebra Standards

- Generalize patterns using explicitly defined and recursively defined functions.
- Understand and perform transformations, such as arithmetically combining, composing, and inverting commonly used functions, using technology to perform such operations on more complicated symbolic expressions.
- Understand the meaning of equivalent forms of expressions, equations, inequalities, and relations.
- Write equivalent forms of equations, inequalities, and systems of equations and solve them with fluency—mentally or with paper and pencil in simple cases and using technology in all cases.
- Use symbolic algebra to represent and explain mathematical relationships.
- Use symbolic expressions, including iterative and recursive forms, to represent relationships arising from various contexts.

Geometry Standard

- Use geometric models to gain insights into, and answer questions in, other areas of mathematics.

Measurement Standard

- Understand and use formulas for the area, surface area, and volume of geometric figures, including cones, spheres, and cylinders.

Lesson Overview

Within the context of this lesson, students will:

- Represent polynomials with the area of geometric figures.
- Complete squares geometrically and algebraically.
- Use the method of completing the square to solve quadratic equations.
- Develop the Quadratic Formula from completing the square.

Essential Questions

These key questions are addressed in the lesson:

1. What is a perfect square trinomial?
2. What is a factor?
3. How can you complete the square to solve a quadratic equation?
4. When should you use the Quadratic Formula to solve a quadratic equation?

Show The Way

Warm Up

13

Place the following questions or an applicable subset of these questions on the board before students enter class. Students should begin working as soon as they are seated.

Factor each expression.

1. $5x + 15$ $5(x + 3)$
2. $7x^5 - 7x^4 + 21x^2 - 7x$ $7x(x^4 - x^3 + 3x - 1)$
3. $x^2 + 8x + 12$ $(x + 2)(x + 6)$
4. $x^2 - 10x + 25$ $(x - 5)^2$
5. $x^2 - 16$ $(x + 4)(x - 4)$
6. $-2x^3 - 4x^2 + 6x$ $-2x(x + 3)(x - 1)$
7. $x^2 + 4x + 4$ $(x + 2)^2$
8. $x^3 - 3x^2 + 3x$ $x(x^2 - 3x + 3)$

Solve each expression for x using the Quadratic Formula $x = \dfrac{-b \pm \sqrt{b^2 - 4ac}}{2a}$.

9. $8x^2 - 2x - 1$ $x = -0.25$ or $x = 0.5$
10. $x^2 - 1.5x - 4.5$ $x = -1.5$ or $x = 3$
11. $10x^2 + 51x + 5$ $x = -5$ or $x = -0.1$
12. $x^2 - 11x + 28$ $x = 4$ or $x = 7$

Motivator

Begin the lesson with the motivator to get students thinking about the topic of the upcoming problem. This lesson is about the history of mathematics. The motivating questions are about ancient mathematics.

Ask the students the following questions to get them interested in the lesson.

- When do you think people first began to think about mathematics?
- What tools did early mathematicians have to work with?
- When do you think calculators were invented? How could you find out if you are correct?
- Why do you think people were interested in mathematics in ancient times?
- What types of problems do you think ancient mathematicians tried to solve?

Explore Together

Problem 1

Students will model polynomials using areas of rectangular regions.

Grouping

Ask for a student volunteer to read the Scenario and Problem 1 aloud. Have a student restate the problem. Pose the Guiding Questions below to verify student understanding. Have students work together in small groups to complete parts (A) through (D); then call the class back together to have the students discuss and present their work for parts (A) through (D) of Problem 1.

Guiding Questions

- What information is given in this problem?
- What is a polynomial? What is a binomial? What is a trinomial?
- What is a perfect square?
- What is a factor?
- What does it mean to factor a polynomial?
- How can you find the area of a square?
- How can you find the area of a rectangle?

Notes Students will model polynomials using the areas of rectangular regions to work toward a visual understanding of completing the square to solve a quadratic function. Completing the square is typically a very abstract concept for students. This visual approach will make the concept much more understandable for most students.

Common Student Errors

Students may be confused in part (B) about the meaning of a figure that is more in the shape of a square. You may need to explain that this means that the shape is closer to a square than the figure in part (A) was. If they are confused, you can ask them what piece would be needed to make the shape a square. Then point out that this is asked in part (D).

SCENARIO When you used the Quadratic Formula in Lesson 8.6, did you wonder where the formula came from or why it works? Quadratic equations have been around since ancient times. Al-Kwārizī, an Islamic mathematician who lived from 780 A.D. to 850 A.D., wrote a text that included methods for solving quadratic equations. In his book, Al-Kwārizī used geometric figures to justify his methods.

Problem 1 Creating a Perfect Square

In previous lessons, you learned how to factor a trinomial of the form $a^2 + 2ab + b^2$ as the perfect square $(a + b)^2$. Now we will construct a perfect square by using a geometric model.

A. Consider the expression $x^2 + 10x$. This sum can be represented geometrically by the figure below. Write the area of each piece in the center of the piece.

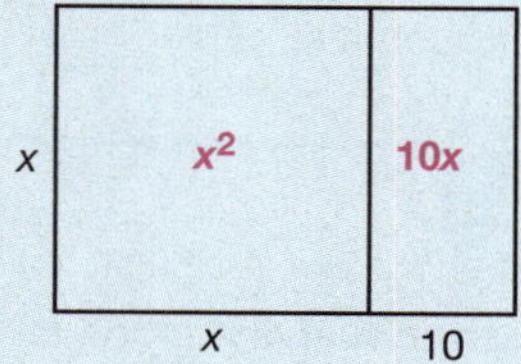

B. Let's rebuild this figure so that it is more in the shape of a square by splitting the rectangle in half and rearranging the pieces. Label the side length of each piece and write the area of each piece in the center of the piece.

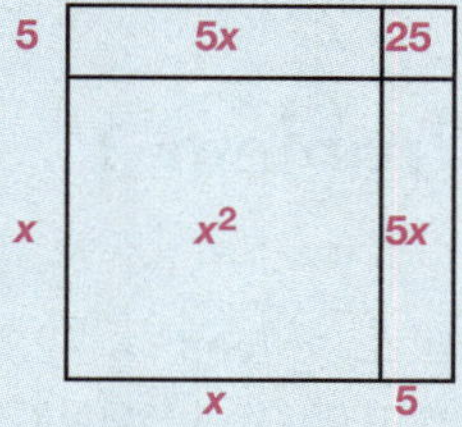

C. Do the two figures represent the same expression? How do you know? Use a complete sentence in your answer.

Yes, because the area of each piece is the same: $x^2 + 10x = x^2 + 5x + 5x$

D. Complete the figure so that it is a square. Label the area of the piece you added.

13

Explore Together

Problem 1

Students will complete the square for the polynomial $x^2 + 2x$ using the area of a larger square divided into squares and rectangles.

Grouping

Ask for a student volunteer to read parts (E) and (F) aloud. Have a student restate the problems. Have students work together in small groups to complete parts (E) through (G). Then, call the class back together to have the students discuss and present their work for parts (E) through (G).

Take Note

A **perfect square trinomial** is a trinomial that can be factored as a square of a difference or sum of binomials:

$a^2 + 2ab + b^2 = (a + b)^2$ or
$a^2 - 2ab + b^2 = (a - b)^2$.

Investigate Problem 1

Grouping

Ask for a student volunteer to read Question 1 aloud. Have a student restate the problem. Pose the Guiding Questions below to verify student understanding. Have students work together in small groups to complete Question 1, then call the class back together to have the students discuss and present their work for Question 1. Next, complete Question 2 together as a whole class

Guiding Questions

- How did you decide you could complete a square in part (G)?
- What are you asked to do in Question 1?
- What would be the first step you would do if you were to complete the square for the expression $x^2 + 12x$?
- Is the expression that you find when you complete the square equal to the original expression? How is it different?

Problem 1 Creating a Perfect Square

E. Add this area to your expression. What is the expression?

$x^2 + 10x + 25$

F. Factor this expression.

$x^2 + 10x + 25 = (x + 5)^2$

G. Use complete sentences to explain how to *complete the square* on the expression $x^2 + bx$ where b is an integer.

Sample Answer: To complete the square, add the square of half of b to the expression.

Investigate Problem 1

1. Use a geometric figure to complete the square (or create a perfect square trinomial) for each expression. Then **factor** the trinomial.

$x^2 + 8x$

4	$4x$	16
x	x^2	$4x$
	x	4

$x^2 + 8x + 16 = (x + 4)^2$

$x^2 + 5x$

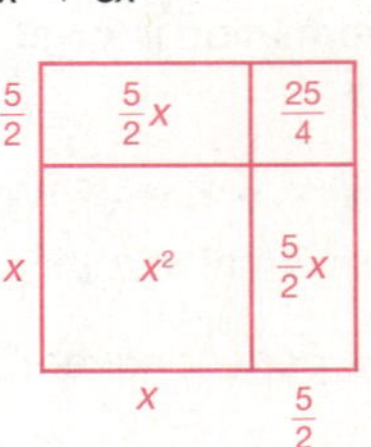

$x^2 + 5x + \frac{25}{4} = \left(x + \frac{5}{2}\right)^2$

2. You can use the idea of completing the square to help you solve quadratic equations. Consider the equation $x^2 + 10x - 11 = 0$. The trinomial in this equation cannot be factored as it is written. But how can we make $x^2 + 10x$ a perfect square trinomial? Use a complete sentence in your answer.

Sample Answer: Add 25 to the binomial.

Because we know that we can make $x^2 + 10x$ into a perfect square trinomial, we will get this expression by itself on one side of the equation $x^2 + 10x - 11 = 0$.

$x^2 + 10x = \boxed{11}$

Common Student Errors

Students may not immediately recognize that the expression they write and factor when they complete the square for the original expression is not equal to the original expression.

Explore Together

Investigate Problem 1

Students will use the method of completing the square to solve trinomials.

Grouping

Ask for a student volunteer to read Question 3 aloud. Have a student restate the problem. Pose the Guiding Questions below to verify student understanding. Have students work together in small groups to complete Question 3, then call the class back together to have the students discuss and present their work for Question 3.

Guiding Questions

- How is an equation different from an expression?
- When we add a value to one side of an equation, why do we always have to add the same value to the other side of the equation?
- What do we mean by the phrase one side of an equation?
- What does it mean to solve an equation?
- What does a solution to an equation represent?
- How can one equation have more than one possible solution for a variable?
- What degree would an equation have if the variable has 2 different solutions possible? Why?
- How can you solve the equation $x^2 = 49$ by extracting square roots?

Common Student Errors

Students will often forget to add the value to complete the square to both sides of the equation. Also, they frequently will make arithmetic errors in their solutions. Remind the students to check their answers.

Investigate Problem 1

Now, add the number to each side of the equation that will make $x^2 + 10x$ a perfect square trinomial. Complete the steps below.

$x^2 + 10x + \boxed{25} = 11 + \boxed{25}$ Add same number to each side.

$x^2 + 10x + \boxed{25} = \boxed{36}$ Add.

$(\boxed{x + 5})^2 = \boxed{36}$ Factor trinomial.

We can extend the method of extracting square roots to solve this equation. Complete each step below.

$x + 5 = \boxed{\sqrt{36}}$ or $x + 5 = \boxed{-\sqrt{36}}$

$x + 5 = \boxed{6}$ or $x + 5 = \boxed{-6}$

$x = \boxed{1}$ or $x = \boxed{-11}$

Check your work by substituting each solution into the original equation. Show your work.

$x = 1$:

$1^2 + 10(1) - 11 \stackrel{?}{=} 0$

$1 + 10 - 11 \stackrel{?}{=} 0$

$11 - 11 = 0$

$x = -11$:

$(-11)^2 + 10(-11) - 11 \stackrel{?}{=} 0$

$121 - 110 - 11 \stackrel{?}{=} 0$

$121 - 121 = 0$

3. Solve each equation by completing the square and then extracting square roots. Show all your work.

$x^2 + 8x - 9 = 0$

$x^2 + 8x = 9$

$x^2 + 8x + 16 = 9 + 16$

$x^2 + 8x + 16 = 25$

$(x + 4)^2 = 25$

$x + 4 = \sqrt{25}$ or $x + 4 = -\sqrt{25}$

$x + 4 = 5$ or $x + 4 = -5$

$x = 1$ or $x = -9$

$x^2 + 20x + 36 = 0$

$x^2 + 20x = -36$

$x^2 + 20x + 100 = -36 + 100$

$x^2 + 20x + 100 = 64$

$(x + 10)^2 = 64$

$x + 10 = \sqrt{64}$ or $x + 10 = -\sqrt{64}$

$x + 10 = 8$ or $x + 10 = -8$

$x = -2$ or $x = -18$

Explore Together

Investigate Problem 1

Students will work through an informal algebraic proof to develop the Quadratic Formula.

Grouping

Ask for a student volunteer to read Question 4 aloud. Have a student restate the problem. Pose the Guiding Questions below to verify student understanding. Have students work together in small groups to complete Question 4. Then, call the class back together to have the students discuss and present their work for Question 4.

Guiding Questions

- Why do you think it might be inconvenient to solve the equation $53x^2 + 104x - 51 = 0$ by completing the square?
- What is a quadratic equation?
- What is the degree of a quadratic equation?
- What does the graph of a quadratic equation look like?
- What is the Quadratic Formula?
- What do we find with the Quadratic Formula?
- What does it mean to solve an equation for x?
- Why is a restricted to not equal zero in this situation?
- What is a radical?

Key Formative Assessments

- How can you draw a geometric figure to complete the square for a polynomial?
- How can you solve a quadratic equation by completing the square?
- How should you solve a quadratic equation that is not easy to solve by completing the square?
- What other methods have you learned previously to solve quadratic equations?

Investigate Problem 1

$x^2 - 14x + 13 = 0$

$x^2 - 14x = -13$

$x^2 - 14x + 49 = -13 + 49$

$x^2 - 14x + 49 = 36$

$(x - 7)^2 = 36$

$x - 7 = \sqrt{36}$ or $x - 7 = -\sqrt{36}$

$x - 7 = 6$ or $x - 7 = -6$

$x = 13$ or $x = 1$

4. The method that you used to solve the equations in Question 2 is not the best method for *all* quadratic equations. For instance, you would probably not want to use this method to solve the equation $53x^2 + 104x - 51 = 0$. That is why we have the Quadratic Formula. Consider the equation $ax^2 + bx + c = 0$, where $a \neq 0$. Complete the steps below to discover how the *Quadratic Formula* was formed by completing the square.

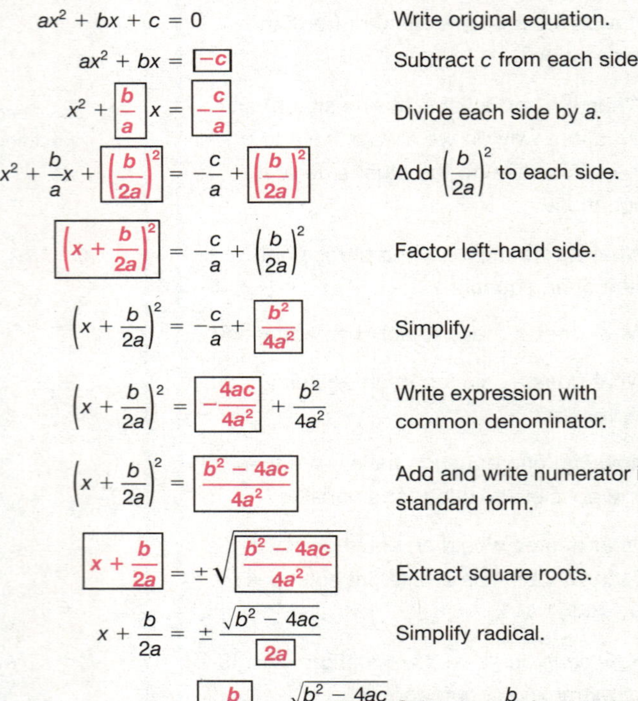

$ax^2 + bx + c = 0$ — Write original equation.

$ax^2 + bx = -c$ — Subtract c from each side.

$x^2 + \frac{b}{a}x = -\frac{c}{a}$ — Divide each side by a.

$x^2 + \frac{b}{a}x + \left(\frac{b}{2a}\right)^2 = -\frac{c}{a} + \left(\frac{b}{2a}\right)^2$ — Add $\left(\frac{b}{2a}\right)^2$ to each side.

$\left(x + \frac{b}{2a}\right)^2 = -\frac{c}{a} + \left(\frac{b}{2a}\right)^2$ — Factor left-hand side.

$\left(x + \frac{b}{2a}\right)^2 = -\frac{c}{a} + \frac{b^2}{4a^2}$ — Simplify.

$\left(x + \frac{b}{2a}\right)^2 = -\frac{4ac}{4a^2} + \frac{b^2}{4a^2}$ — Write expression with common denominator.

$\left(x + \frac{b}{2a}\right)^2 = \frac{b^2 - 4ac}{4a^2}$ — Add and write numerator in standard form.

$x + \frac{b}{2a} = \pm\sqrt{\frac{b^2 - 4ac}{4a^2}}$ — Extract square roots.

$x + \frac{b}{2a} = \pm\frac{\sqrt{b^2 - 4ac}}{2a}$ — Simplify radical.

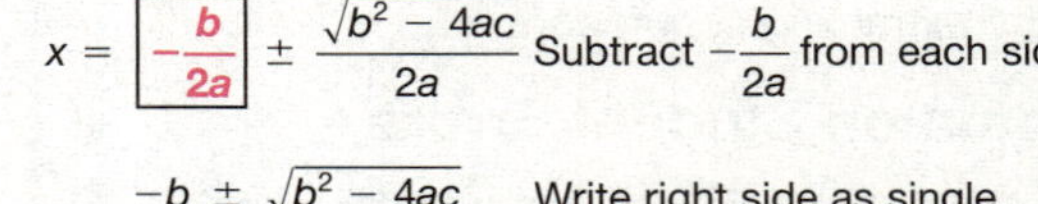

$x = -\frac{b}{2a} \pm \frac{\sqrt{b^2 - 4ac}}{2a}$ — Subtract $-\frac{b}{2a}$ from each side.

$x = \frac{-b \pm \sqrt{b^2 - 4ac}}{2a}$ — Write right side as single fraction.

Wrap Up

13

Close

- Review all key terms and their definitions. Include the terms *perfect square trinomial, factor, complete the square,* and *Quadratic Formula.*
- You may also want to review any other vocabulary terms that were discussed during the lesson, which may include *polynomial, binomial, trinomial, quadratic equation, expression, area, square, rectangle, construct, isolate, geometric figure, extract square roots,* and *radical.*
- Remind the students to write the key terms and their definitions in the notes section of their notebooks. You may also want the students to include examples.
- Have the students list all the ways they can think of to solve quadratic equations on the front board. Their list should include at least the following: factoring and setting the factors equal to zero, graphing and finding x-intercepts, completing the square, the Quadratic Formula, and extracting square roots.
- Have the students work in small groups to summarize each of these methods of solving quadratic equations. Have them list advantages, disadvantages, and appropriate uses of each as well as explain each process.

Ties to the Cognitive Tutor Software

In the Cognitive Tutor software, students apply the Pythagorean Theorem to a number of different real-world situations. These applications help students to understand that the Pythagorean Theorem applies in cases where there is an object that is a right triangle but also in more abstract cases, where distances in orthogonal dimensions are known.

Follow Up

Assignment

Use the Assignment for Lesson 13.3 in the Student Assignments book. See the Teacher's Resources and Assessments book for answers.

Assessment

See the Assessments provided in the Teacher's Resources and Assessments book for Chapter 13.

Open-Ended Writing Task

Ask the students to write an exit slip on a piece of paper before leaving the class session. Their exit slip should explain how to solve a quadratic equation by completing the square. They should also list any questions that they still have about the concept of solving quadratic equations by completing the square. Then, you can answer those questions at the beginning of the next class session.

13

Reflections

Insert your reflections on the lesson as it played out in class today.

What went well?

What did not go as well as you would have liked?

How would you like to change the lesson in order to improve the things that did not go well and capitalize on the things that did go well?

Notes

13.4 Learning to Be a Teacher

Vertex Form of a Quadratic Equation

Learning By Doing Lesson Map

Get Ready

Objectives

In this lesson, you will:

- Write quadratic equations in standard form.
- Write quadratic equations in factored form.
- Write quadratic equations in vertex form.

Key Terms

- standard form
- factored form
- vertex form

NCTM Content Standards

Grades 9–12 Expectations

Algebra Standards

- Generalize patterns using explicitly defined and recursively defined functions.
- Understand and perform transformations, such as arithmetically combining, composing, and inverting commonly used functions, using technology to perform such operations on more complicated symbolic expressions.
- Interpret representations of functions of two variables.
- Understand the meaning of equivalent forms of expressions, equations, inequalities, and relations.
- Use symbolic algebra to represent and explain mathematical relationships.
- Use a variety of symbolic representations, including recursive and parametric equations, for functions and relations.
- Identify essential quantitative relationships in a situation and determine the class or classes of functions that might model the relationships.
- Use symbolic expressions, including iterative and recursive forms, to represent relationships arising from various contexts.

Lesson Overview

Within the context of this lesson, students will be asked to:

- Write equations in standard form, in factored form, and in vertex form when given one of these forms of the equation.
- Graph quadratic functions.
- Identify the direction a parabola will open based on the coefficient of the x^2 term for a quadratic equation written in standard form.
- Identify the x-intercepts of a parabola for a quadratic equation written in factored form.
- Identify the vertex of a parabola for a quadratic equation written in vertex form.

Essential Questions

The following key questions are addressed in this lesson:

1. What is a quadratic equation?
2. What will the graph of a quadratic equation look like?
3. How can you write a quadratic equation in standard form?
4. How can you write a quadratic equation in factored form?
5. How can you write a quadratic equation in vertex form?

Show The Way

Warm Up

Place the following questions or an applicable subset of these questions on the board before students enter class. Students should begin working as soon as they are seated.

Tell whether each trinomial opens upward or downward.

1. $8x^2 - 2x - 1$ **Up**
2. $-7x^2 + 13x - 4$ **Down**
3. $0.03x^2 - 8x - 12$ **Up**
4. $2x^2 - 10x + 25$ **Up**
5. $x^2 - 16$ **Up**
6. $-2x^3 - 4x^2 + 6x$ **Down**
7. $-\frac{1}{2}x^2 + 4x - 4$ **Down**
8. $0.75x^2 - 3x + 1.25$ **Up**

Solve each equation by completing the square.

9. $x^2 - 2x - 1 = 0$ **$(x - 1)^2 = 2$; So, $x = 1 + \sqrt{2}$ or $x = 1 - \sqrt{2}$**
10. $x^2 + 2x - 1 = 0$ **$(x + 1)^2 = 2$; So, $x = -1 + \sqrt{2}$ or $x = -1 - \sqrt{2}$**
11. $x^2 + 6x + 5 = 0$ **$(x + 3)^2 = 4$; So, $x = -1$ or $x = -5$**
12. $x^2 - 8x = 0$ **$(x - 4)^2 = 16$; So, $x = 0$ or $x = 8$**

Motivator

Begin the lesson with the motivator to get students thinking about the topic of the upcoming problem. This lesson is about becoming a student studying to become a math teacher. The motivating questions are about becoming a math teacher.

Ask the students the following questions to get them interested in the lesson.

- What do you think you would have to do to become a math teacher?
- What kinds of classes do you think a prospective teacher would need to take?
- What colleges or universities do you know of that have strong education departments?
- How could you find a college or university to attend to major in mathematics education?
- What do you think are the advantages of being a math teacher?

Explore Together

Problem 1

Students will investigate three ways to write and represent quadratic equations.

Grouping

Ask for a student volunteer to read the Scenario and Problem 1 aloud. Have a student restate the problem. Pose the Guiding Questions below to verify student understanding. Have students work together in small groups to complete parts (A) through (C). Then, call the class back together to have the students discuss and present their work for parts (A) through (C).

Guiding Questions

- What is a parabola?
- Name an object that has the shape of a parabola in the natural world.
- Are rainbows examples of parabolas that open up or down?
- What type of coefficient is in the x^2 term of a quadratic equation written in standard form for a parabola that opens up?
- What type of coefficient is in the x^2 term of a quadratic equation written in standard form for a parabola that opens down?
- If you were to model the shape of a rainbow with a quadratic equation, would the coefficient of the x^2 term be positive or negative?
- What is the standard form of a quadratic equation? In standard form, what does a represent? What does b represent? What does c represent?
- What is the factored form of a quadratic equation? What is the vertex form of a quadratic equation?
- What is the vertex of a quadratic equation?

SCENARIO A college student is in school to learn how to be a teacher. An assignment for one of his classes is to create an interactive project that helps students learn about the different shapes of quadratic functions. He plans to make an interactive animation of water coming out of a garden hose so that the water's path is in the shape of a parabola. In the animation, students can change the numbers in a quadratic equation and see how the shape of the water's path changes.

13

Problem 1 Choosing a Quadratic Form

It has been a while since the college student has worked with quadratic equations, so he looks over material from old notes. According to his notes, there are three forms of a quadratic equation: the *standard form,* the *factored form,* and the *vertex form.* He has an example in his notes of a function represented by each of the forms:

Standard form: $y = x^2 - 4x + 3$

Factored form: $y = (x - 3)(x - 1)$

Vertex form: $y = (x - 2)^2 + (-1)$

A. How would you show that all of the equations represent the same function? Use a complete sentence in your answer.

Sample Answer: Simplify the factored form and the vertex form and compare them to the standard form.

Use your method to show that all of the equations represent the same function. Show all your work.

Factored form: $y = (x - 3)(x - 1) = x^2 - x - 3x + 3 = x^2 - 4x + 3$
Vertex form: $y = (x - 2)^2 + (-1) = x^2 - 4x + 4 - 1 = x^2 - 4x + 3$

B. Without graphing the equation, what can you determine about the graph by looking at the equation in standard form? Use complete sentences in your answer.

Sample Answer: You can tell whether the parabola opens upward or downward. You can determine the y-intercept.

C. Without graphing the equation, what can you determine about the graph by looking at the equation in factored form? Use a complete sentence in your answer.

You can determine the x-intercepts.

Note You may want to introduce the terminology of concave up for a parabola that opens up and concave down for a parabola that opens down.

Explore Together

Problem 1

13

Students will graph a parabola, identify the vertex, and investigate the vertex form of a quadratic equation.

Grouping

Ask for a student volunteer to read part (D) of Problem 1 aloud. Have a student restate the problem. Pose the Guiding Questions below to verify student understanding. Have students work together in small groups to complete parts (D) and (E). Then, call the class back together to have the students discuss and present their work for parts (D) and (E).

Guiding Questions

- How can you graph a quadratic function?
- What does it mean for the equation to be a function?
- What do you expect to be true for the number of different y-values for each x-value because we are told the equation is a function?
- How can you choose the appropriate bounds for this graph?
- Which equation for the quadratic function will you choose to use for your graph? Will you use a combination of equations? Explain.
- Do you expect the vertex to be the absolute maximum or absolute minimum y-value for this function? Explain.

Problem 1 Choosing a Quadratic Form

D. Create a graph of the quadratic function on the grid below. First, choose your bounds and intervals. Be sure to label your graph clearly.

Variable quantity	Lower bound	Upper bound	Interval
x	−6	9	1
y	−6	9	1

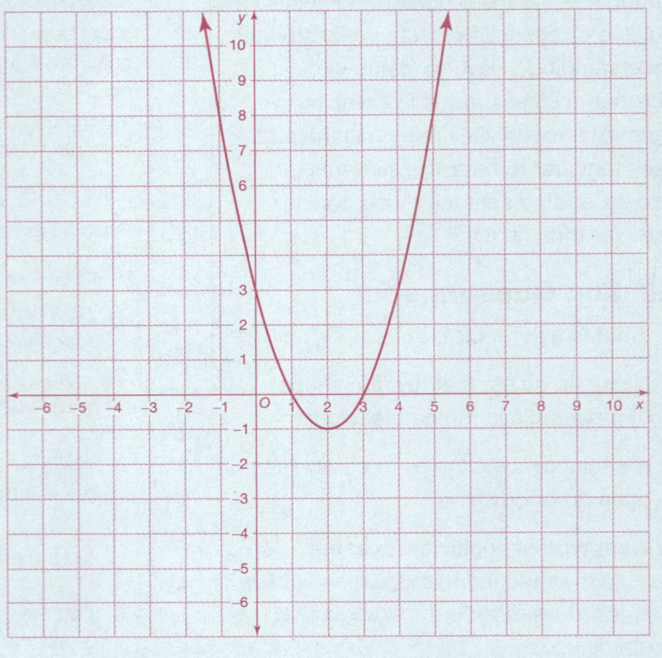

E. What is the vertex of the graph? Is it obvious why the last form is called the vertex form? Use complete sentences in your answer.

The vertex is (2, −1). Yes, because the x-coordinate is after the subtraction sign by the x and the y-coordinate is the constant in the function.

Investigate Problem 1

1. Just the Math: Vertex Form of a Quadratic Equation The **vertex form of a quadratic equation** is an equation of the form $y = a(x - h)^2 + k$ where (h, k) is the vertex of the graph of the equation.

Explore Together

Investigate Problem 1

Students will continue to investigate the vertex form of a quadratic equation.

Just the Math

Students will formally investigate the vertex form of a quadratic equation.

Ask for a student volunteer to read Question 1 aloud. Have a student restate the problem. Pose the Guiding Questions below to verify student understanding. Have the students complete Question 1 individually.

Guiding Questions

- What does h represent in the vertex form of a quadratic equation? What does k represent?
- How can you identify the vertex for a quadratic equation written in vertex form?

Common Student Errors

Students will frequently incorrectly identify the sign of the vertex for a quadratic equation written in vertex form. For instance, if the equation includes $(x + 1)^2$, then the correct vertex form is $[x - (-1)]^2$ and $h = -1$, not $h = 1$. Likewise, if the equation is $(x - 1)^2 - 5$, then the vertex form is $(x - 1)^2 + (-5)$ and $k = -5$, not $k = 5$.

Grouping

Ask for a student volunteer to read Question 2 aloud. Have a student restate the problem. Pose the Guiding Questions below at the right to verify student understanding. Have students work together in small groups to complete Questions 2 and 3. Then, call the class back together to have the students discuss and present their work for Questions 2 and 3.

Investigate Problem 1

1. Identify the vertex of the graph of each equation.

$y = (x - 3)^2 + 9$ — **(3, 9)**

$y = (x - 1)^2 - 2$ — **(1, –2)**

$y = (x + 4)^2 + 8$ — **(–4, 8)**

$y = (x + 7)^2 - 5$ — **(–7, –5)**

2. How is the vertex form of a quadratic equation different from the standard form? Use a complete sentence in your answer.

Sample Answer: The vertex form contains a perfect square.

In Problem 1, you wrote an equation in standard form by simplifying the expressions in the vertex form of the equation. How do you think an equation that is written in standard form could be written in vertex form? Use a complete sentence to explain your reasoning.

Sample Answer: Because the vertex form contains a perfect square, you should complete the square on the x^2-term and the x-term.

3. Consider the quadratic equation from Problem 1 in standard form: $y = x^2 - 4x + 3$. What number should you add to $x^2 - 4x$ to make a perfect square trinomial?

4

Add this number to each side of the quadratic equation in standard form.

$y + 4 = x^2 - 4x + 4 + 3$

Why did you have to add the number to each side of the equation? Use a complete sentence in your answer.

Sample Answer: Because you have to maintain the property of equality.

Now factor your perfect square trinomial.

$y + 4 = (x - 2)^2 + 3$

Finally, write your equation in vertex form.

$y = (x - 2)^2 - 1$

4. Write the equation $y = x^2 + 2x + 5$ in vertex form. Show your work. Then identify the vertex of the graph of the equation.

Sample Answer:

$$y = x^2 + 2x + 5$$
$$y + 1 = x^2 + 2x + 1 + 5$$
$$y + 1 = (x + 1)^2 + 5$$
$$y = (x + 1)^2 + 4$$

Vertex: (−1, 4)

Guiding Questions

- What is a perfect square?
- How can you complete a square for a quadratic equation?
- If you add a value to one side of an equation, what do you have to do to the other side of the equation? Why?

Explore Together

Investigate Problem 1

Students will graph a quadratic equation given in standard form by first writing the equation in factored and vertex forms.

Grouping

Have students work individually to solve Question 4. Then, call the class back together to have the students discuss and present their work for Question 4.

Common Student Errors

Some students will recognize in Question 4 that 5 can be written as $1 + 4$ and use this fact to complete the square. You may want to mention this alternate solution approach to students after they have completed the problem.

Key Formative Assessments

- Why would we want to write an equation in a different form than the form in which it is initially given?
- How can you write a quadratic equation that is in standard form in vertex form? How can you write a quadratic equation in factored form?

Grouping

Ask for a student volunteer to read Question 5 aloud. Have a student restate the problem. Pose the Guiding Questions below to verify student understanding. Have students work together in small groups to complete Question 5.

Guiding Questions

- What will you do first to solve this problem?
- What information can you determine from the standard form of this equation?
- What information do you expect to find for your graph from the factored form of the equation?
- What information do you expect to find for your graph from the vertex form of the equation?

Investigate Problem 1

5. The college student has to determine which form of the equation to use in the interactive project. He will make his decision after creating several graphs and comparing the three forms of each function. Write each equation in the other two forms. Then graph each equation in the given form.

Standard form: $y = x^2 + 2x - 15$

Factored form: $y = (x + 5)(x - 3)$

Vertex form: $y = (x + 1)^2 - 16$

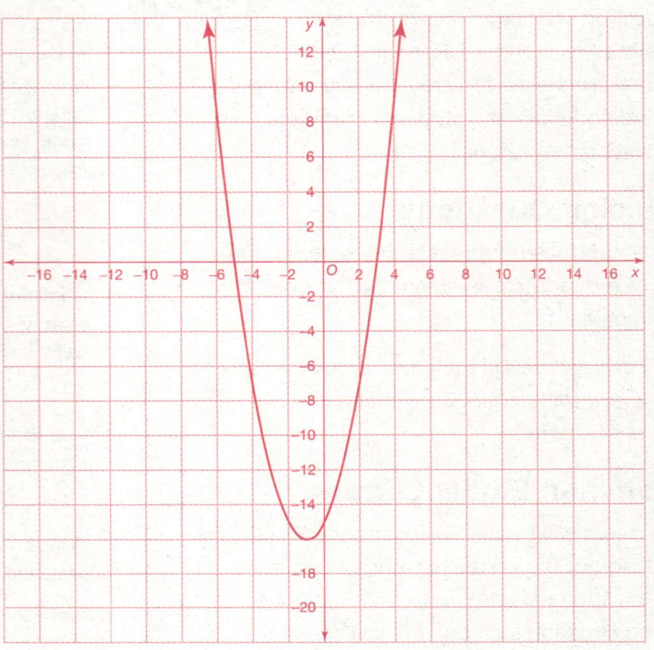

Key Formative Assessments

- Does your parabola open up or down?
- What are the x-intercepts for this equation?
- What is the vertex for your equation?
- Do these answers make sense for the factored and vertex forms of the equation? If not, what does that tell you about your graph and your equations?

Explore Together

Investigate Problem 1

Students will graph a quadratic equation given in factored form by first writing the equation in standard and vertex forms.

Grouping

Students will be working together in small groups to complete Question 5.

Key Formative Assessments

- Does your parabola open up or down?
- What are the x-intercepts for this equation?
- What is the vertex for your equation?
- Do these answers make sense for the factored and vertex forms of the equation? If not, what does that tell you about your graph and your equations?

Common Student Errors

Some students will have difficulty connecting the different forms of quadratic equations with their advantages and disadvantages. These students will be tempted to solve for the other forms of the equation, but ignore them and only focus on the standard form when creating the graph of the parabola. To prevent this, you can walk around the room and monitor for this problem. If you find any students ignoring the different forms of the equation, prompt them with probing questions to help them realize that by solving for the equation in the various forms, they have done much of the work toward graphing already. Ask them what they can determine from the factored form, what they can determine from the vertex form, and what they can determine from the standard form.

Investigate Problem 1

Standard form: $y = x^2 - 6x + 8$

Factored form: $y = (x - 2)(x - 4)$

Vertex form: $y = (x - 3)^2 - 1$

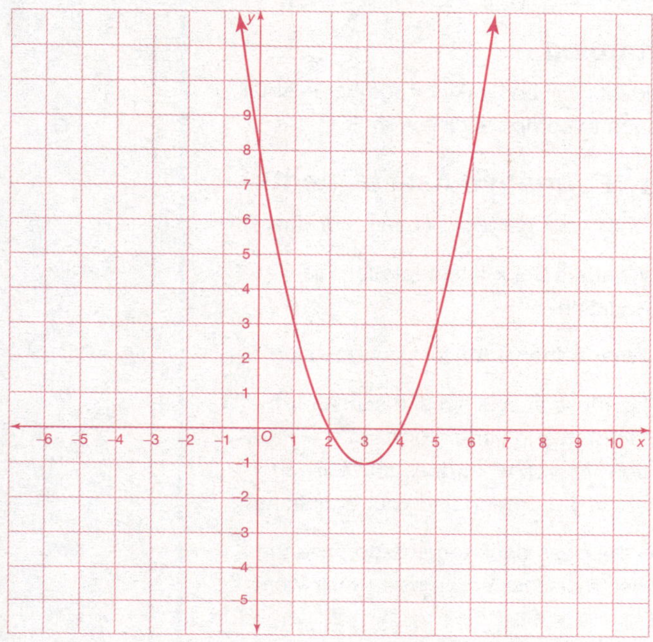

Explore Together

Investigate Problem 1

Students will graph a quadratic equation given in vertex form by first writing the equation in standard and factored forms.

Grouping

Students will be working together in small groups to complete Question 5.

Key Formative Assessments

- Does your parabola open up or down?
- What are the x-intercepts for this equation?
- What is the vertex for your equation?
- Do these answers make sense for the factored and vertex forms of the equation? If not, what does that tell you about your graph and your equations?

Call the class back together to have the students discuss and present their work for Question 5.

Grouping

Ask for a student volunteer to read Question 6 aloud. Have a student restate the problem. Assign Question 6 to be completed individually as part of a homework assignment. This question can be collected and graded as a writing assignment. Discuss this question at the beginning of the next class session.

Key Formative Assessments

- How can you write a quadratic equation in standard form?
- How can you write a quadratic equation in factored form?
- How can you write a quadratic equation in vertex form?
- Why do we need each of these forms of quadratic equations?

Investigate Problem 1

Standard form: $y = x^2 + 8x + 7$

Factored form: $y = (x + 1)(x + 7)$

Vertex form: $y = (x + 4)^2 - 9$

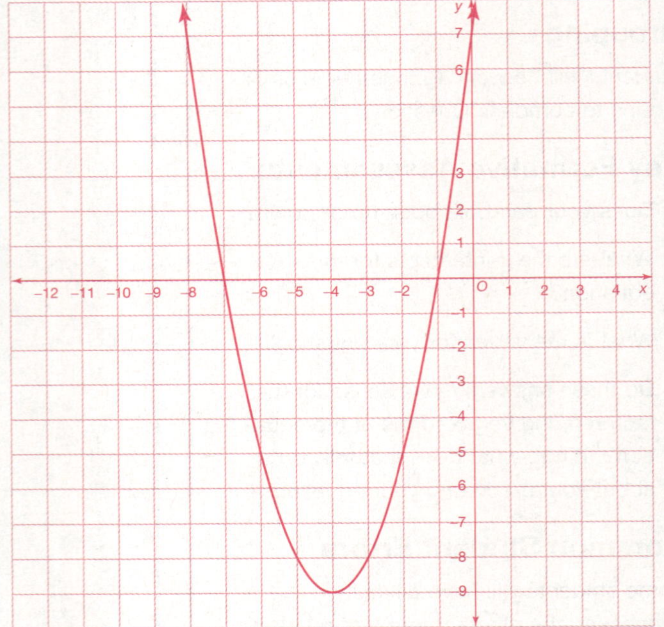

6. Which form of a quadratic equation do you think the college student should choose for his interactive project? He has the following options:

In the standard form $y = ax^2 + bx + c$, students would be able to change a, b, and c in the animation.

In the factored form $y = a(x - p)(x - q)$, students would be able to change a, p, and q in the animation.

In the vertex form $y = a(x - h)^2 + k$, students would be able to change a, h, and k in the animation.

Explain your reasoning. Use complete sentences in your answer.

Answers will vary.

Wrap Up

Close

- Review all key terms and their definitions. Include the terms *standard form, factored form,* and *vertex form.*
- You may also want to review any other vocabulary terms that were discussed during the lesson, which may include *parabola, function, quadratic equation, quadratic function, vertex, completing the square, x-intercept, y-intercept, concave up, concave down, absolute maximum, absolute minimum,* and *perfect square.*
- Remind the students to write the key terms and their definitions in the notes section of their notebooks. You may also want the students to include examples.
- Have the students summarize the standard form of a quadratic equation. They should include the general form of the equation, how to write an equation in standard form if given the factored or vertex form, the meaning of the values a, b, and c, the advantages, the disadvantages, and what can easily be determined about a function when written in standard form.
- Have the students summarize the factored form of a quadratic equation. They should include the general form of the equation, how to write an equation in factored form if given the standard or vertex form, how to find the x-intercepts, the advantages, the disadvantages, and what can easily be determined about a function when written in factored form.
- Have the students summarize the vertex form of a quadratic equation. They should include the general form of the equation, how to write an equation in vertex form if given the standard or factored form, how to find the vertex, the advantages, the disadvantages, and what can easily be determined about a function when written in vertex form.
- Ask the students to explain why we need to be able to write quadratic functions using each of these forms of the equation.

Ties to the Cognitive Tutor Software

Through the algebraic transformation problems in the Cognitive Tutor software, students are asked to relate specific parameters in a function to specific aspects of the graphical representation of the function. These activities help students understand that, for instance, adding a constant to a function transforms the graph of the function vertically, regardless of the form of the function. The idea that a function can be considered a single object that can be transformed, rather than just a collection of points helps provide the basis for a strong understanding of mathematical function.

Follow Up

Assignment

Use the Assignment for Lesson 13.4 in the Student Assignments book. See the Teacher's Resources and Assessments book for answers.

Assessment

See the Assessments provided in the Teacher's Resources and Assessments book for Chapter 13.

13

Open-Ended Writing Task

Ask the students to compare and contrast the standard form, factored form, and vertex form of a quadratic equation.

Reflections

Insert your reflections on the lesson as it played out in class today.

What went well?

___

___

What did not go as well as you would have liked?

___

___

How would you like to change the lesson in order to improve the things that did not go well and capitalize on the things that did go well?

___

___

___

___

___

Notes

13.5

Screen Saver

Graphing by Using Parent Functions

13

Learning By Doing Lesson Map

Get Ready

Objectives

In this lesson, you will:

- Use the graph of a parent function to describe the graph of a function.
- Identify the parent function given a function.
- Write equations of functions based on the graphs of functions.

Key Terms

- parent function
- reflection
- translation

Materials

- Colored pencils
- Transparency graph grid overlays

NCTM Content Standards

Grades 9–12 Expectations

Algebra Standards

- Generalize patterns using explicitly defined and recursively defined functions.
- Understand and perform transformations, such as arithmetically combining, composing, and inverting commonly used functions, using technology to perform such operations on more complicated symbolic expressions.
- Understand and compare the properties of classes of functions, including exponential, polynomial, rational, logarithmic, and periodic functions.
- Interpret representations of functions of two variables.
- Use symbolic algebra to represent and explain mathematical relationships.
- Identify essential quantitative relationships in a situation and determine the class or classes of functions that might model these relationships.

Geometry Standard

- Understand and represent translations, reflections, rotations, and dilations of objects in the plane by using sketches, coordinates, vectors, function notation, and matrices.

Lesson Overview

Within the context of this lesson, students will be asked to:

- Graph a parent function and related functions.
- Describe the graph of a function in relation to the graph of the parent function.
- Identify the parent function of a given function.
- Write the equations of functions based on the graphs of functions.

Essential Questions

The following key questions are addressed in this lesson:

1. What is a parent function?
2. What is a translation?
3. What is a reflection?

Show The Way

13

Warm Up

Place the following questions or an applicable subset of these questions on the board before students enter class. Students should begin working as soon as they are seated.

Write each equation in standard form, factored form, and vertex form.

	Standard Form	Factored Form	Vertex Form
1.	$x^2 + 12x + 35$	$(x + 5)(x + 7)$	$(x + 6)^2 - 1$
2.	$x^2 + 2x - 8$	$(x + 4)(x - 2)$	$(x + 1)^2 - 9$
3.	$x^2 - 6x + 8$	$(x - 2)(x - 4)$	$(x - 3)^2 - 1$
4.	$3x^2 - 12x - 36$	$3(x - 6)(x + 2)$	$3(x - 2)^2 - 48$

Motivator

Begin the lesson with the motivator to get students thinking about the topic of the upcoming problem. This lesson is about a creating a computer screen saver. The motivating questions are about screen savers.

Ask the students the following questions to get them interested in the lesson.

- What is a screen saver?
- What does it save?
- What would happen if a computer screen were to have the same image for a long time without changing?
- What different types of screen savers have you seen?
- How do you think screen savers are made?

Explore Together

Problem 1

Students will graph a parabola, find its vertex, and write the equation for the parabola in vertex form.

Grouping

Ask for a student volunteer to read the Scenario and Problem 1 aloud. Have a student restate the problem. Pose the Guiding Questions below to verify student understanding. Have students work together in small groups to complete parts (A) and (B). When they are finished, call the class back together to have the students discuss and present their work.

Guiding Questions

- What type of mathematical figure will be moved around the screen in this screen saver?
- What do you think the graph of $y = x^2$ will look like ?
- What is a parabola?
- How can you tell that the function will have a parabolic shape when you graph it?
- What part of the equation of the function $y = x^2$ allows you to determine whether the parabola will open up or down? Which way does $y = x^2$ open?
- What is a vertex? How can you find the vertex for a parabola?
- What does the h represent in the vertex form of quadratic equation? What does the k represent? What does the a represent?
- If a parabola has a vertex at the point (5, 3), and $a = 2$, what is the vertex form of the equation? If a parabola has a vertex at the point (–4, 8) and $a = 3$, what is the vertex form of the equation? If a parabola has a vertex at the point (–2, –5), and $a = 9$, what is the vertex form of the equation?

SCENARIO Your friend is learning how to create a screen saver by moving a mathematical figure around on the screen. She chooses to use a parabola and will start with the graph of $y = x^2$. Each step of her program will require her to display the graph of a different quadratic function to show the movement.

Problem 1 Right, Left, Up, and Down

A. First, your friend must decide how the parabola will move around on the screen. Create the graph of $y = x^2$ on the grid below.

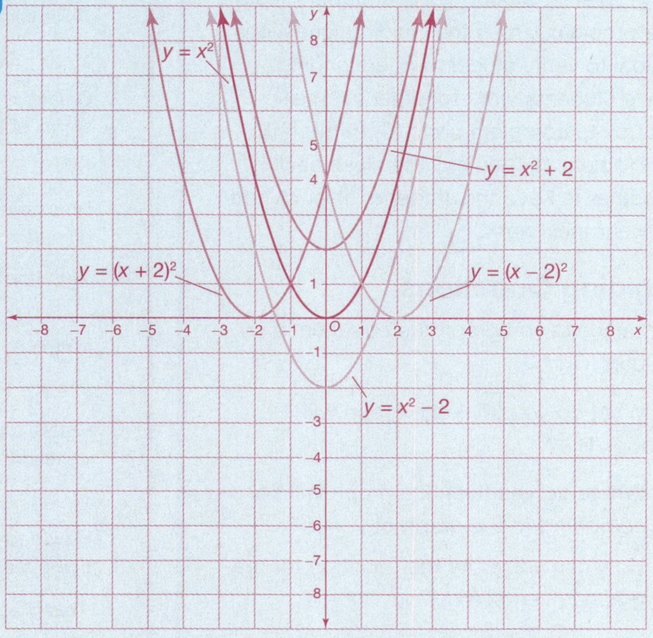

B. Create a graph of $y = x^2 + 2$ on the grid in part (A). What is the vertex of the graph of $y = x^2 + 2$?

(0, 2)

How does this graph compare to the graph of $y = x^2$? Use a complete sentence in your answer.

This graph is two units above the graph of $y = x^2$.

Write $y = x^2 + 2$ in vertex form.

$y = (x - 0)^2 + 2$

Notes You may want to have the students use a different color to create each graph in parts (A) through (E). Colored pencils are convenient for student use.

Explore Together

Problem 1

Students will graph and investigate the equations for parabolas with the parent function of $y = x^2$.

Grouping

Ask for a student volunteer to read part (C) of Problem 1 aloud. Have a student restate the problem. Pose the Guiding Questions below to verify student understanding. Have students work together in small groups to complete parts (C) through (E) of Problem 1. Then call the class back together to have the students discuss and present their work.

Guiding Questions

- How can you graph the equation in part (C)?
- What do you think the graph will look like?
- Where do you think the graph will be compared to the graph of $y = x^2$?
- How can you write the equation for the graph in vertex form?

Key Formative Assessments

- What function did you graph first?
- What other functions did you graph in parts (B) through (E)?
- How are they related? In what ways are they similar? In what ways are they different?
- How is writing the equation of a quadratic function in standard form different from writing the equation in vertex form?
- How can you find the vertex from a quadratic function written in vertex form?
- How can you write a quadratic equation in vertex form from the graph of the function?
- How can you write a quadratic equation in vertex form if it is given in standard form?

Problem 1 Right, Left, Up, and Down

C. Create a graph of $y = x^2 - 2$ on the grid in part (A). What is the vertex of the graph of $y = x^2 - 2$?

(0, –2)

How does this graph compare to the graph of $y = x^2$? Use a complete sentence in your answer.

This graph is two units below the graph of $y = x^2$.

Write $y = x^2 - 2$ in vertex form.

$y = (x - 0)^2 + (-2)$

D. Create a graph of $y = (x + 2)^2$ on the grid in part (A). What is the vertex of the graph of $y = (x + 2)^2$?

(–2, 0)

How does this graph compare to the graph of $y = x^2$? Use a complete sentence in your answer.

This graph is two units to the left of the graph of $y = x^2$.

Write $y = (x + 2)^2$ in vertex form.

$y = [x - (-2)]^2 + 0$

E. Create a graph of $y = (x - 2)^2$ on the grid in part (A). What is the vertex of the graph of $y = (x - 2)^2$?

(2, 0)

How does this graph compare to the graph of $y = x^2$? Use a complete sentence in your answer.

This graph is two units to the right of the graph of $y = x^2$.

Write $y = (x - 2)^2$ in vertex form.

$y = (x - 2)^2 + 0$

Investigate Problem 1

1. How can you obtain the graph of $y = x^2 + n$ when n is positive by using the graph of $y = x^2$? Use a complete sentence in your answer.

Sample Answer: The graph of $y = x^2$ can be moved up $|n|$ units to obtain the graph of $y = x^2 + n$.

Investigate Problem 1

Ask for a student volunteer to read Question 1 aloud. Have a student restate the problem. Have students complete Question 1 individually.

Explore Together

Investigate Problem 1

Students will investigate ways to graph functions based on a parent function.

Call the class back together to have the students discuss and present their work for Question 1.

Grouping

Have students work together in small groups to complete Questions 2 and 3. Then, call the class back together to have the students discuss and present their work for Questions 2 and 3.

Common Student Errors

Most students will quickly understand that adding a positive constant to a quadratic term will result in a vertical shift upward for a graph. Similarly, they will usually recognize that adding a negative constant or subtracting a positive constant will move the graph downward vertically. However, the need to subtract a positive number in the parentheses with the variable x to move a function to the right or adding a positive number in the parentheses to move a function to the left is not easy for most students to understand. Use the Key Formative Assessments to help the students solidify their understanding.

Key Formative Assessments

- How would you write the equation of the function that would have a graph of $y = x^2$ shifted up vertically by 5 units?
- How would you write the equation of the function that would have a graph of $y = x^2$ shifted down vertically by 7 units?
- How would you write the equation of the function that would have a graph of $y = x^2$ shifted horizontally to the left by 4 units?
- How would you write the equation of the function that would have a graph of $y = x^2$ shifted horizontally to the right by 3 units?

Investigate Problem 1

How can you obtain the graph of $y = x^2 + n$ when n is negative by using the graph of $y = x^2$? Use a complete sentence in your answer.

Sample Answer: The graph of $y = x^2$ can be moved down $|n|$ units to obtain the graph of $y = x^2 + n$.

2. How can you obtain the graph of $y = (x - n)^2$ when n is positive by using the graph of $y = x^2$? Use a complete sentence in your answer.

Sample Answer: The graph of $y = x^2$ can be moved $|n|$ units to the right to obtain the graph of $y = (x - n)^2$.

How can you obtain the graph of $y = (x - n)^2$ when n is negative by using the graph of $y = x^2$? Use a complete sentence in your answer.

Sample Answer: The graph of $y = x^2$ can be moved $|n|$ units to the left to obtain the graph of $y = (x - n)^2$.

3. Explain how you can use the graph of $y = x^2$ to draw the graph of $y = (x - 1)^2 + 4$. Use a complete sentence in your answer.

Sample Answer: Move the graph of $y = x^2$ one unit to the right and four units up to obtain the graph of $y = (x - 1)^2 + 4$.

Explain how you can use the graph of $y = x^2$ to draw the graph of $y = (x + 2)^2 + 3$. Use a complete sentence in your answer.

Sample Answer: Move the graph of $y = x^2$ two units to the left and three units up to obtain the graph of $y = (x + 2)^2 + 3$.

Explain how you can use the graph of $y = x^2$ to draw the graph of $y = (x - 3)^2 - 1$. Use a complete sentence in your answer.

Sample Answer: Move the graph of $y = x^2$ three units to the right and one unit down to obtain the graph of $y = (x - 3)^2 - 1$.

Explain how you can use the graph of $y = x^2$ to draw the graph of $y = (x + 5)^2 - 2$. Use a complete sentence in your answer.

Sample Answer: Move the graph of $y = x^2$ five units to the left and two units down to obtain the graph of $y = (x + 5)^2 - 2$.

Use complete sentences to explain how you found your answers to Question 3.

Sample Answer: Identify the vertex of the given function. Then determine how to get to the vertex from (0, 0), the vertex of $y = x^2$.

- How would you write the equation of the function that would have a graph of $y = x^2$ shifted down vertically by 2 units and horizontally to the left by 8 units?

Explore Together

3

Investigate Problem 1

Students will identify the equation and vertex for given movements of the function $y = x^2$.

Grouping

Ask for a student volunteer to read Question 4 aloud. Have a student restate the problem. Have the students complete Question 4 individually. If class-time is limited, you may assign Question 4 as part of a homework assignment. Question 4 can also be used as a warm up review activity to start the next class period.

Call the class back together to have the students discuss and present their work for Question 4.

Investigate Problem 1

4. Your friend's parabola will move as described below in the table. She will later add intermediate steps to make the movement smooth. Your friend will start with the graph of $y = (x - 5)^2 + 8$. In each step, write the equation that will graph the parabola described. Then identify the vertex.

Step	Movement	Equation	Vertex
0	none	$y = (x - 5)^2 + 8$	(5, 8)
1	left two units, up two units	$y = (x - 3)^2 + 10$	(3, 10)
2	left four units down ten units	$y = (x + 1)^2 + 0$	(−1, 0)
3	right three units, down ten units	$y = (x - 2)^2 - 10$	(2, −10)
4	right three units, up ten units	$y = (x - 5)^2 + 0$	(5, 0)

Problem 2

Students will be introduced to reflections and translations for the graphs of functions.

Grouping

Ask for a student volunteer to read Problem 2 aloud. Have a student restate the problem. Pose the Guiding Questions below to verify student understanding. Have students complete part (A) individually.

Guiding Questions

- What do you think it might mean to "squish" a graph vertically?
- How do you think you might write an equation for a graph that is only half as tall as the original function?
- What do you think it might mean to "stretch" a graph vertically?
- How do you think you might write an equation for a graph that is a stretch of another?
- What do you think it might mean to "flop" a graph?
- How do you think you might write the equation of a "flop" of another function?

Problem 2 Squish, Stretch, and Flop

Your friend is also considering other changes to the parabola as it moves around on the screen.

A. Begin with the graph of $y = x^2$. Create a graph of $y = x^2$ on the grid below.

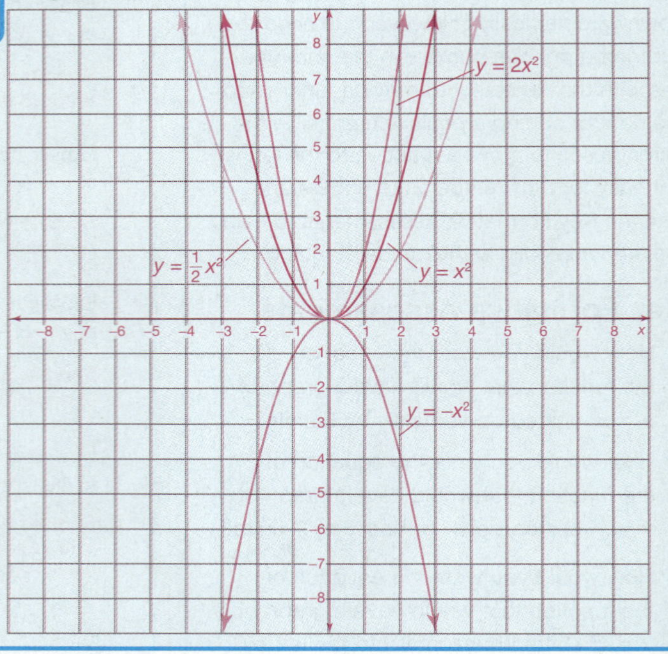

Notes You may want to have the students use a different color to create each graph in parts (A) through (E) of Problem 2 also. Colored pencils are convenient for student use.

Explore Together

Problem 2

Students will create graphs of dilations and reflections on the quadratic equation $y = x^2$.

Grouping

Ask for a student volunteer to read part (B) aloud. Have a student restate the problem. Pose the Guiding Questions below to verify student understanding. Have students work together in small groups to complete parts (B) through (E). Have the students summarize what they discovered in parts (B) through (E).

Guiding Questions

- What do you predict will be the effect on the y-value of the function by multiplying the x^2 factor by 2?
- If each y-value in the function $y = 2x^2$ is twice as large as the corresponding y-value for the function $y = x^2$, how do you think the graph of $y = 2x^2$ will appear compared to the graph of $y = x^2$?
- How can you check your prediction?
- What type of function looks like a parabola that opens up?
- What type of function looks like a parabola that opens down?
- How do you think you can flip a graph so that it is a mirror image around the x-axis? How might you write the equation of such a function?

Investigate Problem 2

Grouping

Ask for a student volunteer to read Question 1 aloud. Have a student restate the problem. Have students work together in small groups to complete Questions 1 and 2. Call the class back together to have the students discuss and present their work for Questions 1 and 2.

Problem 2 Squish, Stretch, and Flop

B. Now, create a graph of $y = 2x^2$ on the grid in part (A). How does this graph compare to the graph of $y = x^2$? Use a complete sentence in your answer.

Sample Answer: The graph of $y = 2x^2$ is narrower than the graph of $y = x^2$.

C. Create a graph of $y = \frac{1}{2}x^2$ on the grid in part (A). How does this graph compare to the graph of $y = x^2$? Use a complete sentence in your answer.

Sample Answer: The graph of $y = \frac{1}{2}x^2$ is wider than the graph of $y = x^2$.

D. Create a graph of $y = -x^2$ on the grid in part (A). How does this graph compare to the graph of $y = x^2$? Use a complete sentence in your answer.

Sample Answer: The graph of $y = -x^2$ is a mirror image of the graph of $y = x^2$.

E. Describe the effect that a constant being multiplied by x^2 has on the graph of $y = x^2$. Use complete sentences in your answer.

Sample Answer: When the constant is greater than one, the graph gets narrower. When the constant is less than one and positive, the graph gets wider. When the constant is -1, the graph is reflected over the x-axis.

Investigate Problem 2

1. How can you obtain the graph of $y = ax^2$ by using the graph of $y = x^2$ when $|a|$ is greater than 1? Use a complete sentence in your answer.

Sample Answer: The graph of $y = x^2$ can be made narrower to obtain the graph of $y = ax^2$.

How can you obtain the graph of $y = ax^2$ by using the graph of $y = x^2$ when $|a|$ is less than 1? Use a complete sentence in your answer.

Sample Answer: The graph of $y = x^2$ can be made wider to obtain the graph of $y = ax^2$.

2. How can you obtain the graph of $y = -x^2$ by using the graph of $y = x^2$? Use a complete sentence in your answer.

Sample Answer: The graph of $y = x^2$ can be reflected over the x-axis to obtain the graph of $y = -x^2$.

13

Explore Together

Investigate Problem 2

Students will learn the formal terminology associated with translations and reflections of functions.

Just the Math

Students will correctly identify the parent functions for given functions.

Grouping

Ask for a student volunteer to read Question 3 aloud. Have a student restate the problem. Pose the Guiding Questions below to verify student understanding. Have students work together in small groups to complete Question 3. Call the class back together to have the students discuss and explain their work for Question 3.

Guiding Questions

- Why is $y = x^2$ considered the most basic quadratic function?
- What is a parent function?
- How can you use a parent function to determine what the graph of a related function will look like?
- Why would we want to be able to predict the shape and appearance of the graph of a function without actually having to calculate sets of ordered pairs for the function?
- How can you find the parent function for an equation without graphing the equation?
- What is the parent function of $y = x^4 - 3$?
- How does the graph of $y = x^4 - 3$ relate to the graph of the parent function $y = x^4$?

Investigate Problem 2

3. **Just the Math: Parent Functions** The graph of $y = x^2$ is called a **parent function** because the function $y = x^2$ represents the most basic quadratic function.

In Problem 1 you worked with *translations*. **Translations** are shifts (left, right, up, or down) of the graph of the parent function. In parts (B) and (C) of Problem 2, the graphs that you drew were either wider or narrower than the graph of the parent function. Finally, in part (D) of Problem 2, you **reflected,** or created the mirror image of the graph of the parent function, over the x-axis.

For each function below, identify the parent function. Then describe how the graph of the given function relates to the graph of the parent function. Use complete sentences in your answer.

$y = 10x^2$

Parent function: **$y = x^2$**

Relationship: **The graph of $y = 10x^2$ is narrower than the graph of $y = x^2$.**

$y = (x - 4)^2$

Parent function: **$y = x^2$**

Relationship: **Translation (shift right 4 units)**

$y = -2x^2$

Parent function: **$y = x^2$**

Relationship: **Reflection (over x-axis) and the graph is narrower**

$y = 3x^2 + 2$

Parent function: **$y = x^2$**

Relationship: **The graph is narrower and translation (shift up 2 units)**

Explore Together

Investigate Problem 2

Students will write the equation of a function given a graph of the function.

Grouping

Ask for a student volunteer to read Question 4 aloud. Have a student restate the problem. Pose the Guiding Questions below to verify student understanding. Have the students work individually to complete Question 4. Call the class back together to have the students discuss and present their work for Question 4.

Guiding Questions

- What is the parent function for the graph in Question 4?
- What is the vertex of the equation?
- How can you write the equation for a function with the parent function $y = x^2$ and the vertex (1, 2)?

Grouping

Ask for a student volunteer to read Question 5 aloud. Have a student restate the problem. Have the students work in small groups to complete Question 5. Call the class back together to have the students discuss and present their work for Question 5.

Note If time is limited, Questions 4 and 5 can be assigned as part of a homework assignment.

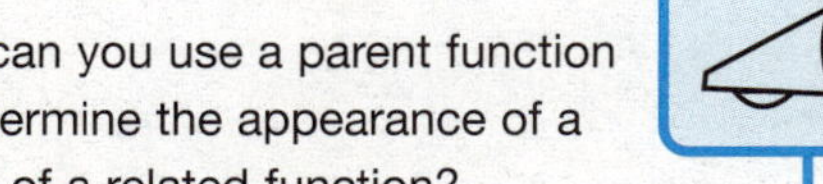

Key Formative Assessments

- What is a parent function?
- How can you use a parent function to determine the appearance of a graph of a related function?
- How can you shift a function vertically? How can you shift a function horizontally?
- How can you reflect a graph across the x-axis? How can you reflect a graph across the y-axis?
- How can you make a graph narrower? How can you make a graph wider?

Investigate Problem 2

4. The graph of $y = x^2$ has been transformed into the graph below. Write an equation for the graph in vertex form.

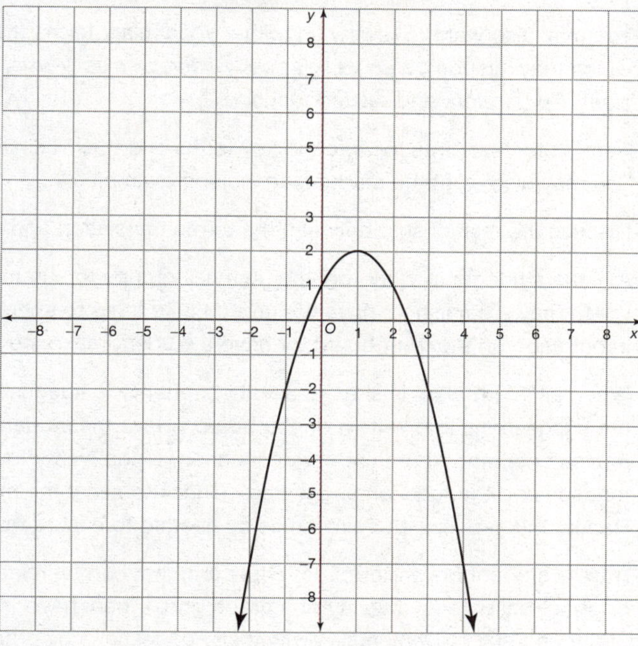

$y = -(x - 1)^2 + 2$

5. Your friend wants to program the movement of the parabola as described below in the table. She will later add intermediate steps to make the movement smooth. Your friend will begin with the graph of $y = x^2$. In each step, write the equation that will graph the parabola described. Then identify the vertex.

Step	Movement	Equation	Vertex
0	none	$y = x^2$	(0, 0)
1	reflection in x-axis	$y = -x^2$	(0, 0)
2	right one unit, down two units	$y = -(x - 1)^2 - 2$	(1, –2)
3	left three units, down two units	$y = -(x + 2)^2 - 4$	(–2, –4)

Wrap Up

13

Close

- Review all key terms and their definitions. Include the terms *parent function, translation,* and *reflection.*
- You may also want to review any other vocabulary terms that were discussed during the lesson, which may include *parabola, x-value, y-value, x-axis, y-axis, horizontal, vertical, wider, narrower, coefficient, flip, stretch,* and *related function.*
- Remind the students to write the key terms and their definitions in the notes section of their notebooks. You may also want the students to include examples.
- Discuss the relationship between the terms translation and reflection.
- Ask the students to work together in small groups to summarize the information learned in this lesson. When they are finished, have the groups take turns to share something that they learned. List the information on the front board, or have a student make the list on the board.
- When all information is shared, ask the students to suggest a function for the class. Choose one of the suggestions and write it on the board. Have the students then identify the parent function and all movements made from the parent function. Then, verify their work by graphing both the parent function and the function suggested. Check to see if the movements they suggested were accurate. Repeat this process as many times as needed to ensure that the students understand.
- Have a student draw a graph of a function that can be related to a parent function at the front board or on a transparency overlay of a graph grid. Then, have the class identify the parent function for the graph and any movements. Verify their suggestions by graphing the function they suggest as well as the parent function on a graphing calculator if available.

Ties to the Cognitive Tutor Software

Through the algebraic transformation problems in the Cognitive Tutor software, students are asked to relate specific parameters in a function to specific aspects of the graphical representation of the function. These activities help students understand that, for instance, adding a constant to a function transforms the graph of the function vertically, regardless of the form of the function. The idea that a function can be considered a single object that can be transformed, rather than just a collection of points helps provide the basis for a strong understanding of mathematical function.

Follow Up

Assignment

Use the Assignment for Lesson 13.5 in the Student Assignments book. See the Teacher's Resources and Assessments book for answers.

Assessment

See the Assessments provided in the Teacher's Resources and Assessments book for Chapter 13.

13

Open-Ended Writing Task

Ask the students to graph a function with a different parent function than $y = x^2$. Then have the students write the equation of the function using the methods learned in this lesson. Have the students then write an explanation in sentences about how their graph is related to the graph of the parent function and how they developed the equation for their function.

Reflections

Insert your reflections on the lesson as it played out in class today.

What went well?

____

____

What did not go as well as you would have liked?

____

____

How would you like to change the lesson in order to improve the things that did not go well and capitalize on the things that did go well?

____

____

____

____

____

Notes

13

13.6 Science Fair

Introduction to Exponential Functions

13

Learning By Doing Lesson Map

Get Ready

Objectives

In this lesson, you will:

- Write and graph exponential functions.
- Identify translations of exponential functions.

Key Terms

- exponential function

NCTM Content Standards

Grades 9–12 Expectations

Algebra Standards

- Understand and perform transformations, such as arithmetically combining, composing, and inverting commonly used functions, using technology to perform such operations on more complicated symbolic expressions.
- Understand and compare the properties of classes of functions, including exponential, polynomial, rational, logarithmic, and periodic functions.
- Use symbolic algebra to represent and explain mathematical relationships.
- Identify essential quantitative relationships in a situation and determine the class or classes of functions that might model the relationships.
- Use symbolic expressions, including iterative and recursive forms, to represent relationships arising from various contexts.
- Draw reasonable conclusions about a situation being modeled.
- Approximate and interpret rates of change from graphical and numerical data.

Lesson Overview

Within the context of this lesson, students will be asked to:

- Write and graph an exponential function to model exponential growth.
- Evaluate and graph various exponential functions.
- Identify transformations of exponential functions.
- Identify the appropriate domain and range for exponential functions.

Essential Questions

The following key questions are addressed in this lesson:

1. What is an exponential function?
2. What are the requirements for an exponential function?
3. How can you evaluate an exponential function?
4. How can you graph an exponential function?

Show The Way

Warm Up

13

Place the following questions or an applicable subset of these questions on the board before students enter class. Students should begin working as soon as they are seated.

Evaluate each power.

1. 1^0 1

2. 6^0 1

3. $(-5)^0$ 1

4. 12^0 1

5. 1^5 1

6. 1^2 1

7. $1^{(-3)}$ 1

8. 1^7 1

9. $(-1)^0$ 1

10. $(-1)^1$ −1

11. $(-1)^2$ 1

12. $(-1)^3$ −1

13. $(-1)^4$ 1

14. $(-1)^5$ −1

15. $(-1)^6$ 1

16. 4^0 1

17. 4^1 4

18. 4^2 16

19. 4^3 64

20. 4^4 256

21. $4^{(-1)}$ $\frac{1}{4}$

22. $4^{(-2)}$ $\frac{1}{16}$

23. $4^{(-3)}$ $\frac{1}{64}$

24. $\left(\frac{1}{4}\right)^{-1}$ 4

25. $\left(\frac{1}{4}\right)^{-2}$ 16

26. $\left(\frac{1}{4}\right)^{-3}$ 64

27. $\left(\frac{1}{4}\right)^{-4}$ 256

Motivator

Begin the lesson with the motivator to get students thinking about the topic of the upcoming problem. This lesson is about a bacteria project submitted to a science fair. The motivating questions are about science fairs.

Ask the students the following questions to get them interested in the lesson.

- Have you ever entered a science fair?
- What project did you make?
- Where was the science fair?
- What is your favorite memory from a science fair?

Explore Together

Problem 1

Students will identify an exponential pattern and write an expression for the relationship as a power of 2.

Grouping

Ask for a student volunteer to read the Scenario and Problem 1 aloud. Have a student restate the problem. Pose the Guiding Questions below to verify student understanding. Have students work together in small groups to complete parts (A) through (E) of Problem 1.

Guiding Questions

- What is a science fair?
- What is a bacteria?
- What information is given in this problem?
- What are you asked to do in part (A) of Problem 1?
- Do you notice a pattern in the data?
- How can you find the number of bacteria that will exist in 6 hours from the beginning if the pattern continues?
- In what part of the problem are you asked to complete the third column of the table?
- Is the number of bacteria increasing or decreasing as time increases?
- If you were to graph the points given with the number of hours on the horizontal *x*-axis and the number of bacteria on the vertical *y*-axis, do you think that the points would be in a straight line? Why or why not?
- Do you think the rate of growth of the bacteria is a linear function? Why or why not?

Common Student Errors

Some students will not immediately recognize the pattern for this relationship because they will expect it to be either a linear or quadratic relationship. They may notice that the number 2 is a factor of each value for the number of bacteria and try to write a linear function with a coefficient of 2 to represent the relationship. Also, some students may forget how to find the prime factorization of a number. You may need to remind them of this skill.

SCENARIO You decide to enter the science fair at your school. Your project will demonstrate how conditions such as temperature, light, food source, and so on, affect the growth of bacteria. From your science class, you know that there are four phases of bacterial growth: lag, log, stationary, and death. In the lag phase the bacteria become accustomed to their surroundings; in the log phase the bacteria grow rapidly; in the stationary phase the number of growing bacteria is equal to the number of dying bacteria; and in the death phase the bacteria die off.

Problem 1 Growing Bacteria

In your first experiment, you record the following data for the log phase of the bacterial growth.

Time since beginning of log phase (in hours)	Number of bacteria	Prime factorization of number of bacteria
0	1	2^0
1	2	2^1
2	4	2^2
3	8	2^3
4	16	2^4
5	32	2^5

A. Assuming that the bacterial growth continues in this manner, how many bacteria will there be six hours after the beginning of the log phase? Use a complete sentence to explain how you found your answer.

There will be 64 bacteria after 6 hours. The number of bacteria is doubling after each hour.

B. Is there a pattern in the numbers of bacteria? If so, what is the pattern? Use a complete sentence in your answer.

Sample Answer: The numbers of bacteria can all be written as a power of 2.

C. Find the prime factorization of each number of bacteria. Record your results in the third column of the table.

13

Explore Together

Investigate Problem 1

Students will graph and investigate an exponential growth function.

Grouping

Call the class back together to have the students discuss and present their work for parts (A) through (E) of Problem 1.

Notes Students often have a very limited understanding of the benefits and applications of graphing functions. While it is too early to introduce detailed calculus concepts, it is not too early to introduce students to the general concept of rates of change. You can discuss the rate of bacterial growth with the students by asking the following Key Formative Assessments.

Key Formative Assessments

- Where is the number of bacteria smallest?
- As time increases, what happens to the number of bacteria?
- Is the slope a constant value for this function?
- How many bacteria are present after 1 hour? How many bacteria are present after 2 hours? How many more bacteria are present after 2 hours than after 1 hour?
- How many bacteria are present after 4 hours? How many bacteria are present after 5 hours? How many more bacteria are present after 5 hours than after 4 hours?
- Does the number of bacteria grow faster as the time changes from 1 hour to 2 hours or does it grow faster as the time changes from 4 hours to 5 hours?
- Do you think that the rate of bacterial growth is faster when the graph is flatter or when the graph is steeper?

Problem 1 Growing Bacteria

D. Write a function that gives the population, or number of bacteria, in terms of time. Let x be the amount of time in hours and let y be the number of bacteria.

$y = 2^x$

E. Create a graph that shows the population as a function of time on the grid below. First, choose your bounds and intervals. Be sure to label your graph clearly.

Variable quantity	Lower bound	Upper bound	Interval
Time	0	7.5	0.5
Population	0	150	10

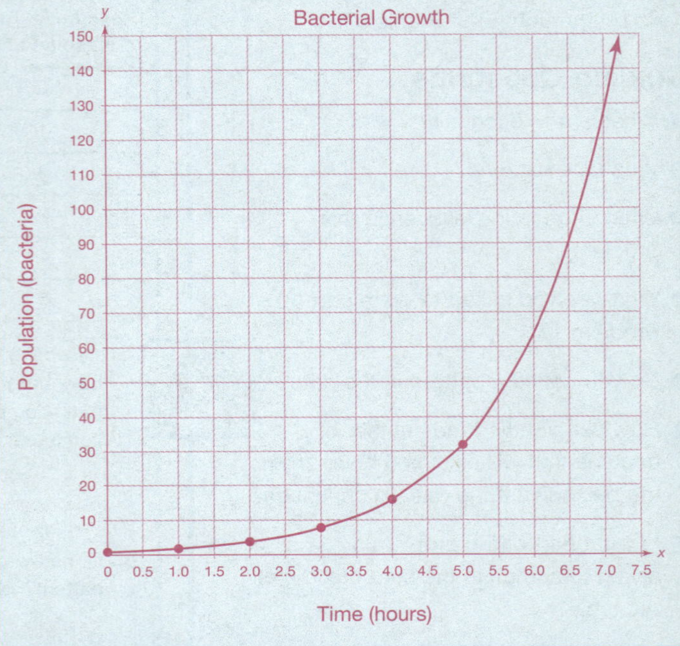

Investigate Problem 1

1. What is the domain of your function in the problem situation? Use a complete sentence in your answer.

   Sample Answer: The domain is all real numbers from zero to the time when the log phase ends.

Investigate Problem 1

Ask for a student volunteer to read Question 1 aloud. Have a student restate the problem. Pose the Guiding Questions on the next page to verify student understanding. Complete Question 1 together as a class.

Explore Together

Investigate Problem 1

Students will evaluate exponential functions.

Guiding Questions

- What is a domain? What is a range?
- What 2 quantities are being compared in this problem?
- What times are appropriate for this situation? What number of bacteria are possible?

Just the Math

Students will be formally introduced to the terminology for exponential functions.

Grouping

Ask for a student volunteer to read Question 2 aloud. Have a student restate the problem. Pose the Guiding Questions below to verify student understanding. Have students work together in small groups to complete Questions 2 and 3. Then, call the class back together to have the students discuss and present their work for Questions 2 and 3.

Guiding Questions

- What is an exponent?
- Why do you think an exponential function has the term exponential as part of its title?
- What is a function?
- What is an exponential function?
- What does the notation $f(x)$ represent?
- How do we read the notation $f(x)$?
- What does the a represent in an exponential function?
- Why must a be a positive value?
- How can you evaluate an exponential function?

Common Student Errors

Students often struggle to evaluate exponential functions when the exponent is a negative value. You may need to review evaluating negative exponents with the students. Additionally, some students will struggle to evaluate powers with an exponent of zero. Students sometimes confuse an exponent of zero with a factor of zero when multiplying. Finally, evaluating functions involving fractions will be challenging for several students. Many students have difficulty evaluating powers of fractions. If the students have too much difficulty with Question 3, you may want to complete it together as a class.

Investigate Problem 1

What does the domain of your function represent in the problem situation?

Sample Answer: The domain represents the length of the log phase.

What factors do you think will affect the length of the log phase? Use complete sentences in your answer.

Sample Answer: The length of the log phase will be affected by whether the bacteria prefer warmer or colder temperatures, whether the bacteria prefer light or no light, and the amount of food available.

What is the range of your function in the problem situation? Use a complete sentence in your answer.

Sample Answer: The range is all real numbers from one to the number of bacteria that are there when the log phase ends.

2. **Just the Math: Exponential Functions** The function that you wrote in part (D) is an example of an *exponential function*. An **exponential function** is a function of the form $f(x) = a^x$ where $a > 0$ and $a \neq 1$. Why do you think that 1 is excluded as a base? Use a complete sentence in your answer.

Sample Answer: It is excluded because $f(x) = 1^x = 1$, which is a constant function.

3. Complete the table of values below for three different exponential functions.

x	$f(x) = 2^x$	$f(x) = 3^x$	$f(x) = \left(\frac{1}{2}\right)^x$
−3	$\frac{1}{8}$	$\frac{1}{27}$	8
−2	$\frac{1}{4}$	$\frac{1}{9}$	4
−1	$\frac{1}{2}$	$\frac{1}{3}$	2
0	1	1	1
1	2	3	$\frac{1}{2}$
2	4	9	$\frac{1}{4}$
3	8	27	$\frac{1}{8}$

Explore Together

Investigate Problem 1

Students will graph and compare exponential functions.

Grouping

Ask for a student volunteer to read Question 4 aloud. Have a student restate the problem. Pose the Guiding Questions below to verify student understanding. Have students work together in small groups to complete Question 4. Then call the class back together to have the students discuss and present their work for Question 4.

Guiding Questions

- What do you predict the graph of the function $y = 2^x$ will look like?
- What do you predict the graph of the function $y = 3^x$ will look like?
- What do you predict the graph of the function $y = \left(\frac{1}{2}\right)^x$ will look like?
- Are negative x-values possible for these functions?
- Are negative y-values possible for these functions?
- What are appropriate bounds for the x-values based on the table in Question 3?
- What are appropriate bounds for the y-values based on the table in Question 3?

Grouping

Ask for a student volunteer to read Question 5 aloud. Have a student restate the problem. Have students work together in small groups to complete Questions 5 through 7.

Common Student Errors

Some students will incorrectly expect the functions to be parabolic in nature. You may need to ask the students to compare a quadratic function such as $y = 3x^2 + 5x - 7$ to an exponential function to help them notice the differences between quadratic functions and exponential functions.

Investigate Problem 1

4. Graph each of the functions in Question 3 together on the grid.

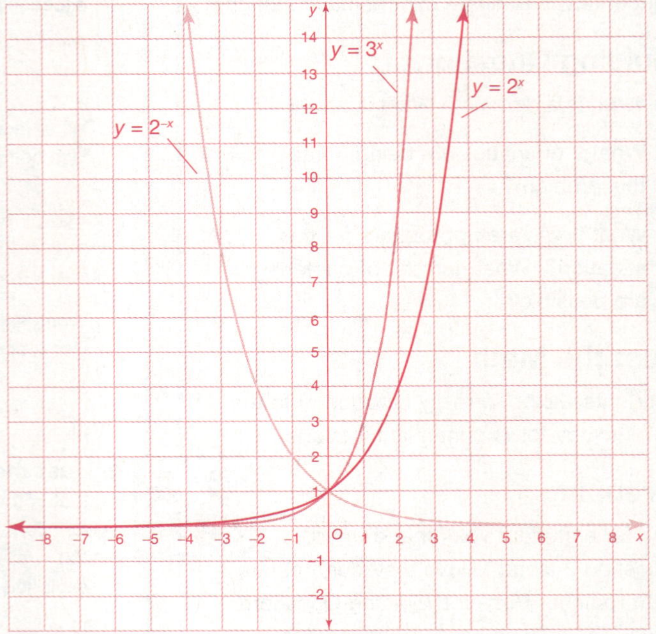

5. Describe the graph of $f(x) = a^x$ when $a > 1$. Use a complete sentence in your answer.

Sample Answer: The graph rises as you move from left to right.

Describe the graph of $f(x) = a^x$ when $0 < a < 1$. Use a complete sentence in your answer.

Sample Answer: The graph falls as you move from left to right.

6. What do think happens to the graph of $f(x) = a^x$ $(a > 1)$ as a gets larger? Use a complete sentence in your answer.

Sample Answer: The graph rises more quickly as a becomes larger.

What do you think happens to the graph of $f(x) = a^x$ $(0 < a < 1)$ as a gets smaller? Use a complete sentence in your answer.

Sample Answer: The graph falls more quickly as a becomes smaller.

Explore Together

Investigate Problem 1

Students will evaluate, investigate, and compare exponential functions.

Call the class back together to have the students discuss and present their work for Questions 5 through 7.

Grouping

Ask for a student volunteer to read Question 8 aloud. Have a student restate the problem. Pose the Guiding Questions below to verify student understanding. Have students work together in small groups to complete Questions 8 and 9. Then, call the class back together to have the students discuss and present their work for Questions 8 and 9.

Guiding Questions

- How do $f(x)$, $g(x)$, and $h(x)$ differ in Question 8?
- How do you expect the function values for $f(x)$, $g(x)$, and $h(x)$ to differ? Why?
- How do $f(x)$, $g(x)$, and $h(x)$ differ in Question 9?
- How do you expect the function values for $f(x)$, $g(x)$, and $h(x)$ to differ? Why?
- How can you test your predictions for Questions 8 and 9?

Key Formative Assessments

- What is an exponential function?
- How can you evaluate an exponential function?
- How can you graph an exponential function?
- What is the appearance of the graph of an exponential function if the value for a is greater than 1? What is the appearance of the graph of an exponential function if the value for a is between zero and 1?

Investigate Problem 1

7. What is the domain of an exponential function of the form $f(x) = a^x$? Use a complete sentence in your answer.

The domain is all real numbers.

What is the range of an exponential function of the form $f(x) = a^x$? Use a complete sentence in your answer.

The range is all real numbers greater than 0.

8. Complete the table of values below for the functions $f(x) = 2^x$, $g(x) = 2^x + 1$, and $h(x) = 2^x - 3$.

x	$f(x) = 2^x$	$g(x) = 2^x + 1$	$h(x) = 2^x - 3$
0	1	2	−2
1	2	3	−1
2	4	5	1
3	8	9	5
4	16	17	13

How do you think the graphs of g and h are related to the graph of f? Use complete sentences to explain.

Sample Answer: The graph of f can be shifted up one unit to obtain the graph of g because each y-value of g is one more unit than the corresponding y-value of f. The graph of f can be shifted down three units to obtain the graph of h because each y-value of h is three units less than the corresponding y-value of f.

9. Complete the table of values below for the functions $f(x) = 2^x$, $g(x) = 2^{x+2}$, and $h(x) = 2^{x-1}$.

x	$f(x) = 2^x$	$g(x) = 2^{x+2}$	$h(x) = 2^{x-1}$
0	1	4	$\frac{1}{2}$
1	2	8	1
2	4	16	2
3	8	32	4
4	16	64	8

How do you think the graphs of g and h are related to the graph of f? Use complete sentences to explain.

Sample Answer: The graph of f can be shifted left two units to obtain the graph of g because each y-value of g corresponds to the y-value of f whose x-value is two units more. The graph of f can be shifted right one unit to obtain the graph of h because each y-value of h corresponds to the y-value of f whose x-value is one unit less.

Wrap Up

Close

13

- Review all key terms and their definitions. Include the term *exponential function.*
- You may also want to review any other vocabulary terms that were discussed during the lesson, which may include *prime number, prime factorization, evaluate, function, linear function, quadratic function, slope, exponent, powers, growth rate, exponential growth, transformations,* and *translations.*
- Remind the students to write the key terms and their definitions in the notes section of their notebooks. You may also want the students to include examples.
- Ask the students to summarize the characteristics of an exponential function.
- Ask the students how exponential functions differ from other functions that they have worked with in the past.
- Ask the students to summarize the rate of growth of the bacteria based on their work in this lesson.
- Ask the students to explain the meaning of a rate of change.
- Ask the students to explain why the rate of change is a constant value for a straight line, but not for a quadratic function or for an exponential function.
- Have the students agree on the requirements and a general notation for exponential functions. For instance, $f(x) = a^x$, where $a > 0$ and $a \neq 1$.

Ties to the Cognitive Tutor Software

Although exponential functions can be mathematically complex, there are many real-world situations that can help students understand the general idea of exponential growth and decay. The Cognitive Tutor software provides students with a number of different application contexts.

Follow Up

Assignment

Use the Assignment for Lesson 13.6 in the Student Assignments book. See the Teacher's Resources and Assessments book for answers.

Assessment

See the Assessments provided in the Teacher's Resources and Assessments book for Chapter 13.

Open-Ended Writing Task

Ask the students to write a short paragraph comparing the similarities and explaining the differences between a linear function, a quadratic function, and an exponential function. The students should provide examples and graph their examples as part of their response.

Reflections

Insert your reflections on the lesson as it played out in class today.

13

What went well?

What did not go as well as you would have liked?

How would you like to change the lesson in order to improve the things that did not go well and capitalize on the things that did go well?

Notes

13

13.7 Money Comes and Money Goes

Exponential Growth and Decay

Learning By Doing Lesson Map

Get Ready

Objectives

In this lesson, you will:

- Write and use an exponential growth model.
- Write and use an exponential decay model.

Key Terms

- simple interest
- compound interest
- exponential growth model
- exponential growth
- growth rate
- growth factor
- exponential decay model
- exponential decay
- decay rate
- decay factor

NCTM Content Standards

Grades 9–12 Expectations

Algebra Standards

- Understand and compare the properties of classes of functions, including exponential, polynomial, rational, logarithmic, and periodic functions.
- Use symbolic algebra to represent and explain mathematical relationships.
- Judge the meaning, utility, and reasonableness of the results of symbol manipulations, including those carried out by technology.
- Identify essential quantitative relationships in a situation and determine the class or classes of functions that might model the relationships.
- Use symbolic expressions, including iterative and recursive forms, to represent relationships arising from various contexts.
- Draw reasonable conclusions about a situation being modeled.
- Approximate and interpret rates of change from graphical and numerical data.

Lesson Overview

Within the context of this lesson, students will be asked to:

- Write, evaluate, and graph an exponential growth model to represent the balance in a savings account after interest is accrued.
- Write, evaluate, and graph an exponential decay model to represent the value of a car after it is purchased.
- Use the models to predict values.

Essential Questions

The following key questions are addressed in this lesson:

1. What is exponential growth?
2. What is exponential decay?
3. What is simple interest?
4. What is compound interest?

Show The Way

Warm Up

13

Place the following questions or an applicable subset of these questions on the board before students enter class. Students should begin working as soon as they are seated.

Complete the table by writing equivalent fractions, decimals, and percents.

	Fraction	Decimal	Percent
1.	$\frac{5}{9}$	$0.\overline{5}$	$55.\overline{5}\%$
2.	$\frac{14}{25}$	0.56	56%
3.	$\frac{1}{5}$	0.2	20%
4.	$\frac{2}{3}$	$0.\overline{6}$	$66.\overline{6}\%$
5.	$\frac{5}{8}$	0.625	62.5%
6.	$\frac{1}{100}$	0.01	1%
7.	$\frac{7}{100}$	0.07	7%
8.	$\frac{3.4}{100} = \frac{17}{500}$	0.034	3.4%
9.	$\frac{157}{100}$	1.57	157%
10.	$\frac{100}{100} = 1$	1	100%

Motivator

Begin the lesson with the motivator to get students thinking about the topic of the upcoming problem. This lesson is about savings, depreciation, and interest. The motivating questions are about earning and losing money.

Ask the students the following questions to get them interested in the lesson.

- What is interest? How can you earn interest for money that you deposit in a bank?
- Why do banks pay interest for money that you deposit?
- What do banks do with the money that you deposit?
- Do you think that banks spend more money on interest for deposited money or collect more money in interest on personal and business loans?
- What is a stock? How can you earn money with a stock that you buy?
- Can you lose money with a stock that you buy? If so, how?

Explore Together

Problem 1

Students will calculate simple interest for deposits into a savings account.

Grouping

Ask for a student volunteer to read the Scenario and Problem 1 aloud. Have a student restate the problem. Pose the Guiding Questions below to verify student understanding. Complete parts (A) and (B) together as a class. Then, have students work together in small groups to complete parts (C) through (E).

Take Note

The **simple interest** formula is $I = Prt$ where I is the interest earned in dollars, P is principal (the amount of money put into the account) in dollars, r is the interest rate in decimal form, and t is the time in years.

Guiding Questions

- What is interest?
- What is simple interest?
- How can you calculate the amount of interest earned?
- If the amount of time is given in months, such as 6 months, how can you find the amount of time in years?
- How can you calculate the amount of money in the savings account after interest has accrued?

Common Student Errors

Students frequently have difficulty remembering the equation for simple interest as $I = Prt$. It may help to give the students the suggestion that interest is their part and PaRT has the letters *Prt* from the formula. Because interest is the part paid to them or that they will pay, this can be easy to remember.

SCENARIO The value of money, cars, homes, and so on can change, for the better or for the worse. For instance, you could put your money into a savings account and earn interest on your money. You could also buy stock in a company and the value of your stock could rise or fall.

Problem 1 Compound Can Be Better than Simple

A. Suppose that you invest \$100 into a savings account that earns 3% interest for 3 years. Find the amount of simple interest that you earned at the end of the 3 year period. Show your work and use a complete sentence in your answer.

$I = 100(0.03)3 = 9$; The amount of interest earned after 3 years is \$9.

How much money do you have altogether? Use a complete sentence in your answer.

I have \$109.

B. Find the amount of simple interest earned by depositing \$100 into an account that earns 3% interest for 1 year. Show your work and use a complete sentence in your answer.

$I = 100(0.03)1 = 3$; The amount of interest earned after 1 year is \$3.

How much money do you have altogether? Use a complete sentence in your answer.

I have \$103.

C. Take the total amount of money that you have from part (B) and put it back into the same account for another year. Find the amount of simple interest that you earned after this year. Show your work and use a complete sentence in your answer.

$I = 103(0.03)(1) = 3.09$; The amount of interest earned after another year is \$3.09.

How much money do you have altogether? Use a complete sentence in your answer.

I have \$106.09.

Explore Together

Problem 1

Students will compare account balances where the interest is calculated yearly and where the interest is calculated once for the entire time of deposit.

Call the class back together to have the students discuss and present their work for parts (C) through (E).

Key Formative Assessments

- What is principal?
- What is an interest rate?
- How is time measured for calculating interest?
- Which is greater, simple interest or compound interest? Why?
- What types of banking accounts can earn interest?
- When would you have to pay interest to a bank?

Take Note

The amount of money in an account is often called the balance of the account.

Investigate Problem 1

Grouping

Ask for a student volunteer to read Question 1 aloud. Have a student restate the problem. Pose the Guiding Questions below to verify student understanding. Have students work together in small groups to complete Questions 1 through 3.

Guiding Questions

- What is a balance?
- What information do you need to calculate the balance for an account?
- How can you calculate the balance for an account?

Problem 1 Compound Can Be Better than Simple

D. Take the total amount of money that you have from part (C) and put it back into the same account for one more year. Find the amount of simple interest that you earned after this year. Show your work and use a complete sentence in your answer.

$I = 106.09(0.03)(1) \approx 3.18$; The amount of interest earned after another year is $3.18.

How much money do you have altogether? Use a complete sentence in your answer.

I have $109.27.

E. How does the total amount of money that you have in part (A) compare to the total amount of money that you have in part (D)?

The amount in part (D) is greater than the amount in part (A) by $.27.

In parts (B) through (D), the interest was **compounded.** In other words, the interest is earned not only on the principal, but on interest that was previously earned.

Investigate Problem 1

1. Create a table of values that shows the balance of the account in terms of time. The third and fourth columns of the table will be completed later.

Time (years)	Balance (dollars)	Balance in terms of previous balance	Balance in terms of original balance
0	100.00	100	100
1	103.00	1.03(100)	1.03(100)
2	106.09	1.03(103)	$1.03^2(100)$
3	109.27	1.03(106.09)	$1.03^3(100)$

2. For successive years, what percent of the previous balance is the new balance? Show your work. Round to the nearest whole percent if necessary.

Year 0 to year 1: $\frac{103}{100} = 1.03$; 103%

Year 1 to year 2: $\frac{106.09}{103} = 1.03$; 103%

Year 2 to year 3: $\frac{109.27}{106.09} \approx 1.03$; 103%

- How can you represent the balance of the account in terms of time?
- How can you calculate what percent of a previous value a new value is?
- How can you write a decimal as a percent?

Explore Together

Investigate Problem 1

Students will write and investigate an exponential growth model.

Notes Students may struggle with Questions 2 and 3. Remind the students to use the percent equation, $\frac{\text{New balance}}{\text{Previous balance}} \cdot 100$, for Questions 2 and 3.

Call the class back together to have the students discuss and present their work for Questions 1 through 3.

Take Note

In later courses, you will learn that compoundings can take place more often than yearly. When you compound interest yearly, you should use the exponential growth model in Question 6.

Grouping

Ask for a student volunteer to read Question 4 aloud. Have a student restate the problem. Pose the Guiding Questions below to verify student understanding. Have students work together in small groups to complete Questions 4 and 5. Then, call the class back together to have the students discuss and present their work for Questions 4 and 5.

Guiding Questions

- What percent of the original balance was in the account at the end of zero years?
- By what can you multiply 100% to find the balance at the end of 1 year? Why do you multiply 100 percent by 1.03 to find the balance at the end of 1 year?
- By what can you multiply 100% to find the balance at the end of 2 years?
- What pattern do you notice?

Just the Math

The students will be formally introduced to the terminology for exponential growth in Question 6.

Investigate Problem 1

What do you notice about the percents? Use a complete sentence in your answer.

Sample Answer: The percents are all the same and they are greater than 100%.

3. Write each balance in the table in terms of the previous balance. Record your answers in the third column of the table.

4. Write each balance in the table in terms of the original balance of the account. Record your answers in the fourth column of the table.

5. Write a function for the account balance y in dollars in terms of the time t in years since the account was opened.

   $y = 100(1.03)^t$

6. **Just the Math: Exponential Growth Model**
   The function that you wrote in Question 5 is an *exponential growth model*. The general form of an **exponential growth model** is the equation $y = C(1 + r)^t$ where C is the original amount (of money, people, and so on) before any growth occurs, r is the **growth rate** in decimal form, $(1 + r)$ is the **growth factor,** t is the amount of time, and y is the new amount. What is the growth rate of your model in Question 5? What is the growth factor? Show your work and use a complete sentence in your answer.

   $1.03 = 1 + 0.03$; The growth rate is 0.03. The growth factor is 1.03.

7. You invest $500 into a savings account that earns 2% interest, compounded yearly. Write a model for the account balance y after t years.

   $y = 500(1.02)^t$

Grouping

Ask for a student volunteer to read Question 6 aloud. Have a student restate the problem. Pose the Guiding Questions on the next page to verify student understanding. Complete Question 6 together as a class. Then, have the students work together in small groups to complete Questions 7 and 8.

13

Explore Together

Investigate Problem 1

Students will graph and investigate another exponential growth model.

Guiding Questions

- How can you find the growth rate for our situation?
- How can you find the growth factor for our situation?
- What does the 1 in the growth factor represent?
- What does the *C* represent in the exponential growth model? What does *t* represent? What does *r* represent?
- What is meant by the phrase "the new amount?"

Call the class back together to have the students discuss and present their work for Questions 7 and 8.

Grouping

Ask for a student volunteer to read Question 9 aloud. Have a student restate the problem. Pose the Guiding Questions below to verify student understanding. Have students work together in small groups to complete Questions 9 and 10.

Guiding Questions

- What 2 quantities are being considered in this situation?
- Which is the independent variable? Which is the dependent variable?
- Which axis of the graph will you use to represent the amount of time?
- Which axis of the graph will you use to represent the account balance?
- What are the units for the time of deposit? What are the units for the account balance?
- How can you use a graph to determine the amount of time required to have a given account balance?

Investigate Problem 1

8. Complete the table of values that shows the account balance as a function of the number of years the account is open.

Quantity Name	Time	Account balance
Unit	years	dollars
Expression	t	$500(1.02)^t$
	0	500.00
	1	510.00
	2	520.20
	5	552.04
	10	609.50
	15	672.93

9. Create a graph that shows the account balance as a function of time on the grid below. First, choose your bounds and intervals. Be sure to label your graph

Variable quantity	Lower bound	Upper bound	Interval
Time	0	15	1
Account balance	0	750	50

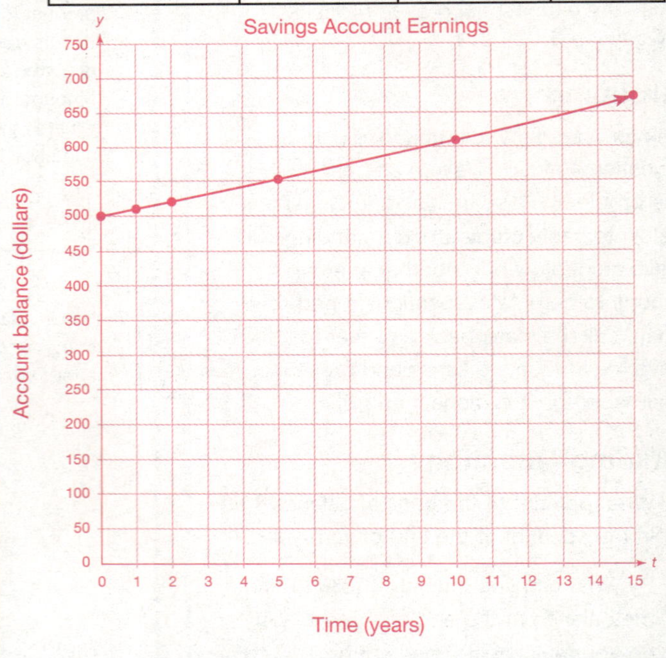

- How can you use a graph to determine the account balance for a given time of deposit?

Common Student Errors

Students may incorrectly assume that the graph will be a linear relationship. Ask them if they expect the relationship to be linear. Have them explain their reasoning. Try to lead them with probing questions to realize that the graph is an exponential function.

Explore Together

Investigate Problem 1

Students will compare account balances between an account earning interest compounded yearly.

Call the class back together to have the students discuss and present their work for Questions 9 and 10.

Grouping

Ask for a student volunteer to read Question 11 aloud. Have a student restate the problem. Pose the Guiding Questions below to verify student understanding. Have the students work individually to complete Question 11. Call the class back together to have the students discuss and present their work for Question 11.

Guiding Questions

- What formula did you use to find the balance in Lupe's account?
- How did you know to use this formula?
- What formula did you use to find the balance in Domingo's account?
- How did you know to use this formula?
- In which account would you invest your money? Why?

Investigate Problem 1

10. Use your graph to determine how long you have to keep the account open in order to have a balance of $575. Use a complete sentence in your answer.

The account has to be open for about seven years in order to have a balance of $575.

Use your graph to determine how long you have to keep the account open in order to have a balance of $650. Use a complete sentence in your answer.

The account has to be open for about thirteen years in order to have a balance of $650.

Use your graph to determine the account balance after three years. Use a complete sentence in your answer.

The account balance after three years is approximately $530.

11. Suppose that Lupe invests $1500 into a savings account that earns 4% simple interest. Domingo invests the same amount of money into a savings account that earns 4% interest compounded yearly. If neither Lupe nor Domingo has removed any money from their account after 10 years, how much money is in each account? Which account has earned more interest? Show your work and use complete sentences to explain your reasoning.

After 10 years, Lupe's account has a balance of $2100 and Domingo's account has a balance of about $2220.37. Domingo's account has earned more interest.

Explore Together

Problem 2

Students will investigate exponential decay.

Grouping

Ask for a student volunteer to read Problem 2 aloud. Have a student restate the problem. Pose the Guiding Questions below to verify student understanding. Have students work together in small groups to complete parts (A) and (B).

Guiding Questions

- Is the cost of a new car or a used car more? Why?
- What words do we use to describe an object that is less valuable as time increases?
- Do you think that the value of a car depreciates more between when it is first purchased and 1 year later or between 5 years after it is purchased and 6 years after it was purchased? Why?
- Do you think that the rate of depreciation is a constant value?
- Do you think that the value of a car as time increases will be a linear function? Why?
- What 2 quantities are being discussed in Problem 2?
- What unit will you use for time? What unit will you use for the value of the car?
- What is the smallest possible amount of time after a car is purchased? What is the largest amount of time given in the table for Problem 2? What bounds for time are reasonable for this situation?
- What is the largest possible value for a car that you purchase for $15,000? What is the smallest possible value for the car?

Problem 2 Losing Value

Whenever you buy an item like a car or a computer, the value of the item depreciates, or decreases, because once you start using the item, it is no longer in its original condition.

The value of a new car from the time it is bought over a five year period is shown in the table at the right.

Time (years)	Value (dollars)
0	15,000.00
1	9000.00
2	5400.00
3	3240.00
4	1944.00
5	1166.40

A. Create a graph that shows the car's value as a function of time on the grid below. First, choose your bounds and intervals. Be sure to label your graph clearly.

Variable quantity	Lower bound	Upper bound	Interval
Time	0	7.5	0.5
Value	0	15,000	1000

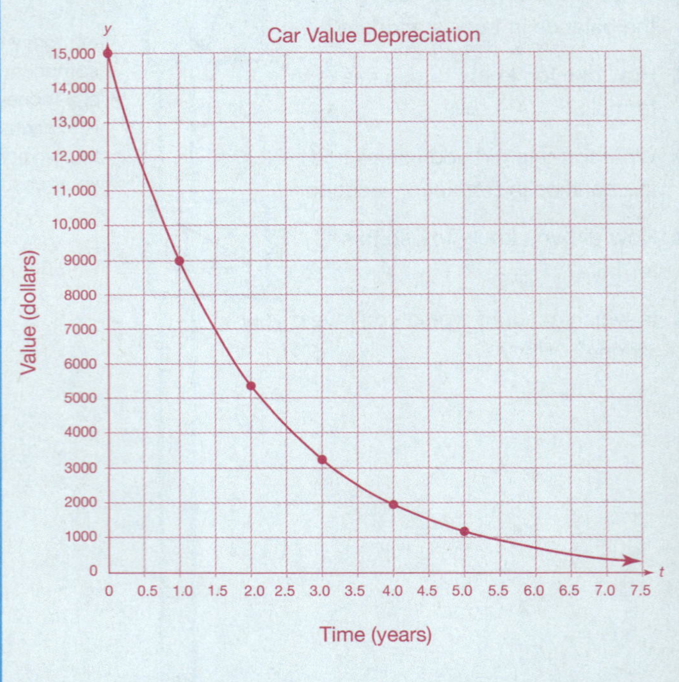

Explore Together

Problem 2

Students will investigate the exponential decay of the value for a new car over time after it is purchased.

Call the class back together to have the students discuss and present their work for parts (A) and (B).

Key Formative Assessments

- How is the graph in part (A) of Problem 2 similar to the graph in Question 9 of Problem 1? How are the graphs different?
- Do you think the trend for the value of the car will continue in the pattern shown as the years continue beyond 5 years?
- Do you expect the value of the car to ever be equal to zero dollars?
- Do you expect the value of the car to ever be negative?
- What value for the car does it approach but never equal?

Notes You may want to introduce the concept of a limit to more advanced students. The limit of the value is zero dollars, because the value will approach, but never equal zero dollars.

Investigate Problem 2

Grouping

Ask for a student volunteer to read Question 1 aloud. Have a student restate the problem. Have students work together in small groups to complete Questions 1 through 4.

Problem 2 Losing Value

B. What kind of function would you use to model your graph? Use complete sentences to explain your reasoning.

Sample Answer: Because the graph falls towards zero as you move from left to right, but does not reach zero, the graph can be modeled by an exponential function.

Investigate Problem 2

1. For successive years, what percent of the previous value is the new value? Show your work.

Year 0 to year 1: $\frac{9000}{15,000} = 0.6$; 60%

Year 1 to year 2: $\frac{5400}{9000} = 0.6$; 60%

Year 2 to year 3: $\frac{3240}{5400} = 0.6$; 60%

Year 3 to year 4: $\frac{1944}{3240} = 0.6$; 60%

Year 4 to year 5: $\frac{1166.4}{1944} = 0.6$; 60%

What do you notice about the percents? Use a complete sentence in your answer.

Sample Answer: The percents are all the same and they are less than 100%.

2. Write each value in the table in terms of the previous value. Record your answers in the third column of the table.

Time (years)	Value (dollars)	Value in terms of previous value	Value in terms of original value
0	15,000.00	15,000	15,000
1	9000.00	0.6(15,000)	0.6(15,000)
2	5400.00	0.6(9000)	0.6^2(15,000)
3	3240.00	0.6(5400)	0.6^3(15,000)
4	1944.00	0.6(1944)	0.6^4(15,000)
5	1166.40	0.6(1944)	0.6^5(15,000)

Explore Together

Investigate Problem 2

3

Students will compare their exponential growth and exponential decay models.

Call the class back together to have the students discuss and present their work for Questions 1 through 4.

Just the Math

The students will be formally introduced to the terminology for exponential decay in Question 5.

Ask for a student volunteer to read Question 5 aloud. Have a student restate the problem. Pose the Guiding Questions below to verify student understanding.

Guiding Questions

- What is a model?
- What is an exponential decay model?
- If something is worth 90% of its original value, what percent of its value did it lose?
- If something is worth 90% of its original value, what was the decay rate?
- How can you calculate the decay rate? How can you calculate the decay factor?

Grouping

Have the students work in small groups to complete Questions 5 through 7. Call the class back together to have the students discuss and present their work for Questions 5 through 7.

Key Formative Assessments

- What is simple interest? What is compound interest?
- How can you calculate the amount of simple interest? How can you calculate the amount of compound interest?
- What is exponential growth? What is exponential decay?
- What is a growth factor? What is a growth rate? What is a decay factor? What is a decay rate?

Investigate Problem 2

3. Write each value in the table in terms of the original value of the car. Record your answers in the fourth column of the table.

4. Write a function for the value y of the car in dollars in terms of the time t in years since the car was bought.

$y = 15{,}000(0.6)^t$

5. Just the Math: Exponential Decay Model The function that you wrote in Question 4 is an *exponential decay model*. The general form of an **exponential decay model** is the equation $y = C(1 - r)^t$ where $0 < r < 1$, C is the original amount before any decay occurs, r is the **decay rate** in decimal form, $(1 - r)$ is the **decay factor,** t is the amount of time, and y is the new amount. What is the decay rate of your model in Question 4? Show your work and use a complete sentence in your answer.

0.6 = 1 – 0.4; The decay rate is 40%.

6. According to this model, will the value of the car ever be $0? Why or why not? Use complete sentences in your answer.

Sample Answer: No, because a power with a positive base can never be zero.

Is this reasonable? Explain your reasoning. Use a complete sentence in your answer.

Sample Answer: No, because at some point, the car will not run and will be worth nothing as a car.

7. How are the exponential growth model and exponential decay model similar? How are they different? Use complete sentences in your answer.

Sample Answer: Both models have an initial value and have a base raised to an exponent. In the exponential growth model, the base is greater than one, and in the exponential decay model, the base is less than one.

Wrap Up

Close

- Review all key terms and their definitions. Include the terms *simple interest, compound interest, exponential growth, growth rate, growth factor, exponential decay, decay rate,* and *decay factor.*
- You may also want to review any other vocabulary terms that were discussed during the lesson, which may include *interest, principal, rate, balance, deposit, withdrawal, accrued, independent variable, dependent variable,* and *depreciate.*
- Remind the students to write the key terms and their definitions in the notes section of their notebooks. You may also want the students to include examples.
- Ask the students the following questions:
  - How are simple interest and compound interest similar? How are they different?
  - How can you calculate simple interest? How can you calculate compound interest?
  - Is interest an example of exponential growth or decay? Why?
  - What else do you think can be modeled with an exponential growth model?
  - What do you think can be modeled with an exponential decay model other than the value of a car?
  - What does the graph of an exponential growth model look like?
  - What does the graph of an exponential decay model look like?
  - What is a growth rate? What is a growth factor?
  - What is a decay rate? What is a decay factor?

Ties to the Cognitive Tutor Software

Although exponential functions can be mathematically complex, there are many real-world situations that can help students understand the general idea of exponential growth and decay. The Cognitive Tutor software provides students with a number of different application contexts.

Follow Up

Assignment

Use the Assignment for Lesson 13.7 in the Student Assignments book. See the Teacher's Resources and Assessments book for answers.

Assessment

See the Assessments provided in the Teacher's Resources and Assessments book for Chapter 13.

Open-Ended Writing Task

Ask the students to write a short summary comparing exponential growth and exponential decay.

13

Reflections

Insert your reflections on the lesson as it played out in class today.

What went well?

What did not go as well as you would have liked?

How would you like to change the lesson in order to improve the things that did not go well and capitalize on the things that did go well?

Notes

13.8 Camping

Special Topic: Logic

Learning By Doing Lesson Map

Get Ready

Objectives

In this lesson, you will:

- Prove a statement using a direct proof.
- Prove a statement using an indirect proof.
- Find a counterexample.

Key Terms

- logical reasoning
- proof
- direct proof
- counterexample
- indirect proof

NCTM Content Standards

Grades 9–12 Expectations

Number and Operations Standards

- Use number-theory arguments to justify relationships involving whole numbers.
- Judge the effects of such operations as multiplication, division, and computing powers and roots on the magnitude of quantities.

Algebra Standards

- Use symbolic algebra to represent and explain mathematical relationships.
- Draw reasonable conclusions about a situation being modeled.

Lesson Overview

Within the context of this lesson, students will be asked to:

- Construct direct proofs to prove algebraic relationships.
- Construct an indirect proof to prove an algebraic statement.
- Develop counterexamples to disprove algebraic relationships.

Essential Questions

The following key questions are addressed in this lesson:

1. What is logical reasoning?
2. What is a proof?
3. What is a direct proof?
4. What is an indirect proof?
5. What is a counterexample?

Show The Way

Warm Up

13

Place the following questions or an applicable subset of these questions on the board before students enter class. Students should begin working as soon as they are seated.

Graph each equation. You may consider using parent functions to help you. Then determine whether the equation is a function. If it is a function, state the type of function and the parent function.

1. $12x - 6y = 24$

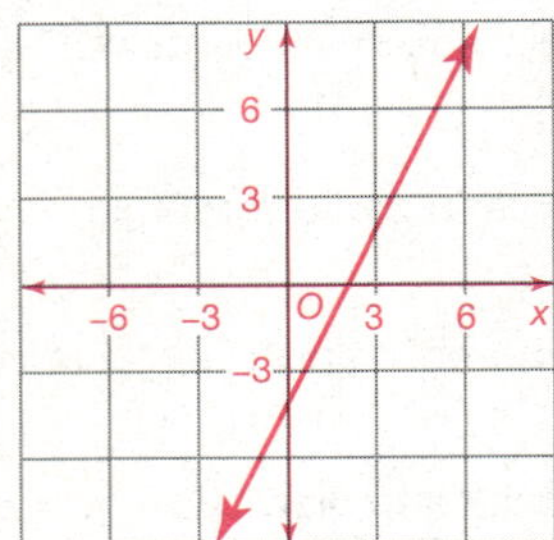

This is a linear function.
The parent function is $y = x$.

2. $x = 3$

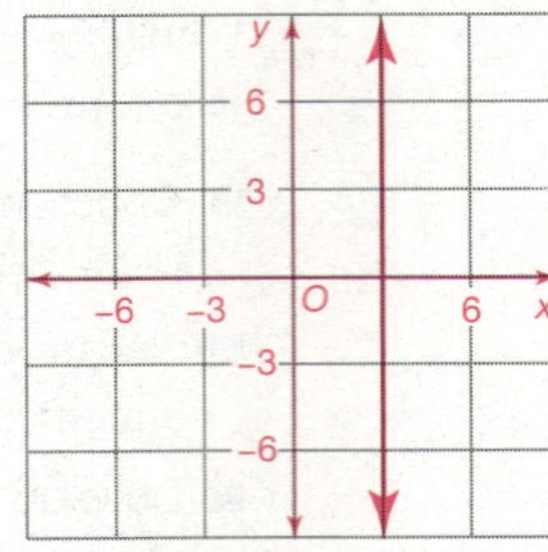

This is not a function.
It fails the vertical line test.

3. $y = (x + 4)^2 - 9$

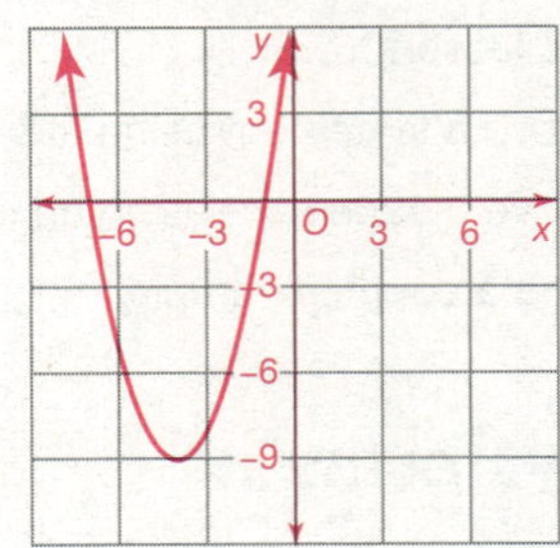

This is a quadratic function.
The parent function is $y = x^2$.

4. $y = (x - 3)(x + 1)$

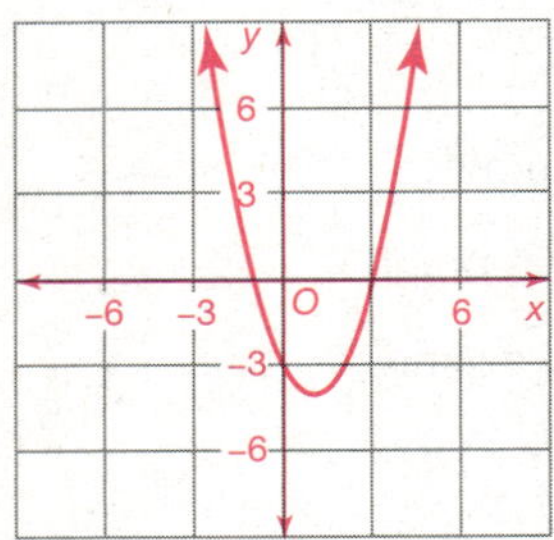

This a quadratic function.
The parent function is $y = x^2$.

5. $y = x^3 - 4$

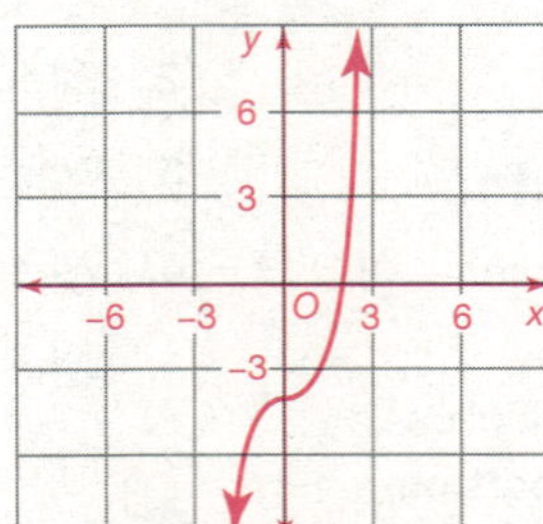

This a cubic function.
The parent function is $y = x^3$.

6. $y = (2^x) + 1$

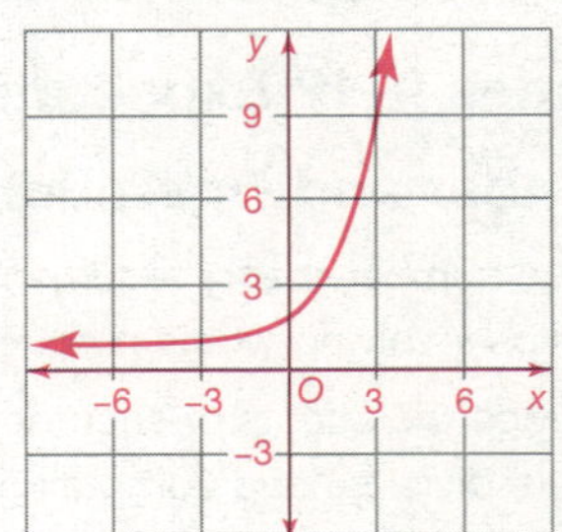

This an exponential function.
The parent function is $y = a^x$.
Where $a \neq 0$, and $a > 1$.

Motivator

Begin the lesson with the motivator to get students thinking about the topic of the upcoming problem. This lesson is about writing assembly instructions for a tent. The motivating questions are about assembly instructions.

Ask the students the following questions to get them interested in the lesson.

- What are assembly instructions?
- Why must assembly instructions be clear and easy to follow?
- What products have you purchased that had to be assembled?
- Have you ever had to assemble a product in which the instructions were hard to follow or incorrect?
- Have you ever returned a product to the store because the instructions were hard to follow or incorrect?

Explore Together

Problem 1

Students will consider a logical order for the steps required to assemble a tent.

Ask for a student volunteer to read the Scenario and Problem 1 aloud. Have a student restate the problem. Pose the Guiding Questions below to verify student understanding.

Guiding Questions

- What is an instruction sheet?
- What is a writer?
- Why is the order of the instructions important?
- How can you determine an appropriate order for the steps?
- What does the phrase "the steps are in no particular order" mean?
- What is a tent fly-cover?
- What does it mean to be waterproof?

13

SCENARIO A camping tent manufacturer has a writer on staff to create instruction sheets for the company's tents, one of which is shown below. The instruction sheets must include a list of all of the tent parts, diagrams, and written assembly instructions. Not only must the assembly instructions be in the correct order, but they must also be clearly written.

Problem 1 What's Your Order

The writer has the following parts list and diagram.

Parts

1 tent body
1 tent fly-cover
7 tent stakes
4 tent poles

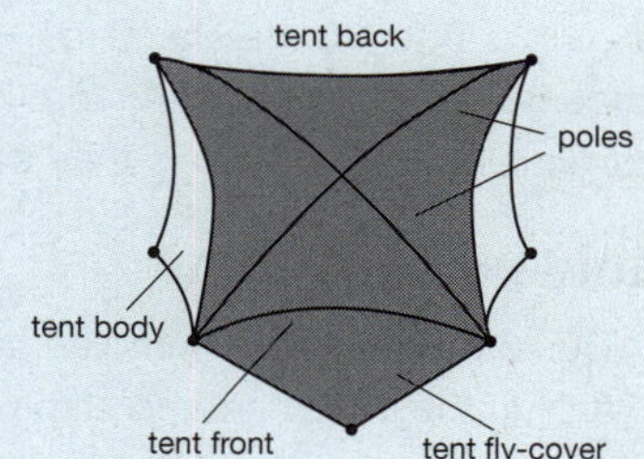

During a meeting with the tent designer, the writer comes up with the following steps, in no particular order.

a) Attach tent fly-cover over top of tent and put stake through loop and pound stake into the ground.

b) Assemble two pairs of tent poles.

c) Put stakes through the two loops at tent front and pound the stakes into the ground.

d) Select a site to place tent.

e) Pull tent fabric tight on one side of tent and put the stake through the loop at tent side and pound the stake into the ground.

f) Unpack tent and lay it on the ground where you want it so that the tent floor is on the bottom.

g) Pull tent fabric tight toward the back and put stakes through the two loops at tent back and pound the stakes into the ground.

h) Thread both tent poles through loops on top of tent.

i) Pull tent fabric tight on other side of tent and put a stake through the loop at tent side and pound the stake into the ground.

13

Explore Together

Problem 1

Students will develop a logical order for the steps required to assemble a tent.

Grouping

Ask for a student volunteer to read part (A) aloud. Pose the Guiding Questions below to verify student understanding. Have the students complete parts (A) and (B) individually. Then, discuss their solutions.

Guiding Questions

- Could the tent be assembled successfully in the order that the steps are given on the previous page?
- How can you determine which step should be first and last?

Notes The order given in part (C) is a sample answer only. For instance, step (b) can occur anytime before the threading of the poles through the loops on top of the tent in step (h). This is addressed in part (D).

Grouping

Ask for a student volunteer to read part (C) aloud. Pose the Guiding below Questions to verify student understanding. Have the students complete parts (C) and (D) individually or together in small groups. Call the class back together to have the students discuss and present their work for parts (C) and (D).

Take Note

When you write a logical series of steps to demonstrate a statement, you are writing a **proof.**

Guiding Questions

- How can you determine the order of the steps?
- Why do the directions for part (C) suggest that we use the letters rather than the words to list the order of the steps?

Problem 1 What's Your Order

A. What do you think should be the first step in the instructions? Why? Use complete sentences in your answer.

Sample Answer: The first step should be selecting the site for the tent because you have to know where you are putting the tent before you can start pounding the stakes into the ground.

B. What do you think should be the last step in the instructions? Why? Use complete sentences in your answer.

Sample Answer: The last step should be the tent fly-cover because it lays on top of the tent body.

C. Write the order of the steps for the instructions. You can identify the order by using the letters shown in Problem 1.

Sample Answer: d, f, c, g, e, i, b, h, a

D. Could any of the steps be moved in your order so that the tent is still properly assembled? If so, identify the steps and explain why and where they could be moved. Use complete sentences in your answer.

Sample Answer: Yes. You could move the assembly of the tent poles (step b) anywhere before threading the poles through the loops on top of the tent (step h).

Investigate Problem 1

1. When you completed the correct order of the tent assembly steps in Problem 1, you were using **logical reasoning.** You also use logical reasoning to simplify an expression or to prove a mathematical statement. To prove a mathematical statement means to use logical steps to show that the statement is true. Consider the statement

$(a + b)(a^2 - ab + b^2) = a^3 + b^3.$

The steps below, in no particular order, can be used to prove that this statement is true.

$= a^3 + b^3$	Additive Inverse
$= a(a^2) - a(ab) + a(b^2) + a^2b - b(ba) + b(b^2)$	Commutative Prop. of Mult.
$= a^3 - a^2b + ab^2 + a^2b - ab^2 + b^3$	Commutative Prop. of Mult.
$= a(a^2 - ab + b^2) + b(a^2 - ab + b^2)$	Distributive Prop.
$= a^3 - a^2b + a^2b + ab^2 - ab^2 + b^3$	Commutative Prop. of Addition
$= a(a^2) - a(ab) + a(b^2) + b(a^2) - b(ab) + b(b^2)$	Distributive Prop.
$= a^3 - a^2b + ab^2 + a^2b - b^2a + b^3$	Product of Powers

Investigate Problem 1

Ask for a student volunteer to read Question 1 aloud. Have a student restate the problem. Pose the Guiding Questions on the next page to verify student understanding. Have students work together in small groups to complete Question 1.

Explore Together

Investigate Problem 1

Students will construct direct proofs for algebraic relationships.

Take Note

Recall that you learned some of the rules of algebra in Lessons 4.4 and 4.5 and you used properties of exponents in Chapter 9.

Guiding Questions

- What is logic?
- What is logical reasoning?
- What does it mean to prove a mathematical statement to be true?
- What is an additive inverse?
- What is the commutative property of multiplication? What is the commutative property of addition?
- What is the distributive property?
- What is the product of powers?
- Does the proof make sense with the steps in the order that they were originally given?
- What is the goal of a proof?

Call the class back together to have the students discuss and present their work for Question 1.

Just the Math

Students will be introduced to the concept of direct proof in Question 2 and counterexamples in Question 3.

Grouping

Ask for a student volunteer to read Question 2 aloud. Have a student restate the problem. Have students work together in small groups to complete Question 2. Call the class back together to have the students discuss and present their work for Question 2. Then, ask for a student volunteer to read Question 3 aloud. Have a student restate the problem.

Investigate Problem 1

Write a proof that shows the steps in the correct order.

$(a + b)(a^2 - ab + b^2)$
$= a(a^2 - ab + b^2) + b(a^2 - ab + b^2)$
$= a(a^2) - a(ab) + a(b^2) + b(a^2) - b(ab) + b(b^2)$
$= a(a^2) - a(ab) + a(b^2) + a^2b - b(ba) + b(b^2)$
$= a^3 - a^2b + ab^2 + a^2b - b^2a + b^3$
$= a^3 - a^2b + ab^2 + a^2b - ab^2 + b^3$
$= a^3 - a^2b + a^2b + ab^2 - ab^2 + b^3$
$= a^3 + b^3$

2. **Just the Math: Direct Proof** The proof in Question 1 is a **direct proof.** Not only do you have to put the steps in the correct order, but you also have to use the rules of algebra. Use a direct proof to show that the following statement is true. Show all your work and give your reason for each step.

$(a^2 - b^2)(a^2 + b^2) = a^4 - b^4$

$(a^2 - b^2)(a^2 + b^2)$	
$= (a^2 - b^2)(a^2) + (a^2 - b^2)(b^2)$	Distributive Prop.
$= a^2(a^2) - b^2(a^2) + a^2(b^2) - b^2(b^2)$	Distributive Prop.
$= a^4 - b^2a^2 + a^2b^2 - b^4$	Product of Powers
$= a^4 - a^2b^2 + a^2b^2 - b^4$	Comm. Prop. of Mult.
$= a^4 - b^4$	Additive Inverse

3. **Just the Math: Counterexample** Your friend claims that the statement $(m + n)^2 = m^2 + n^2$ is true. You claim that it is false. Give an example that uses real numbers to show that the statement is false. An example that shows that a statement is not true is called a **counterexample.**

Sample Answer: Let $m = 1$ and $n = 2$.

$(m + n)^2 = (1 + 2)^2 = 3^2 = 9$

$m^2 + n^2 = 1^2 + 2^2 = 1 + 4 = 5$

$9 \neq 5$

Do you think that the statement $a - (b - c) = (a - b) - c$ is true? If so, give a proof. If not, give a counterexample.

Sample Answer: The statement is not true. Let $a = 1$, $b = 2$, and $c = 3$.

$1 - (2 - 3) = 1 - (-1) = 1 + 1 = 2$

$(1 - 2) - 3 = (-1) - 3 = -4$

$2 \neq -4$

Have the students work together in small groups to complete Question 3. Call the class back together to have the students discuss and present their work for Question 3.

13

Explore Together

Investigate Problem 1

Students will construct indirect proofs for algebraic relationships.

Just the Math

Students will be formally introduced to the concept of indirect proof in Question 4.

Take Note

To say that an even integer has the form $2n$ where n is an integer is the same as saying that an even integer is an integer that is divisible by 2.

Ask for a student volunteer to read Question 4 aloud. Have a student restate the problem. Pose the Guiding Questions below to verify student understanding. Have the students work in small groups to complete Question 4.

Guiding Questions

- What is an indirect proof?
- How is an indirect proof different from a direct proof?
- What type of integer has 2 as a factor?
- What type of integer is not divisible by 2?
- Why can we say that every odd integer is 1 more than an even integer? Can you find a counterexample to contradict it?
- If you are unable to think of a counterexample, is it possible that the statement is false?
- If you can think of a counterexample, is it possible that the statement is true?

Common Student Errors

The concept of indirect proof is difficult for many students to understand. It may help to relate the process to our judicial system. A trial begins with the assumption that a person is not guilty. Then evidence is presented that either proves or disproves the person's guilt. Then a judge or jury determines whether the person is guilty of committing the crime.

Key Formative Assessments

- What is a proof?
- How can you prove an algebraic statement to be correct? Explain all 3 ways.

Investigate Problem 1

4. Just the Math: Indirect Proof In some cases, it is better to prove that a statement is true by using an **indirect proof.** To indirectly prove that a statement is true, you begin by assuming that the conclusion, or end result, of the statement is false. If you can show that this assumption leads to an impossible result, then you have indirectly proven that the statement must be true.

Consider the following statement: If p^2 is an even integer, then p must be an even integer.

To indirectly prove that this statement is true, we will assume that the conclusion, p is an even integer, is false. So, assume that p is an odd integer.

If p is an odd integer, then it has the form $2n + 1$ where n is an integer (an even integer has the form $2n$ where n is an integer). Complete the following steps to finish the proof.

$p^2 = (\underline{2n + 1})^2$ Substitute $2n + 1$ for p.

$p^2 = \underline{4n^2 + 4n + 1}$ Find the product.

Does the expression $4n^2 + 4n$ represent an even number or an odd number? Use a complete sentence to explain your reasoning.

Sample Answer: The expression $4n^2 + 4n$ is an even number because $4n^2 + 4n = 2(2n^2 + 2n)$ and the product of 2 and any number is even.

Is the expression $4n^2 + 4n + 1$ an even number or an odd number? Use a complete sentence to explain your reasoning.

Sample Answer: The expression $4n^2 + 4n + 1$ is an odd number because the sum of an even number and 1 is an odd number.

What can you conclude about p^2? Use a complete sentence in your answer.

The expression p^2 must be odd.

How does this answer complete the indirect proof? Use a complete sentence in your answer.

Sample Answer: The result that p^2 is odd is impossible because we know that p^2 is even.

Wrap Up

Close

- Review all key terms and their definitions. Include the terms *logical reasoning, proof, direct proof, indirect proof,* and *counterexample.*
- You may also want to review any other vocabulary terms that were discussed during the lesson, which may include *instruction sheet, logic, even, odd, integer, additive inverse, commutative property of addition, commutative property of multiplication, distributive property,* and *product of powers.*
- Remind the students to write the key terms and their definitions in the notes section of their notebooks. You may also want the students to include examples.
- Ask the students the following:
  - What is the purpose of a proof?
  - What is a direct proof?
  - When can you use a direct proof?
  - What is an indirect proof?
  - When can you use an indirect proof?
  - What is a counterexample?
  - When can you use a counterexample?
- Together with the class, develop a strategy for constructing a direct proof, an indirect proof, and a counterexample.

Follow Up

Assignment

Use the Assignment for Lesson 13.8 in the Student Assignments book. See the Teacher's Resources and Assessments book for answers.

Assessment

See the Assessments provided in the Teacher's Resources and Assessments book for Chapter 13.

Open-Ended Writing Task

Ask the students to write a short description of a direct proof, an indirect proof, and a counterexample. Have the students include the characteristics of each.

13

Reflections

Insert your reflections on the lesson as it played out in class today.

What went well?

What did not go as well as you would have liked?

How would you like to change the lesson in order to improve the things that did not go well and capitalize on the things that did go well?

Notes

Glossary

absolute deviation

The absolute deviation is the absolute value of the difference between a data value and the mean of the data set.

Example

The mean of the data set 2, 3, 9, and 13 is 6.75.

The data value 2 has an absolute deviation of 4.75.

The data value 3 has an absolute deviation of 3.75.

The data value 9 has an absolute deviation of 2.25.

The data value 13 has an absolute deviation of 6.25.

absolute value

The absolute value of a number is the distance between zero and the point that represents the number on a real number line. The absolute value of a number is always greater than or equal to zero.

Example

$|5| = 5$ because 5 is 5 units from 0 on the number line. $|-3| = 3$ because -3 is 3 units from 0 on the number line.

additive identity

The additive identity is a number such that when you add it to a second number, the sum is the second number.

Example

The additive identity is the number 0. $0 + 2 = 2$

additive inverse

The additive inverse of a number is the number such that the sum of the given number and its additive inverse is 0 (the additive identity).

Example

The additive inverse of 5 is -5 because $5 + (-5) = 0$.

algebraic equation

An algebraic equation is an expression used to generalize a problem situation, formed by writing an equals sign between two algebraic expressions.

Example

The statement $10 = 2x + 3$ is an equation.

algebraic expression

An algebraic expression is a mathematical phrase consisting of numbers, variables, and operations.

Example

If one pizza costs $7 and a pizza shop charges a $2.50 delivery charge, the cost of buying one or more pizzas can be represented by the algebraic expression $7p + 2.50$, where p is the number of pizzas purchased.

area

The area of a figure is the number of square units needed to cover the figure.

Example

The area of the rectangle is 18 square units.

The area of the triangle is 10 square units.

The area of circle is about 19.63 square units.

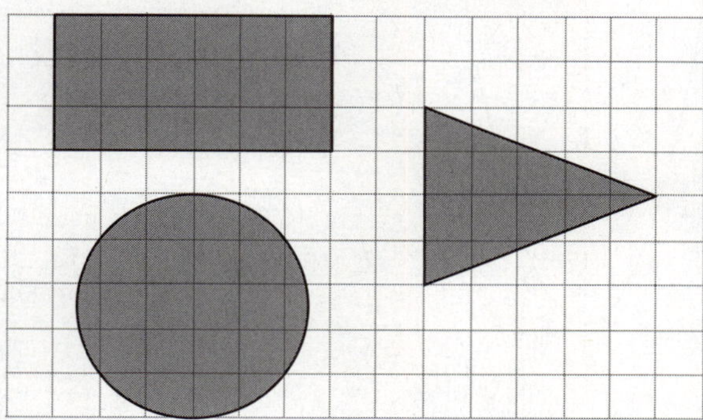

area model

An area model is a method for multiplying two polynomials. Each polynomial represents a side length. The area of the rectangle is the product of the polynomials.

Example

The product of the polynomials $3x$ and $4x + 1$ can be represented by using an area model. The product is equal to the area, which is $12x^2 + 3x$.

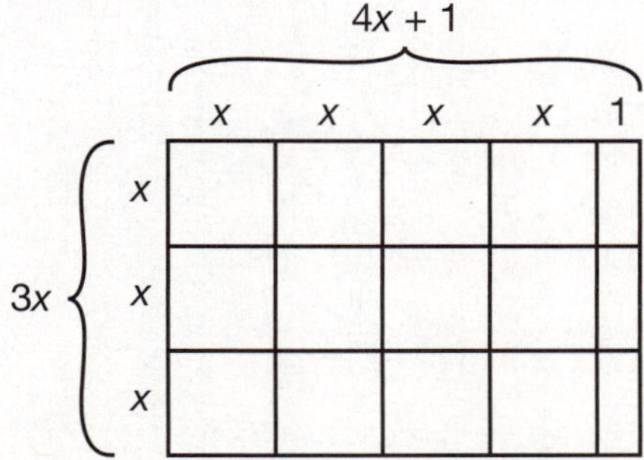

area of a square

The area of a square is equal to the side length of the square multiplied by itself: $A = (s)(s) = s^2$.

Example

In square *ABCD*, the length of each side is 12 centimeters. So, the area of the square is $A = (12)(12) = 144$ square centimeters.

axis of symmetry

An axis of symmetry is a line that passes through a figure and divides the figure into two symmetrical parts that are mirror images of each other.

Example

Line *K* is the axis of symmetry of the parabola.

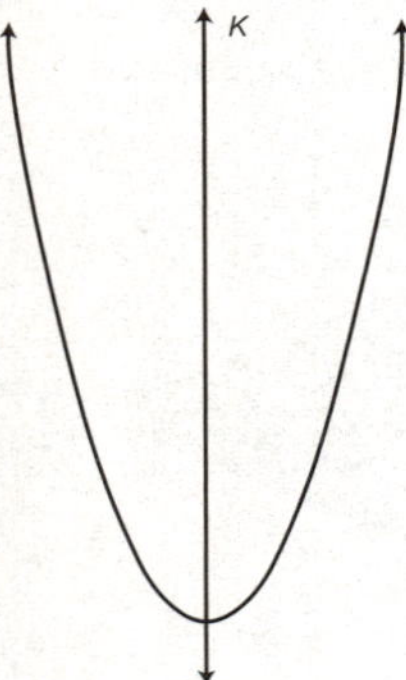

balance

A balance is the amount of money in an account.

Example

Carla opens a checking account with $200. She writes a check for $50. The balance of her checking account is $150.

bar graph

A bar graph is a graph that uses parallel bars to represent data. The heights of the bars represent quantities from the data set.

Example

The bar graph represents John's earnings over a four-week period.

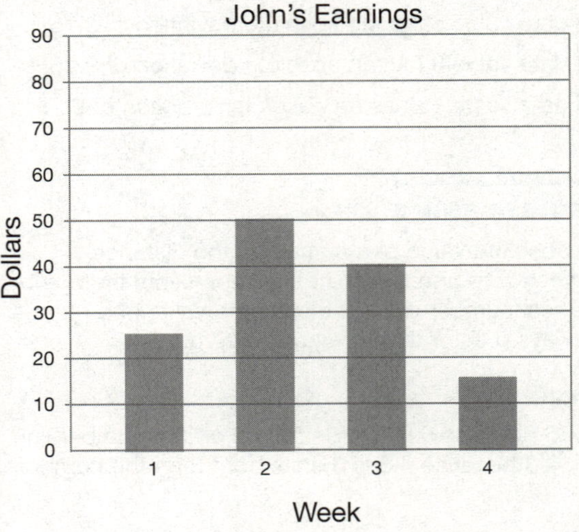

base of a power

The base of a power is the repeated factor in a power.

Example

In the expression 3^5, the number 3 is the base. $3^5 = (3)(3)(3)(3)(3) = 243$.

biased sample

A biased sample is a sample that does not accurately represent all of a population.

Example

A survey is conducted asking students their favorite class. Only students in the math club are surveyed. The sample of students is a biased sample.

bimodal

A data set is bimodal if two values occur in the data set the same number of times.

Example

The data set 1, 1, 3, 3, 4, 5, 7 is bimodal.

binomial

A binomial is a polynomial with exactly two terms.

Example

The polynomial $3x + 5$ is a binomial.

bounds

The lower and upper bounds determine the portion of a graph that you will see. The data that you are graphing should be greater than the lower bounds and less than the upper bounds.

Example

In this graph, the lower bound for the x-axis is 0 and the upper bound for the x-axis is 300. The lower bound for the y-axis is 0 and the upper bound for the y-axis is 225.

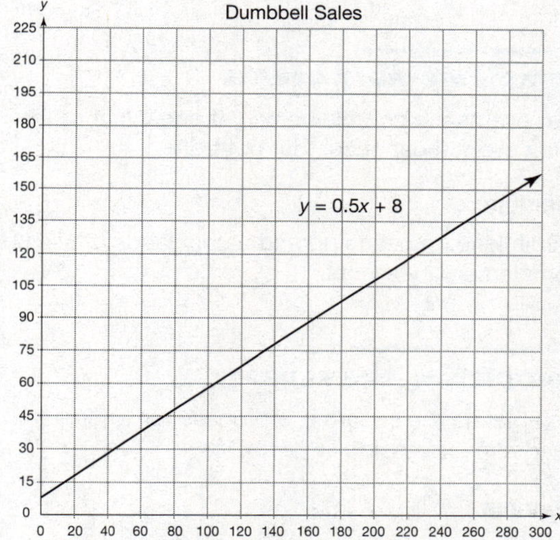

box-and-whisker plot

A box-and-whisker plot is a visual display of data that organizes the data values into four groups using the upper and lower bounds, the median, and the upper and lower quartiles.

Example

The box-and-whisker plots compare the test scores from two algebra classes.

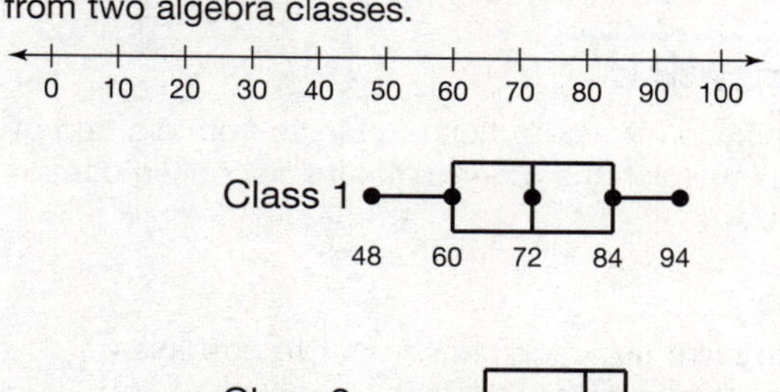

Cartesian coordinate system

A Cartesian coordinate system is a method of representing the location of a point using an ordered pair of real numbers of the form (x, y).

Example

Point J is represented by the ordered pair (5, 4).

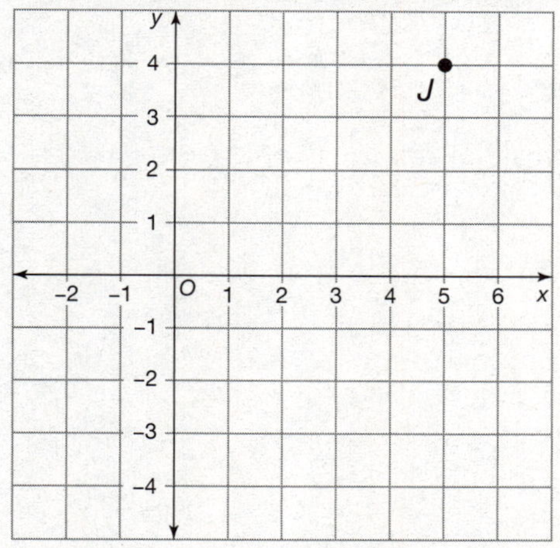

closure property

The closure property states that a set of numbers is closed under an operation if the result of the operation on two numbers in the set is a number in the set.

Example

The set of whole number numbers is closed under addition. The sum of any two whole numbers is always another whole number.

Glossary

coefficient

In a term containing a number multiplied by one or more variables, the number is the coefficient of the term.

Example

The term $3x^5$ has a coefficient of 3.

combination

A combination is a selection of objects from a group of objects for which the order of the items chosen does not matter.

Example

Choosing two pizza toppings from five possible toppings is a combination.

combination of distinct objects taken *r* at a time

A combination of n distinct objects taken r at a time is the number of groupings of n objects taken r at a time and is given by ${}_nC_r = \frac{n!}{(n-r)!r!}$.

Example

You choose two pizza toppings from five possible toppings.

$$
\begin{aligned}
{}_nC_r &= \frac{n!}{(n-r)!r!} \\
&= \frac{5!}{(5-2)!2!} \\
&= \frac{5!}{3!2!} \\
&= \frac{120}{(6)(2)} \\
&= \frac{120}{12} \\
&= 10
\end{aligned}
$$

combine like terms

To combine like terms is to add or subtract like terms.

Example

The expression $5x + 3x$ can be simplified to $8x$ by combining like terms.

commission

A commission is a fee or a percent of sales that are paid to a sales representative or an agent for services rendered.

Example

A sales person is to receive a 5% commission on her sales. Suppose that she sells $500 worth of merchandise. Her commission will be $25: 5% of 500 = (0.05)(500) = 25.

common factor

A common factor is a whole number that is a factor of two or more integers or expressions.

Example

The whole number 2 is a common factor of 6 and 8.

complementary events

Two events are complementary if one event or the other event must occur, but not both.

Example

A coin landing heads up and a coin landing tails up are complementary events.

completing the square

Completing the square is a process for writing a quadratic expression in vertex form.

Example

To write the equation $y = x^2 + 10x + 2$ in vertex form, perform the following steps.

$y = x^2 + 10x + 2$

$y = x^2 + 10x + 25 + 2 - 25$

$y = (x + 5)^2 - 23$

composite number

A composite number is a whole number greater than 1 that is divisible by 1, itself, and at least one other positive number.

Example

The first five composite numbers are 4, 6, 8, 9, and 10.

compound event

A compound event is an event that is made up of two or more simple events.

Example

Choosing two socks from a drawer is a compound event.

compound inequality

A compound inequality is formed by two inequalities that are connected by the word "and" or the word "or."

Example

The statement $x > 5$ or $x > -5$ is a compound inequality.

compound interest

Compound interest is interest on both the principal and previously earned interest.

Example

Sonya opens a savings account with $100. She earns $4 in interest the first year. The compound interest y is found by using the equation $y = 100(1 + 0.04)^t$, where t is the time in years.

congruent figures

Two figures are congruent if they have the same size and the same shape.

Example

Triangle *ABC* and triangle *DEF* are congruent triangles.

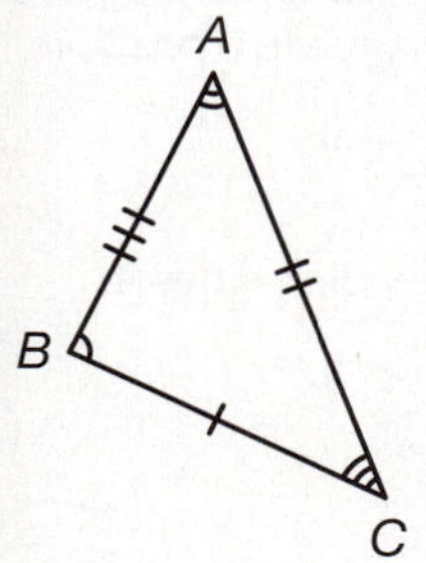

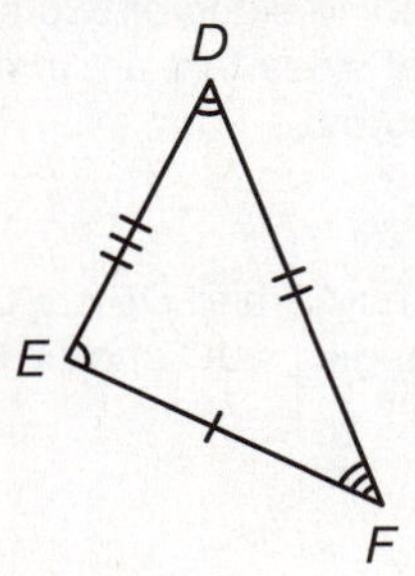

conjecture

A conjecture is a possible explanation that can be tested by further investigation.

Example

Stating that more people in a school are right-handed is a conjecture.

constant ratio

A constant ratio is a ratio that has a constant value.

Example

A recipe for fruit punch specifies 1 cup of cranberry juice for every 2 cups of apple juice. The ratio of the amount of cranberry juice to the amount of apple juice is a constant ratio.

converse

The converse of an if-then statement is the statement that results from interchanging the hypothesis (the "if" part) and the conclusion (the "then" part) of the original statement.

Example

The converse of the statement "If $a = 0$ or $b = 0$, then $ab = 0$" is "If $ab = 0$, then $a = 0$ or $b = 0$."

converse of the Pythagorean theorem

The converse of the Pythagorean Theorem states that if a, b, and c are sides of a triangle and $a^2 + b^2 = c^2$, then the triangle is a right triangle.

Example

In triangle *ABC*, the lengths of the sides are 5 inches, 10 inches, and 12 inches. To determine whether the triangle is a right triangle, substitute the side lengths into the Pythagorean theorem.

$$a^2 + b^2 = c^2$$
$$5^2 + 10^2 \stackrel{?}{=} 12^2$$
$$25 + 100 \stackrel{?}{=} 144$$
$$125 \neq 144$$

The expression is false, so the triangle is not a right triangle.

coordinate plane

A coordinate plane is a plane formed by the intersection of a vertical real number line and a horizontal real number line. The vertical number line is the y-axis and the horizontal number line is the x-axis. The number lines intersect at right angles and the point of intersection is the origin.

Example

The origin is labeled on the coordinate plane below.

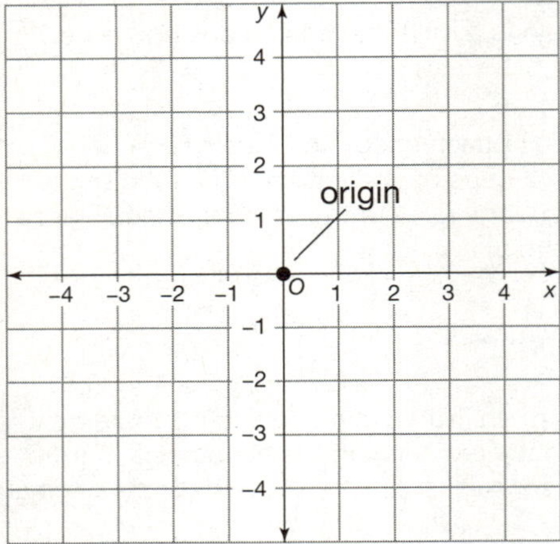

correlation

A correlation exists between the x- and y-values of a data set when it is appropriate to use a line of best fit to approximate a collection of points.

Example

A correlation does not exist for this data set.

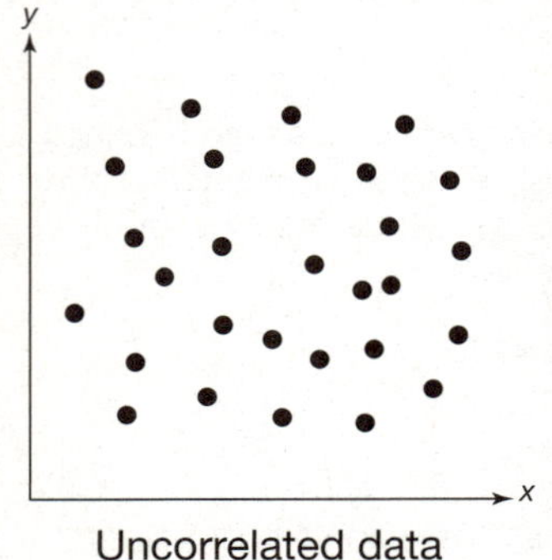
Uncorrelated data

correlation coefficient

The correlation coefficient, r, indicates how close the data are to forming a straight line. If the value of r is between 0 and 1, the linear regression equation has a positive slope. If the value is between 0 and -1, the linear regression equation has a negative slope. The closer r is to 1 or -1, the closer the data are to being in a straight line.

Example

The correlation coefficient for the data is –0.9935. The value is negative so the equation has a negative slope. The value is close to –1 so the data is very close to forming a straight line.

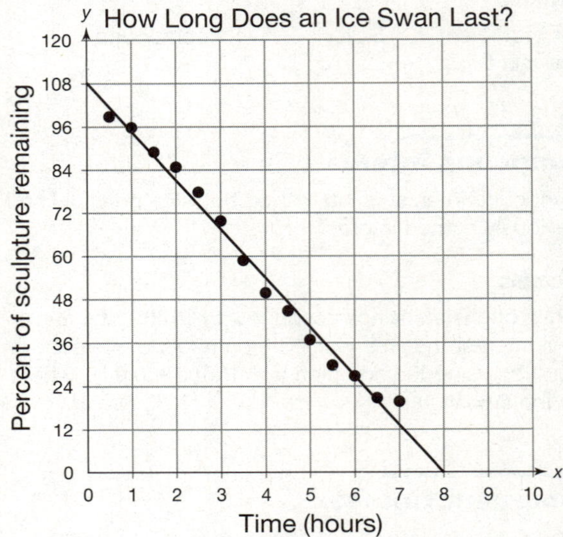

corresponding sides

Corresponding sides of two similar or congruent figures are pairs of sides that are in the same relative position in both figures.

Example

Side lengths AB and DE are corresponding sides in similar triangles ABC and DEC.

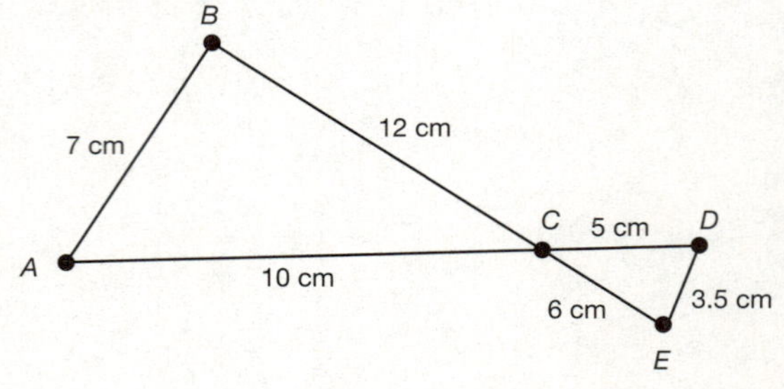

counterexample

A counterexample is an example that shows that a statement is not true.

Example

Your friend claims that you add fractions by adding the numerators and adding the denominators.

A counterexample is $\frac{1}{2} + \frac{1}{2}$. The sum of these two fractions is 1. Your friend's method results in $\frac{2}{4}$, or $\frac{1}{2}$. Your friend's method is incorrect.

cube root

The cube root of a given number is a number that, when cubed, equals the given number.

Example

The cube root of 8 is 2.

decay factor

The decay factor is the expression $(1 - r)$ in the exponential decay equation.

Example

Your uncle bought a car for $20,000. The value of the car decreases 30% each year. A model for the value of the car after t years is $y = 20{,}000(1 - 0.3)^t$.

The decay factor is 0.70.

decay rate

The decay rate is the variable r in the exponential decay equation, written in decimal form.

Example

Your uncle bought a car for $20,000. The value of the car decreases 30% each year. A model for the value of the car after t years is $y = 20{,}000(1 - 0.3)^t$.

The decay rate is 0.30.

degree of a polynomial

The degree of a polynomial in one variable is the exponent of that variable with the largest numerical value.

Example

The polynomial $2x^3 + 5x^2 - 6x + 1$ has a degree of 3.

demand

The demand is the amount of a good that is used.

Example

The demand for oil in the world can be modeled by the equation $y = 360.1x + 15.179$, where x is the time, in years, since 1965 and y is the demand for oil, in millions of barrels.

dependent events

Dependent events are events in which the outcome of one event affects the outcome of the other event.

Example

Choosing a sock from a drawer, keeping the sock, and choosing another sock from the drawer are dependent events.

dependent variable

A dependent variable, or output value of a function, is a variable whose value is determined by an independent variable, or input value of a function.

Example

In the relationship between driving time and distance traveled, distance is represented by the dependent variable d because the value of d depends on the value of the driving time t.

deviation

The deviation of a data value is the difference between a data value and the mean of the data set.

Example

The mean of the data set 2, 3, 9, and 13 is 6.75.

The data value 2 has a deviation of –4.75.

The data value 3 has a deviation of –3.75.

The data value 9 has a deviation of 2.25.

The data value 13 has a deviation of 6.25.

dilation

A dilation is a transformation of a figure in which the figure stretches or shrinks with respect to a fixed point. The scale factor of a dilation is the ratio of a side length of the dilated figure to the original figure. An enlargement or reduction of a photo is an example of a dilation.

Example

The original light hexagon is dilated to produce the dark hexagon by a scale factor of 2.

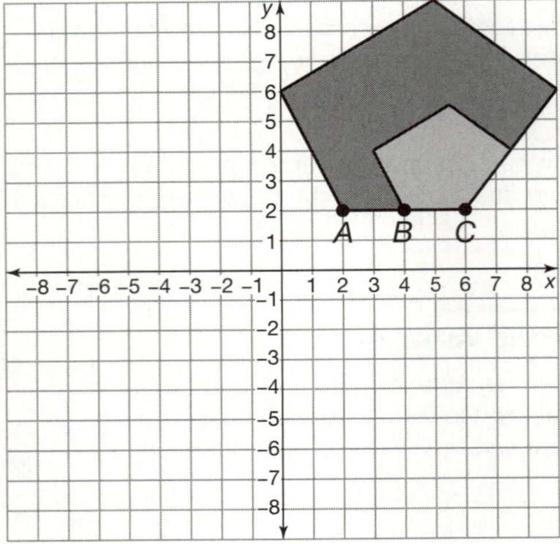

direct proof

To prove that a statement is true by using a direct proof, put all the steps in order using mathematical rules.

Example

To prove the statement $(a - b)(a^2 + ab + b^2) = a^3 - b^3$, use the following direct proof.

$(a - b)(a^2 + ab + b^2)$

$= a(a^2 + ab + b^2) - b(a^2 + ab + b^2)$

Distributive property

$= a(a^2) + a(ab) + a(b^2) - b(a^2) - b(ab) - b(b^2)$

Distributive Property of Multiplication Over Addition

$= a^3 + a^2b + ab^2 - a^2b - ab^2 - b^3$

Product of Powers

$= a^3 - b^3$

Additive Inverse

direct variation

Direct variation is the relationship between two quantities x and y having a constant ratio; one quantity varies directly with the other.

Example

The relationship between the variables x and y in the equation $y = 2x$ is a direct relationship.

discriminant

In a quadratic equation $ax^2 + bx + c = 0$, the discriminant is equal to the expression $b^2 - 4ac$.

Example

In the quadratic equation $x^2 + 2x - 24 = 0$, the discriminant is equal to $2^2 - 4(1)(-24)$, or 100.

distance formula

The distance formula can be used to find the distance between two points.

The distance between points (x_1, y_1) and (x_2, y_2) is

$d = \sqrt{(x_2 - x_1)^2 + (y_2 - y_1)^2}$.

Example

To find the distance between the points (–1, 4) and (2, –5), substitute the coordinates into the Distance Formula.

$d = \sqrt{(x_2 - x_1)^2 + (y_2 - y_1)^2}$

$d = \sqrt{(2 + 1)^2 + (-5 - 4)^2}$

$d = \sqrt{3^2 + (-9)^2}$

$d = \sqrt{9 + 81}$

$d = \sqrt{90}$

$d \approx 9.49$

So, the distance between the points (–1, 4) and (2, –5) is approximately 9.49 units.

distribution

A distribution is the way in which the data are distributed, such as being spread out or clustered together.

Example

symmetric

Skewed to the Right

distributive property

The distributive property states that for any numbers a, b, and c it is true that $a(b + c) = ab + ac$.

Example

The distributive property can be used to write the expression $2(x + 4)$ as $2x + 8$.

domain of a function

The domain of a function is the set of all input values for the function.

Example

The domain of the function $y = 2x$ is the set of all real numbers.

equivalent equations

Two equations are equivalent if they have the same solution or solutions.

Example

The equations $x + y = 1$ and $2x + 2y = 2$ are equivalent.

estimation

Estimation is the process of finding the approximate value of an expression, often done through rounding.

Example

To estimate 697 + 309, round 697 to 700 and round 309 to 300. Then you can estimate that 697 + 309 is approximately 700 + 300, or 1000.

evaluate an expression

To evaluate an expression, find the value of the expression by replacing each variable with a given value and then simplifying the result.

Example

To evaluate $3x + 6$ when $x = 5$, replace the x by 5, and then simplify.

$$(3)(5) + 6 = 15 + 6$$
$$= 21$$

event

A simple event is a collection of outcomes of an experiment. An outcome is one possible result of an experiment.

Example

Flipping a coin or rolling a number cube is a simple event.

excluded value

An excluded value is a number that causes the denominator of a rational expression to equal zero.

Example

The excluded value for the rational expression $\frac{2x + 1}{x - 3}$ is $x = 3$.

experiment

An experiment is a test to demonstrate a known truth.

Example

You want to find the probability of a coin landing heads up. One experiment is flipping a coin 100 times and recording the results.

experimental probability

An experimental probability is a probability that is based on repeated trials of an experiment.

Example

You want to find the probability of a coin landing heads up. You flip a coin 100 times and record the results. The coin lands heads up 55 times and lands tails up 45 times. The experimental probability of the coin landing heads up is $\frac{55}{100}$.

exponent of a power

The exponent of a power is the number of times that the factor is repeated.

Example

In the expression 10^3, the number 3 is the exponent. This indicates that the base 10 is used as a factor 3 times: $10^3 = (10)(10)(10) = 1000$.

exponential decay model

The general form of an exponential decay model is the equation $y = C(1 - 4)^t$ where $0 < r < 1$, C is the original amount before any decay occurs, r is the decay rate in decimal form, $(1 - r)$ is the decay factor, t is the amount of time, and y is the new amount.

Example

Your uncle bought a car for $20,000. The value of the car decreases 30% each year. A model for the value of the car after t years is $y = 20{,}000(1 - 0.3)^t$.

exponential function

An exponential function is a function of the form $f(x) = a^x$ where $a > 0$ and $a \neq 1$.

Example

The function $y = 2^x$ is an exponential function.

exponential growth model

The general form of an exponential growth model is the equation $y = C(1 + r)^t$ where C is the original amount before any growth occurs, r is the growth rate in decimal form, $(1 + r)$ is the growth factor, t is the amount of time, and y is the new amount.

Example

You invest $1000 into a savings account that earns 2.5% interest, compounded annually. A model for the account balance after t years is $y = 1000(1 + 0.025)^t$.

expression

An expression is any symbolic mathematical phrase that may include constants, variables, and operators.

Examples

Three expressions are shown below.

$5y$ $\quad 4x - 2 \quad$ $6^3 + 8$

extracting square roots

Extracting square roots is the process of recognizing that $x = \sqrt{b}$ and $x = -\sqrt{b}$ satisfy the equation $x^2 = b$.

Example

The equation $x^2 = 10$ has two solutions; $x = \sqrt{10}$ and $x = -\sqrt{10}$.

extreme

An extreme is the greatest and least value of a data set. The greatest value is called the upper extreme and the least value is called the lower extreme.

Example

The upper extreme of the numbers 1, 5, 7, 10, 14 is 14. The lower extreme is 1.

extremes of a proportion

The extremes of a proportion are the two outside quantities of a proportion.

Example

In the proportion 3 dimes: 5 quarters:: 15 dimes: 25 quarters, the extremes are the outside quantities 3 dimes and 25 quarters.

factor an expression

To factor an expression is to use the distributive property in reverse.

Example

The expression $2x + 4$ can be factored as $2(x + 2)$.

factor of a number

A factor of a number is a number that evenly divides the given number with no remainder.

Example

The number 24 has eight factors: 1, 2, 3, 4, 6, 8, 12, and 24.

factor of a polynomial

A factor of a polynomial is a polynomial that evenly divides the given polynomial with no remainder.

Example

The polynomial $x^2 + 5x + 6$ has two factors: $x + 2$ and $x + 3$.

factorial

The factorial of a number n is the product of all of the positive integers less than or equal to n. The factorial of n is expressed using the notation $n!$

Example

$5! = 5 \cdot 4 \cdot 3 \cdot 2 \cdot 1 = 120$

factoring a polynomial

Factoring is the process of expressing a polynomial as the product of monomials and binomials.

Example

The polynomial $x^2 + 5x + 6$ can be written in factored form as $(x + 2)(x + 3)$.

factoring out a common factor

Factoring out a common factor is a method of factoring a polynomial where the polynomial is written as the product of the common factor of all terms on the remaining polynomial.

Example

$2x^3 - 6x^2 + 8x = 2x(x^2 - 3x + 4)$

fair game

A fair game is a game in which each player has an equally likely chance of winning.

Example

You are playing a two-player game with your friend using two six-sided number cubes. In each round, one player is the roller and one player is the recorder. If the sum of the numbers on a roller's turn is even, the roller gets one point. If the sum of the numbers on a roller's turn is odd, the recorder gets one point. The winner of the round is the player who has the most points after 20 rolls.

The probability that the roll results in an even number is $\frac{18}{36}$, or $\frac{1}{2}$.

The probability that the roll results in an odd number is $\frac{18}{36}$, or $\frac{1}{2}$.

After 20 rolls, you can expect 10 even rolls and 10 odd rolls which results in 10 points for the roller and 10 points for the recorder. This game is fair.

favorable outcome

A favorable outcome is a specific outcome chosen for a particular event.

Example

You are finding the probability of rolling an even number with a six-sided number cube. There are three favorable outcomes: rolling a 2, 4, or 6.

first quartile

The first quartile, Q_1, is the median of the lower half of the data.

Example

In the data set 13, 17, 23, 24, 25, 29, 31, 45, 46, 53, 60, the median, 29, divides the data into two halves. The first quartile, 23, is the median of the lower half of the data.

13 17 23 24 25 29 31 45 46 53 60

lower quartile (23) | median (29) | upper quartile (46)

FOIL pattern

The FOIL pattern is a method for finding the product of two binomials by multiplying the first terms, the outer terms, the inner terms, and the last terms.

Example

$(x - 1)(2x + 5) = 2x^2 + 5x - 2x - 5 = 2x^2 + 3x - 5$

function

A function is a relation in which for every input there is exactly one output.

Example

The equation $y = 2x$ is a function. Every value of x has exactly one corresponding y-value.

function notation

Function notation is a notation used to write functions such that the dependent variable is replaced with the name of the function.

Example

In the function $f(x) = 0.75x$, f is the name of the function, x represents the domain, and $f(x)$ represents the range.

fundamental counting principle

The fundamental counting principle states that if you have m choices for one event and n choices for another event, then the number of choices for both events is $m \cdot n$.

Example

You flip a coin and then roll a six-sided number cube. There are 2 choices for flipping a coin and 6 choices for rolling the number cube. The number of choices for both events is $2 \cdot 6$, or 12.

geometric probability

A geometric probability is the ratio of a desired length, area, or volume to the total length, area, or volume.

Example

A circular dartboard has a radius of 6.75 inches. The radius of the circular bull's-eye is 0.5 inch. To find the probability of hitting the bull's-eye, find the ratio of the area of the bull's-eye to the total area of the dart board.

$$\text{Probability} = \frac{\text{Area of bull's–eye}}{\text{Area of dart board}} \approx \frac{3.14(0.5)^2}{3.14(6.75)^2}$$

$$= \frac{(0.5)^2}{(6.75)^2} = \frac{0.25}{45.5625} = 0.0055$$

So, the probability of hitting the bull's-eye is 0.0055, or 0.55%.

graph

A graph is a visual representation of the relationship between two sets of values as points in a coordinate plane.

Example

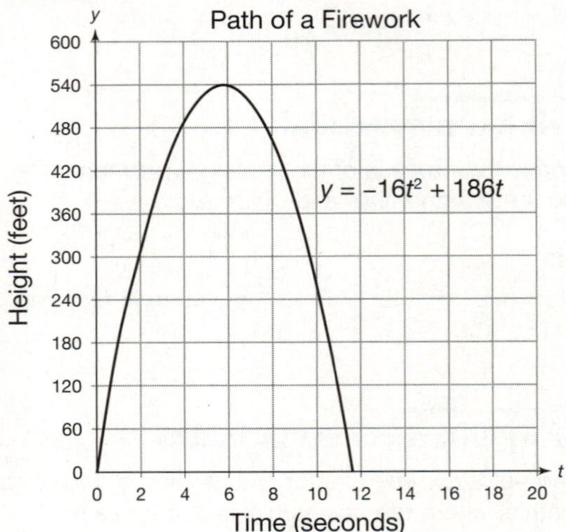

graph of an inequality

The graph of an inequality in one variable is the set of all points that make the inequality true.

Example

The inequality $x > 5$ is represented by the graph shown.

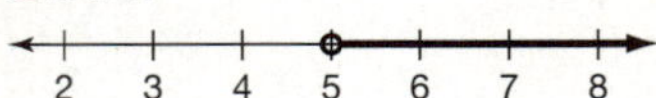

greatest common factor

The greatest common factor of two whole numbers is the largest whole number that is a factor of both numbers. The greatest common factor is abbreviated as GCF.

Example

The whole number 3 is the greatest common factor of 6 and 9.

gross pay

Gross pay is the total amount of money an employee earns before any taxes or deductions are subtracted.

Example

Nadia's gross pay was $2400 per month.

growth factor

The growth factor is the expression $(1 + r)$ in the exponential growth equation.

Example

You invest $1000 into a savings account that earns 2.5% interest, compounded annually. A model for the account balance after t years is $y = 1000(1 + 0.025)^t$.

The growth factor is 1.025.

growth rate

The growth rate is the variable r in the exponential growth equation, written in decimal form.

Example

You invest $1000 into a savings account that earns 2.5% interest, compounded annually. A model for the account balance after t years is $y = 1000(1 + 0.025)^t$.

The growth rate is 0.025.

half-plane

A half-plane is the graph of a linear inequality or half a coordinate plane.

Example

The shaded portion of the graph is a half-plane.

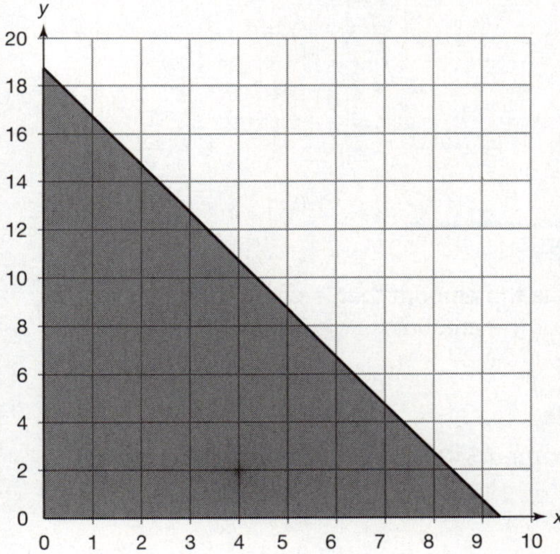

hypotenuse of a right triangle

In a right triangle, the hypotenuse is the side of the triangle that is opposite the right angle.

Example

In triangle ABC, angle A is the right angle, so side BC is the hypotenuse. In triangle DEF, angle F is the right angle, so side DE is the hypotenuse.

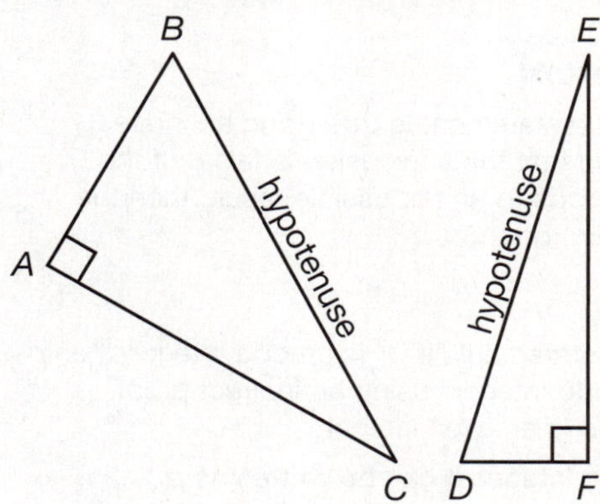

income

Income is the amount of money that a company earns.

Example

A company makes $10,000 before expenses are considered. The company's expenses are $8000. The company's income is $10,000.

independent events

Independent events are two events in which the outcome of the first event does not affect the probability of the second event.

Example

Choosing a sock from a drawer, replacing the sock, and choosing another sock from the drawer are independent events.

independent variable

An independent variable, or input value, is a variable whose value is not determined by another variable.

Example

In the relationship between driving time and distance traveled, time is represented by the independent variable t because the value of t does not depend on any variable.

Glossary

index

An index is a number used to indicate what root is to be determined. It is placed above and to the left of the radical sign.

Example

In the radical expression $\sqrt[4]{16}$, 4 is the index.

indirect proof

To prove that a statement is true using an indirect proof, assume that the conclusion is false. If the assumption leads to an impossible result, then the statement must be true.

Example

To prove the statement "If p^2 is an odd integer, then p must be an odd integer" using an indirect proof, assume that p is an even integer.

If p is an even integer, it can be written as $p = 2n$.

$p^2 = (2n)^2$
$p^2 = 4n^2$

The expression $4n^2$ is even because it can be written as $2(2n^2)$. If $4n^2$ is even then p^2 must be even. This is impossible because p^2 is given as odd.

So, the statement "If p^2 is an odd integer, then p must be an odd integer" must be true.

inequality

An inequality is a statement that one expression is less than, less than or equal to, greater than or equal to, or greater than another expression. The inequality symbols are $<$, $>$, $\leq$, and $\geq$.

Example

The statement $y < 2x + 5$ is an inequality.

inequality symbol

An inequality symbol is one of the symbols for less than ($<$), greater than ($>$), less than or equal to ($\leq$), or greater than or equal to ($\geq$).

input value

An input value is the first coordinate of an ordered pair in a relation.

Example

The ordered pair (5, 8) has an input value of 5.

integer

An integer is any of the numbers –4, –3, –2, 1, 0, 1, 2, 3, 4, 5, ... Integers include all of the whole numbers and their additive inverses.

Examples

The numbers –12, 0, and 30 are integers.

intercept

An intercept is the point where a graph intersects the x- or y-axis.

Example

In the graph below, the points x-intercept is 1 and the y-intercept is –3.

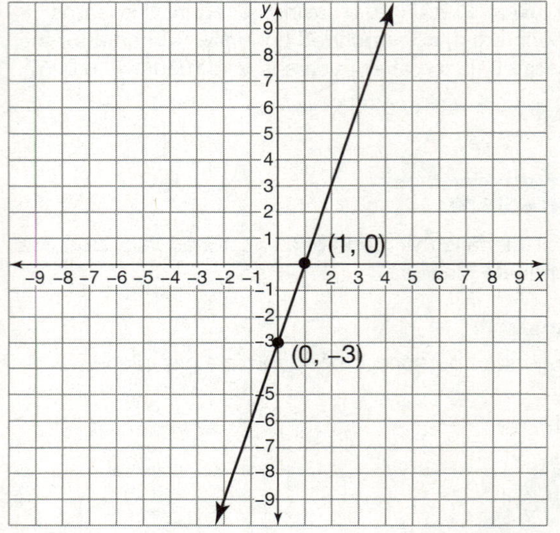

interest

Interest is the amount that is charged for borrowing money or the amount that is earned from saving money.

Example

Derek earned $13.58 in interest from his savings account.

Glossary

interquartile range

The interquartile range, IQR, is the difference between the upper and lower quartiles.

Example

In the data set 13, 17, 23, 24, 25, 29, 31, 45, 46, 53, 60, the median, 29, divides the data into two halves. The first quartile, 23, is the median of the lower half of the data. The third quartile, 46, is the median of the upper half of the data. The interquartile range is 46 – 23, or 23.

inverse operations

Inverse operations are operations that undo each other.

Example

The operations of addition and subtraction are inverse operations.

irrational number

An irrational number is a number that cannot be written as $\frac{a}{b}$, where a and b are integers.

Example

The number π is an irrational number.

label

A label is a written description that identifies an object.

Example

In the graph, the label on the x-axis is "Time (hours)" and the label on the y-axis is "Earnings (dollars)."

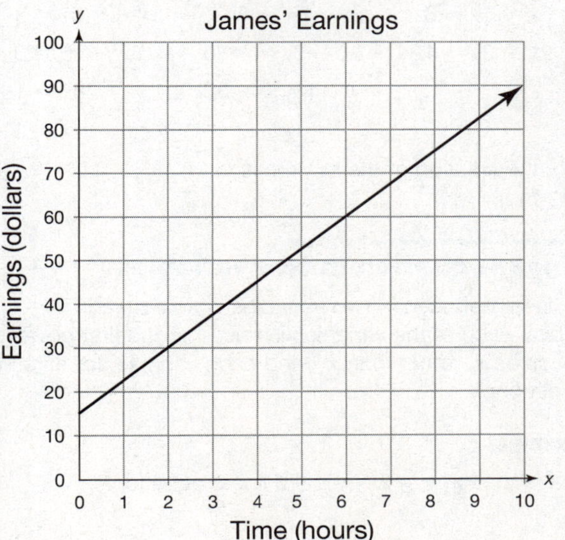

least squares method

The least squares method is a method used to find the equation of the line of best fit.

Example

Most graphing calculators use the least squares method to calculate the line of best fit.

legs of a right triangle

In a right triangle, the legs are the two sides of the triangle that form the right angle.

Example

In triangle ABC, angle A is the right angle, so sides AB and AC are the legs of the triangle.

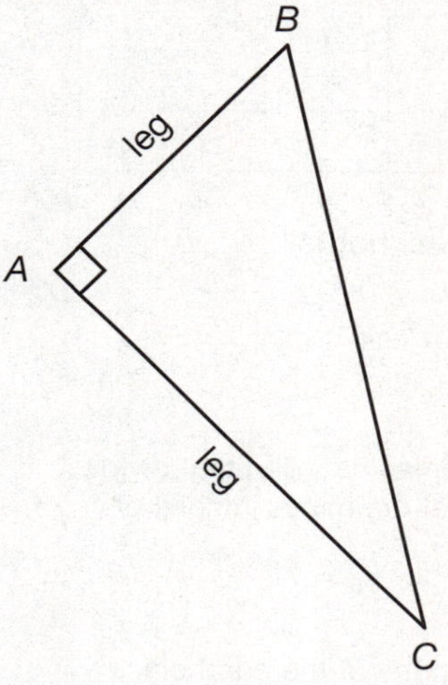

Glossary

Glossary

line of best fit

A line of best fit is a line that is used to approximate the data as close to as many points as possible.

Example

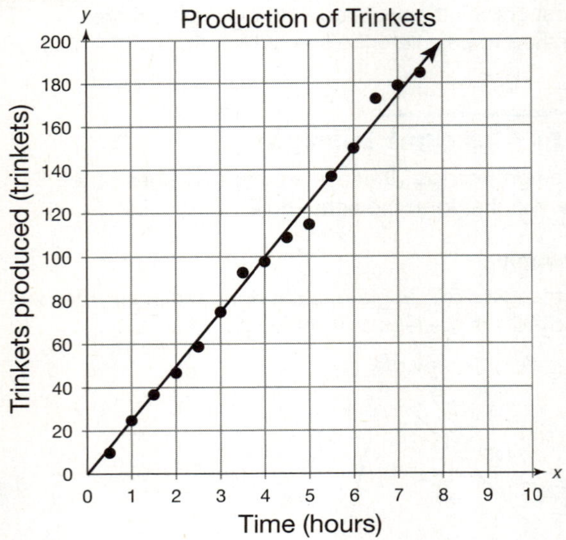

line of symmetry

A line of symmetry is an imaginary line that divides a graph into two parts that are mirror images of each other.

Example

Line *K* is the line of symmetry of the parabola.

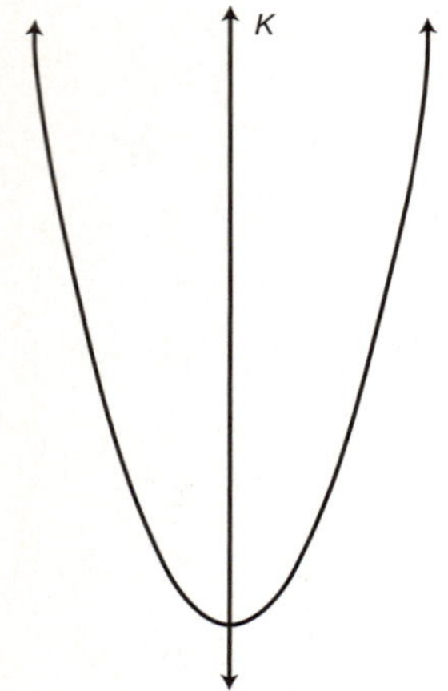

line plot

A line plot is a graph that has an X above each number on a number line that appears in the data set.

Example

linear combination

A linear combination is an equation that is the result of adding two equations to each other.

Example

The linear combination of the equations $2x + 3y = 5$ and $x - 3y = -1$ is $3x = 4$.

linear combinations method

The linear combinations method is a process to solve a system of equations by adding two equations to each other resulting in an equation with one variable.

Example

Solve the following system of linear equations using linear combinations.

$6x - 5y = 3$
$2x + 2y = 12$

To solve the system by using linear combinations, first multiply the second equation by –3. Then, add the equations and solve for the remaining variable. Finally, substitute $y = 3$ into the first equation and solve for x.

$6x - 5y = 3$	$6x - 5y = 3$	$6x - 5(3) = 3$
$2x + 2y = 12$	$-6x - 6y = -36$	$6x - 15 = 3$
	$-11y = -33$	$6x = 18$
	$y = 3$	$x = 3$

So, the solution of the system is (3, 3).

linear equation in two variables

A linear equation in two variables is an equation in which each of the variables is raised to the first power (such as x, rather than x^2) and each variable appears at most once.

Example

The equation $y = 2x + 1$ is a linear equation.

linear factor

A linear factor is a factor that is a linear expression.

Example

The polynomial $x^2 + 5x + 6$ has two linear factors: $x + 2$ and $x + 3$.

linear function

A linear function is a function whose graph is a non-vertical straight line.

Example

$f(x) = 3x + 2$ is a linear function.

linear inequality in two variables

A linear inequality in two variables is any inequality that can be written in one of these forms: $ax + by > c$, $ax + by < c$, $ax + by \geq c$, or $ax + by \leq c$.

Example

The inequality $3x + 4y > 7$ is a linear inequality.

linear regression equation

A linear regression equation is the equation of the line of best fit.

Example

The linear regression equation for the data is $y = 25x$.

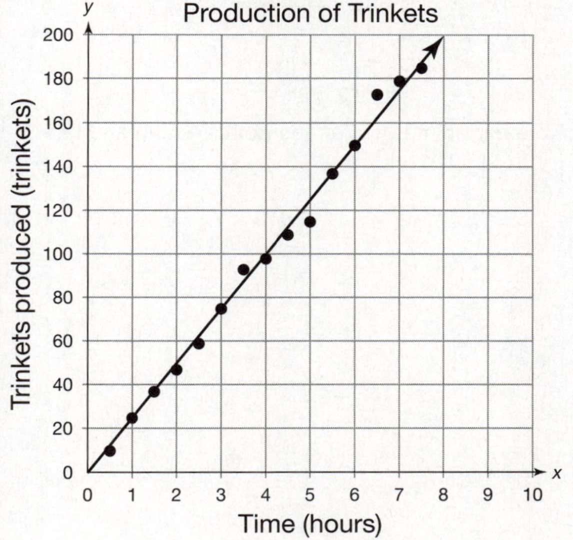

linear system

A linear system is two or more linear equations in the same variables.

Example

The equations $y = 3x + 7$ and $y = -4x$ are a system of linear equations.

literal equation

A literal equation is an equation that contains two or more variables to represent known quantities.

Example

The equations $I = Prt$ and $A = lw$ are literal equations.

maximum point

The maximum point of the graph of a function is the ordered pair on the graph with the greatest y-coordinate.

Example

The ordered pair (4, 2) is the maximum point of the graph of the function $y = -\frac{1}{2}x^2 + 4x - 6$.

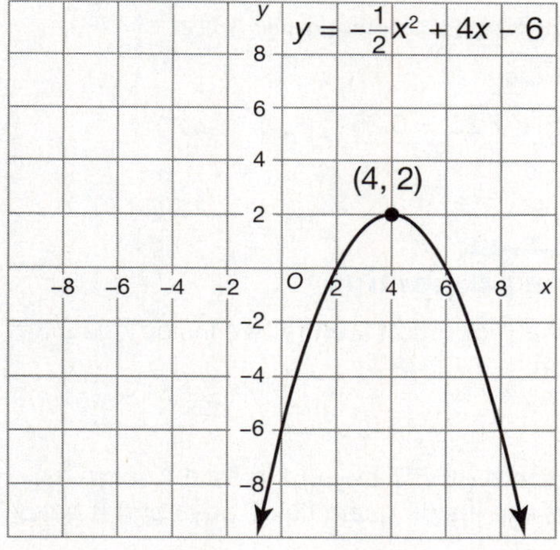

mean

The mean of a data set is the sum of all of the values of the data set divided by the number of values in the data set. The mean is also called the average.

Example

The mean of the numbers 7, 9, 13, 4, and 7 is $\frac{7 + 9 + 13 + 4 + 7}{5}$, or 8.

mean absolute deviation

The mean absolute deviation is the average of the absolute deviations for all data values.

Example

The mean of the data set 2, 3, 9, and 13 is 6.75.

The data value 2 has an absolute deviation of 4.75.

The data value 3 has an absolute deviation of 3.75.

The data value 9 has an absolute deviation of 2.25.

The data value 13 has an absolute deviation of 6.25.

The mean absolute deviation is

$\frac{4.75 + 3.75 + 2.25 + 6.25}{4}$, or 4.25.

mean deviation

The mean deviation is the average of the deviations for all data values.

Example

The mean of the data set 2, 3, 9, and 13 is 6.75.

The data value 2 has a deviation of –4.75.

The data value 3 has a deviation of –3.75.

The data value 9 has a deviation of 2.25.

The data value 13 has a deviation of 6.25.

The mean deviation is

$\frac{-4.75 - 3.75 + 2.25 + 6.25}{4}$, or 0.

means of a proportion

The means of a proportion are the two inside quantities of a proportion.

Example

In the proportion 4 girls: 7 boys:: 8 girls: 14 boys, the means are the inside quantities 7 boys and 8 girls.

measure of central tendency

A measure of central tendency is a single value that represents a typical value in a data set.

Example

The mean, median, and mode are the most common measures of central tendency.

median

The median of a data set that is arranged in numerical order is either the middle value (when the number of data values is odd), or the average of the two middle values (when the number of data values is even).

Example

The median of the numbers 1, 5, 7, 10, 14 is 7.

midpoint

The midpoint of a segment is the point that divides the segment into two congruent segments.

Example

Because point *B* is the midpoint of segment *AC*, segment *AB* is congruent to segment *BC*.

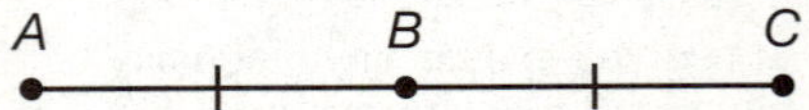

midpoint formula

The midpoint formula can be used to find the midpoint between two points. The midpoint between (x_1, y_1) and (x_2, y_2) is $\left(\frac{x_1 + x_2}{2}, \frac{y_1 + y_2}{2}\right)$.

Example

To find the midpoint between the points (–1, 4) and (2, –5), substitute the coordinates into the Midpoint Formula.

$$\left(\frac{x_1 + x_2}{2}, \frac{y_1 + y_2}{2}\right) = \left(\frac{-1 + 2}{2}, \frac{4 - 5}{2}\right)$$

$$= \left(\frac{1}{2}, \frac{-1}{2}\right)$$

So, the midpoint between the points (–1, 4) and (2, –5) is $\left(\frac{1}{2}, -\frac{1}{2}\right)$.

Glossary

minimum point

The minimum point of the graph of a function is the ordered pair on the graph with the least y-coordinate.

Example

The ordered pair (1, –4) is the minimum point of the graph of the function $y = \frac{2}{3}x^2 - \frac{4}{3}x - \frac{10}{3}$.

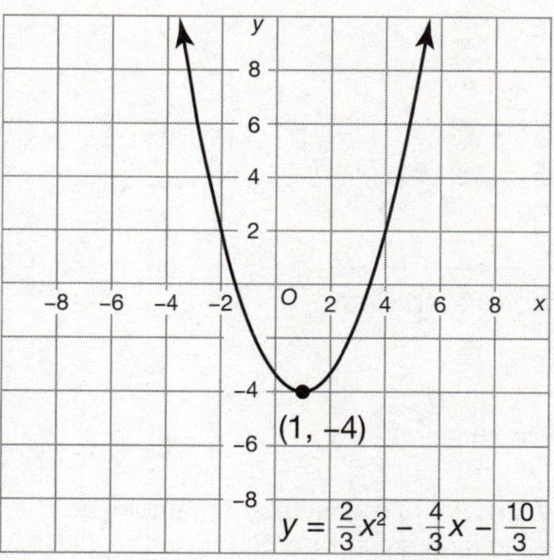

mode

The mode is the number (or numbers) that occurs most often in a data set. If there is no number that occurs most often, the data set has no mode.

Example

The mode of the numbers 1, 3, 3, 5, 8, 10, 17 is 3.

model of a data set

A model of a data set is a line of best fit and its equation.

Example

The equation of the line of best fit is $y = 25x$.

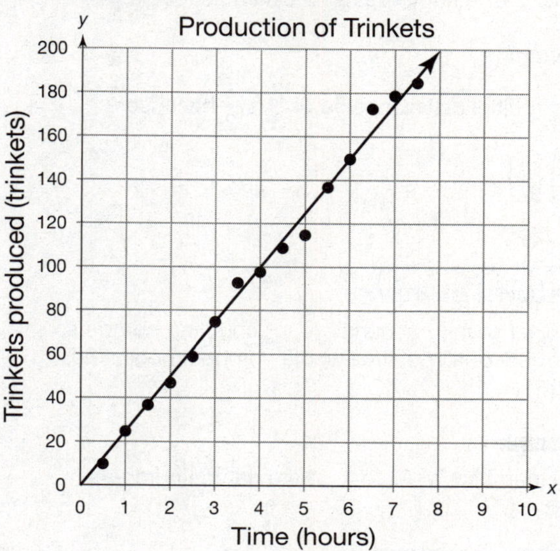

monomial

A monomial is an expression that consists of a single term that is either a constant, a variable, or a product of a constant and one or more variables. A monomial is a polynomial with one term.

Example

The expressions $5x$, 7, $-2xy$, and $13x^3$ are monomials.

multiplicative identity

The multiplicative identity is a number such that when you multiply it by a second number, the product is the second number.

Example

The multiplicative identity is the number 1.

$1 \cdot 5 = 5$

Glossary

multiplicative inverse

The multiplicative inverse of a number $\frac{a}{b}$ is the number $\frac{b}{a}$. The product of any nonzero number and its multiplicative inverse is 1. The multiplicative inverse of a number is also called its reciprocal.

Example

The multiplicative inverse of $\frac{2}{3}$ is $\frac{3}{2}$ because $\left(\frac{2}{3}\right)\left(\frac{3}{2}\right) = 1$.

natural number

The set of natural numbers, or counting numbers, consists of all positive whole numbers beginning with 1.

Example

The numbers 1, 2, 3, 4, ... are natural numbers.

negative exponent

A negative exponent is an exponent in a power. A power of a whole number with a negative exponent represents a number that is less than 1.

Example

In the expression x^{-2}, –2 is a negative exponent.

negative square root

A negative square root is one of the two square roots of a positive number.

Example

The negative square root of 25 is –5.

negatively correlated

Points are negatively correlated when the line of best fit has a negative slope.

Example

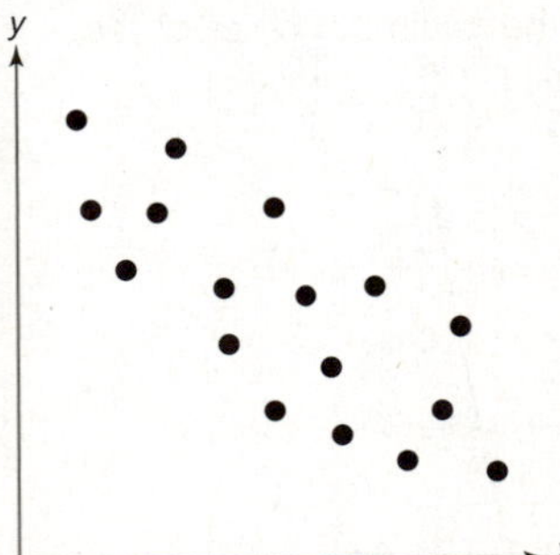

Negatively correlated data

net pay

Net pay is the amount of money that an employee earns after deductions are subtracted from the employee's gross pay.

Example

An employee earns $2400 per month of gross pay. Deductions of $432 in taxes and $164 in insurance are subtracted from this amount. So, the employee's net pay is $2400 – $432 – $164 = $1804.

*n*th percentile

The *n*th percentile for a data set is the value for which *n* percent of the numbers in the set are less than that value.

Example

A student scored 987 points on a standardized test. The score is the 60th percentile because 60% of the test scores are less than 987.

*n*th term

The *n*th term is a convention used to generate or represent terms of a sequence.

Example

A sequence is given by $a_n = 7n$.

The first term is $a_1 = 7(1) = 7$.

The second term is $a_2 = 7(2) = 14$.

number line

A number line is a line on which a unique point is assigned to every real number.

Example

The point on the number line below corresponds to the rational number 1.5.

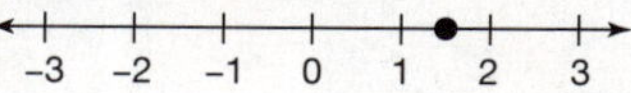

numerical expression

A numerical expression is a mathematical phrase consisting of numbers and operations to be performed.

Example

$2(1 + 5) - 6$ is a numerical expression.

odds against an event

The odds against an event are the ratio of the number of unfavorable outcomes to the number of favorable outcomes.

Example

The odds against rolling a number that is a multiple of 3 with a six-sided number cube is $\frac{4}{2}$.

odds in favor of an event

The odds in favor of an event are the ratio of the number of favorable outcomes to the number of unfavorable outcomes.

Example

The odds of rolling a number that is a multiple of 3 with a six-sided number cube is $\frac{2}{4}$.

one-step equation

A one-step equation is an equation that requires only one operation to solve.

Example

The equation $x + 5 = 8$ is a one-step equation. It can be solved by subtracting 5 from each side of the equation.

opposite of a number

The opposite of a number is the number that is the same distance from zero but on the opposite side of zero on a number line.

Example

The opposite of −3 is 3. Both numbers are 3 units from 0 on the number line.

order of operations

The order of operations is a set of rules that ensure that the result of combining numbers and operations, such as addition and multiplication, is the same every time. The order of operations is:

1. Evaluate expressions inside grouping symbols such as ( ) or [ ].
2. Evaluate powers.
3. Multiply and divide from left to right.
4. Add and subtract from left to right.

Example

To evaluate the expression $(3 + 4)^2 + 5 \cdot 2$, perform the operations in this order. Evaluate expressions inside the parentheses first.

$$\begin{aligned}(3 + 4)^2 + 5 \cdot 2 &= 7^2 + 5 \cdot 2\\ &= 49 + 5 \cdot 2\\ &= 49 + 10\\ &= 59\end{aligned}$$

Glossary

ordered pair

An ordered pair is a pair of numbers of the form (x, y) that represents a unique position in the coordinate plane. The first number in the ordered pair is the x-coordinate and the second number is the y-coordinate.

Example

The ordered pairs (4, 2) and (–2, –3) are shown in the coordinate plane.

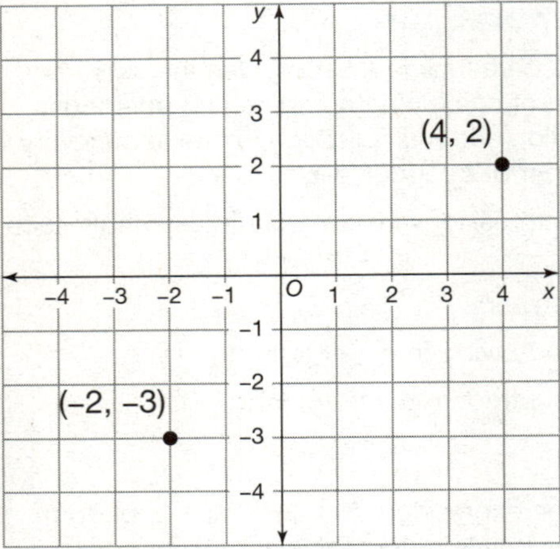

origin

The origin is the point where the x-axis and y-axis intersect in the coordinate plane. The ordered pair that represents the origin is (0, 0).

Example

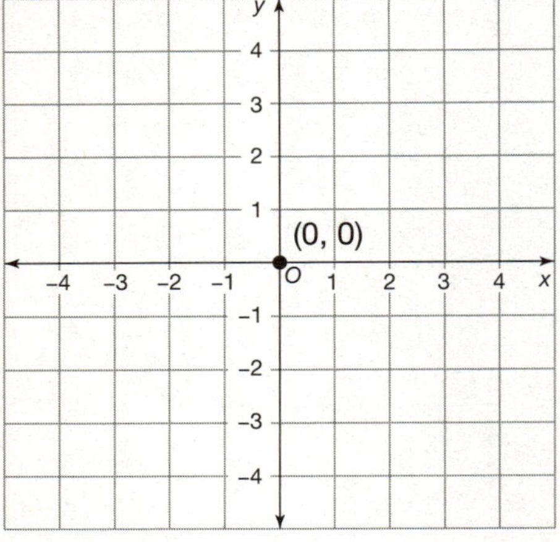

outcome

An outcome is a possible result of an event.

Example

Flipping a coin has two outcomes: heads and tails.

outlier

An outlier is a data value that is much less or much greater than the other values in the data set.

Example

The data set 1, 1, 3, 3, 4, 4, 5, 1000 has outlier of 1000.

output value

An output value is the second coordinate of an ordered pair in a relation.

Example

The ordered pair (5, 8) has an output value of 8.

parabola

A parabola is the U-shaped graph of a quadratic function of the form $y = ax^2 + bx + c$, where $a \neq 0$.

Example

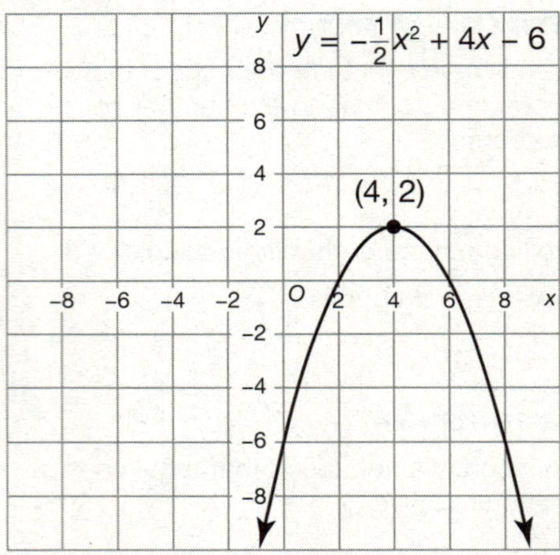

parallel

Two lines in the same plane are parallel to each other if they do not intersect.

Example

Lines m and n are parallel.

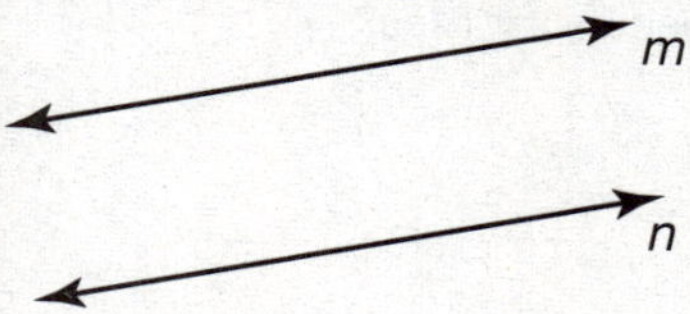

parallelogram

A parallelogram is a quadrilateral in which both pairs of opposite sides are parallel.

Example

In parallelogram $ABCD$, opposite sides AB and CD are parallel; opposite sides AD and BC are parallel.

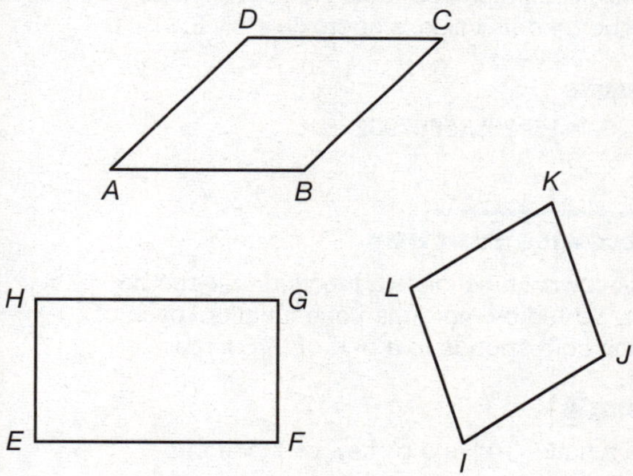

parent function

A parent function is the most basic function of a family of functions.

Example

The parent linear function is $y = x$. The parent quadratic function is $y = x^2$.

pattern

A pattern is an ordered sequence of numbers, shapes, or other objects that are arranged according to a rule.

Examples

The pattern a, b, a, b, a, b, a, b, ... is the sequence of alternating letters a and b.

The pattern 0, 1, 4, 9, 16, 25, 36, 49, ... is the sequence of the squares of whole numbers.

percent

A percent is a ratio whose denominator is 100. One percent of a quantity is $\frac{1}{100}$ of the quantity.

Example

You buy a notebook for $4.00 and pay a sales tax of 7%. The sales tax is equal to $\frac{7}{100}$ of $4.00, or $.28.

percent equation

A percent equation is an equation of the form $a = \frac{p}{100}b$, where p is the percent and the numbers being compared are a and b. A percent equation can also be written in the form $a = pb$ where p is the percent in decimal form and the numbers that are being compared are a and b.

Example

A store is having a sale. All merchandise is 75% of the original price. The equation $y = \frac{75}{100}x$ is a percent equation where x represents the original price and y represents the discounted price.

percentile

A percentile divides a data set into 100 equal parts.

perfect square

A perfect square is a whole number whose square root is also a whole number.

Example

The numbers 1, 4, 9, and 16 are perfect squares.

perfect square trinomial

A perfect square trinomial is a trinomial of the form $ax^2 + 2ab + b^2$ or $ax^2 - 2ab + b^2$. A perfect square trinomial can be written as the square of a binomial.

Example

The trinomial $x^2 + 6x + 9$ is a perfect square trinomial and can be written as $(x + 3)$ squared.

perimeter of a square

The perimeter of a square is the sum of all the side lengths.

Example

The perimeter of square *ABCD* is (4)(5) or 20 centimeters.

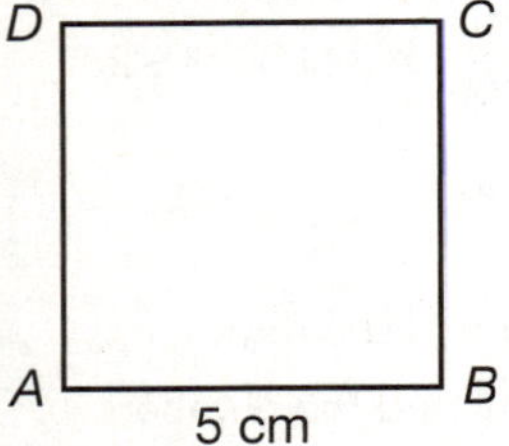

permutation

A permutation is an arrangement of a set of items for which the order of the items is important.

Example

Selecting the order in which students will give presentations is a permutation.

perpendicular

Perpendicular lines are two lines that intersect to form a right angle. The slopes of perpendicular lines are negative reciprocals of each other.

Example

Lines *m* and *k* are perpendicular lines.

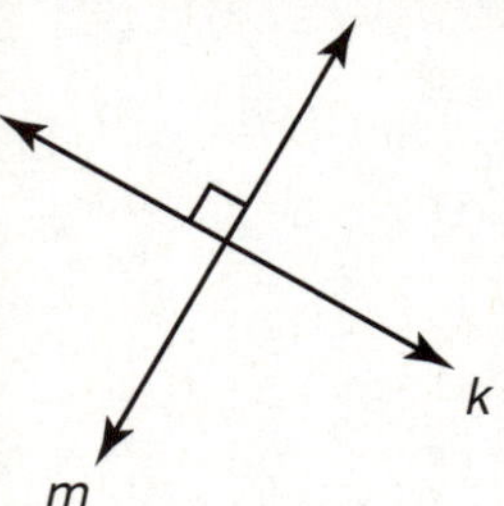

pi

π is the symbol that is used to represent the ratio of a circle's circumference to its diameter. π is an irrational number and its value is approximately 3.14.

Example

$\pi = 3.14159265358979323846...$

piecewise function

A piecewise function is a function that can be represented by more than one function, each of which corresponds to a part of the domain.

Example

The function $f(x)$ is a piecewise function.

$$f(x) = \begin{cases} x + 5 & x \leq -2 \\ -2x + 1 & -2 < x \leq 2 \\ 2x - 9 & x > 2 \end{cases}$$

point of intersection

The point of intersection is the location on a graph where two lines or functions intersect indicating that the values at that point are the same.

Example

The point of intersection is the point (100, 300).

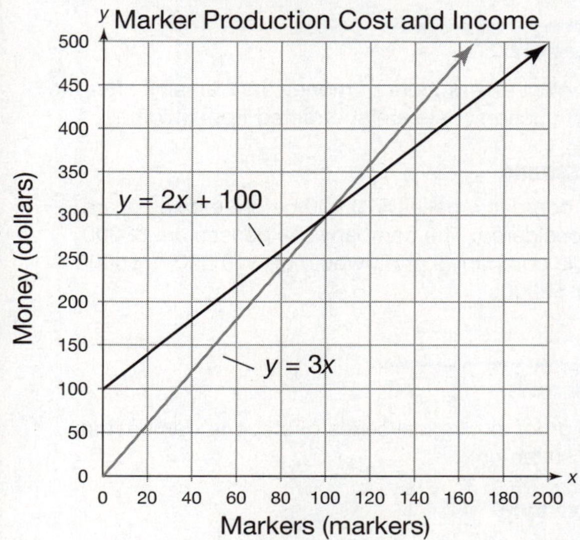

point-slope form of a linear equation

The point-slope form of a linear equation that passes through the point (x_1, y_1) and has slope m is $y - y_1 = m(x - x_1)$.

Example

A line passing through the point (1, 2) with a slope of $\frac{1}{2}$ can be written in point-slope form as $y - 2 = \frac{1}{2}(x - 1)$.

polynomial

A polynomial is an expression of the form

$a_0 + a_1x + a_2x^2 + ... + a_nx^n$

where the coefficients $(a_0, a_1, a_2, ...)$ are real numbers or complex numbers and the exponents are nonnegative integers.

Example

The expression $3x^3 + 5x^2 - 6x + 1$ is a polynomial.

population

A population is the entire group of people or itmes from which information is gathered.

Example

The population of math students in your school is every student in your school who is enrolled in a math class.

positive exponent

A positive exponent is an exponent in a power.

Example

In the expression x^2, 2 is a positive exponent.

positive square root

A positive square root is one of the two square roots of a positive number.

Example

The positive square root of 25 is 5.

positively correlated

Points are positively correlated when the line of best fit has a positive slope.

Example

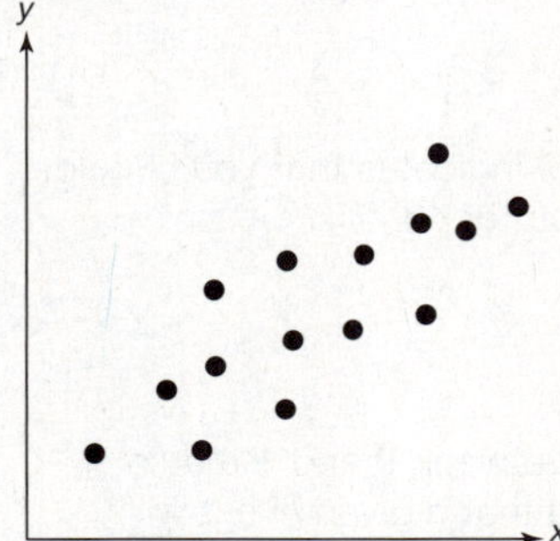

Positively correlated data

power

A power is an expression in which a number or variable is raised to an exponent. A power is a notation used to represent repeated multiplication.

Example

The expression 2^5 is a power.

Glossary

prime factorization

The prime factorization of a number is the representation of the number as a product of prime numbers.

Example

The prime factorization of 24 is $2 \cdot 2 \cdot 2 \cdot 3$ or $2^3 \cdot 3$.

prime number

A prime number is a whole number that is greater than 1 and has exactly two whole number factors, 1 and itself.

Example

The first five prime numbers are 2, 3, 5, 7, and 11.

principal

The principal is an amount of money that is borrowed or invested, represented by P.

Example

Sharon deposits $500 into a new checking account. The principal is $500.

principal square root

The positive square root is called the principal square root.

Example

An expression such as $\sqrt{49}$ indicates that you should find the principal square root of 49.

probability

A probability is a number between 0 and 1 that is a measure of the likelihood that a given event will occur. The probability of an event when all outcomes are equally likely is equal to the number of desired outcomes divided by the number of possible outcomes.

Example

The probability of rolling an even number with a six-sided number cube is $\frac{3}{6}$, or $\frac{1}{2}$.

product

A product is the result of multiplying one quantity by another.

Example

The product of 2 and 6 is 12.

profit

Profit is the amount of money that remains after the production costs are subtracted from income.

Example

A company makes $10,000 before expenses are considered. The company's expenses are $8000. The company's profit would be $10,000 − $8000 or $2000.

proof

A proof is a logical series of steps to demonstrate a statement.

Example

To prove the statement $(a - b)(a^2 + ab + b^2) = a^3 - b^3$, use the following direct proof.

$(a - b)(a^2 + ab + b^2)$

$= a(a^2 + ab + b^2) - b(a^2 + ab + b^2)$

Distributive property

$= a(a^2) + a(ab) + a(b^2) - b(a^2) - b(ab) - b(b^2)$

Distributive Property of Multiplication Over Addition

$= a^3 + a^2b + ab^2 - a^2b - ab^2 - b^3$

Product of Powers

$= a^3 - b^3$

Additive Inverse

proportion

A proportion is an equation that states that two ratios are equal.

Examples

The equation $\frac{4}{8} = \frac{1}{2}$ is a proportion. The equation $\frac{x}{12} = \frac{5}{60}$ is a proportion.

Glossary

Pythagorean Theorem

The Pythagorean Theorem states that if a and b are the legs of a right triangle, and c is the hypotenuse, then the sum of the squares of the lengths of the legs equals the square of the length of the hypotenuse: $a^2 + b^2 = c^2$.

Example

In triangle ABC, angle C is the right angle, so side BC is a leg, side AC is a leg, and side AB is the hypotenuse.

$$a^2 + b^2 = c^2$$
$$3^2 + b^2 = 8^2$$
$$9 + b^2 = 64$$
$$b^2 = 64 - 9$$
$$b^2 = 55$$
$$b = \sqrt{55}$$

So, the length of side AC is $\sqrt{55}$, or approximately 7.42 centimeters.

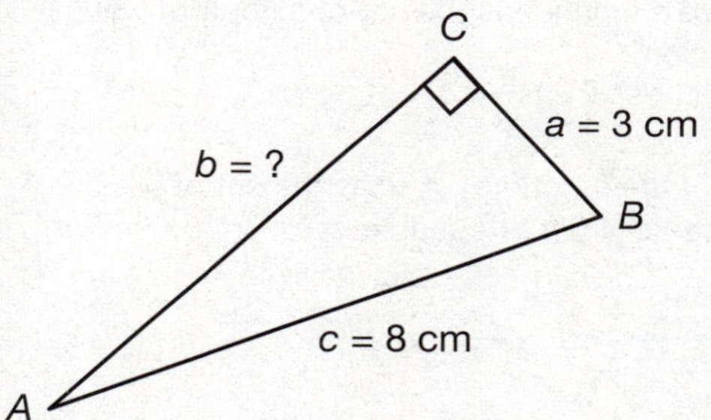

quadratic formula

The Quadratic Formula is a formula used to find the solutions of a quadratic equation. For a quadratic equation of the form $ax^2 + bx + c = 0$, the solutions can be found using the Quadratic Formula $x = \dfrac{-b \pm \sqrt{b^2 - 4ac}}{2a}$.

Example

To use the Quadratic Formula to find the solutions of $x^2 + 2x - 24 = 0$, use $a = 1$, $b = 2$, and $c = -24$.

$$x = \frac{-2 \pm \sqrt{(2)^2 - 4(1)(-24)}}{2(1)}$$
$$= \frac{-2 \pm \sqrt{100}}{2}$$
$$= \frac{-2 \pm 10}{2}$$
$$x = \frac{-2 + 10}{2} \quad \text{or} \quad x = \frac{-2 - 10}{2}$$
$$= 4 \quad \text{or} \quad = -6$$

So, the solutions of $x^2 + 2x - 24 = 0$ are –6 and 4.

quadratic function

A quadratic function is a function that can be written in the form $f(x) = ax^2 + bx + c$, where a, b, and c are real numbers and a is not equal to zero.

Example

The equations $y = x^2 + 2x + 5$ and $y = -4x^2 - 7x + 1$ are quadratic functions.

quartile

The three quartiles divide a data set into four equal parts. The middle quartile is the median. The other two values are the upper quartile and the lower quartile.

Example

In the data set 13, 17, 23, 24, 25, 29, 31, 45, 46, 53, 60, the median, 29, divides the data into two halves. The first quartile, 23, is the median of the lower half of the data. The third quartile, 46, is the median of the upper half of the data.

13 17 23 24 25 29 31 45 46 53 60

lower quartile (23) — median (29) — upper quartile (46)

Glossary

quotient

A quotient is the number that results from the division of one number by another. The quotient is the answer of a division problem.

Example

The quotient of the division problem $96 \div 12 = 8$ is the number 8.

radical symbol

A radical symbol is the symbol used to write a square root.

Example

The square root of 25 can be written using a radical symbol as $\sqrt{25}$.

radicand

A radicand is the quantity under a radical sign in an expression.

Example

In the radical expression $\sqrt{25}$, 25 is the radicand.

radius

The radius is the distance from the center of a circle to a point on the circle.

Example

In the circle, O is the center and the length of segment OA is the radius.

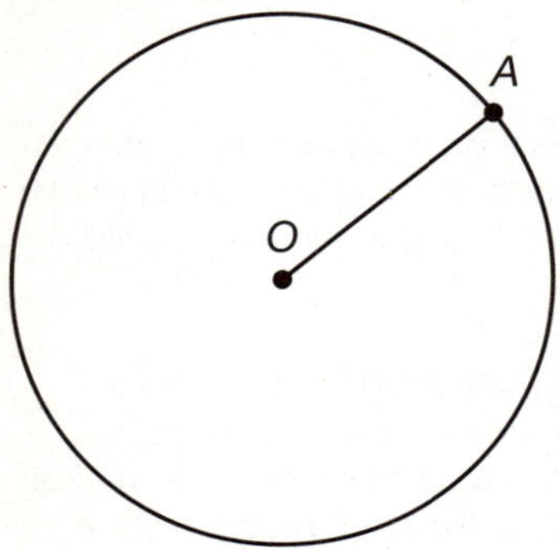

randomly chosen

To be randomly chosen means that no particular rule was used to choose a person.

Example

Choosing 100 fans at random to participate in a survey from crowd of 5000 people is an example of being randomly choosen.

range of a data set

The range of a data set is the difference between the greatest number and the least number in the data set.

Example

The range of the numbers 1, 5, 7, 10, 14 is $14 - 1$, or 13.

range of a function

The range of a function is the set of all output values for the function.

Example

The range of the function $y = x^2$ is the set of all numbers greater than or equal to zero.

rate

A rate is a ratio in which the units of the quantities being compared are different.

Example

A car uses 20 gallons of gasoline to drive 600 miles. The car's fuel consumption rate is

$\frac{600 \text{ miles}}{20 \text{ gallons}} = \frac{30 \text{ miles}}{1 \text{ gallon}}$ or 30 miles per gallon.

rate of change

Rate of change is a comparison of two quantities with different units that are changing.

Example

The speed of a car, measured in miles per hour, is a rate of change.

Glossary

ratio

A ratio is a way to compare two quantities that are measured in the same units by using division. The ratio of two numbers *a* and *b*, with the restriction that *b* cannot equal zero, can be written in three ways.

a to *b*

a : *b*

$\frac{a}{b}$

Example

In Central High School, there are 4 boys for every 5 girls. The ratio of boys to girls can be written as 4 boys to 5 girls, 4 boys: 5 girls, or $\frac{4 \text{ boys}}{5 \text{ girls}}$.

rational exponent

A rational exponent is an exponent that is a rational number.

Example

In the expression $x^{2/3}$, $\frac{2}{3}$ is a rational exponent

rational expression

A rational expression is an expression that can be written as the quotient of two nonzero polynomials.

Example

The expression $\frac{2x + 1}{x - 3}$ is a rational expression.

rational number

A rational number is a number that can be written as the quotient of two integers.

Example

The number 0.5 is a rational number because it can be written as the fraction $\frac{1}{2}$.

real number

The real numbers consist of all rational numbers and irrational numbers. Real numbers can be represented on the real number line.

Example

The numbers -3, 11.4, $\frac{1}{2}$, and $\sqrt{5}$ are real numbers.

real number system

The real number system is comprised of the real numbers together with their properties and operations.

reciprocals

Two non-zero numbers are reciprocals if their product is 1.

Example

The fractions $\frac{2}{3}$ and $\frac{3}{2}$ are reciprocals.

reflection

A reflection is a transformation in which a figure is reflected, or flipped, in a given line called the line of reflection.

Example

The triangle on the right is a reflection of the triangle on the left.

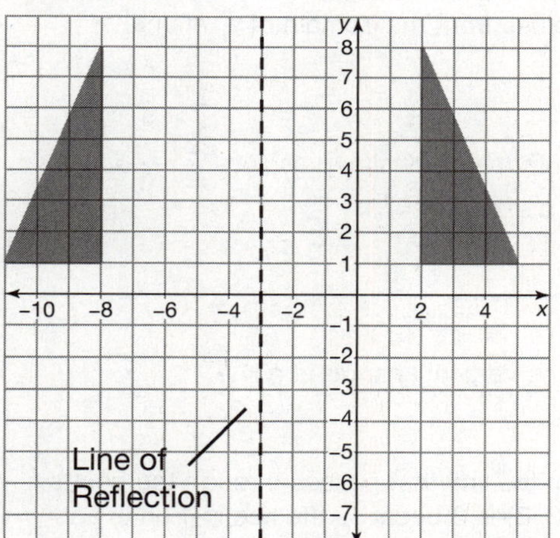

relation

A relation is any set of ordered pairs.

Example

The set of points {(0, 1), (1, 8), (2, 5), (3, 7)} is a relation.

Glossary

remainder

The remainder is the whole number left over in a division problem if the divisor does not divide the dividend evenly.

Example

When 17 is divided by 3, the remainder is 2.

repeating decimal

A repeating decimal is a decimal with one or more digits that repeat indefinitely. A repeating decimal can be represented by placing a bar over the repeating digits.

Example

The fraction $\frac{1}{3}$ can be written as the repeating decimal $0.\overline{3}$.

restricting the domain

Restricting the domain is a process of eliminating excluded values from the domain of a rational expression.

Example

The domain of the rational expression $\frac{2x + 1}{x - 3}$ is all real numbers except $x = 3$.

rise

The rise is the vertical change of a line.

Example

The slope of the line that passes through the points (1, 4) and (3, 8) is 2 because the rise is 4 units and the run is 2 units.

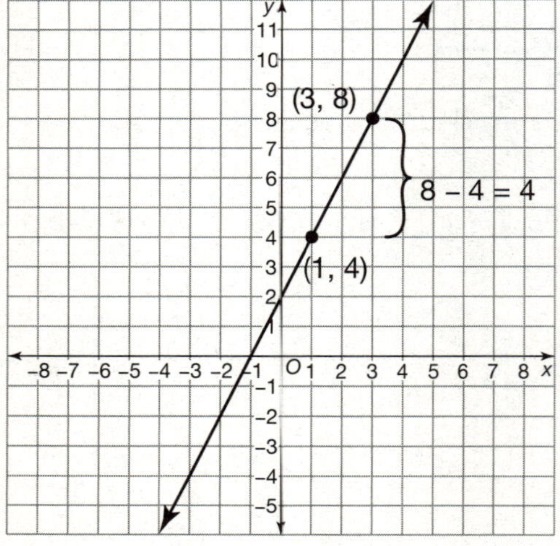

root of a number

A root of a number or *n*th root of a number *b* is a solution of the equation $x^n = b$.

Example

The fourth root of 16 is 2 because $2^4 = 16$.

run

The run is the horizontal change of a line.

Example

The slope of the line that passes through the points (1, 4) and (3, 8) is 2 because the rise is 4 units and the run is 2 units.

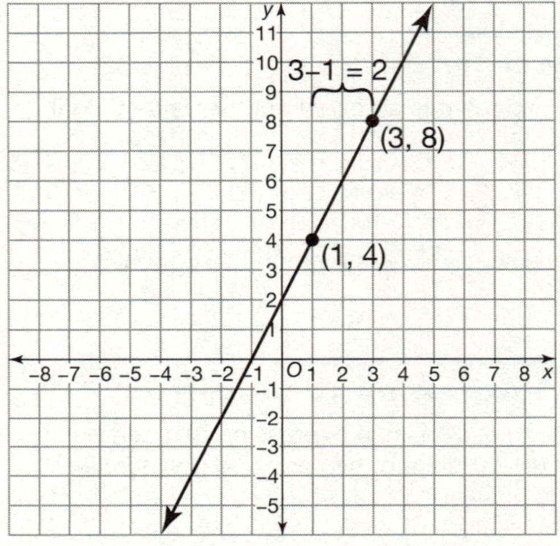

sample

A sample is a group of items that are selected at random from a larger group of items called the population.

Example

If the population of a study concerning health care includes everyone born in the United States from 1995 to 2005, then everyone born on May 22 of each year from 1995 to 2005 is a sample.

sample size

The sample size is the number of people that are surveyed.

Example

A newspaper reporter surveys 100 people. The sample size is 100.

sample space

A sample space of a random experiment is the set of all possible outcomes of the experiment.

Example

The sample space for flipping a coin twice consists of four outcomes: Heads-Heads, Heads-Tails, Tails-Heads, and Tails-Tails.

sample standard deviation

The sample standard deviation, s, is the square root of the sample variance.

Example

The mean of the data set 2, 3, 9, and 13 is 6.75.

The data value 2 has a deviation of –4.75.

The data value 3 has a deviation of –3.75.

The data value 9 has a deviation of 2.25.

The data value 13 has a deviation of 6.25.

The sample variance is equal to $\frac{80.75}{3}$, or approximately 26.92.

The sample standard deviation is the square root of 26.92, or approximately 5.19.

sample variance

The sample variance, s^2, is found by dividing the variance by one less than the number of data values in the set.

Example

The mean of the data set 2, 3, 9, and 13 is 6.75.

The data value 2 has a deviation of –4.75.

The data value 3 has a deviation of –3.75.

The data value 9 has a deviation of 2.25.

The data value 13 has a deviation of 6.25.

The variance is $(-4.75)^2 + (-3.75)^2 + 2.25^2 + 6.25^2$, or 80.75.

The sample variance is equal to $\frac{80.75}{3}$, or approximately 26.92.

sampling method

A sampling method is a method for selecting people or items from a population for a survey. Types of sampling methods are random sampling, stratified sampling, systematic sampling, convenience sampling, and self-selected sampling.

Example

Choosing 100 fans at random to participate in a survey from a crowd of 5000 people is an example of random sampling.

scatter plot

A scatter plot is a graph in the coordinate plane in which values of x and y are plotted as points (x, y).

Example

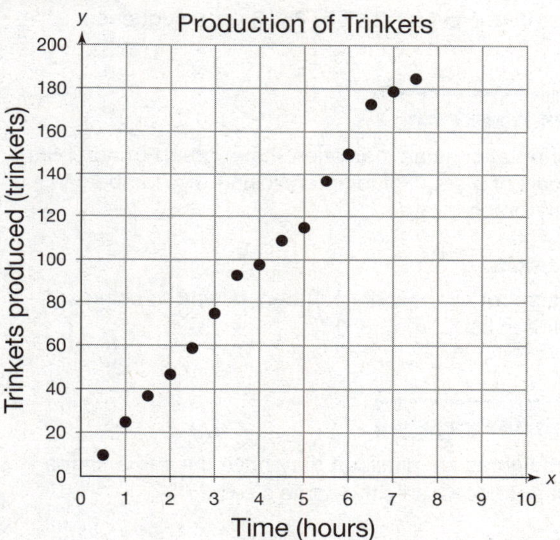

scientific notation

Scientific notation is a way of writing very large or very small numbers. A number written in scientific notation has the form $c \times 10^n$, where c is greater than or equal to 1 and less than 10 and n is an integer.

Example

The number 1,000,000,000 can be written in scientific notation as 1×10^9.

Glossary

second quartile

The second quartile, Q_2, is the median of a data set.

Example

In the data set 13, 17, 23, 24, 25, 29, 31, 45, 46, 53, 60, the median, 29, is the second quartile.

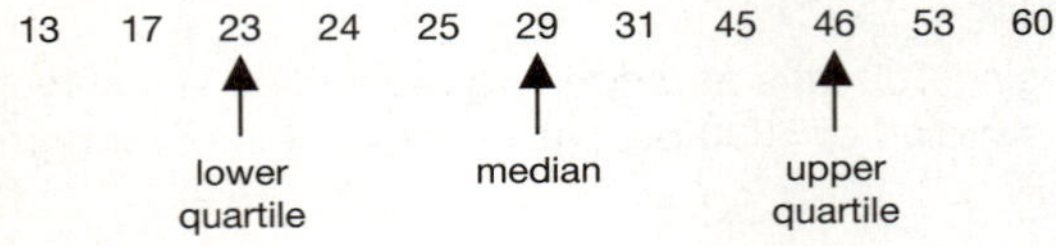

sequence

A sequence is an ordered set of objects or numbers.

Example

The numbers 1, 1, 2, 3, 5, 8, 13 are a sequence.

set notation

Set notation is an indication that a group of numbers is part of a set, including enclosing the numbers in curly braces: { }.

Example

The set of even whole numbers is written using set notation as {2, 4, 6, 8, 10, ...}.

similar figures

Two figures are similar if they have the same shape, but not necessarily the same size.

Example

Triangle *ABC* is similar to triangle *DEF*

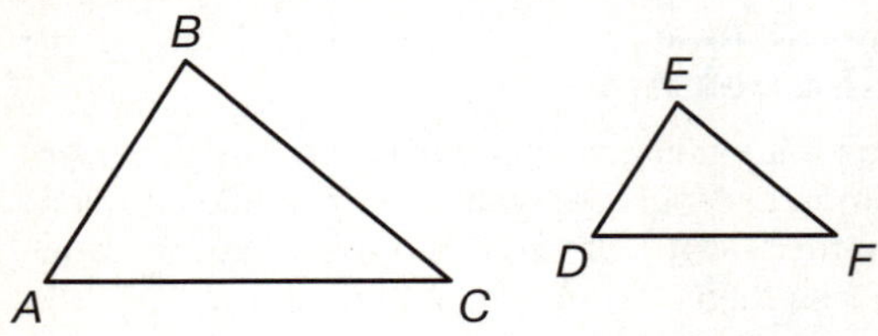

simple event

A simple event is a collection of outcomes of an experiment. An outcome is one possible result of an experiment.

Example

Flipping a coin or rolling a number cube is a simple event.

simple interest

Simple interest is when interest is paid only as a percent of the principal. To find simple interest, multiply the principal P by the annual interest rate r written as a decimal and the time t in years: $I = Prt$.

Example

Tonya deposits $200 in a 3-year certificate of deposit that earns 4% interest. The amount of interest that Tonya earns can be found using the simple interest formula.

$I = (200)(0.04)(3)$

$I = 24$

Tonya earns $24 interest.

simplify an expression

To simplify an expression, rewrite the expression as an equivalent yet briefer expression that is easier to work with.

Example

The expression $2(4 + 5)$ can be simplified as 18.

slope

The slope of a non-vertical line is the ratio of the vertical change to the horizontal change.

Example

The slope of the line that passes through the points (3, –6) and (2, –4) is –2 because the vertical change is –2 units and the horizontal change is 1 unit.

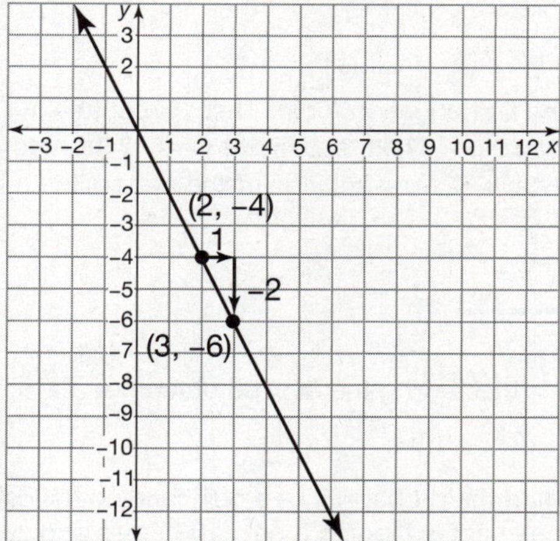

slope-intercept form of a linear equation

The slope-intercept form of a linear equation is $y = mx + b$, where m is the slope of the line and b is the y-intercept of the line.

Example

The linear equation $y = 2x + 1$ is written in slope-intercept form. The slope of the line is 2 and the y-intercept is 1.

solution of a linear system

A solution of a linear system is an ordered pair (x, y) that is a solution of both equations in the system.

Example

The equations $y = 3x + 7$ and $y = -4x$ are a system of equations. The solution to the system of equations is the intersection point (–1, 4).

solution of an equation

The solution of an equation is a number that, when substituted for the variable, makes the equation true.

Example

The solution of the equation $3x + 4 = 25$ is 7 because 7 makes the equation true: $3(7) + 4 = 25$, or $25 = 25$.

solve an inequality

To solve an inequality, find the values of the variable that make the inequality true.

Example

The inequality $x + 5 > 6$ can be solved by subtracting 5 from each side of the inequality. The solution is $x > 1$. Any number greater than 1 will make the inequality $x + 5 > 6$ true.

square of a binomial difference

The square of a binomial difference $(a - b)^2$ is equal to $a^2 - 2ab + b^2$.

Example

$(x - 3)^2 = x^2 - 6x + 9$

square of a binomial sum

The square of a binomial sum $(a + b)^2$ is equal to $a^2 + 2ab + b^2$.

Example

$(x + 3)^2 = x^2 + 6x + 9$

square root

A number b is a square root of a if $b^2 = a$.

Example

The square roots of 25 are 5 and –5.

standard form of a linear equation

The standard form of a linear equation is $ax + by = c$, where a, b, and c are constants and a and b are not both zero.

Example

The linear equation $2x + 3y = 5$ is written in standard form.

Glossary

standard form of a number

A number in standard form is a number written as a numeral. In standard form, the position of the digit represents the place value of the digit.

Example

The number 243 is written in standard form. The digit 2 is in the hundreds place, the digit 4 is in the tens place, and the digit 3 is in the ones place.

standard form of a polynomial

A polynomial in standard form is written with the terms in descending order, starting with the term with the greatest degree and ending with the term with the least degree.

Example

The polynomial $2x^2 + x^4 + 1 + 5x - 2x^3$ can be written in standard form as $x^4 - 2x^3 + 2x^2 + 5x + 1$.

stem-and-leaf plot

A stem-and-leaf plot is a visual display of data that is organized by digits. Each data value is separated into a stem and a leaf. The leading digits of the data value are represented by the stem and the last digit is represented by the leaf.

Example

A stem-and-leaf plot can be drawn to represent test scores.

55, 62, 73, 75, 76, 79, 80, 83, 86, 87, 87, 88, 88, 89, 89, 89

The tens' place represents the stem and the ones' place represents the leaves.

Stems	Leaves
1	
2	
3	
4	
5	5
6	2
7	3 5 6 9
8	0 3 6 7 7 8 8 9 9 9

Key: 7 | 3 = 73

substitution method

The substitution method is a process of solving a system of equations by substituting a variable in one equation by an equal expression.

Example

Solve the following system of equations by using the substitution method:

$x - 3y = 4$

$2x + 5y = -14$

First, solve the first equation for x. Then, substitute in the second equation. Next, substitute $y = -2$ into the equation $x - 3y = 4$. The solution of the system is $(-2, -2)$.

success

A success is a trial having the desired outcome.

Example

You want to find the probability of a coin landing heads up. One experiment is flipping a coin 100 times and recording the results. Each coin flip that lands heads up is a successful trial.

sum

A sum is the result of adding one quantity to another.

Example

The sum of 26 and 13, 26 + 13, is the number 39.

supply

The supply is the amount of a good that is available.

Example

The supply of oil in the world can be modeled by the equation $y = 324.3x + 15.856$, where x is the time, in years, since 1965 and y is the supply of oil, in millions of barrels.

survey

A survey is an investigation of a characteristic of a population to gather information.

Example

A newspaper takes a survey to determine which mayoral candidate is favored.

Glossary

system of linear equations

A system of linear equations is two or more linear equations in the same variables.

Example

The equations $y = 3x + 7$ and $y = -4x$ are a system of equations.

system of linear inequalities

A system of linear inequalities is two or more linear inequalities in the same variables.

Example

The inequalities $y > 3x + 7$ and $y < -4x$ are a system of linear inequalties.

tax rate

A tax rate is the percent used to calculate the amount of money taken out of your gross pay.

Example

John lives in a township that has a tax rate of 2% so 2% of John's pay is taken out for local taxes.

term

A term is a member of a sequence. The first term is the first object or number in the sequence; the second term is the second object or number in the sequence; and so on.

Example

In the sequence 2, 4, 6, 8, 10, the first term is 2, the second term is 4, and the third term is 6.

terms of an expression

The terms of an expression are the parts that are added together. A term may be a number, a variable, or a product of a number and a variable or variables.

Example

The polynomial $2x + 3y + 5$ has three terms: $2x$, $3y$, and 5.

theoretical probability

A theoretical probability is a probability that is based on knowing all of the possible outcomes that are equally likely to occur.

Example

The theoretical probability of a coin landing heads up is $\frac{1}{2}$.

third quartile

The third quartile, Q_3, is the median of the upper half of the data .

Example

In the data set 13, 17, 23, 24, 25, 29, 31, 45, 46, 53, 60, the median, 29, divides the data into two halves. The third quartile, 46, is the median of the upper half of the data.

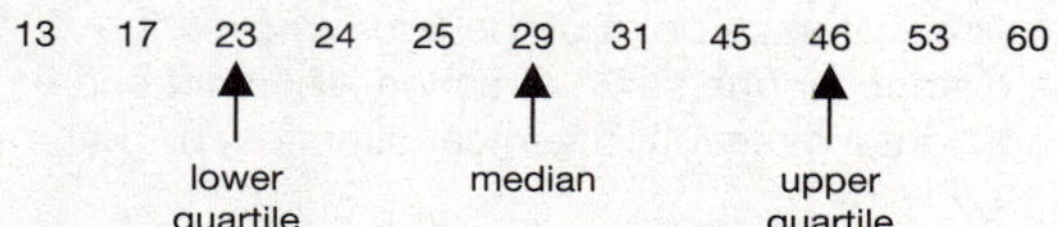

tolerance

A tolerance is the amount by which a quantity is allowed to vary from the normal or target quantity.

Example

A centimeter ruler has a tolerance of $\frac{1}{10}$ centimeter, or 1 millimeter.

Glossary

transformation

A transformation is an operation that maps, or moves a figure, called the preimage, to form a new figure, called the image. Three types of transformations are reflections, rotations, and translations.

Example

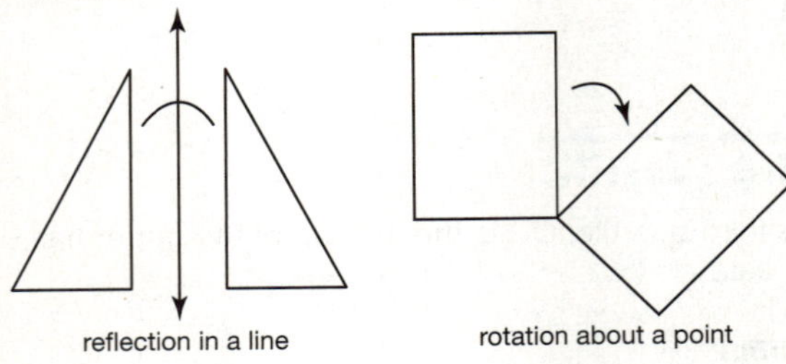

translation

A translation is a transformation in which a figure is shifted so that each point of the figure moves the same distance in the same direction. The shift can be in a horizontal direction, a vertical direction, or both.

Example

The top trapezoid is a vertical translation of the bottom trapezoid by 5 units.

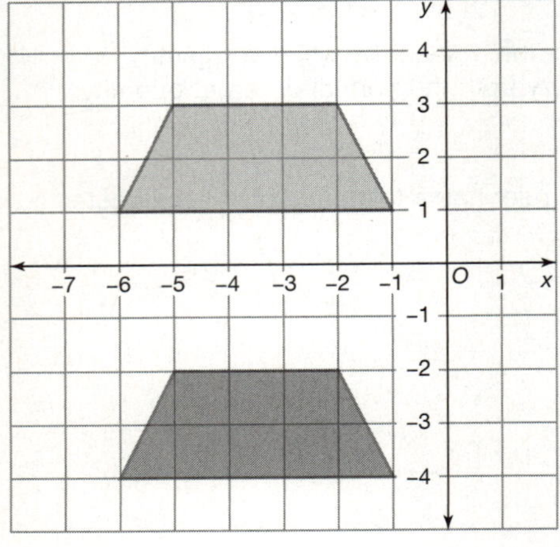

trapezoid

A trapezoid is a quadrilateral with exactly one pair of parallel sides. The parallel sides are called bases and the nonparallel sides are called legs. The perpendicular distance between the bases is the height of the trapezoid.

Example

Quadrilateral *ABCD* is a trapezoid. The height is 4 meters, the length of base *AD* is 12 meters, and the length of base *BC* is 6 meters.

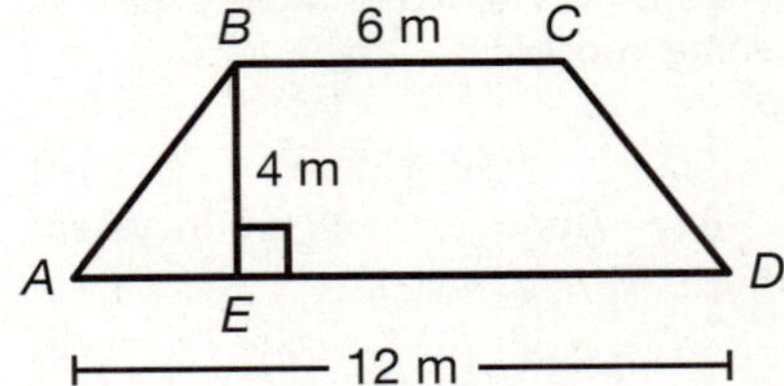

tree diagram

A tree diagram is a visual display that represents the outcomes for a series of events.

Example

The tree diagram shows the sample space for flipping a coin three times.

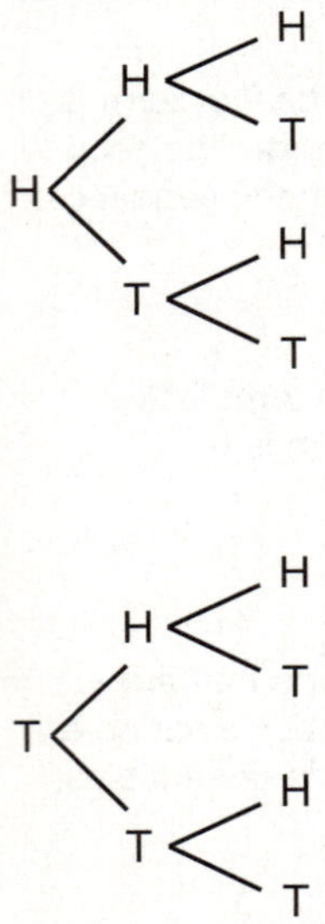

H – H – H
H – H – T
H – T – H
H – T – T
T – H – H
T – H – T
T – T – H
T – T – T

trial

A trial is one instance of an experiment.

Example

You want to find the probability of a coin landing heads up. One experiment is flipping a coin 100 times and recording the results. Each coin flip is a trial.

triangle

A triangle is a three-sided polygon that is formed by joining three points called vertices with line segments.

Example

In triangle *ABC* below, vertices *A*, *B*, and *C* are joined by segments *BA*, *AC*, and *CB*.

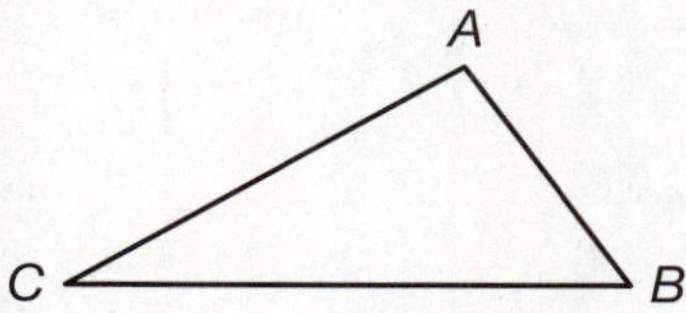

trimodal

A data set is trimodal if three values occur in the data set the same number of times.

Example

The data set 1, 1, 3, 3, 4, 4, 5, 7 is trimodal.

trinomial

A trinomial is a polynomial that consists of three terms.

Example

The polynomial $5x^2 - 6x + 9$ is a trinomial.

two-step equation

A two-step equation is an equation that requires two steps to solve.

Example

The equation $2x + 1 = 5$ is a two-step equation. It can be solved by subtracting 1 from both sides of the equation and then dividing both sides of the equation by 2.

undefined division

An undefined division is division by zero.

Example

The expression $6 \div 0$ is undefined.

unit

A unit is a standard measurement of one, such as one inch, one pound, or one second.

Examples

A unit of money is one dollar. A unit of distance is one foot.

unit rate

A unit rate is the rate per one given unit.

Example

The rate $\frac{150 \text{ miles}}{3 \text{ hours}}$ can be written as the unit rate of 50 miles per hour:

$$\frac{(150 \div 3) \text{ miles}}{(3 \div 3) \text{ hours}} = \frac{50 \text{ miles}}{1 \text{ hour}}.$$

value of an expression

The value of an expression is the result that is obtained when the indicated operations are carried out.

Example

The value of $9^{-1/2}$ is $\frac{1}{3}$. The value of $3x^2 + 6$ when $x = 2$ is 18.

variable

A variable is a letter or symbol that represents an unspecified member of a set.

Example

In the expression $2x + 3$, the letter "x" is a variable.

Glossary

Venn diagram

A Venn diagram uses circles to show how elements among sets of numbers or objects are related.

Example

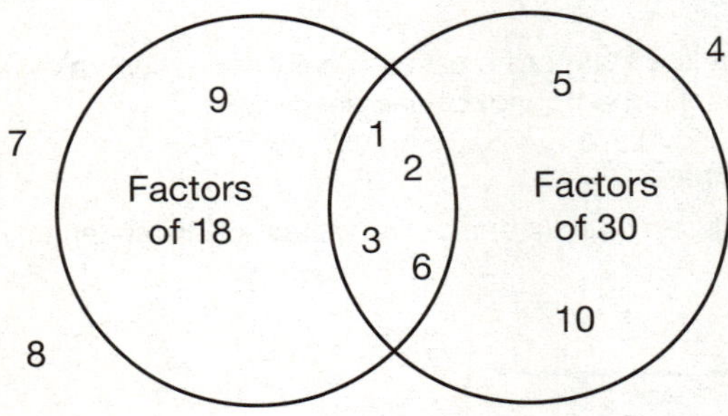

vertex form of a quadratic equation

The vertex form of a quadratic equation is an equation of the form $y = a(x - h)^2 + k$ where (h, k) is the vertex of the graph of the equation.

Example

The quadratic equation $y = 2(x - 5)^2 + 10$ is written in vertex form. The vertex of the graph is the point (5, 10).

vertex of a parabola

The vertex of a parabola, which lies on the axis of symmetry, is the highest or lowest point on the parabola.

Example

The vertex of the graph of $y = \frac{2}{3}x^2 - \frac{4}{5}x - \frac{10}{3}$ is the point (1, –4), the minimum point on the parabola.

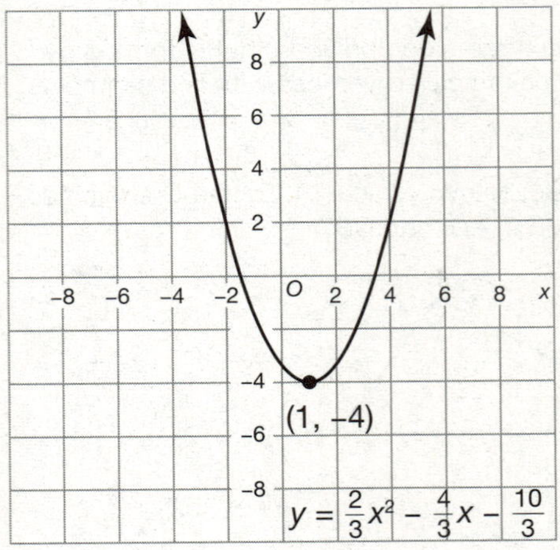

vertical line

A vertical line is a line of the form $x = a$, where a is a real number.

Example

The line represented by the equation $x = 5$ is a vertical line.

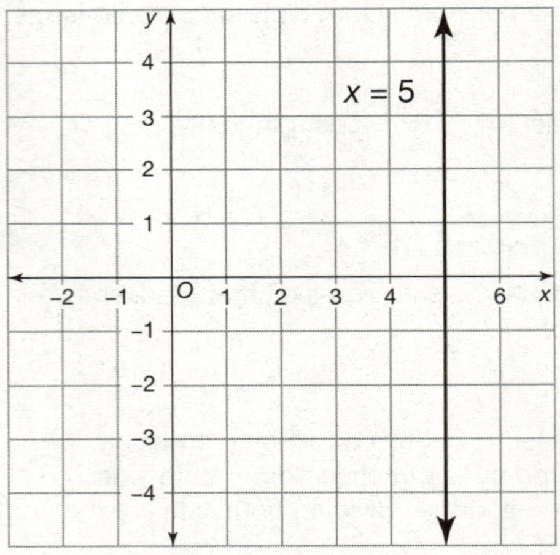

Glossary

vertical line test

The vertical line test is a method of determining whether an equation is a function. It states that an equation is a function if you can pass a vertical line through any part of the graph of the equation and the line intersects the graph at most one time.

Example

The equation $y = 3x^2$ is a function, because the graph of the function passes the Vertical Line Test.

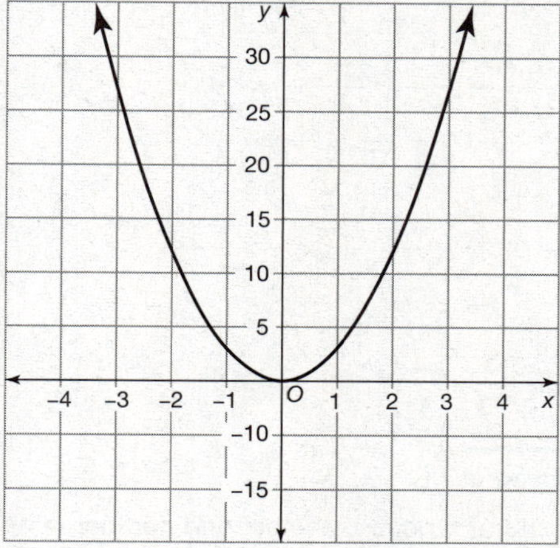

The equation $x^2 + y^2 = 9$ is not a function, because the graph of the function fails the Vertical Line Test.

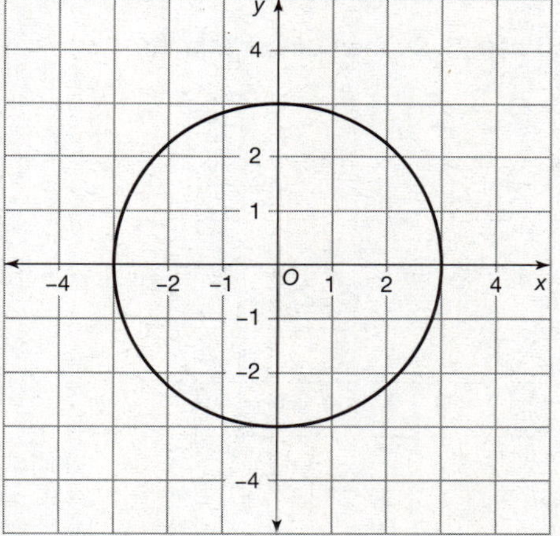

vertical motion model

A vertical motion model is an equation of the form $y = -16t^2 + vt + h$, where t is the time in seconds that the object has been moving, v is the initial velocity (speed) of the object in feet per second, h is initial height of the object in feet, and y is the height of the object in feet at time t seconds.

Example

A rock is thrown in the air at a velocity of 10 feet per second from a cliff that is 100 feet. The height of the rock is modeled by the equation $y = -16t^2 + 10t + 100$.

whole number

A whole number is any counting number or zero.

Example

The numbers 0, 1, 2, 3, ... are whole numbers.

x-axis

The x-axis is the horizontal number line in a Cartesian coordinate system.

Example

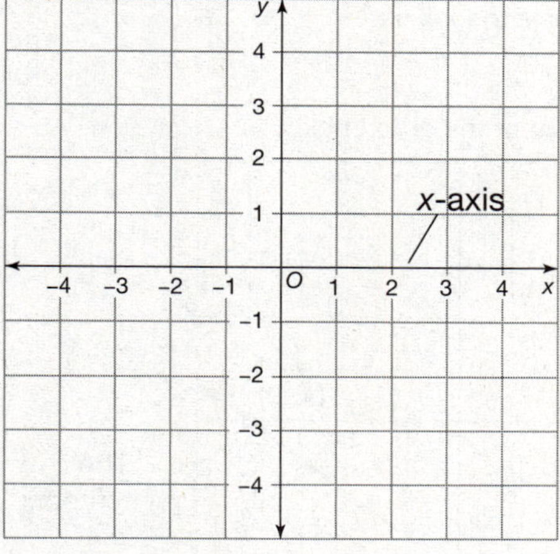

Glossary

x-coordinate

The x-coordinate of a point is the first number in an ordered pair. It indicates the distance of the point from the y-axis.

Example

In the ordered pair (5, 2), the number 5 is the x-coordinate.

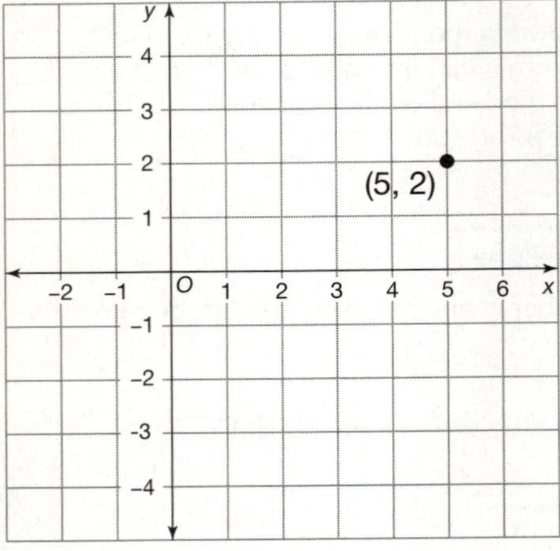

x-intercept

The x-intercept is the x-coordinate of the point where a graph crosses the x-axis.

Example

The x-intercept of the graph below is 4.

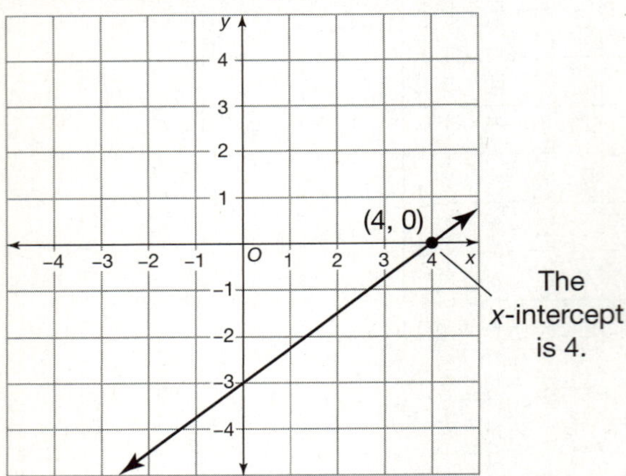

y-axis

The y-axis is the vertical number line in a Cartesian coordinate system.

Example

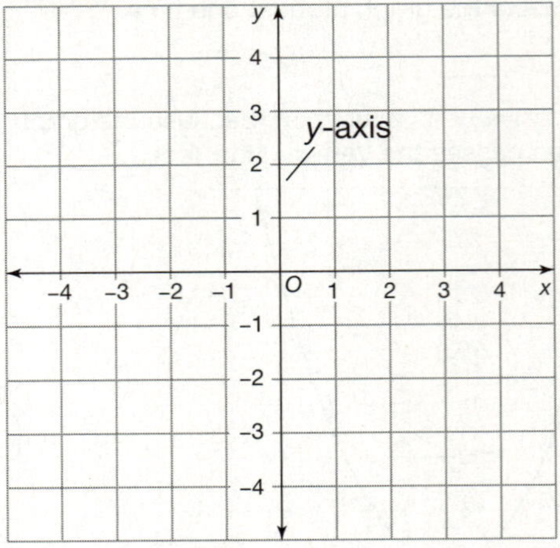

y-coordinate

The y-coordinate of a point is the second number in an ordered pair. It indicates the distance of the point from the x-axis.

Example

In the ordered pair (5, 2), the number 2 is the y-coordinate.

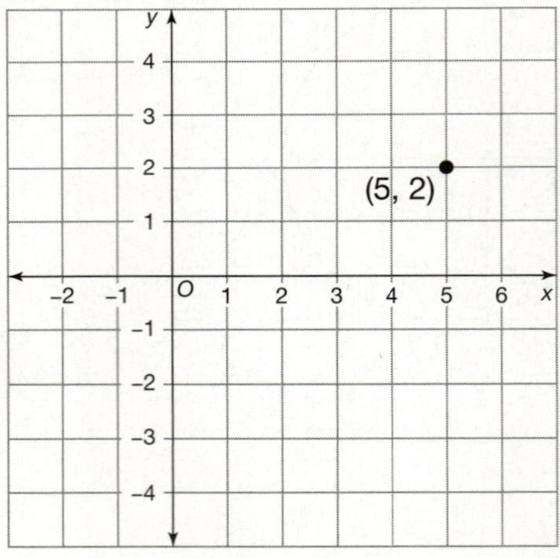

Glossary

y-intercept

The y-intercept is the y-coordinate of the point where a graph crosses the y-axis.

Example

The y-intercept of the graph below is –3.

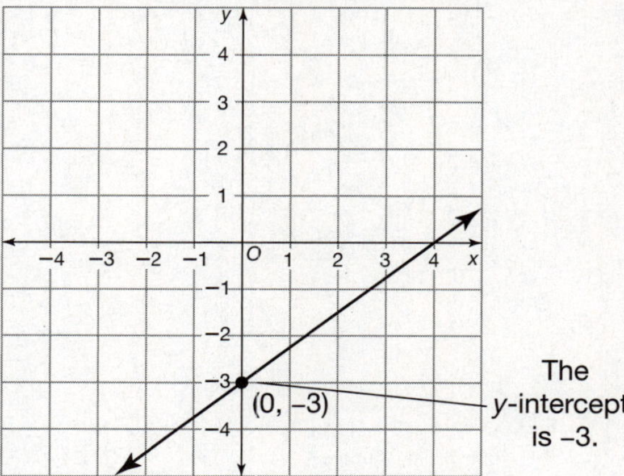

zero exponent

A zero exponent is an exponent in a power.
A non-zero power with a zero exponent is equal to 1.

Example

The expression x^0 contains a zero exponent.

Glossary

Index

B

C

D

E

Index

F

G

H

I

L

M

N

O

P

Q

R

S

Index

T

U

V

W

X

Y

Z